LEWIS *and* CLARK
Trail Maps

Lewis and Clark Trail Maps

A Cartographic Reconstruction, Volume III

Columbia River to the Pacific Ocean, and Further Columbia, Marias, and Yellowstone Explorations (Washington/Oregon/ Idaho/Montana)—Outbound 1805; Return 1806.

Martin Plamondon II

Washington State University Press
Pullman, Washington

Washington State University Press
PO Box 645910
Pullman, Washington 99164-5910
Phone: 800-354-7360
Fax: 509-335-8568
E-mail: wsupress@wsu.edu
Web site: wsupress.wsu.edu

First printing 2004

Library of Congress Cataloging-in-Publication Data

Plamondon, Martin.
Lewis and Clark trail maps, a cartographic reconstruction, Volume III Columbia River to the Pacific Ocean, and further Columbia, Marias, and Yellowstone explorations (Washington/Oregon/Idaho/Montana)—Outbound 1805; return 1806 / by Martin Plamondon II.
p. cm.
Includes index.
ISBN 0-87422-265-6 (hdb.)—ISBN 0-87422-266-4 (pbk.)—ISBN 0-87422-267-2 (spiral)
1. Lewis and Clark National Historic Trail—Maps. 2. Lewis and Clark Expedition (1804–1806)—Maps. 3. Cartography—Northwestern States—History—Maps. I. Title.

G1417.L4 P5 2004
912.78—dc21 v. 3

Cover art: John Ford Clymer, *The Salt Makers,* 1975.
Courtesy of Doris Clymer and the Clymer Museum of Art, Ellensburg, Washington.

WSU Press acknowledges the assistance provided to Martin Plamondon II
by the Lewis and Clark Trail Heritage Foundation,
the Governor's Washington Lewis and Clark Trail Committee,
two anonymous grants in the name of the Columbia Gorge
Interpretive Center (Stevenson, Washington),
Captain and Mrs. J.P. Brooks (Camas, Washington),
VIAs Multimedia Productions (Lolo, Montana),
the Daughters of the Pioneers of Washington,
the Hillside Senior Center (McMinnville, Oregon),
and Marten Plamondon Sr.

Contents

Index Maps and Legend, Volume III

Lewis and Clark Trail Maps, Volume III

Index, Volume III

Dedication

William Clark, the Man
Marten and Teresa Plamondon
Evelyn Plamondon

Volume I was dedicated to "General William Clark," and *Volume II* to "William Clark, the Explorer." *Volume III* honors "William Clark, the Man," for his personal understanding, respect, equal-standing, and dignity when in council with American Indians, and for his great partnering in leadership with Captain Meriwether Lewis.

Volume III also is dedicated to my parents—Marten Plamondon, for his encouragement and his companionship on field trips, and to my mother, Teresa Plamondon. She taught me at an early age to respect and honor our Indian people.

Finally, *Volume III* is dedicated to Evelyn, my wife of 36 years. I shall always be grateful for her constant support, research assistance, and for helping me think through problems and mysteries in the Corps of Discovery story.

Acknowledgments

Over thirty years of mapping the Lewis and Clark trail now come to a close. There were times when I despaired mightily that I would ever get this horse into the barn. But it ultimately has been accomplished, with the help of people along the way who are legion. Regretfully, to recognize them all is perhaps impossible; human memory is insufficient to the daunting challenge. I believe most, however, are gratefully acknowledged in *Volumes I & II*, and here in *Volume III*. To those few I may have missed, I offer my sincerest apologies.

There is one who cannot be passed over lightly—Glen Lindeman, Editor-in-Chief of Washington State University Press. Glen and I have been friends for fourteen years; both of us serve on the Governor's Washington Lewis and Clark Trail Committee. Glen is a capable man with a tremendous love of history and a special interest in Lewis and Clark. During the five years that we have labored over publishing the *Trail Maps*, I cannot ever remember a cross or thoughtless word between us. Our mutual friendship and respect has grown over the years. Such a friendship truly is one of the treasures acquired in a lifetime of work.

Also at the WSU Press, I wish to acknowledge Mary Read, Jenni Lynn, Caryn Lawton, Jean Taylor, Nancy Grunewald, Diana Whaley, Arline Lyons, and Lynne Crawford. These fine people have become my friends, too. And, let us not forget the lithographers and pressmen of WSU's University Publishing. They are the folks who have scanned and printed the original 18" by 24" maps down to a 9" by 12" size, allowing the *Lewis and Clark Trail Maps* to be affordable to the public. The work is technical and critical, and their printing of all three volumes of the *Trail Maps* has been superb.

Special recognition likewise is due to the U.S. Geological Survey of the Department of the Interior, probably the preeminent mapping agency in the world. In compiling the *Lewis and Clark Trail Maps*, thousands of USGS quad maps served as a basic source for many types of topographical information. We Americans are truly fortunate to have a government agency that provides such accurate, highly detailed maps.

For unique assistance and a personal tour in helping locate the Corps' return trail across southeast Washington, an appreciative thank-you goes to: Walt Gary, a retired WSU Agricultural Extension Agent; George Touchette, a former Columbia County commissioner; Steve Plucker, a local Lewis and Clark researcher whose Walla Walla County farm sits directly on the Corps' route; and Gary Lentz, a Ranger at Lewis and Clark Trail State Park, likewise situated right on the expedition's cutoff path. Gary Lentz and Walt Gary also are fellow members of the Governor's Washington Lewis and Clark Trail Committee.

Other memorable hours were spent in discussion with Rex Ziak of Naselle, Washington, as well as Gary Lentz, sharing our compatible views regarding Clark's first sighting of the Pacific Ocean and rambles on Cape Disappointment, as well as the Corps' campsite locations in storm-pounded niches on the Washington shoreline. Rex recently authored *In Full View* (2002), a beautifully illustrated chronicle of the Corps' time on the lower Columbia. He, too, is on the Governor's Washington Lewis and Clark Trail Committee.

Additional essential information regarding the lower Columbia's mouth, coastal forts, and navigation came from Steve Wang, Chief of Interpretive Services, Washington State Parks; the University of Washington map library; and Jeffrey Smith, Associate Curator of the Columbia River Maritime Museum, Astoria, Oregon.

Special recognition goes to David Nicandri, Director of the Washington State Historical Society—and again to Lentz and Plucker—regarding investigations into when and where the Corps first sighted the great Cascade peaks. The year 2003 always will have a special place in my memory when this small circle of investigators struggled mightily among Wallula Gap cliffs in trying to resolve when Clark first viewed Mount Hood. Steve Plucker, too, provided verification that Mount Adams is visible in northern Walla Walla County, from hilly places above the Snake River. Meriwether Lewis might have been the first to sight a Cascade peak (Mount Adams) while hiking into the uplands from the Corps' extended October 15, 1805, campsite, with two days yet to travel down the Snake River.

Curly Bear Wagner, historian of the Blackfeet tribe, shared appreciated insights and unique knowledge about the Marias River country in Montana. Likewise, George Aguilar Sr., an upper Chinook on Oregon's Warm Springs reservation, provided information about native food preservation and its uses.

The *Lewis and Clark Trail Maps* comprise an undertaking of massive proportions, compiled with many thousands of bits of data. Despite the most diligent efforts, there are bound to be some errors. This cartographer offers his apologies, hoping that none will suffer confusion or inconvenience due to this. In this regard, David Saxe of Auburn, New Hampshire, provided appreciated corrections concerning several Universal Transverse Mercator readings.

And finally, I wish to thank the many past and present staff members and volunteers at the Fort Clatsop National Memorial, including Cynthia Orlando, Curt Johnson, Janice Eldridge, Jill Harding, and Sandra Reinebach of the Fort Clatsop Historical Association.

To everyone, my gratitude is without limits on this 200th anniversary of when the Corps of Volunteers for Northwest Discovery turned the bows of their watercraft into the Missouri River to begin the epic journey to western waters.

Martin Plamondon II
Vancouver, Washington
May 14, 2004

A Tribute to Martin Plamondon II for Reconstructing William Clark and Meriwether Lewis's Traverse Notes into Corps of Discovery Trail Maps

Glen W. Lindeman, Editor-in-Chief,
Washington State University Press

Traverse *To determine the locations of survey-points by measuring bearings and distances between successive points.*—Alan H. Hartley, *Lewis & Clark Lexicon of Discovery* (WSU Press, 2004)

On a fine summer day in 1999, it was my distinct pleasure to have three generations of Martin Plamondons visit my WSU Press office. Martin Plamondon II, with father and son of similar name, presented a proposal to publish special volumes of Lewis and Clark maps. I was surprised to hear that years of work already had gone into *Volume I* and it virtually was ready for publication.

Martin and I knew each other as fellow members of the Governor's Washington Lewis and Clark Trail Committee. Being a life-long Washingtonian, Martin wished to have the maps published in his home state—a decision for which WSU will be forever grateful. The distinguished value of these atlases in Western exploration history was readily apparent to the Press staff and to our seven-member editorial board of history and science professors, administrators, and archivists. A deal was cordially and quickly struck.

Martin's roots in western North America go deep. He is of the Plamondon clan, whose earliest regional immigrant, Simon Plamondon, arrived when Canadian fur companies held sway in old Oregon after the mid 1810s. Simon Plamondon traded among the Columbia tribes, originally out of the British post at Astoria, before settling in the late 1820s at the Hudson's Bay Company's agricultural colony near Cowlitz Landing (Toledo, Washington). Martin's more immediate branch of Plamondon kindred were Canadian Plains residents, who emigrated to the Pacific Northwest in an early decade of the twentieth century. At that time they joined their relatives, the immigrants of earlier days.

Martin and I, close in age, grew up in adjacent parts of southwest Washington—I in Lewis County, and he in Clark County. (In fact, beginning in the 1880s, Lindemans on Clackamas Prairie resided but a few miles from Plamondons on Cowlitz Prairie.) These rural lowlands are a delight to farm kids—where beautiful, natural prairie openings are bordered by the unforgettable tattered geometry of dark-green stands of giant Douglas fir.

Somewhere, sometime in this setting, Martin developed a keen passion for exploration history. The Columbia country's geography can cast pervasive spells on youngsters with a historical bent. Regional names are constant reminders of a not-too-distant era of grand exploration, international struggle, and imperial vision. The Columbia River itself is named for the European discoverer of the New World, while community names (Vancouver, Astoria, Clarkston, Lewiston) memorialize great seekers of Pacific waters.

Pervasive, too, are scores upon scores of Indian names still marking the Northwest landscape. For example, on the Columbia—from the mouth of the Snake River to the sea—this broad Indian nomenclature, much recorded by Lewis and Clark, identifies rivers (Walla Walla, Cowlitz, Chinook), natural features (Wallula Gap, Klickitat Valley, Tillamook Head), towns (Umatilla, Celilo, Cathlamet), and counties (Multnomah, Wahkiakum, Clatsop), to name just a few. In the Northwest states, a good sprinkling of French Canadian names, too, are reminders of those early decades when the HBC's red British banner was far more prevalent than the Stars and Stripes. (Not until the 1840s did Oregon Trail pioneers finally tip the balance to the United States.)

In his career, Martin worked for 28 years in Vancouver, Washington, at the Clark County map office, rising to chief cartographer. Only a background such as this can provide one with the skills to undertake a massive reconstruction of the Corps' traverse notes. Martin's home, well stocked with historical narratives and map-making tools, reflects other aspects of exploration history as well. Inside and out, there are rock and mineral samples from along Lewis and Clark's trek, as well as plantings of natural shrubs, herbs, and flowers—some described in Lewis's flora descriptions.

Martin's historical mapping efforts began early when illustrating an unpublished novel he wrote about the Lewis and Clark Expedition. With time—as family life and career allowed—the map project gradually but relentlessly expanded, until after two decades it was his main focus when Martin walked into my office.

Martin's three-volume atlas clearly is a vast body of work. It includes 530 trail maps, 73 index maps, 17 special maps, and 7 sketches and charts—for a total of 627 illustrations. This is all "hand-crafted," not computer-generated. Martin felt that handwork had more flexibility. He estimated it took about three days to complete a map. At this rate, by basic arithmetic, about 1,880 full 8-hour work days were expended on the effort—equivalent to over seven years of constant effort, with only weekends off. Martin referenced literally thousands of USGS maps and historical sources and illustrations, while relying mostly on Reuben G. Thwaites for the journal quotes placed on the maps. The exquisite Gary E. Moulton volumes did not start appearing until well after Martin's *Volume I* was in progress. Nevertheless, they have been of significant help in compiling much of the *Trail Maps*. Moulton, a masterful geographer himself, wrote extensive footnotes that are a rich source of place, time, and setting for the Corps' trek.

Regrettably in the past few years, a triad of medical concerns has kept Martin mostly at home. Ironically, this has been a boon to the Lewis and Clark story, allowing Martin to fully dedicate his time to completing *Volumes II & III* within a half-decade, as compared to the 25 years that it took for *Volume I* when the important demands of family and workplace cut deeply into his days. In the past year when Martin felt the strain of map-making amid his health challenges, both of us noted how William Clark must have felt, too, when writing under adversity—in wind, rain and snow, starving and freezing, beaten from traveling, ill and afflicted, and suffering clouds of mosquitoes. Duty and stoic determination brought Clark through, as it has Martin, an admirer of the famous surveyor-explorer.

During *Volume III*'s editing, Martin and I exchanged maps in Umatilla, Oregon, near where Clark climbed a barren desert cliff to view the distant Cascade Range. Umatilla is the exact halfway point in the 340-mile distance between Martin's Vancouver home and Pullman's WSU Press office. Always liking the open road, Martin and I rendezvoused a half dozen times at a crossroads truck-stop café. The rumbling of long-haul semi-trucks on the nearby Interstate seemed congruous to our own mission of publishing travelers' accounts. In the parking lot, we traded maps practically worth their weight in gold. As a university-trained historian, I knew the Lewis and Clark story well, but working with Martin and his maps took my understanding of the Corps' story to ever-higher levels.

Martin includes a special focus on native groups throughout the trail maps. In community service, too, he has partnered with today's Chinook and Cowlitz peoples on many issues, including tribal recognition in the halls of government. Lewis and Clark, of course, occupy a key place in the Indian history of America. The explorers' peaceful demeanor was memorable, and the extensive journals themselves provide a unique window into native life 200 years ago, when the tribes held sway across the West. This is of irreplaceable value.

However, indigenous people's views of the Corps of Discovery understandably are mixed. Lewis and Clark, indeed, were harbingers of great change, unfortunately some of it negative, that came to native peoples in the 19th century. (Please excuse my diversion here, but I have always found it intriguing how American western expansion is reminiscent of another vast northwest conquest two millennia earlier in Europe. Beginning in 58 BC, armies under Julius Caesar, also a fine frontier chronicler, marched into the homelands of Europe's many Celtic and Briton tribes, essentially in France, Belgium, and England. In a century's time, as in the American experience, Roman soldiers, cavalry, and officials subdued the great tribal warrior classes. Along with Roman forts and colonial outposts came a new language, economy, religion, and governing structure—in essence, a wholesale reordering of society itself. For the tribal peoples, life had profoundly changed forever.)

This progression was much the same for American Indians in the 19th century. Nevertheless, Martin has felt special gratification today that many American Indian groups are graciously participating in the Lewis and Clark Bicentennial. In the following map pages, the reader will note the extensive original Indian occupation along the Columbia's shorelines. In this region, the Corps' journalists depicted one of the highest concentrations of native peoples in America north of Mexico.

Also, *Volume III* begins with a special "journal" chart, which is modified from one that Martin found so useful for quickly referencing those days that a journalist wrote an entry.

On September 26, 1806, three days after the Corps of Discovery had returned to St. Louis, Christy's tavern held an evening "dinner & Ball" honoring Lewis and Clark. Attendees prodigiously drank 18 toasts, beginning with one for President Thomas Jefferson, and ending with, "Captains Lewis and Clark—Their perilous services endear them to every American heart." It seems appropriate that a 19th toast—a modern day one—is due to Martin Plamondon (and Moulton, Thwaites, Coues, and Biddle), who have so ably provided us with the writings, mapping, and surveying of the Lewis and Clark Expedition.

JOURNAL ENTRY DATES *

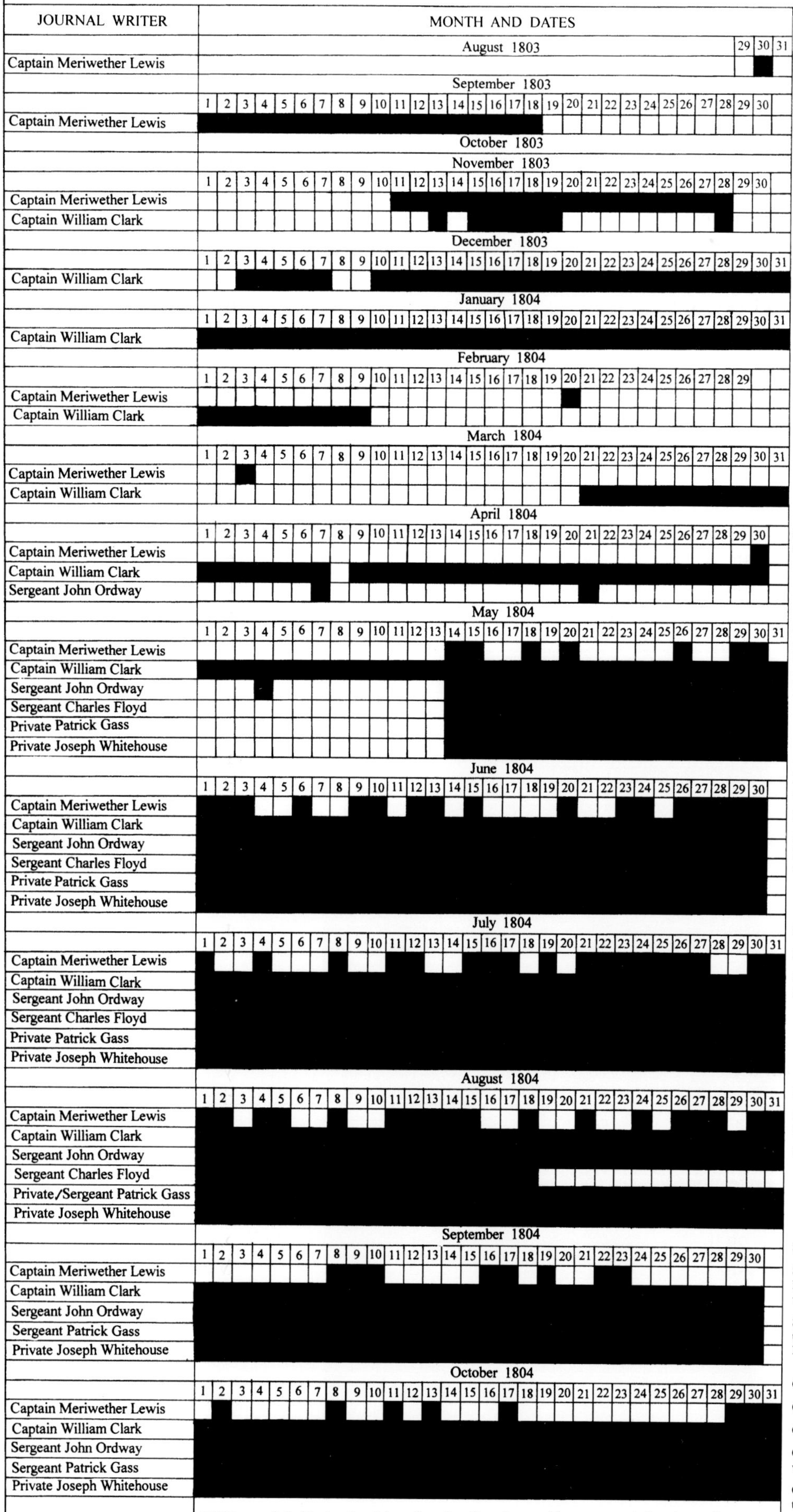

JOURNAL WRITER	MONTH AND DATES
	August 1803 (29, 30, 31)
Captain Meriwether Lewis	
	September 1803 (1–30)
Captain Meriwether Lewis	
	October 1803
	November 1803 (1–30)
Captain Meriwether Lewis	
Captain William Clark	
	December 1803 (1–31)
Captain William Clark	
	January 1804 (1–31)
Captain William Clark	
	February 1804 (1–29)
Captain Meriwether Lewis	
Captain William Clark	
	March 1804 (1–31)
Captain Meriwether Lewis	
Captain William Clark	
	April 1804 (1–30)
Captain Meriwether Lewis	
Captain William Clark	
Sergeant John Ordway	
	May 1804 (1–31)
Captain Meriwether Lewis	
Captain William Clark	
Sergeant John Ordway	
Sergeant Charles Floyd	
Private Patrick Gass	
Private Joseph Whitehouse	
	June 1804 (1–30)
Captain Meriwether Lewis	
Captain William Clark	
Sergeant John Ordway	
Sergeant Charles Floyd	
Private Patrick Gass	
Private Joseph Whitehouse	
	July 1804 (1–31)
Captain Meriwether Lewis	
Captain William Clark	
Sergeant John Ordway	
Sergeant Charles Floyd	
Private Patrick Gass	
Private Joseph Whitehouse	
	August 1804 (1–31)
Captain Meriwether Lewis	
Captain William Clark	
Sergeant John Ordway	
Sergeant Charles Floyd	
Private/Sergeant Patrick Gass	
Private Joseph Whitehouse	
	September 1804 (1–30)
Captain Meriwether Lewis	
Captain William Clark	
Sergeant John Ordway	
Sergeant Patrick Gass	
Private Joseph Whitehouse	
	October 1804 (1–31)
Captain Meriwether Lewis	
Captain William Clark	
Sergeant John Ordway	
Sergeant Patrick Gass	
Private Joseph Whitehouse	

* This graph indicates the dates when the Corps' record keepers wrote in a journal, orderly book, or log. "Black" spaces show when notations were recorded. "Blank" spaces indicate no entries were written or are *known* to exist for those days. Even the briefest daily entries are included here, inasmuch as they provide an account of personal activity.

NOTES

August 31, 1803—Lewis's party of approximately a dozen men leaves Pittsburgh, proceeding down the Ohio River in a new keelboat. His first entry is dated August 30, 1803, but evidence indicates it probably was meant for the day he left Pittsburgh, August 31.

Lewis stops keeping a daily journal, September 19 through November 10.

October 14–26, 1803—Lewis arrives at the Falls of the Ohio and shortly meets Clark at Clarksville. No journal record describes this key juncture in the assembling of the Corps of Discovery.

November 14, 1803—The party, with one keelboat and two pirogues, encamps at the Ohio's mouth. A week later they proceed up the Mississippi River.

December 12–13, 1803—Clark establishes Camp River Dubois (Wood River), on the Mississippi shore opposite to the mouth of the Missouri River.

December 1803–March 1805—Lewis apparently does not keep a "daily" journal. In late spring and summer 1804, however, he records flora, fauna and other scientific descriptions, detachment orders, weather statistics, and especially celestial observations. This latter, fairly extensive writing activity is marked on this chart for this period.

February 20 and March 3, 1804—Detachment orders place Sergeant Ordway in charge when the captains leave Camp Dubois for a time. They delineate the military courtesy that the men are to render Ordway when the commanding officers are absent.

Lewis spends most of the winter in St. Louis on official business and making preparations for the journey.

Ordway's April and early May entries are new standing orders, entered into the Orderly Book.

May 14, 1804—The Corps of Discovery leaves Camp River Dubois, crossing the Mississippi to begin the momentous journey up the Missouri.

May 16–20, 1804—Extended layover at the village of St. Charles, waiting for Lewis's arrival from St. Louis.

June 1–2, 1804—The Corps stops two days at the Osage River, clearing trees to allow the captains to take celestial sightings.

June 13, 1804—Camped at the Grand River, taking observations.

June 26–28, 1804—A three day delay at the Kansas River to clear trees and build a redoubt, while the captains make celestial observations.

July 21, 1804—Expedition passes the Platte River.

July 22–26, 1804—Extended stay at Camp White Catfish.

July 30–August 2, 1804—Meeting with tribesmen at "Council Bluff."

August 13–19, 1804—At Fish Camp near an Omaha village.

August 20, 1804—Sergeant Floyd is buried on a bluff just south of modern-day Sioux City, Iowa.

August 28–31, 1804—Yankton Sioux council.

September 4, 1804—Expedition camps at the Niobrara River.

September 20, 1804—A day is spent working around the "Grand Detour." Sandy ground at this evening's encampment is rapidly washed away by the current; the party hastily relocates to safety.

September 24–27, 1804—In the Bad River locality, the Corps has tense confrontation with the Teton Sioux near today's Pierre, South Dakota.

October 1, 1804—The Cheyenne River is reached.

October 8–11, 1804—Among the Arikara villages.

October 18, 1804—Expedition passes Cannonball River.

October 26–31, 1804—Encampment at Mandan and Minitari villages.

October 30, 1804—Clark reaches farthest upriver point in 1804, then turns back to establish winter quarters a short distance downstream.

EXPLORATIONS OF LEWIS AND CLARK 1804 - 1806
CARTOGRAPHIC RECONSTRUCTION

JOURNAL ENTRY DATES

J.E.D.---1

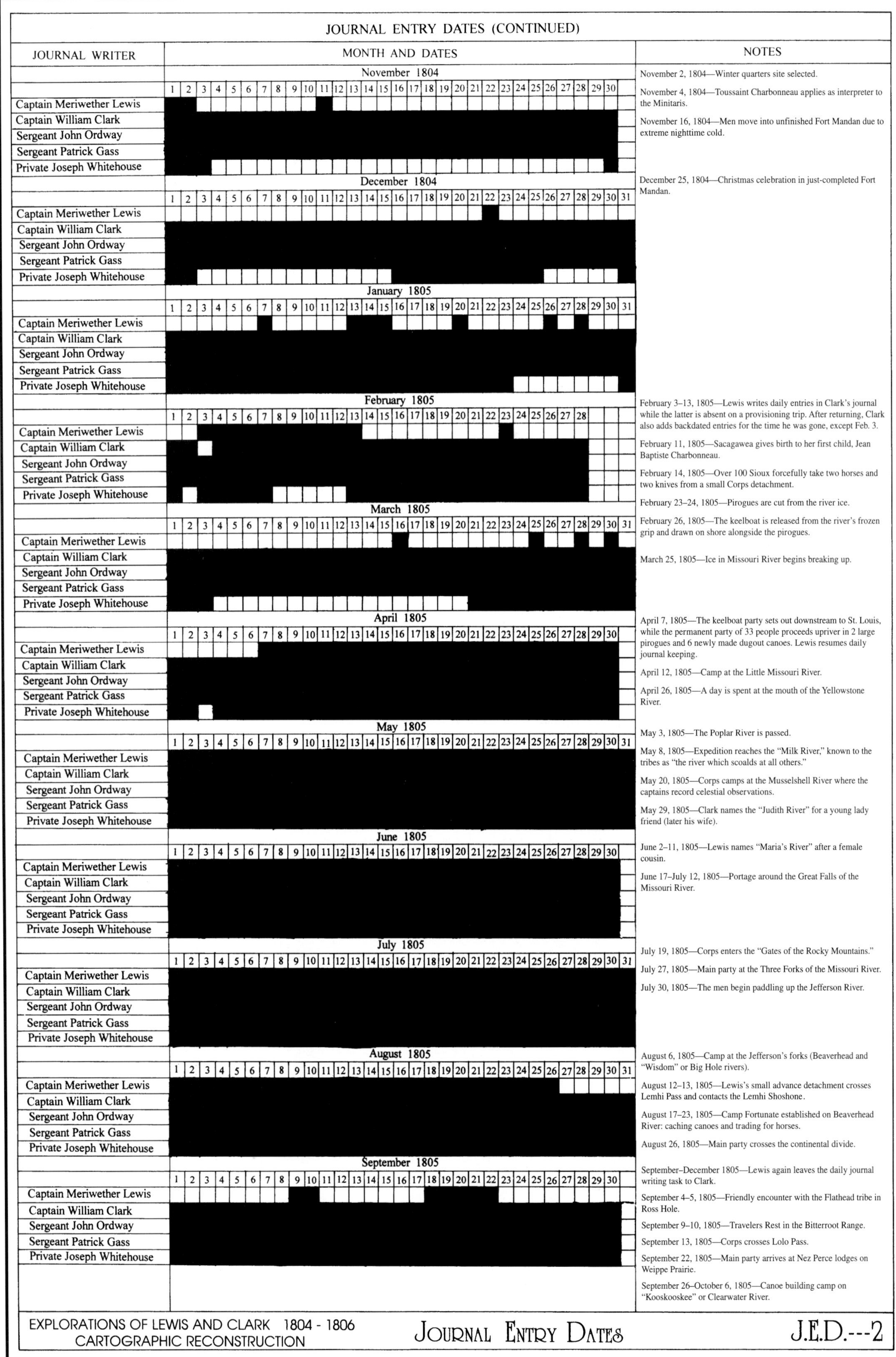
JOURNAL ENTRY DATES (CONTINUED)
JOURNAL WRITER
MONTH AND DATES
NOTES
November 1804
1 2 3 4 5 6 7 8 9 10 11 12 13 14 15 16 17 18 19 20 21 22 23 24 25 26 27 28 29 30
Captain Meriwether Lewis
Captain William Clark
Sergeant John Ordway
Sergeant Patrick Gass
Private Joseph Whitehouse
November 2, 1804—Winter quarters site selected.
November 4, 1804—Toussaint Charbonneau applies as interpreter to the Minitaris.
November 16, 1804—Men move into unfinished Fort Mandan due to extreme nighttime cold.
December 1804
1 2 3 4 5 6 7 8 9 10 11 12 13 14 15 16 17 18 19 20 21 22 23 24 25 26 27 28 29 30 31
Captain Meriwether Lewis
Captain William Clark
Sergeant John Ordway
Sergeant Patrick Gass
Private Joseph Whitehouse
December 25, 1804—Christmas celebration in just-completed Fort Mandan.
January 1805
1 2 3 4 5 6 7 8 9 10 11 12 13 14 15 16 17 18 19 20 21 22 23 24 25 26 27 28 29 30 31
Captain Meriwether Lewis
Captain William Clark
Sergeant John Ordway
Sergeant Patrick Gass
Private Joseph Whitehouse
February 1805
1 2 3 4 5 6 7 8 9 10 11 12 13 14 15 16 17 18 19 20 21 22 23 24 25 26 27 28
Captain Meriwether Lewis
Captain William Clark
Sergeant John Ordway
Sergeant Patrick Gass
Private Joseph Whitehouse
February 3–13, 1805—Lewis writes daily entries in Clark's journal while the latter is absent on a provisioning trip. After returning, Clark also adds backdated entries for the time he was gone, except Feb. 3.
February 11, 1805—Sacagawea gives birth to her first child, Jean Baptiste Charbonneau.
February 14, 1805—Over 100 Sioux forcefully take two horses and two knives from a small Corps detachment.
February 23–24, 1805—Pirogues are cut from the river ice.
February 26, 1805—The keelboat is released from the river's frozen grip and drawn on shore alongside the pirogues.
March 1805
1 2 3 4 5 6 7 8 9 10 11 12 13 14 15 16 17 18 19 20 21 22 23 24 25 26 27 28 29 30 31
Captain Meriwether Lewis
Captain William Clark
Sergeant John Ordway
Sergeant Patrick Gass
Private Joseph Whitehouse
March 25, 1805—Ice in Missouri River begins breaking up.
April 1805
1 2 3 4 5 6 7 8 9 10 11 12 13 14 15 16 17 18 19 20 21 22 23 24 25 26 27 28 29 30
Captain Meriwether Lewis
Captain William Clark
Sergeant John Ordway
Sergeant Patrick Gass
Private Joseph Whitehouse
April 7, 1805—The keelboat party sets out downstream to St. Louis, while the permanent party of 33 people proceeds upriver in 2 large pirogues and 6 newly made dugout canoes. Lewis resumes daily journal keeping.
April 12, 1805—Camp at the Little Missouri River.
April 26, 1805—A day is spent at the mouth of the Yellowstone River.
May 1805
1 2 3 4 5 6 7 8 9 10 11 12 13 14 15 16 17 18 19 20 21 22 23 24 25 26 27 28 29 30 31
Captain Meriwether Lewis
Captain William Clark
Sergeant John Ordway
Sergeant Patrick Gass
Private Joseph Whitehouse
May 3, 1805—The Poplar River is passed.
May 8, 1805—Expedition reaches the "Milk River," known to the tribes as "the river which scoalds at all others."
May 20, 1805—Corps camps at the Musselshell River where the captains record celestial observations.
May 29, 1805—Clark names the "Judith River" for a young lady friend (later his wife).
June 1805
1 2 3 4 5 6 7 8 9 10 11 12 13 14 15 16 17 18 19 20 21 22 23 24 25 26 27 28 29 30
Captain Meriwether Lewis
Captain William Clark
Sergeant John Ordway
Sergeant Patrick Gass
Private Joseph Whitehouse
June 2–11, 1805—Lewis names "Maria's River" after a female cousin.
June 17–July 12, 1805—Portage around the Great Falls of the Missouri River.
July 1805
1 2 3 4 5 6 7 8 9 10 11 12 13 14 15 16 17 18 19 20 21 22 23 24 25 26 27 28 29 30 31
Captain Meriwether Lewis
Captain William Clark
Sergeant John Ordway
Sergeant Patrick Gass
Private Joseph Whitehouse
July 19, 1805—Corps enters the "Gates of the Rocky Mountains."
July 27, 1805—Main party at the Three Forks of the Missouri River.
July 30, 1805—The men begin paddling up the Jefferson River.
August 1805
1 2 3 4 5 6 7 8 9 10 11 12 13 14 15 16 17 18 19 20 21 22 23 24 25 26 27 28 29 30 31
Captain Meriwether Lewis
Captain William Clark
Sergeant John Ordway
Sergeant Patrick Gass
Private Joseph Whitehouse
August 6, 1805—Camp at the Jefferson's forks (Beaverhead and "Wisdom" or Big Hole rivers).
August 12–13, 1805—Lewis's small advance detachment crosses Lemhi Pass and contacts the Lemhi Shoshone.
August 17–23, 1805—Camp Fortunate established on Beaverhead River: caching canoes and trading for horses.
August 26, 1805—Main party crosses the continental divide.
September 1805
1 2 3 4 5 6 7 8 9 10 11 12 13 14 15 16 17 18 19 20 21 22 23 24 25 26 27 28 29 30
Captain Meriwether Lewis
Captain William Clark
Sergeant John Ordway
Sergeant Patrick Gass
Private Joseph Whitehouse
September–December 1805—Lewis again leaves the daily journal writing task to Clark.
September 4–5, 1805—Friendly encounter with the Flathead tribe in Ross Hole.
September 9–10, 1805—Travelers Rest in the Bitterroot Range.
September 13, 1805—Corps crosses Lolo Pass.
September 22, 1805—Main party arrives at Nez Perce lodges on Weippe Prairie.
September 26–October 6, 1805—Canoe building camp on "Kooskooskee" or Clearwater River.
EXPLORATIONS OF LEWIS AND CLARK 1804 - 1806
CARTOGRAPHIC RECONSTRUCTION
JOURNAL ENTRY DATES
J.E.D.---2

JOURNAL ENTRY DATES (CONTINUED)

October 1805

Journal Writer	1	2	3	4	5	6	7	8	9	10	11	12	13	14	15	16	17	18	19	20	21	22	23	24	25	26	27	28	29	30	31
Captain Meriwether Lewis																															
Captain William Clark	■	■	■	■	■	■	■	■	■	■	■	■	■	■	■	■	■	■	■	■	■	■	■	■	■	■	■	■	■	■	■
Sergeant John Ordway	■	■	■	■	■	■	■	■	■	■	■	■	■	■	■	■	■	■	■	■	■	■	■	■	■	■	■	■	■	■	■
Sergeant Patrick Gass	■	■	■	■	■	■	■	■	■	■	■	■	■	■	■	■	■	■	■	■	■	■	■	■	■	■	■	■	■	■	■
Private Joseph Whitehouse	■	■	■	■	■	■	■	■	■	■	■	■	■	■	■	■	■	■	■	■	■	■	■	■	■	■	■	■	■	■	■

November 1805

Journal Writer	1	2	3	4	5	6	7	8	9	10	11	12	13	14	15	16	17	18	19	20	21	22	23	24	25	26	27	28	29	30
Captain Meriwether Lewis																													■	■
Captain William Clark	■	■	■	■	■	■	■	■	■	■	■	■	■	■	■	■	■	■	■	■	■	■	■	■	■	■	■	■	■	■
Sergeant John Ordway	■	■	■	■	■	■	■	■	■	■	■	■	■	■	■	■	■	■	■	■	■	■	■	■	■	■	■	■	■	■
Sergeant Patrick Gass	■	■	■	■	■	■	■	■	■	■	■	■	■	■	■	■	■	■	■	■	■	■	■	■	■	■	■	■	■	■
Private Joseph Whitehouse	■	■	■	■	■	■	■	■	■	■	■	■	■	■	■	■	■	■	■	■	■	■	■	■	■	■	■	■	■	■

December 1805

Journal Writer	1	2	3	4	5	6	7	8	9	10	11	12	13	14	15	16	17	18	19	20	21	22	23	24	25	26	27	28	29	30	31
Captain Meriwether Lewis	■																	■													
Captain William Clark	■	■	■	■	■	■	■	■	■	■	■	■	■	■	■	■	■	■	■	■	■	■	■	■	■	■	■	■	■	■	■
Sergeant John Ordway	■	■	■	■	■	■	■	■	■	■	■	■	■	■	■	■	■	■	■	■	■	■	■	■	■	■	■	■	■	■	■
Sergeant Patrick Gass	■	■	■	■	■	■	■	■	■	■	■	■	■	■	■	■	■	■	■	■	■	■	■	■	■	■	■	■	■	■	■
Private Joseph Whitehouse	■	■	■	■	■	■	■	■	■	■	■	■	■	■	■	■	■	■	■	■	■	■	■	■	■	■	■	■	■	■	■

January 1806

Journal Writer	1	2	3	4	5	6	7	8	9	10	11	12	13	14	15	16	17	18	19	20	21	22	23	24	25	26	27	28	29	30	31
Captain Meriwether Lewis	■	■	■	■	■	■	■	■	■	■	■	■	■	■	■	■	■	■	■	■	■	■	■	■	■	■	■	■	■	■	■
Captain William Clark	■	■	■	■	■	■	■	■	■	■	■	■	■	■	■	■	■	■	■	■	■	■	■	■	■	■	■	■	■	■	■
Sergeant John Ordway	■	■	■	■	■	■	■	■	■	■	■	■	■	■	■	■	■	■	■	■	■	■	■	■	■	■	■	■	■	■	■
Sergeant Patrick Gass	■	■	■	■	■	■	■	■	■	■	■	■	■	■	■	■	■	■	■	■	■	■	■	■	■	■	■	■	■	■	■
Private Joseph Whitehouse	■	■	■	■	■	■	■	■	■	■	■	■	■	■	■	■	■	■	■	■	■	■	■	■	■	■	■	■	■	■	■

February 1806

Journal Writer	1	2	3	4	5	6	7	8	9	10	11	12	13	14	15	16	17	18	19	20	21	22	23	24	25	26	27	28
Captain Meriwether Lewis	■	■	■	■	■	■	■	■	■	■	■	■	■	■	■	■	■	■	■	■	■	■	■	■	■	■	■	■
Captain William Clark	■	■	■	■	■	■	■	■	■	■	■	■	■	■	■	■	■	■	■	■	■	■	■	■	■	■	■	■
Sergeant John Ordway	■	■	■	■	■	■	■	■	■	■	■	■	■	■	■	■	■	■	■	■	■	■	■	■	■	■	■	■
Sergeant Patrick Gass	■	■	■	■	■	■	■	■	■	■	■	■	■	■	■	■	■	■	■	■	■	■	■	■	■	■	■	■
Private Joseph Whitehouse	■	■	■	■	■	■	■	■	■	■	■	■	■	■	■	■	■	■	■	■	■	■	■	■	■	■	■	■

March 1806

Journal Writer	1	2	3	4	5	6	7	8	9	10	11	12	13	14	15	16	17	18	19	20	21	22	23	24	25	26	27	28	29	30	31
Captain Meriwether Lewis	■	■	■	■	■	■	■	■	■	■	■	■	■	■	■	■	■	■	■	■	■	■	■	■	■	■	■	■	■	■	■
Captain William Clark	■	■	■	■	■	■	■	■	■	■	■	■	■	■	■	■	■	■	■	■	■	■	■	■	■	■	■	■	■	■	■
Sergeant John Ordway	■	■	■	■	■	■	■	■	■	■	■	■	■	■	■	■	■	■	■	■	■	■	■	■	■	■	■	■	■	■	■
Sergeant Patrick Gass	■	■	■	■	■	■	■	■	■	■	■	■	■	■	■	■	■	■	■	■	■	■	■	■	■	■	■	■	■	■	■
Private Joseph Whitehouse	■	■	■	■	■	■	■	■	■	■	■	■	■	■	■	■	■	■	■	■	■	■	■	■	■	■	■	■	■	■	■

April 1806

Journal Writer	1	2	3	4	5	6	7	8	9	10	11	12	13	14	15	16	17	18	19	20	21	22	23	24	25	26	27	28	29	30
Captain Meriwether Lewis	■	■	■	■	■	■	■	■	■	■	■	■	■	■	■	■	■	■	■	■	■	■	■	■	■	■	■	■	■	■
Captain William Clark	■	■	■	■	■	■	■	■	■	■	■	■	■	■	■	■	■	■	■	■	■	■	■	■	■	■	■	■	■	■
Sergeant John Ordway	■	■	■	■	■	■	■	■	■	■	■	■	■	■	■	■	■	■	■	■	■	■	■	■	■	■	■	■	■	■
Sergeant Patrick Gass	■	■	■	■	■	■	■	■	■	■	■	■	■	■	■	■	■	■	■	■	■	■	■	■	■	■	■	■	■	■
Private Joseph Whitehouse	■	■																												

May 1806

Journal Writer	1	2	3	4	5	6	7	8	9	10	11	12	13	14	15	16	17	18	19	20	21	22	23	24	25	26	27	28	29	30	31
Captain Meriwether Lewis	■	■	■	■	■	■	■	■	■	■	■	■	■	■	■	■	■	■	■	■	■	■	■	■	■	■	■	■	■	■	■
Captain William Clark	■	■	■	■	■	■	■	■	■	■	■	■	■	■	■	■	■	■	■	■	■	■	■	■	■	■	■	■	■	■	■
Sergeant John Ordway	■	■	■	■	■	■	■	■	■	■	■	■	■	■	■	■	■	■	■	■	■	■	■	■	■	■	■	■	■	■	■
Sergeant Patrick Gass	■	■	■	■	■	■	■	■	■	■	■	■	■	■	■	■	■	■	■	■	■	■	■	■	■	■	■	■	■	■	■

June 1806

Journal Writer	1	2	3	4	5	6	7	8	9	10	11	12	13	14	15	16	17	18	19	20	21	22	23	24	25	26	27	28	29	30
Captain Meriwether Lewis	■	■	■	■	■	■	■	■	■	■	■	■	■	■	■	■	■	■	■	■	■	■	■	■	■	■	■	■	■	■
Captain William Clark	■	■	■	■	■	■	■	■	■	■	■	■	■	■	■	■	■	■	■	■	■	■	■	■	■	■	■	■	■	■
Sergeant John Ordway	■	■	■	■	■	■	■	■	■	■	■	■	■	■	■	■	■	■	■	■	■	■	■	■	■	■	■	■	■	■
Sergeant Patrick Gass	■	■	■	■	■	■	■	■	■	■	■	■	■	■	■	■	■	■	■	■	■	■	■	■	■	■	■	■	■	■

July 1806

Journal Writer	1	2	3	4	5	6	7	8	9	10	11	12	13	14	15	16	17	18	19	20	21	22	23	24	25	26	27	28	29	30	31
Captain Meriwether Lewis	■	■	■	■	■	■	■	■	■	■	■	■	■	■	■	■	■	■	■	■	■	■	■	■	■	■	■	■	■	■	■
Captain William Clark	■	■	■	■	■	■	■	■	■	■	■	■	■	■	■	■	■	■	■	■	■	■	■	■	■	■	■	■	■	■	■
Sergeant John Ordway	■	■	■	■	■	■	■	■	■	■	■	■	■	■	■	■	■	■	■	■	■	■	■	■	■	■	■	■	■	■	■
Sergeant Patrick Gass	■	■	■	■	■	■	■	■	■	■	■	■	■	■	■	■	■	■	■	■	■	■	■	■	■	■	■	■	■	■	■

August 1806

Journal Writer	1	2	3	4	5	6	7	8	9	10	11	12	13	14	15	16	17	18	19	20	21	22	23	24	25	26	27	28	29	30	31
Captain Meriwether Lewis	■	■	■	■	■	■	■	■	■	■	■	■																			
Captain William Clark	■	■	■	■	■	■	■	■	■	■	■	■	■	■	■	■	■	■	■	■	■	■	■	■	■	■	■	■	■	■	■
Sergeant John Ordway	■	■	■	■	■	■	■	■	■	■	■	■	■	■	■	■	■	■	■	■	■	■	■	■	■	■	■	■	■	■	■
Sergeant Patrick Gass	■	■	■	■	■	■	■	■	■	■	■	■	■	■	■	■	■	■	■	■	■	■	■	■	■	■	■	■	■	■	■

September 1806

Journal Writer	1	2	3	4	5	6	7	8	9	10	11	12	13	14	15	16	17	18	19	20	21	22	23	24	25	26	27	28	29	30
Captain Meriwether Lewis																														
Captain William Clark	■	■	■	■	■	■	■	■	■	■	■	■	■	■	■	■	■	■	■	■	■	■	■	■	■	■				
Sergeant John Ordway	■	■	■	■	■	■	■	■	■	■	■	■	■	■	■	■	■	■	■	■	■	■	■							
Sergeant Patrick Gass	■	■	■	■	■	■	■	■	■	■	■	■	■	■	■	■	■	■	■				■							

NOTES

October 10, 1805—Camp at Clearwater-Snake confluence.

October 12, 1805—Corps passes "Drouillards" or Palouse River.

October 16–17, 1805—Meeting with tribes at Snake-Columbia confluence.

October 19, 1805—Umatilla Rapids on the Columbia River.

October 22–23, 1805—Expedition passes mouth of "Twornehiooks" or Deschutes River and portages around Celilo Falls.

October 24–27, 1805—Passage of The Dalles to the "Rock Fort."

October 30–November 2, 1805—Portage of the Cascades Rapids.

November 3, 1805—Mouth of "Quicksand" or Sandy River inspected.

November 6, 1805—"Coweliskee" or Cowlitz River is passed.

November 7, 1805—First sighting of the Pacific Ocean from near Pillar Rock.

November 15–24, 1805—Station Camp, with the Pacific Ocean immediately in view.

November 26, 1805—After moving upriver, the Corps paddles from the Washington to Oregon side and proceeds down the Columbia's south shore.

December 9, 1805—Fort Clatsop construction begins.

December 28, 1805—Five men ordered to find an eligible site on the nearby coast for a salt making camp.

December 30, 1805—Fort Clatsop completed.

January 1, 1806—Lewis orders a formal military guard routine at the stockade; he also resumes regular journal entries.

January 6–9, 1805—Clark leads a party over Tillamook Head to observe the remains of a beached whale at today's Cannon Beach.

Winter 1806—The Corps prepares 338 pairs of moccasins for the homeward trek. The captains make plans for separate detachments to explore new sections of the Rockies and Plains on the return journey.

February 14, 1806—Journal entries indicate that the captains are fully aware of a shorter, better route between Travelers Rest and the Great Falls (to be taken by Lewis, summer 1806).

February 19–21, 1806—Sergeant Ordway closes down the salt camp.

March 23, 1806—Corps leaves Fort Clatsop with five canoes, beginning the long trek home.

March 28, 1806—Hunting on "Elallar" or Deer Island below the Cowlitz River.

March 29, 1806—Cathlapotle village just above the Lewis River's mouth.

March 31–April 5, 1806—"Provision Camp" at the "Seal" or Washougal River, near the Columbia gorge.

April 2, 1806—Last entry of Whitehouse's *known* journal writing.

April 2–3, 1806—Clark and a small detachment explore the lower "Multnomah" or Willamette River.

April 12, 1806—Cascades Rapids portage.

April 15–16, 1806—Return to "Rock Fort" at The Dalles.

April 23–24, 1806—After passing the John Day River, the last two canoes are traded to natives and the combined party proceeds entirely by land with horses.

April 30, 1806—Corps sets out on the Nez Perce or Overland Trail across today's southeast Washington (Alternate Return Route No. 1).

May 4, 1806—Crossing of Alpowa summit to the "Lewis" or Snake River.

May 8, 1806—The party climbs southeast to plateaus above the Clearwater River (Alternate Return Route No. 2).

May 14–June 9, 1806—Corps at Camp Chopunnish on the Clearwater River.

June 10, 1806—Relocating to Weippe Prairie in preparation for entering the Bitterroot Range.

June 17, 1806—Deep snow foils first attempt to cross Bitterroots.

June 24, 1806—Beginning of the second (and successful) attempt to cross the Bitterroots.

June 28–29, 1806—Party approaches Lolo Pass by a different trail (Alternate Return Route No. 3).

June 30–July 2, 1806—At Travelers Rest, the Corps separates into two exploring groups.

July 8, 1806—Crossing Big Hole valley, Clark's party reaches Beaverhead caches at Camp Fortunate (Alternate Return Route No. 4).

July 11, 1806—Lewis's detachment arrives at the Great Falls (Alternate Return Route No. 5).

July 13, 1806—Clark arrives at Three Forks, dispatching Ordway's canoe party to Great Falls, while Clark's horse entourage turns east to Yellowstone River (Alternate Return Route No. 6).

July 16, 1806—Leaving Gass's party (soon joined by Ordway's detachment) to portage around Great Falls, Lewis and three men depart to explore the upper Marias River (Alternate Return Route No. 7).

July 27, 1806—Lewis party clashes with 8 Blackfeet.

July 28, 1806—Following a long horseback ride, Lewis joins the combined Gass-Ordway boat party on the Missouri.

August 11, 1806—Lewis accidentally shot by Cruzatte while hunting.

August 12, 1806—Reunion of all expedition units above the Mandan villages, after which Lewis leaves the main journal writing to Clark.

September 23, 1806—Arrival at St. Louis.

EXPLORATIONS OF LEWIS AND CLARK 1804 - 1806
CARTOGRAPHIC RECONSTRUCTION

JOURNAL ENTRY DATES

J.E.D.---3

Index Maps and Legend, Volume III

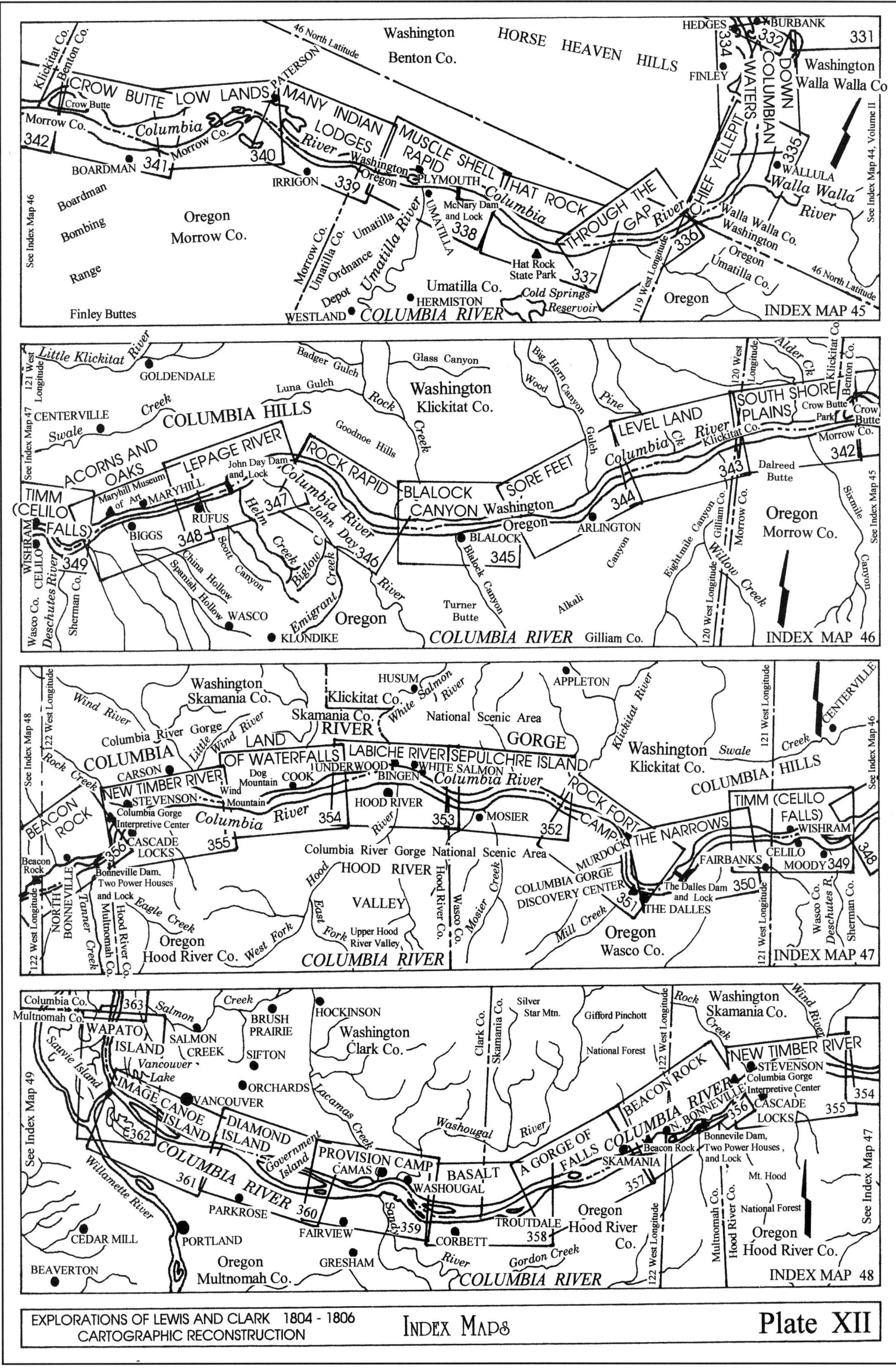

EXPLORATIONS OF LEWIS AND CLARK 1804 - 1806
CARTOGRAPHIC RECONSTRUCTION
INDEX MAPS
Plate XII

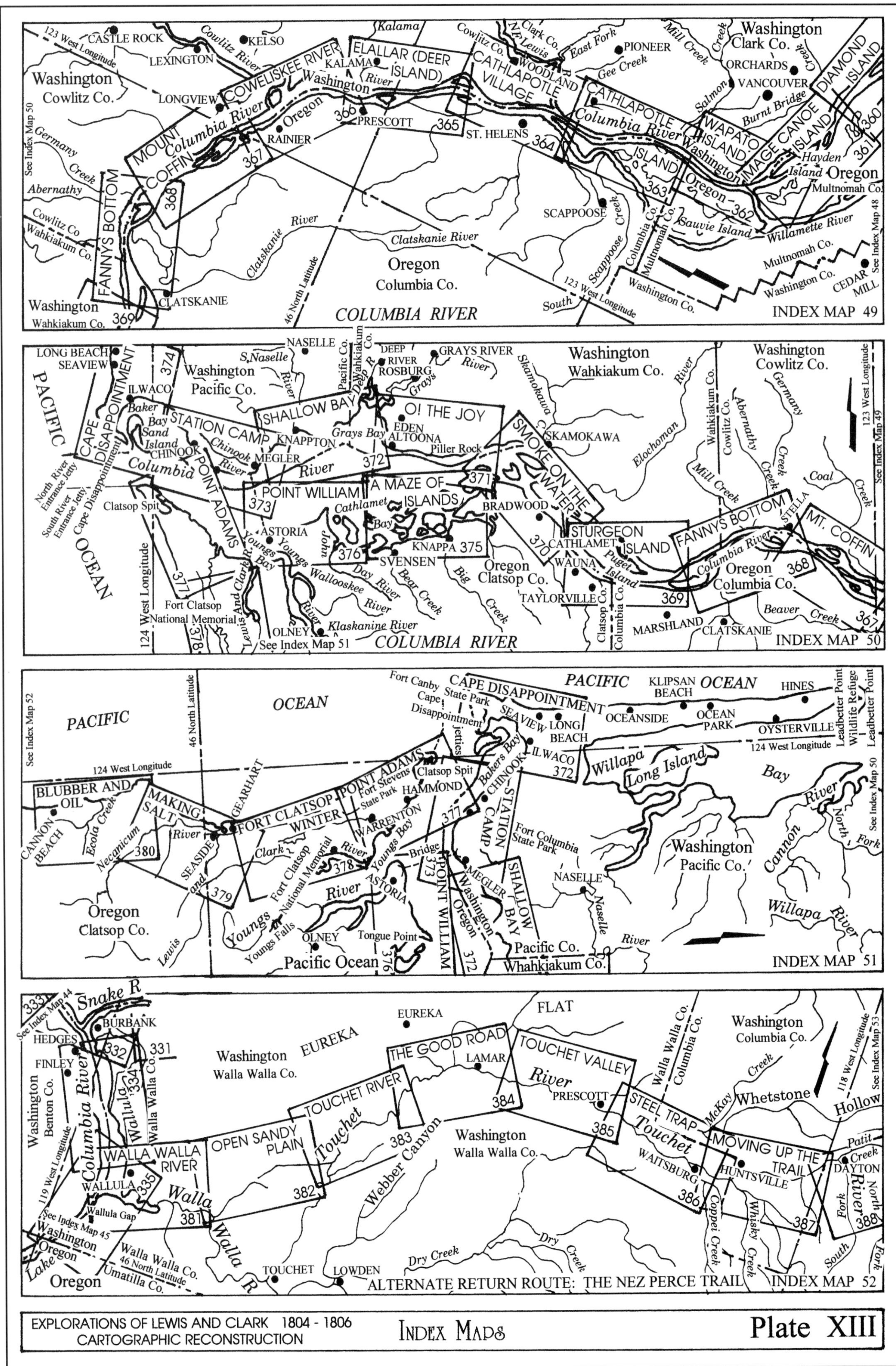

CASTLE ROCK
KELSO
Cowlitz River
LEXINGTON
Washington
Cowlitz Co.
LONGVIEW
COWELISKEE RIVER
ELALLAR (DEER ISLAND)
KALAMA
Kalama
CATHLAPOTLE VILLAGE
WOODLAND
East Fork
PIONEER
Gee Creek
Mill Creek
Washington
Clark Co.
ORCHARDS
VANCOUVER
DIAMOND ISLAND
MOUNT COFFIN
Columbia River
Oregon
RAINIER
366
PRESCOTT
365
ST. HELENS
364
CATHLAPOTLE ISLAND
WAPATO ISLAND
IMAGE CANOE ISLAND
Burnt Bridge
Salmon
360
361
Hayden Island
Oregon
Multnomah Co.
367
368
FANNYS BOTTOM
Germany Creek
Abernathy
Cowlitz Co
Wahkiakum Co.
Clatskanie River
River
Oregon
Columbia Co.
SCAPPOOSE
Scappoose Creek
363
362
Sauvie Island
Willamette River
Columbia Co.
Multnomah Co.
Washington Co.
CEDAR MILL
123 West Longitude
46 North Latitude
See Index Map 50
See Index Map 48
Washington
Wahkiakum Co.
CLATSKANIE
369
COLUMBIA RIVER
INDEX MAP 49
LONG BEACH
SEAVIEW
NASELLE
S.Naselle River
Washington
Pacific Co.
DEEP RIVER
ROSBURG
GRAYS RIVER
Grays River
Washington
Wahkiakum Co.
Washington
Cowlitz Co.
PACIFIC OCEAN
CAPE DISAPPOINTMENT
374
ILWACO
Baker Bay
Sand Island
STATION CAMP
SHALLOW BAY
O! THE JOY
SMOKE ON THE WATER
SKAMOKAWA
Skamokawa
Elochoman River
Abernathy
Germany Creek
Mill Creek
Coal Creek
CHINOOK
KNAPPTON
MEGLER
Grays Bay
EDEN
ALTOONA
Piller Rock
372
371
Columbia River
POINT WILLIAM
A MAZE OF ISLANDS
373
Cathlamet Bay
BRADWOOD
POINT ADAMS
Clatsop Spit
ASTORIA
Youngs Bay
376
KNAPPA
375
SVENSEN
STURGEON ISLAND
CATHLAMET
370
Puget Island
WAUNA
FANNYS BOTTOM
STELLA
MT. COFFIN
Oregon
Clatsop Co.
Oregon
Columbia Co.
368
TAYLORVILLE
369
377
Fort Clatsop National Memorial
378
Wallooskee River
John Day River
Bear Creek
Big Creek
Klaskanine River
OLNEY
See Index Map 51
COLUMBIA RIVER
MARSHLAND
CLATSKANIE
Beaver Creek
367
INDEX MAP 50
124 West Longitude
123 West Longitude
North River Entrance Jetty
South River Entrance Jetty
PACIFIC
OCEAN
PACIFIC
OCEAN
KLIPSAN BEACH
HINES
Fort Canby State Park
Cape Disappointment
CAPE DISAPPOINTMENT
SEAVIEW
LONG BEACH
OCEANSIDE
OCEAN PARK
OYSTERVILLE
Leadbetter Point Wildlife Refuge
124 West Longitude
46 North Latitude
See Index Map 52
BLUBBER AND OIL
MAKING SALT
GEARHART
FORT CLATSOP WINTER
POINT ADAMS
Clatsop Spit
Fort Stevens State Park
HAMMOND
Bakers Bay
ILWACO
372
Willapa Bay
Long Island
CANNON BEACH
Ecola Creek
Necanicum River
380
SEASIDE
379
Clark
WARRENTON
Youngs Bay
378
377
CHINOOK
STATION CAMP
Fort Columbia State Park
Washington
Pacific Co.
Cannon River
North Fork
See Index Map 50
Oregon
Clatsop Co.
Lewis and Clark
Fort Clatsop National Memorial
Youngs River
Youngs Falls
ASTORIA
Bridge
373
POINT WILLIAM
MEGLER
Washington
Oregon
SHALLOW BAY
NASELLE
Naselle River
Willapa River
OLNEY
Tongue Point
376
372
Pacific Ocean
Pacific Co.
Whahkiakum Co.
INDEX MAP 51
333
Snake R
See Index Map 44
BURBANK
HEDGES
332
331
FINLEY
EUREKA
EUREKA
FLAT
Washington
Walla Walla Co.
THE GOOD ROAD
LAMAR
TOUCHET VALLEY
Walla Walla Co.
Columbia Co.
Washington
Columbia Co.
Washington
Benton Co.
Columbia River
Wallula
334
Walla Walla Co.
TOUCHET RIVER
River
384
PRESCOTT
Whetstone Hollow
McKay Creek
118 West Longitude
See Index Map 53
OPEN SANDY PLAIN
Touchet
383
Washington
Walla Walla Co.
STEEL TRAP
Touchet
385
MOVING UP THE TRAIL
Patit Creek
DAYTON
119 West Longitude
WALLA WALLA RIVER
WALLULA
335
382
Webber Canyon
WAITSBURG
HUNTSVILLE
386
387
388
North Fork River
Wallula Gap
See Index Map 45
381
Walla Walla R
Washington
Oregon
Lake
Oregon
Walla Walla Co.
46 North Latitude
Umatilla Co.
TOUCHET
LOWDEN
Dry Creek
Dry Creek
Coppei Creek
Whisky Creek
South Fork
ALTERNATE RETURN ROUTE: THE NEZ PERCE TRAIL
INDEX MAP 52
EXPLORATIONS OF LEWIS AND CLARK 1804 - 1806
CARTOGRAPHIC RECONSTRUCTION
INDEX MAPS
Plate XIII

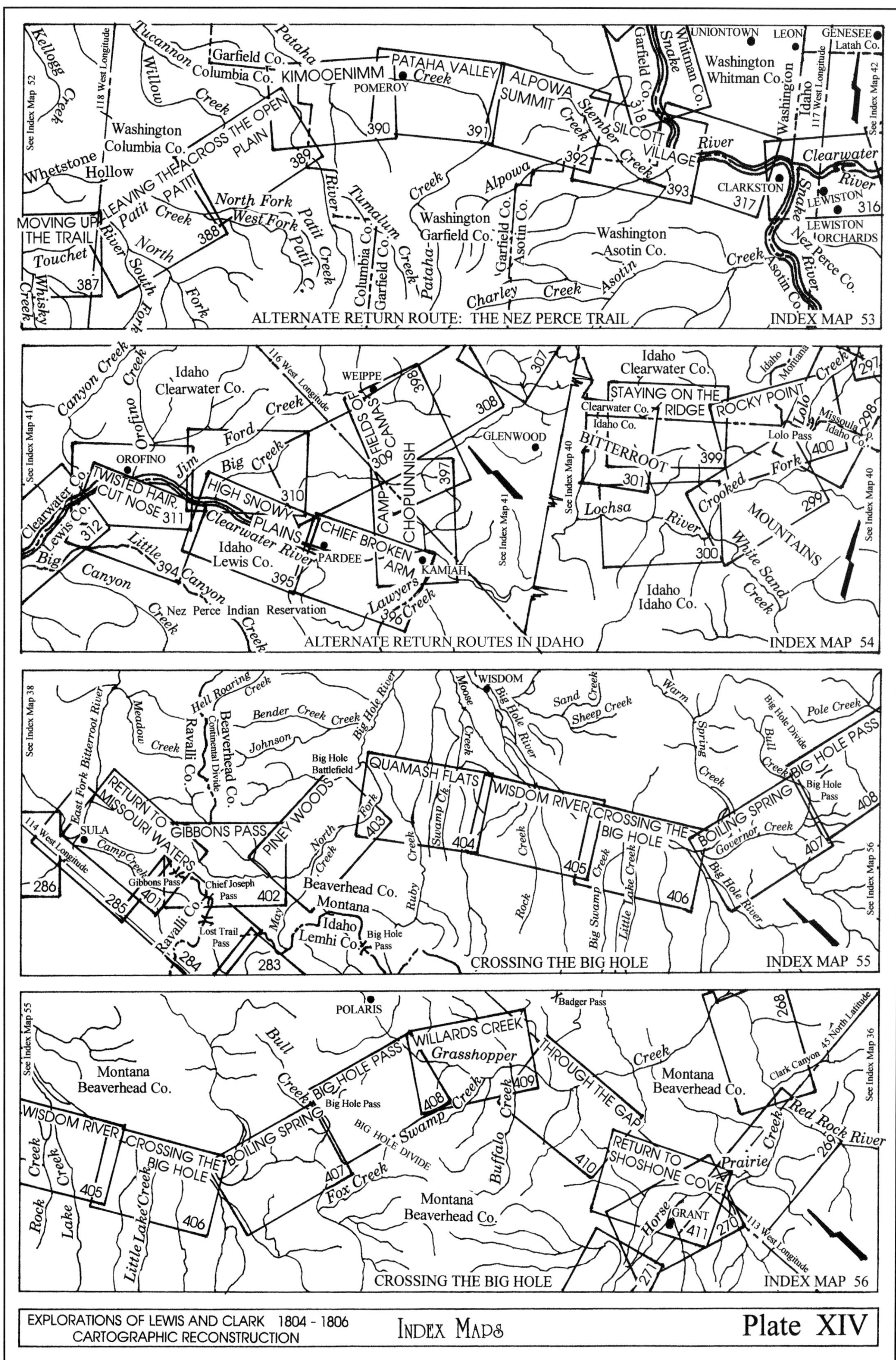
ALTERNATE RETURN ROUTE: THE NEZ PERCE TRAIL
INDEX MAP 53
MOVING UP THE TRAIL
387
LEAVING THE PATIT
388
ACROSS THE OPEN PLAIN
389
KIMOOENIMM
390
PATAHA VALLEY
391
ALPOWA SUMMIT
392
SILCOTT VILLAGE
393
317
316
Washington Columbia Co.
Washington Garfield Co.
Washington Asotin Co.
Washington Whitman Co.
Garfield Co.
Columbia Co.
Latah Co.
POMEROY
UNIONTOWN
LEON
GENESEE
CLARKSTON
LEWISTON
LEWISTON ORCHARDS
Tucannon
Willow
Creek
Kellogg
Pataha
Whetstone Hollow
Touchet
Whisky Creek
Patit Creek
North Fork
West Fork
South Fork
Tumalum
Alpowa
Stember Creek
Snake River
Clearwater River
Nez Perce Co.
Asotin Co.
Charley Creek
Asotin Creek
118 West Longitude
117 West Longitude
See Index Map 52
See Index Map 42
ALTERNATE RETURN ROUTES IN IDAHO
INDEX MAP 54
TWISTED HAIR, CUT NOSE
311
312
HIGH SNOWY PLAINS
310
FIELDS OF CAMAS
309
398
CAMP CHOPUNNISH
397
CHIEF BROKEN ARM
394
395
396
308
307
STAYING ON THE RIDGE
ROCKY POINT
297
298
BITTERROOT
301
399
400
MOUNTAINS
299
300
Idaho Clearwater Co.
Idaho Lewis Co.
Idaho Idaho Co.
Clearwater Co.
Idaho Co.
Missoula Co.
Idaho
Montana
OROFINO
WEIPPE
GLENWOOD
PARDEE
KAMIAH
Lolo Pass
Canyon Creek
Orofino Creek
Jim Ford Creek
Big Creek
Clearwater River
Little Canyon Creek
Big Canyon Creek
Lawyers Creek
Nez Perce Indian Reservation
Lochsa River
Crooked Fork
White Sand Creek
Lolo Creek
116 West Longitude
See Index Map 41
See Index Map 40
CROSSING THE BIG HOLE
INDEX MAP 55
RETURN TO MISSOURI WATERS
GIBBONS PASS
PINEY WOODS
QUAMASH FLATS
WISDOM RIVER
CROSSING THE BIG HOLE
BOILING SPRING
BIG HOLE PASS
286
285
284
283
401
402
403
404
405
406
407
408
SULA
WISDOM
Big Hole Battlefield
Gibbons Pass
Chief Joseph Pass
Lost Trail Pass
Big Hole Pass
Ravalli Co.
Beaverhead Co.
Continental Divide
Beaverhead Co. Montana
Idaho Lemhi Co.
East Fork Bitterroot River
Meadow Creek
Hell Roaring Creek
Bender Creek
Johnson
Big Hole River
Moose Creek
Sand Creek
Sheep Creek
Warm Spring Creek
Pole Creek
Bull Creek
Big Hole Divide
Governor Creek
Camp Creek
North Fork Creek
May Creek
Ruby Creek
Swamp Ck.
Rock Creek
Big Swamp Creek
Little Lake Creek
114 West Longitude
See Index Map 38
See Index Map 56
CROSSING THE BIG HOLE
INDEX MAP 56
WISDOM RIVER
CROSSING THE BIG HOLE
BOILING SPRING
BIG HOLE PASS
WILLARDS CREEK
THROUGH THE GAP
RETURN TO SHOSHONE COVE
405
406
407
408
409
410
411
268
269
270
271
Montana Beaverhead Co.
POLARIS
Badger Pass
Big Hole Pass
GRANT
Bull Creek
Grasshopper Creek
Swamp Creek
Fox Creek
Buffalo Creek
Big Hole Divide
Rock Creek
Lake Creek
Little Lake Creek
Clark Canyon
Red Rock River
Prairie Creek
Horse
45 North Latitude
113 West Longitude
See Index Map 55
See Index Map 36
EXPLORATIONS OF LEWIS AND CLARK 1804 - 1806
CARTOGRAPHIC RECONSTRUCTION
INDEX MAPS
Plate XIV

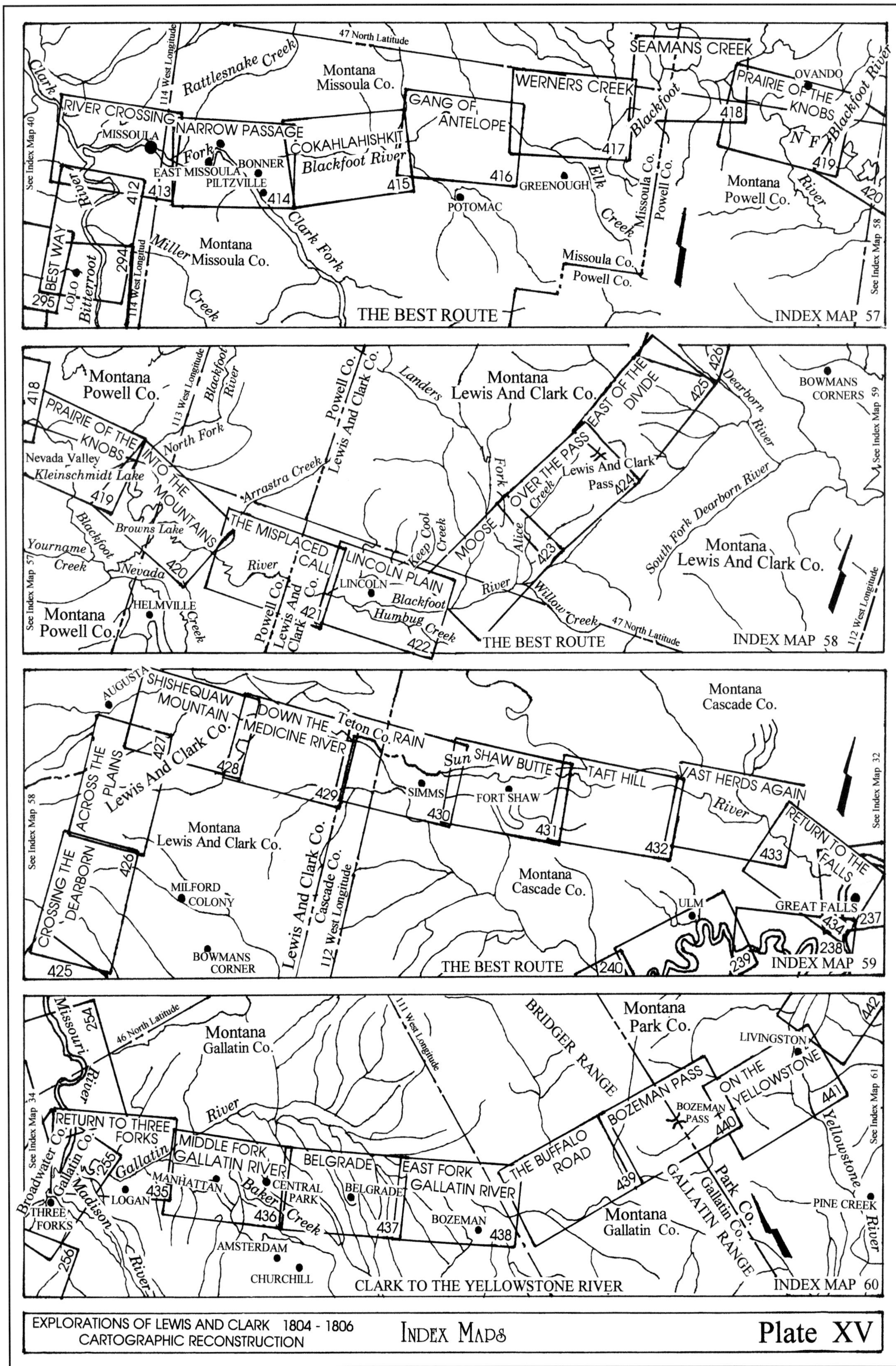

EXPLORATIONS OF LEWIS AND CLARK 1804 - 1806
CARTOGRAPHIC RECONSTRUCTION
INDEX MAPS
Plate XV

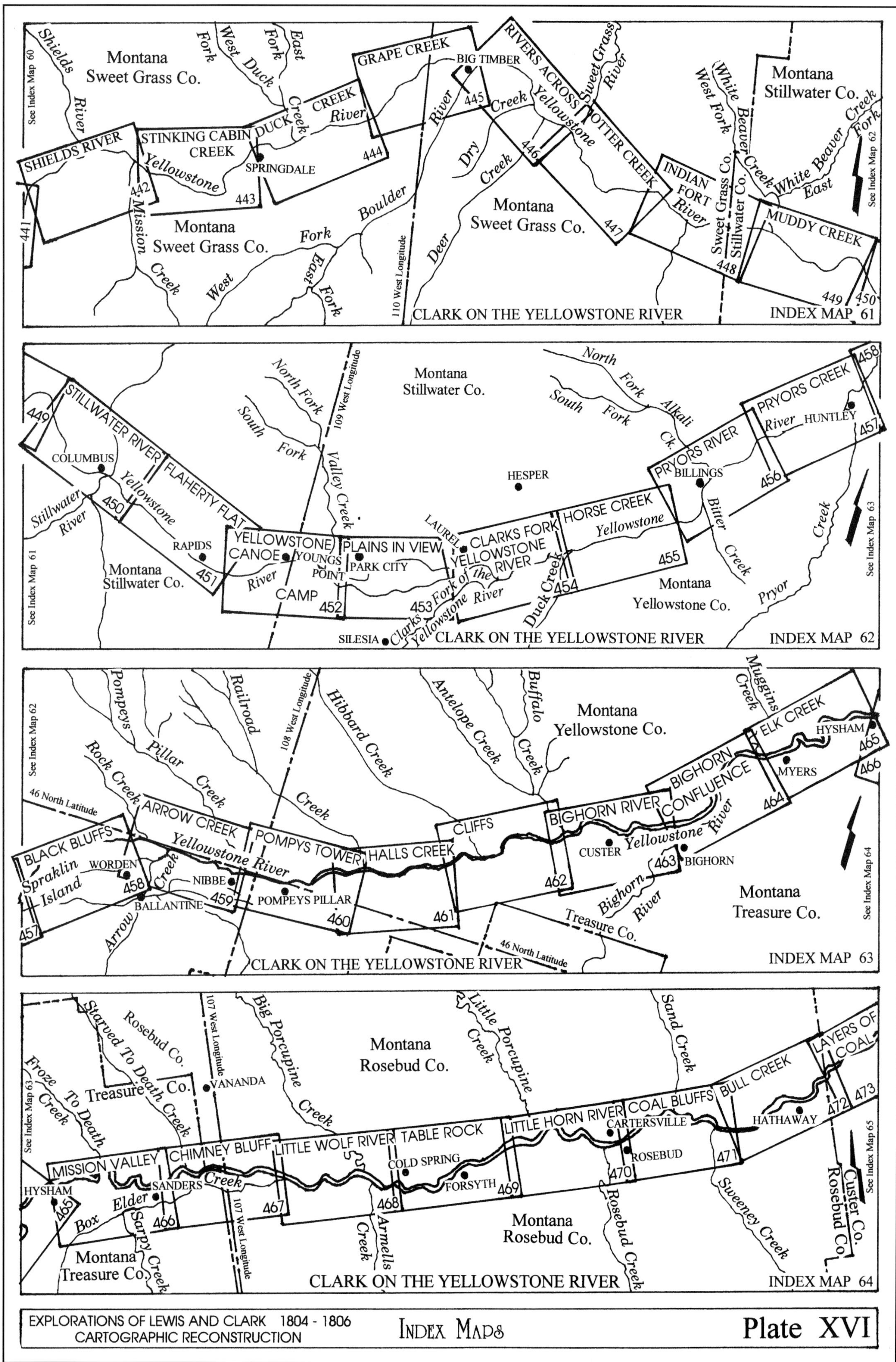

Montana Sweet Grass Co.
Shields River
SHIELDS RIVER
STINKING CABIN CREEK
DUCK CREEK
GRAPE CREEK
BIG TIMBER
RIVERS ACROSS
OTTER CREEK
INDIAN FORT
MUDDY CREEK
SPRINGDALE
Yellowstone
Mission Creek
West Fork Boulder
East Fork
Dry Creek
Deer Creek
Sweet Grass River
White Beaver Creek
East White Beaver Creek Fork
Montana Stillwater Co.
Sweet Grass Co.
Stillwater Co.
110 West Longitude
See Index Map 60
See Index Map 62
441
442
443
444
445
446
447
448
449
450
CLARK ON THE YELLOWSTONE RIVER
INDEX MAP 61
STILLWATER RIVER
FLAHERTY FLAT
COLUMBUS
RAPIDS
Stillwater River
Montana Stillwater Co.
YELLOWSTONE CANOE CAMP
YOUNGS POINT
PLAINS IN VIEW
PARK CITY
LAUREL
CLARKS FORK YELLOWSTONE RIVER
Clarks Fork of the Yellowstone River
HORSE CREEK
Duck Creek
HESPER
PRYORS RIVER
BILLINGS
PRYORS CREEK
HUNTLEY
North Fork
South Fork
Valley Creek
Alkali Ck.
Bitter Creek
Pryor Creek
109 West Longitude
SILESIA
Montana Yellowstone Co.
See Index Map 61
See Index Map 63
449
450
451
452
453
454
455
456
457
458
CLARK ON THE YELLOWSTONE RIVER
INDEX MAP 62
BLACK BLUFFS
Spraklin Island
WORDEN
BALLANTINE
ARROW CREEK
Arrow Creek
Rock Creek
Pompeys Pillar Creek
Railroad Creek
Hibbard Creek
Antelope Creek
Buffalo Creek
Muggins Creek
NIBBE
POMPYS TOWER
POMPEYS PILLAR
HALLS CREEK
CLIFFS
BIGHORN RIVER
BIGHORN CONFLUENCE
CUSTER
BIGHORN
Bighorn River
ELK CREEK
HYSHAM
MYERS
Yellowstone River
46 North Latitude
108 West Longitude
Montana Yellowstone Co.
Montana Treasure Co.
Treasure Co.
See Index Map 62
See Index Map 64
457
458
459
460
461
462
463
464
465
466
CLARK ON THE YELLOWSTONE RIVER
INDEX MAP 63
Froze To Death Creek
Starved To Death Creek
Rosebud Co.
Treasure Co.
VANANDA
107 West Longitude
Big Porcupine Creek
Little Porcupine Creek
Montana Rosebud Co.
Sand Creek
MISSION VALLEY
HYSHAM
CHIMNEY BLUFF
SANDERS
Box Elder Creek
Sarpy Creek
LITTLE WOLF RIVER
TABLE ROCK
COLD SPRING
FORSYTH
Armells Creek
LITTLE HORN RIVER
CARTERSVILLE
ROSEBUD
Rosebud Creek
COAL BLUFFS
BULL CREEK
HATHAWAY
LAYERS OF COAL
Sweeney Creek
Custer Co.
Rosebud Co.
Montana Treasure Co.
See Index Map 63
See Index Map 65
465
466
467
468
469
470
471
472
473
CLARK ON THE YELLOWSTONE RIVER
INDEX MAP 64
EXPLORATIONS OF LEWIS AND CLARK 1804 - 1806
CARTOGRAPHIC RECONSTRUCTION
INDEX MAPS
Plate XVI

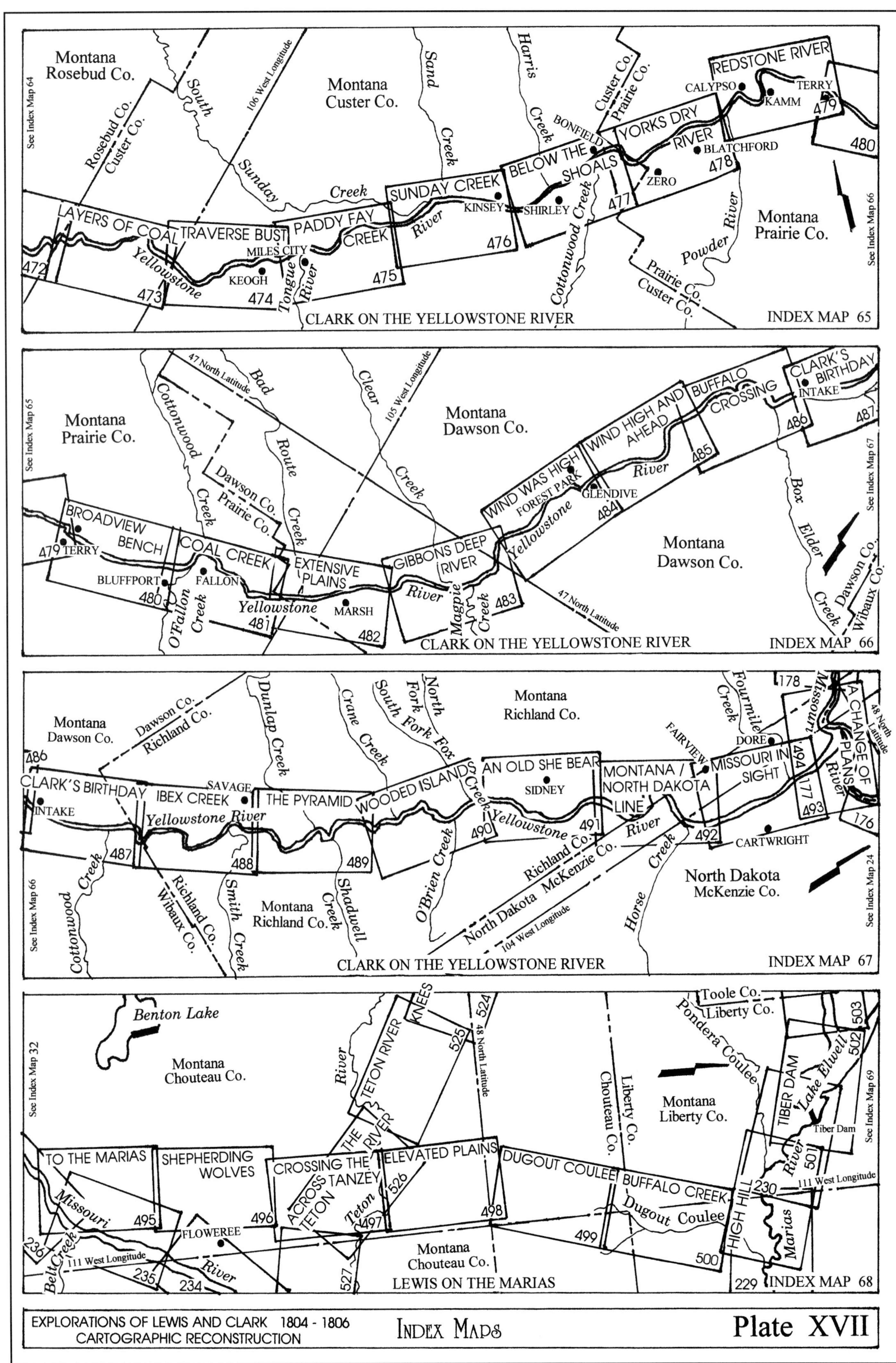

EXPLORATIONS OF LEWIS AND CLARK 1804 - 1806
CARTOGRAPHIC RECONSTRUCTION

INDEX MAPS

Plate XVII

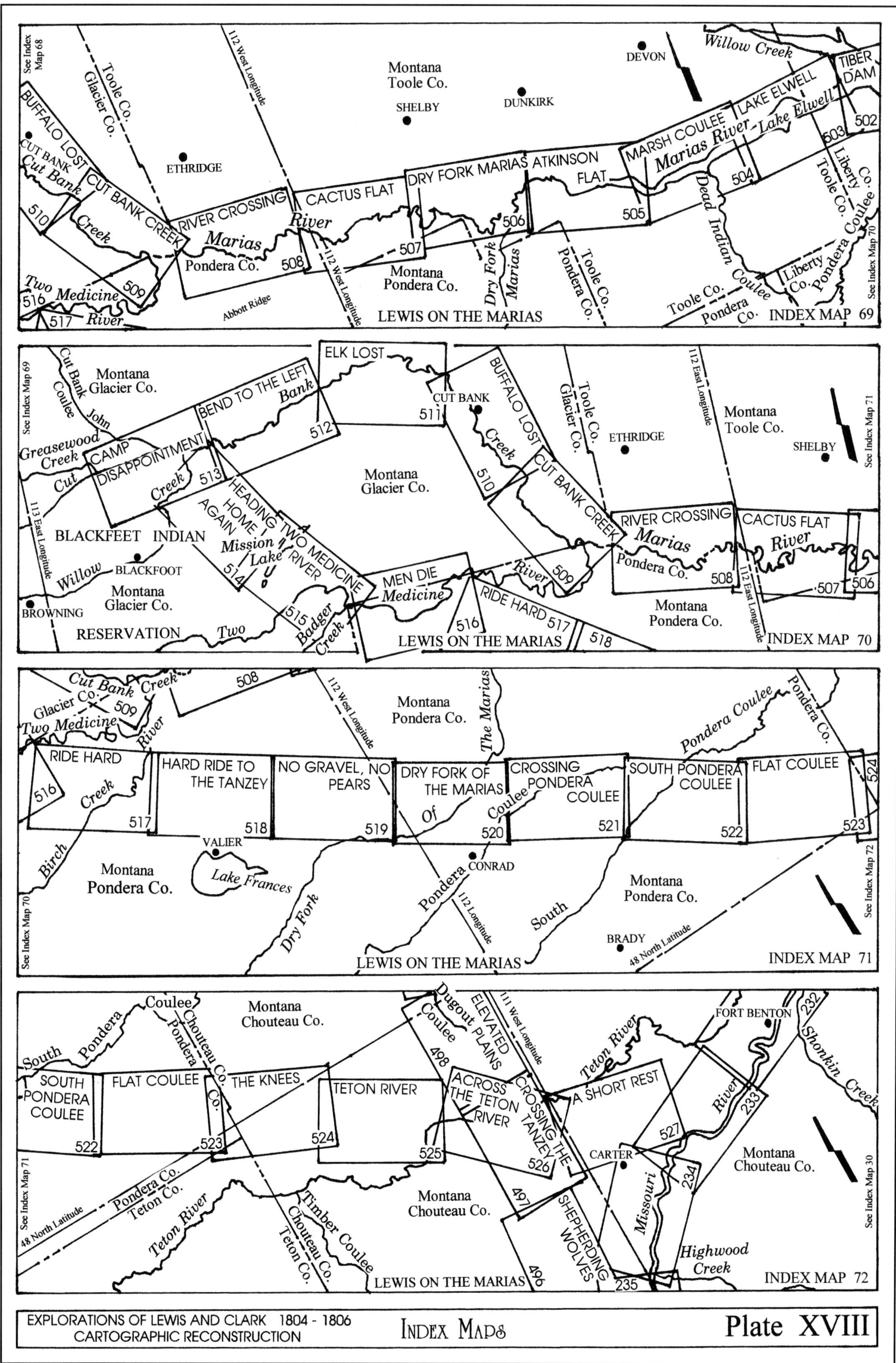

See Index Map 68
Montana
Toole Co.
SHELBY
DUNKIRK
DEVON
Willow Creek
TIBER DAM
502
LAKE ELWELL
Lake Elwell
Marias River
MARSH COULEE
503
504
Liberty Co.
Toole Co.
Toole Co.
Glacier Co.
112 West Longitude
BUFFALO LOST
CUT BANK
Cut Bank
ETHRIDGE
CUT BANK CREEK
Creek
510
RIVER CROSSING
CACTUS FLAT
DRY FORK MARIAS
ATKINSON FLAT
505
River
Marias
506
507
508
509
Pondera Co.
Montana
Pondera Co.
Dry Fork
Marias
Pondera Co.
Toole Co.
Dead Indian Coulee
Pondera Coulee
Liberty Co.
Toole Co.
Pondera Co.
Two Medicine
516
517
River
Abbott Ridge
112 West Longitude
LEWIS ON THE MARIAS
INDEX MAP 69
See Index Map 70
See Index Map 69
Cut Bank Coulee
John
Montana
Glacier Co.
ELK LOST
BEND TO THE LEFT
Bank
512
CUT BANK
511
BUFFALO LOST
Creek
510
Toole Co.
Glacier Co.
ETHRIDGE
112 East Longitude
Montana
Toole Co.
SHELBY
See Index Map 71
Greasewood Creek
CAMP DISAPPOINTMENT
Cut
Creek
513
Montana
Glacier Co.
CUT BANK CREEK
113 East Longitude
BLACKFEET INDIAN
HEADING HOME AGAIN
TWO MEDICINE RIVER
Mission Lake
514
RIVER CROSSING
Marias
CACTUS FLAT
River
Pondera Co.
508
509
507
506
Willow
BLACKFOOT
Montana
Glacier Co.
BROWNING
RESERVATION
Two
515
Badger Creek
MEN DIE
Medicine
River
RIDE HARD
516
517
518
LEWIS ON THE MARIAS
Montana
Pondera Co.
112 East Longitude
INDEX MAP 70
Cut Bank Creek
Glacier Co.
509
508
Two Medicine
River
112 West Longitude
Montana
Pondera Co.
The Marias
Pondera Coulee
Pondera Co.
RIDE HARD
516
Creek
517
HARD RIDE TO THE TANZEY
518
NO GRAVEL, NO PEARS
519
DRY FORK OF THE MARIAS
Of
520
CROSSING PONDERA COULEE
Coulee
521
SOUTH PONDERA COULEE
522
FLAT COULEE
523
524
See Index Map 72
Birch
VALIER
Montana
Pondera Co.
Lake Frances
Dry Fork
Pondera
CONRAD
112 Longitude
South
Montana
Pondera Co.
BRADY
48 North Latitude
See Index Map 70
LEWIS ON THE MARIAS
INDEX MAP 71
South
Pondera
Coulee
Pondera
Chouteau Co.
Montana
Chouteau Co.
Dugout Coulee
ELEVATED PLAINS
111 West Longitude
498
Teton River
FORT BENTON
232
Shonkin Creek
SOUTH PONDERA COULEE
522
FLAT COULEE
523
Co.
THE KNEES
TETON RIVER
524
ACROSS THE TETON RIVER
525
CROSSING THE TANZEY
526
A SHORT REST
527
River
233
CARTER
234
Montana
Chouteau Co.
See Index Map 30
See Index Map 71
Pondera Co.
Teton Co.
48 North Latitude
Teton River
Timber Coulee
Chouteau Co.
Teton Co.
Montana
Chouteau Co.
497
SHEPHERDING WOLVES
Missouri
Highwood Creek
496
235
LEWIS ON THE MARIAS
INDEX MAP 72
EXPLORATIONS OF LEWIS AND CLARK 1804 - 1806
CARTOGRAPHIC RECONSTRUCTION
INDEX MAPS
Plate XVIII

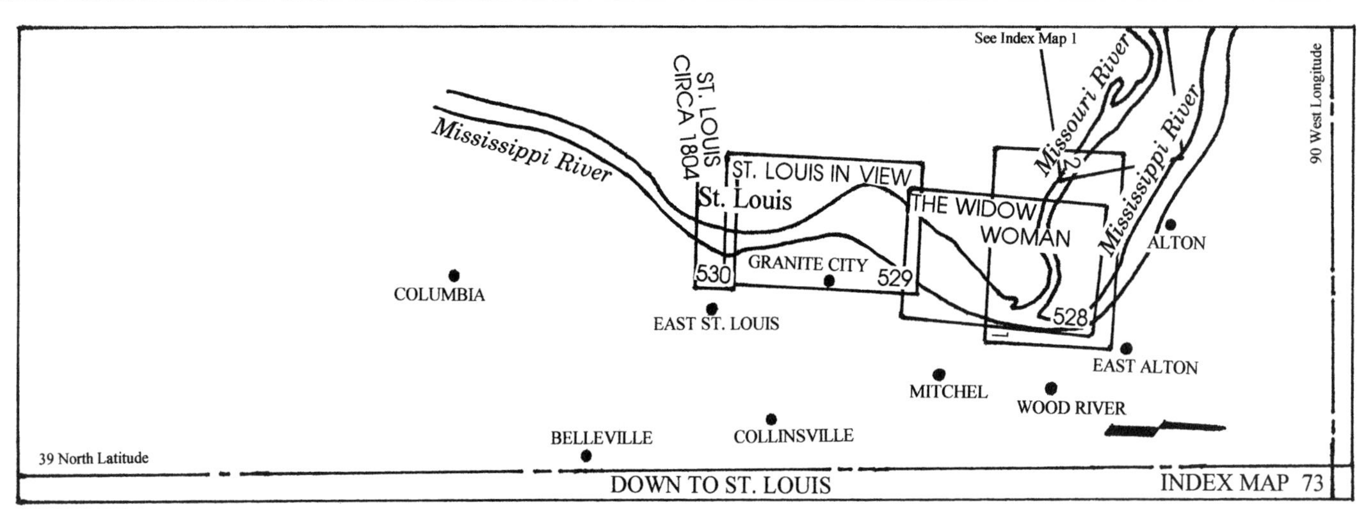

LEGEND

Lines

- Water boundary 1804 - 1806
- Small stream 1804 - 1806
- Water boundary or stream, modern
- Water boundary for reservoir (normal pool elevation)
- Principal parallel or meridian
- State boundary line
- County, reservation, or other jurisdictional boundary
- Major highway, modern
- Contour line (historically reconstructed)
- Depression in topography
- Shading lines 1804 - 1806 (water bodies only)

Symbols

- Interstate highway shield with number
- Federal highway shield with number
- State or local highway
- Airport or landing strip
- Bridge
- Dam (small)
- Spring
- Cave
- Falls
- Rapid
- Point of interest
- Surveyed Traverse Symbol
  End of one bearing/distance and beginning of next. Direction and distance as stated in Clark's journal. Traverse corrected for magnetic declination and adjusted to fit topography. Unusual deviation of bearing is noted.
- Universal Transverse Mercator
  Mapping grid. Grid tics are shown at 1,000 meter intervals along map border. Larger number in notation indicates thousands of meters, smaller number indicates hundreds of thousands of meters.

Fonts

Various font styles have been used to relate certain types of information.

Century Font *Century Font*

The Century font is used to denote feature names existing at the time of Lewis and Clark. Vertical for land features. Italicized for water features. Quotation marks about the name indicate that Lewis and Clark gave the name.

Dot Matrix Font

The Dot Matrix font is used for all notations relating to modern or post Lewis and Clark information. Notations relevant to water are italicized.

Script Roundhand Font

Script Roundhand font is used to convey information from the journals. Short notes that have been altered slightly to fit mapping standards do not have quotation marks. Direct quotes from the journals are enclosed in quotation marks. They are followed by the name of the journal writer and an arrow indicating whether the quote comes from the Outbound travel (⇨) or Return journey (⇦). See further explanation Maps 1 and 2.

Advance Font

The Advance font is used to denote information related to the story but not addressed adequately in the journals.

Cartographer's Note:

Generally an explanation of a mapping detail, a mapping decision, a major adjustment to the surveyed traverse, or an explanation of events as viewed by the cartographer.

All north arrows on The Trail Maps, as well as the index sheets, indicate True North.

Campsite Notations

Camp
May 6, 1805 — Outbound Camp with date of the evening camp. Corps left camp on morning following last date.

Return Camp
June 28, 1806 — Home Bound Camp with date of the evening camp. Corps left camp on morning following last date.

Station Camp
Nov. 15 - 24, 1805 — Named Camp. Important enough to be named by the explorers along with date of encampment.

Notations

453	Lewis and Clark mileposts. Mileposts based on traverse distances. Distances are not scaleable. Outbound route only.
+398	Corrected mileposts. Outbound only.
N 23 W 2 MILES	Traverse notations. Traverse bearings (directions) based on magnetic north, 1804-1806. Notations are from Clark's journal and have been adjusted to fit topography.
VANCOUVER	City or town. Position of note indicates location of town.
American Grizzly Bear (Ursus horribilis horribilis)	Flora and fauna. New to science, first described by Lewis and Clark, or significant mention by journalist. Common name followed by scientific or Latin name in parentheses.

Title Block Notations

Approximate date(s) for post Expedition data shown on map.

Visual Scale.

Dates for which the Expedition was in an area covered by a map. Outbound and return dates are shown.

Project Title.

Map sheet title.

Universal Transverse Mercator (UTM) zone number.

Map sheet number.

Map covers portions of these states.

Contour interval information.

EXPLORATIONS OF LEWIS AND CLARK 1804 - 1806
CARTOGRAPHIC RECONSTRUCTION

LEGEND AND INDEX MAP

Plate XIX

Lewis and Clark Trail Maps, Volume III

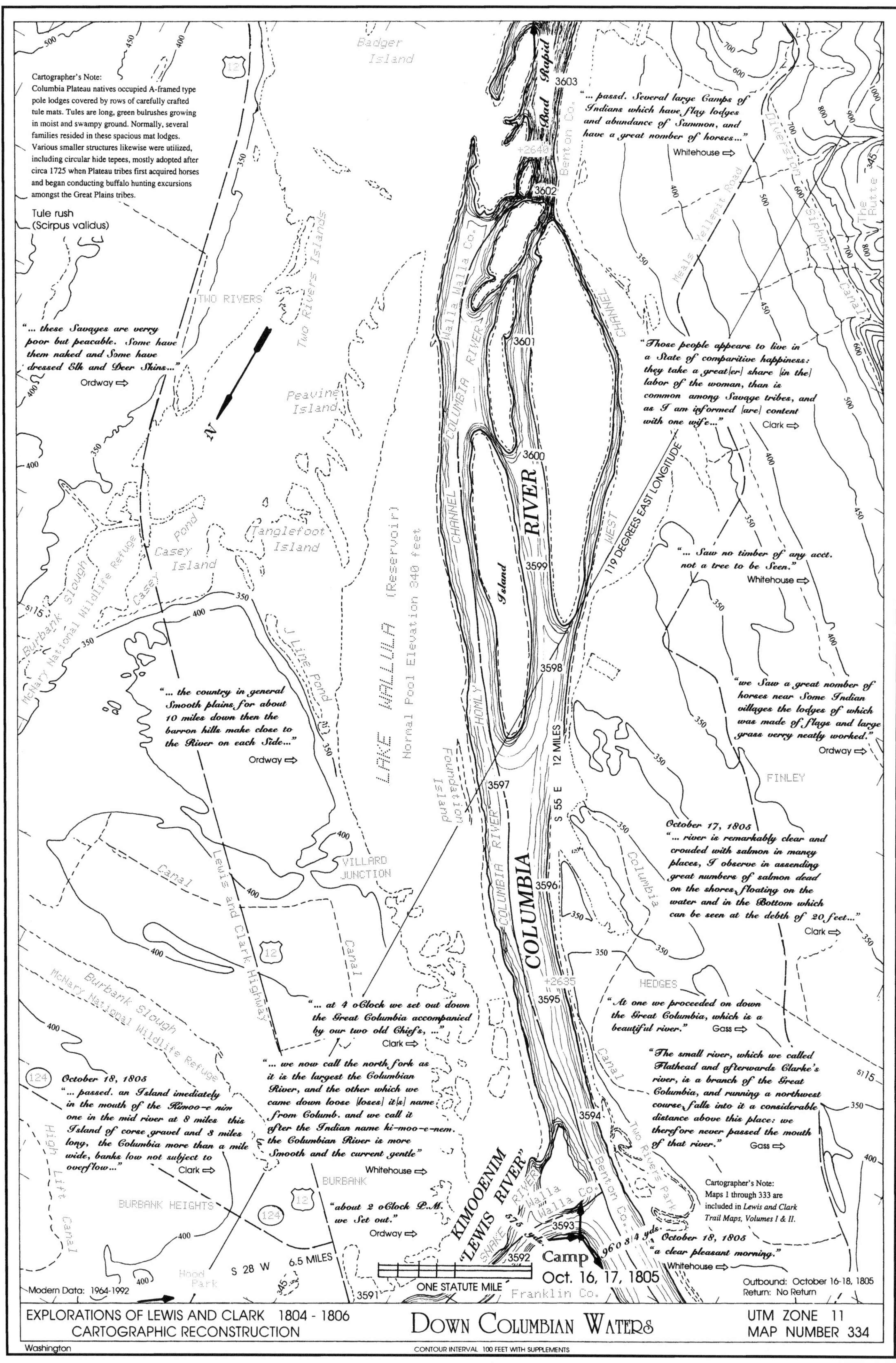

Cartographer's Note:
Columbia Plateau natives occupied A-framed type pole lodges covered by rows of carefully crafted tule mats. Tules are long, green bulrushes growing in moist and swampy ground. Normally, several families resided in these spacious mat lodges. Various smaller structures likewise were utilized, including circular hide tepees, mostly adopted after circa 1725 when Plateau tribes first acquired horses and began conducting buffalo hunting excursions amongst the Great Plains tribes.
Tule rush (Scirpus validus)
Badger Island
Bad Rapid
"... passd. Several large Camps of Indians which have flag lodges and abundance of Sammon, and have a great nomber of horses..." Whitehouse ⇨
Benton Co.
Walla Walla Co.
Meals Yellepit Road
Diversion Siphon Canal
The Butte
TWO RIVERS
Two Rivers Islands
"... these Savages are very poor but peacable. Some have them naked and Some have dressed Elk and Deer Skins..." Ordway ⇨
Peavine Island
COLUMBIA RIVER
CHANNEL
"Those people appears to live in a State of comparitive happiness: they take a great[er] share [in the] labor of the woman, than is common among Savage tribes, and as I am informed [are] content with one wife..." Clark ⇨
119 DEGREES EAST LONGITUDE
WEST CHANNEL
RIVER
Island
Tanglefoot Island
Casey Pond
Casey Island
LAKE WALLULA (Reservoir)
Normal Pool Elevation 340 feet
"... Saw no timber of any acct. not a tree to be Seen." Whitehouse ⇨
Burbank Slough
McNary National Wildlife Refuge
J Line Pond
"... the country in general Smooth plains for about 10 miles down then the barron hills make close to the River on each Side..." Ordway ⇨
HOMLY
"we Saw a great nomber of horses near Some Indian villages the lodges of which was made of flags and large grass very neatly worked." Ordway ⇨
Foundation Island
12 MILES
S 55 E
FINLEY
October 17, 1805
"... river is remarkably clear and crouded with salmon in maney places, I observe in assending great numbers of salmon dead on the shores, floating on the water and in the Bottom which can be seen at the debth of 20 feet..." Clark ⇨
VILLARD JUNCTION
Canal
Lewis and Clark Highway
Columbia
COLUMBIA
HEDGES
"... at 4 oClock we set out down the Great Columbia accompanied by our two old Chiefs, ..." Clark ⇨
"At one we proceeded on down the Great Columbia, which is a beautiful river." Gass ⇨
"... we now call the north fork as it is the largest the Columbian River, and the other which we came down loose [loses] it[s] name from Columb. and we call it after the Indian name ki-moo-e-nem. the Columbian River is more Smooth and the current gentle" Whitehouse ⇨
"The small river, which we called Flathead and afterwards Clarke's river, is a branch of the Great Columbia, and running a northwest course, falls into it a considerable distance above this place: we therefore never passed the mouth of that river." Gass ⇨
October 18, 1805
"... passed. an Island imediately in the mouth of the Kimoo-e nim one in the mid river at 8 miles this Island of corse gravel and 3 miles long, the Columbia more than a mile wide, banks low not subject to overflow..." Clark ⇨
High Lift Canal
BURBANK
BURBANK HEIGHTS
"about 2 oClock P.M. we Set out." Ordway ⇨
KIMOOENIM "LEWIS RIVER"
SNAKE RIVER
Walla Walla Co.
Two Rivers Park
575 yds.
960 3/4 yds.
Camp Oct. 16, 17, 1805
October 18, 1805 "a clear pleasant morning." Whitehouse ⇨
Cartographer's Note: Maps 1 through 333 are included in Lewis and Clark Trail Maps, Volumes I & II.
S 28 W
6.5 MILES
ONE STATUTE MILE
Hood Park
Franklin Co.
Modern Data: 1964-1992
Outbound: October 16-18, 1805
Return: No Return
EXPLORATIONS OF LEWIS AND CLARK 1804 - 1806
CARTOGRAPHIC RECONSTRUCTION
DOWN COLUMBIAN WATERS
UTM ZONE 11
MAP NUMBER 334
Washington
CONTOUR INTERVAL 100 FEET WITH SUPPLEMENTS

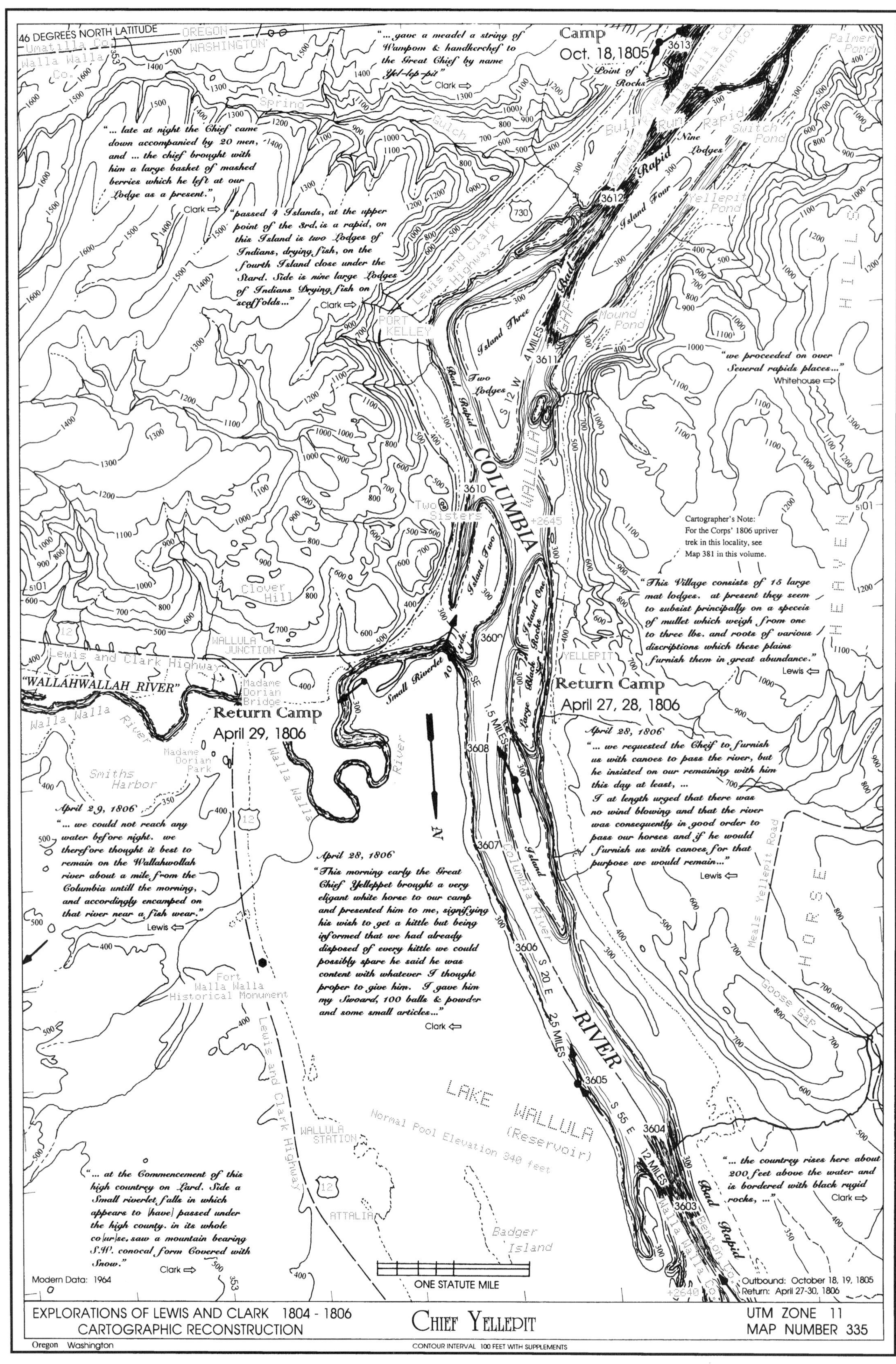
46 DEGREES NORTH LATITUDE
OREGON
WASHINGTON
Umatilla Co.
Walla Walla Co.
"... gave a meadel a string of Wampom & handkerchef to the Great Chief by name Yel-lep-pit"
Clark ⇨
Camp
Oct. 18, 1805
Point of Rocks
3613
Spring
Gulch
Bull Run Rapid
Nine Lodges
Switch Pond
Palmer Pond
Rapid
Island Four
3612
Yellepit Pond
"... late at night the Chief came down accompanied by 20 men, and ... the chief brought with him a large basket of mashed berries which he left at our Lodge as a present."
Clark ⇨
"passed 4 Islands, at the upper point of the 3rd. is a rapid, on this Island is two Lodges of Indians, drying fish, on the fourth Island close under the Stard. Side is nine large Lodges of Indians Drying fish on scaffolds..."
Clark ⇨
Lewis and Clark Highway
730
PORT KELLEY
Island Three
Bad Rapid
Mound Pond
HORSE HEAVEN HILLS
4 MILES
S 12 W
3611
Two Lodges
"we proceeded on over Several rapids places..."
Whitehouse ⇨
COLUMBIA
WALLULA
3610
Two Sisters
+2645
Cartographer's Note: For the Corps' 1806 upriver trek in this locality, see Map 381 in this volume.
Island Two
Island One
Clover Hill
"This Village consists of 15 large mat lodges. at present they seem to subsist principally on a species of mullet which weigh from one to three lbs. and roots of various discriptions which these plains furnish them in great abundance."
Lewis ⇦
12
WALLULA JUNCTION
Lewis and Clark Highway
Black Rocks
YELLEPIT
3609
40 yds.
"WALLAHWALLAH RIVER"
Madame Dorian Bridge
Small Riverlet
SE
Return Camp
April 27, 28, 1806
Return Camp
April 29, 1806
Walla Walla River
Large Island
1.5 MILES
3608
April 28, 1806
"... we requested the Cheif to furnish us with canoes to pass the river, but he insisted on our remaining with him this day at least, ... I at length urged that there was no wind blowing and that the river was consequently in good order to pass our horses and if he would furnish us with canoes for that purpose we would remain..."
Lewis ⇦
Madame Dorian Park
Smiths Harbor
Walla Walla
N
April 29, 1806
"... we could not reach any water before night. we therefore thought it best to remain on the Wallahwollah river about a mile from the Columbia untill the morning, and accordingly encamped on that river near a fish wear."
Lewis ⇦
April 28, 1806
"This morning early the Great Chief Yelleppet brought a very eligant white horse to our camp and presented him to me, signifying his wish to get a kittle but being informed that we had already disposed of every kittle we could possibly spare he said he was content with whatever I thought proper to give him. I gave him my Swoard, 100 balls & powder and some small articles..."
Clark ⇦
3607
Columbia River
Meals Yellepit Road
3606
S 20 E
Fort Walla Walla Historical Monument
2.5 MILES
RIVER
Goose Gap
3605
S 55 E
LAKE WALLULA (Reservoir)
Normal Pool Elevation 340 feet
WALLULA STATION
3604
12 MILES
Lewis and Clark Highway
"... the countrey rises here about 200 feet above the water and is bordered with black rugid rocks, ..."
Clark ⇨
"... at the Commencement of this high countrey on Lard. Side a Small riverlet falls in which appears to [have] passed under the high county. in its whole co[ur]se, saw a mountain bearing S.W. conocal form Covered with Snow."
Clark ⇨
ATTALIA
Badger Island
3603
Bad Rapid
Walla Walla Co.
Benton Co.
+2640
Modern Data: 1964
353
ONE STATUTE MILE
Outbound: October 18, 19, 1805
Return: April 27-30, 1806
EXPLORATIONS OF LEWIS AND CLARK 1804 - 1806
CARTOGRAPHIC RECONSTRUCTION
CHIEF YELLEPIT
UTM ZONE 11
MAP NUMBER 335
Oregon Washington
CONTOUR INTERVAL 100 FEET WITH SUPPLEMENTS

The high Cascade summits were first observed by seafaring explorers/traders in the 1780s–90s. In 1805–6, the locations of four of these great peaks caused confusion for Clark for nearly a half-year. They included:

Mount St. Helens—a beautiful, symmetrical, 9,671-ft. summit (pre-1980 eruption) named by Captain George Vancouver in recognition of Britain's ambassador to Spain.

Mount Hood—an 11,235-ft. pointed peak named for a British admiralty official.

Mount Jefferson, with a 10,405-ft. pointed top, named by William Clark after President Jefferson.

Mount Adams—a rounded, 12,276-ft. bulky summit extending three miles long in a north-south configuration and named after the former U.S. president.

Of these, just Hood and St. Helens were indicated on Nicholas King's special 1803 map provided to the Corps.

Only until the return trip from the coast—at the Willamette River, April 2, 1806—did Clark definitively identify Mount Adams and its volcanic neighbors (G): "I can plainly see Mt. Jefferson which is high and covered with snow S.E.(,) Mt. Hood East, Mt St. Helians [and] a high humped mountain [Mount Adams] to the East of Mt. St. Helians." Clark now was aware of a "high humped" peak (Mount Adams) to the east of St. Helens, but was this Clark's first actual sighting of Mount Adams? Apparently not!

Over five months earlier at Wallula Gap, October 18, 1805, Clark noted (A B): "saw a mountain bearing S. W. conocal form Covered with Snow." Questions surround Clark's initial sighting of a Cascade peak on this date. His entry lacks essential information expected from a surveyor—namely, his vantage point, the estimated distance, and a compass bearing to the peak. Nor did Clark's comment appear in this day's "first draft" journal or traverse entries, but only in his "final" notes. And, Clark was the only diarist mentioning the sighting. Some suggest he somehow caught a glimpse of Mount Hood while busy with other matters (A). However, might it have been conical Mount Jefferson (C) that Clark saw? Perhaps Mount Jefferson was the only volcano potentially visible from within Wallula Gap's walls, possibly before the building of today's McNary Dam, located just downstream?

Most modern-day researchers put Clark's sighting at that evening's campsite (B). Capable regional investigators (including David Nicandri of the Washington State Historical Society, Gary Lentz of Lewis and Clark Trail State Park, and local Lewis and Clark researcher Steve Plucker) have found that Mount Hood (B) is visible from high up on the larboard shore, when looking across the Columbia and over the top of 800-ft. cliffs on the opposite shore. Though an impressive research effort, the visible portion of Mount Hood is quite small and one almost needs to know in advance the direction to sight it. Furthermore, Clark gave no indication that he climbed up the Wallula Gap cliffs to make observations.

Next day at the Umatilla Rapids, October 19, Clark noted (D): "I assended a high clift about 200 feet above the water…descovered a high mountain of emence hight covered with Snow, this must be one of the mountains laid down by Vancouver, as seen from the mouth of the Columbia River, from the course which it bears which is *West* I take it to be Mt. St. Helens…(also) destant about 120 miles a range of mountains in the Derection crossing a conical mountain S. W. toped with snow." The "conical mountain S. W." is Mount Hood, but was the "high mountain of emence hight" really St. Helens? Only Mount Adams is visible in that direction.

On November 3, at Government Island (E), Clark wrote: "the mountain we saw from near the forks proves to be Mount *Hood*." Next day at Sauvie Island (F), Clark recorded: "we had a full view of *Mt. Helien*…which is perhaps the highest pinical in America…it bears N. 25°. E. about 90 miles. This is the mountain I saw from the Muscle Shell (Umatilla) rapid on the 19th. of October last covered with Snow, it rises Something in the form of a Sugar lofe." Clark's compass direction is suspect due to local magnetic anomalies, but the distance indicates the latter actually was Mount Adams, which is visible from this part of Government Island.

Mount Rainier and Saddle Mountain, too, have caused some confusion in the Corps' story. For example, on November 12, while at Point Ellice, Whitehouse noted (H I): "We saw a high mountain which lay on the opposite to where we are encamped covered with snow." It has been suggested that this was Mount Hood (H). However, Mount Hood is not visible from this area. Rather, this was Saddle Mountain (I). On clear days, St. Helens and a north part of Adams are visible from the lower Columbia's north shore (J).

Thus, as related, not till April 2, 1806, at the Willamette River, did Clark geographically sort out these volcanic peaks.

Meriwether Lewis actually might have been the first to sight a Cascade volcano (Mount Adams) while the Corps was still two day's travel up the Snake River, October 15, 1805. On a clear, crisp autumn morning while the Corps dried wet equipment at Pine Tree Rapid till mid-afternoon (see *Volume II*, Map 328), Clark reports that "Lewis assended the hills," seeing in the far distance "a high point to the West." Granted, this and other of Clark's descriptions regarding Lewis's walk this day "on the plains" are rather vague, but today's local residents on nearby Eureka Flat say that Mount Adams is visible from the hills above the Snake River.

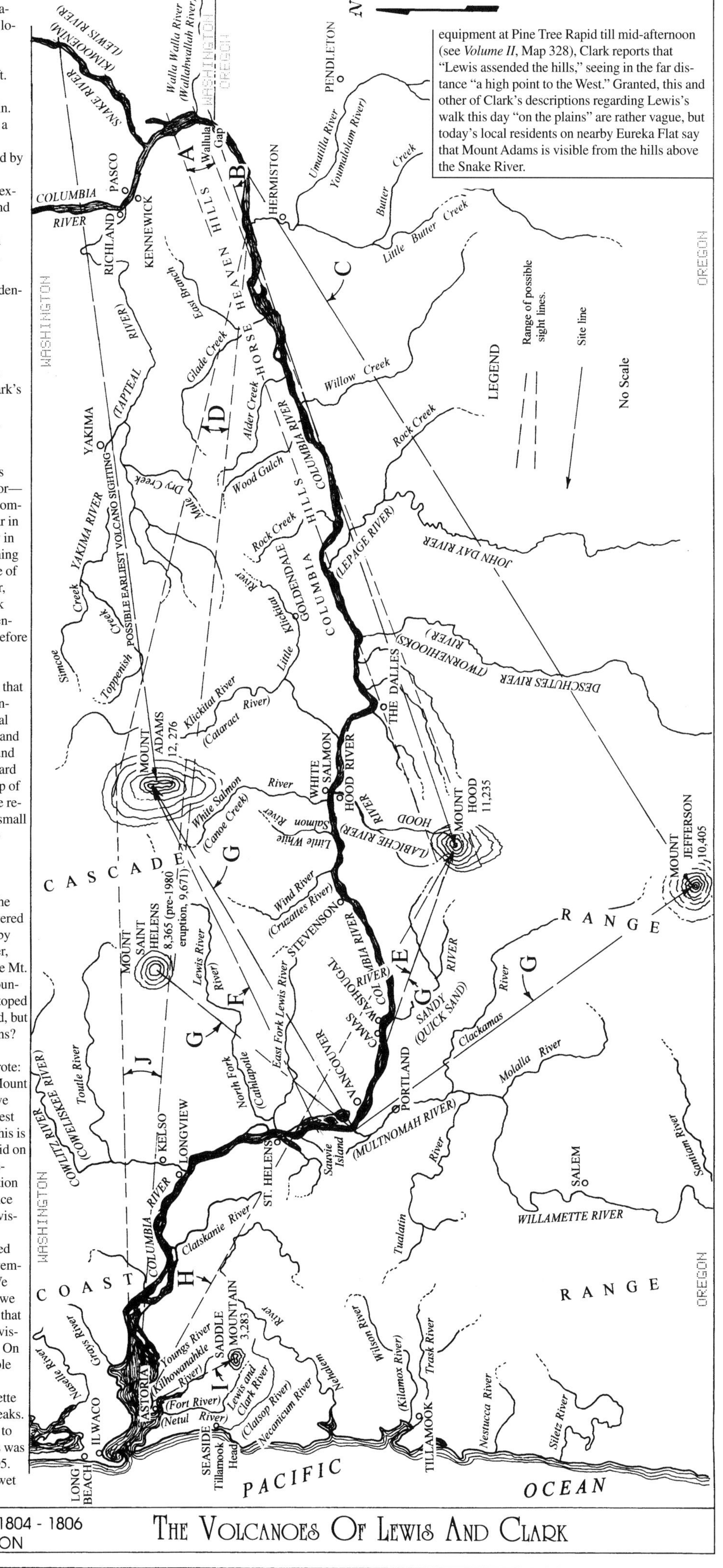

THE VOLCANOES OF LEWIS AND CLARK

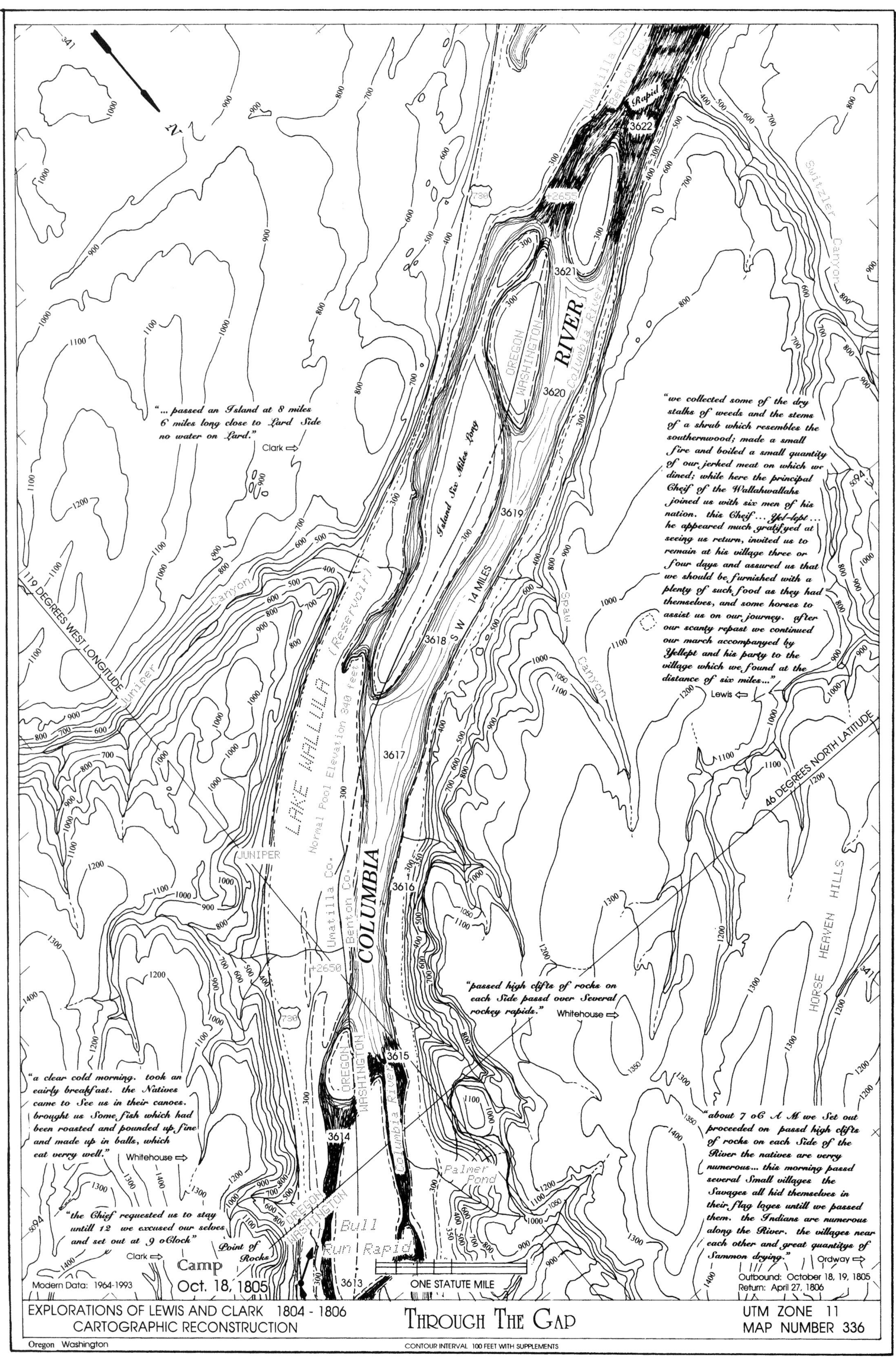
"... passed an Island at 8 miles 6 miles long close to Lard Side no water on Lard."
Clark
"we collected some of the dry stalks of weeds and the stems of a shrub which resembles the southernwood; made a small fire and boiled a small quantity of our jerked meat on which we dined; while here the principal Cheif of the Wallahwallahs joined us with six men of his nation. this Cheif ... Yel-lept ... he appeared much gratifyed at seeing us return, invited us to remain at his village three or four days and assured us that we should be furnished with a plenty of such food as they had themselves, and some horses to assist us on our journey. after our scanty repast we continued our march accompanyed by Yellept and his party to the village which we found at the distance of six miles..."
Lewis
"passed high clifts of rocks on each Side passd over Several rockey rapids."
Whitehouse
"a clear cold morning. took an eairly breakfast. the Natives came to See us in their canoes. brought us Some fish which had been roasted and pounded up fine and made up in balls, which eat verry well."
Whitehouse
"the Chief requested us to stay untill 12 we excused our selves and set out at 9 oClock"
Clark
"about 7 oC A M we Set out proceeded on passd high clifts of rocks on each Side of the River the natives are verry numerous... this morning passd several Small villages the Savages all hid themselves in their flag loges untill we passed them. the Indians are numerous along the River. the villages near each other and great quantitys of Sammon drying."
Ordway
Island Six Miles Long
LAKE WALLULA (Reservoir)
Normal Pool Elevation 340 feet
COLUMBIA RIVER
Columbia River
SW 14 MILES
Rapid
Bull Run Rapid
Point of Rocks
Palmer Pond
Camp Oct. 18, 1805
Switzler Canyon
Spring Canyon
Juniper Canyon
JUNIPER
HORSE HEAVEN HILLS
Umatilla Co.
Benton Co.
OREGON
WASHINGTON
119 DEGREES WEST LONGITUDE
46 DEGREES NORTH LATITUDE
ONE STATUTE MILE
Modern Data: 1964-1993
Outbound: October 18, 19, 1805
Return: April 27, 1806
EXPLORATIONS OF LEWIS AND CLARK 1804 - 1806
CARTOGRAPHIC RECONSTRUCTION
THROUGH THE GAP
UTM ZONE 11
MAP NUMBER 336
Oregon Washington
CONTOUR INTERVAL 100 FEET WITH SUPPLEMENTS

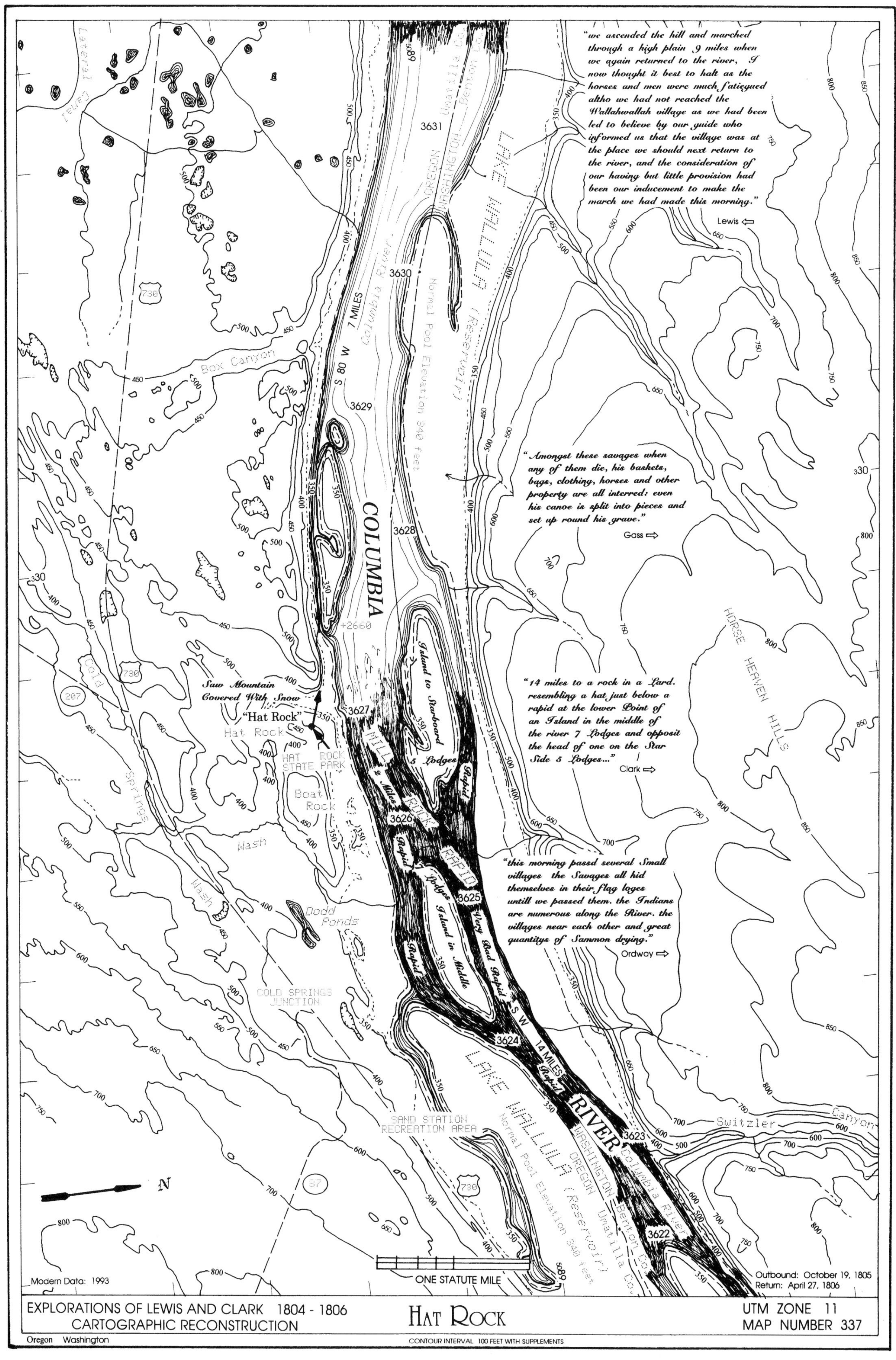
"we ascended the hill and marched through a high plain 9 miles when we again returned to the river, I now thought it best to halt as the horses and men were much fatiegued altho we had not reached the Wallahwallah village as we had been led to believe by our guide who informed us that the village was at the place we should next return to the river, and the consideration of our having but little provision had been our inducement to make the march we had made this morning."
Lewis
"Amongst these savages when any of them die, his baskets, bags, clothing, horses and other property are all interred: even his canoe is split into pieces and set up round his grave."
Gass
"14 miles to a rock in a Lard. resembling a hat just below a rapid at the lower Point of an Island in the middle of the river 7 Lodges and opposit the head of one on the Star Side 5 Lodges..."
Clark
"this morning passd several Small villages the Savages all hid themselves in their flag loges untill we passed them. the Indians are numerous along the River. the villages near each other and great quantitys of Sammon drying."
Ordway
COLUMBIA
RIVER
LAKE WALLULA
LAKE WALLULA
Columbia River
Normal Pool Elevation 340 feet
(Reservoir)
OREGON
WASHINGTON
Umatilla Co.
Benton Co.
S 80 W 7 MILES
S W 14 MILES
Saw Mountain Covered With Snow
"Hat Rock"
Hat Rock
HAT ROCK STATE PARK
Boat Rock
Island to Starboard
5 Lodges
Island in Middle
7 Lodges
Rapid
Very Bad Rapid
ROCK RAPID
Box Canyon
Cold Springs Wash
Dodd Ponds
COLD SPRINGS JUNCTION
SAND STATION RECREATION AREA
HORSE HEAVEN HILLS
Switzler Canyon
Lateral Canal
730
207
37
N
ONE STATUTE MILE
Modern Data: 1993
Outbound: October 19, 1805
Return: April 27, 1806
EXPLORATIONS OF LEWIS AND CLARK 1804 - 1806
CARTOGRAPHIC RECONSTRUCTION
HAT ROCK
UTM ZONE 11
MAP NUMBER 337
Oregon Washington
CONTOUR INTERVAL 100 FEET WITH SUPPLEMENTS

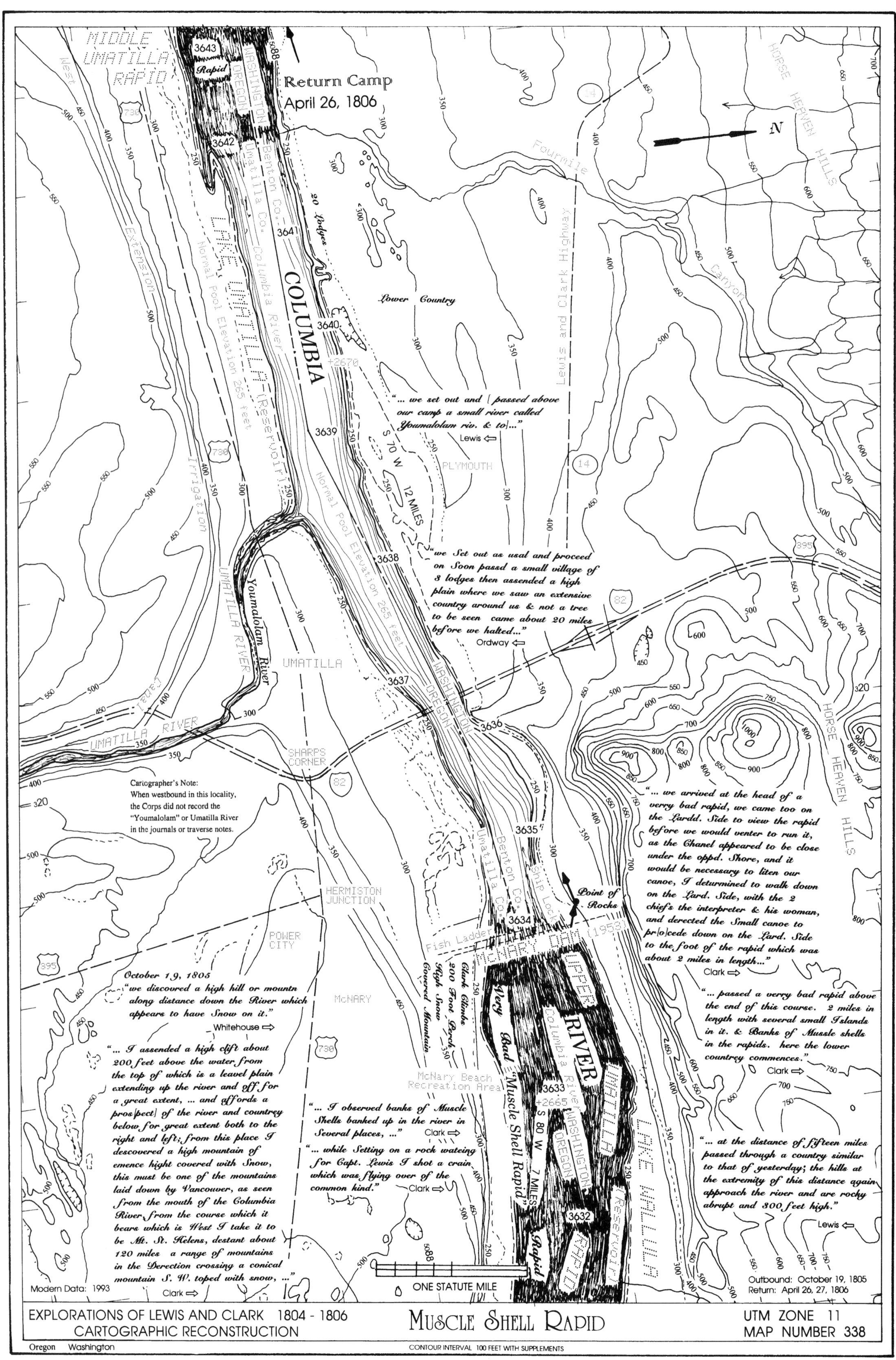

MIDDLE UMATILLA RAPID
Return Camp
April 26, 1806
COLUMBIA
20 Lodges
Lower Country
LAKE UMATILLA (Reservoir)
Normal Pool Elevation 265 feet
Benton Co.
Umatilla Co.
Columbia River
WASHINGTON
OREGON
Rapid
Fourmile
HORSE HEAVEN HILLS
Canyon
Lewis and Clark Highway
"... we set out and [passed above our camp a small river called Youmalolam riv. & to]..."
Lewis ⇦
S 70 W
12 MILES
PLYMOUTH
"we Set out as usal and proceed on Soon passd a small village of 8 lodges then assended a high plain where we saw an extensive country around us & not a tree to be seen came about 20 miles before we halted..."
Ordway ⇦
Youmalolam River
UMATILLA RIVER
UMATILLA
SHARPS CORNER
Extension
Irrigation
Cartographer's Note:
When westbound in this locality, the Corps did not record the "Youmalolam" or Umatilla River in the journals or traverse notes.
HERMISTON JUNCTION
POWER CITY
McNARY
"... we arrived at the head of a verry bad rapid, we came too on the Lardd. Side to view the rapid before we would venter to run it, as the Chanel appeared to be close under the oppd. Shore, and it would be necessary to liten our canoe, I determined to walk down on the Lard. Side, with the 2 chiefs the interpreter & his woman, and derected the Small canoe to pr[o]cede down on the Lard. Side to the foot of the rapid which was about 2 miles in length..."
Clark ⇨
Point of Rocks
Ship Lock
Fish Ladder
McNARY DAM (1953)
"... passed a verry bad rapid above the end of this course. 2 miles in length with several small Islands in it. & Banks of Mussle shells in the rapids. here the lower countrey commences."
Clark ⇨
October 19, 1805
"we discovered a high hill or mountn along distance down the River which appears to have Snow on it."
Whitehouse ⇨
Clark Climbs 200 Foot Rock High Snow Covered Mountain
Very Bad "Muscle Shell Rapid"
UPPER RIVER
UMATILLA
Columbia River
S 80 W
7 MILES
Rapid
LAKE WALLULA
(Reservoir)
RAPID
"... I assended a high clift about 200 feet above the water from the top of which is a leavel plain extending up the river and off for a great extent, ... and affords a pros[pect] of the river and countrey below for great extent both to the right and left; from this place I descovered a high mountain of emence hight covered with Snow, this must be one of the mountains laid down by Vancouver, as seen from the mouth of the Columbia River, from the course which it bears which is West I take it to be Mt. St. Helens, destant about 120 miles a range of mountains in the Derection crossing a conical mountain S. W. toped with snow, ..."
Clark ⇨
"... I observed banks of Muscle Shells banked up in the river in Several places, ..."
Clark ⇨
"... while Setting on a rock wateing for Capt. Lewis I shot a crain which was flying over of the common kind."
Clark ⇨
McNary Beach Recreation Area
"... at the distance of fifteen miles passed through a country similar to that of yesterday; the hills at the extremity of this distance again approach the river and are rocky abrupt and 300 feet high."
Lewis ⇦
ONE STATUTE MILE
Modern Data: 1993
Outbound: October 19, 1805
Return: April 26, 27, 1806
EXPLORATIONS OF LEWIS AND CLARK 1804 - 1806
CARTOGRAPHIC RECONSTRUCTION
MUSCLE SHELL RAPID
UTM ZONE 11
MAP NUMBER 338
Oregon Washington
CONTOUR INTERVAL 100 FEET WITH SUPPLEMENTS

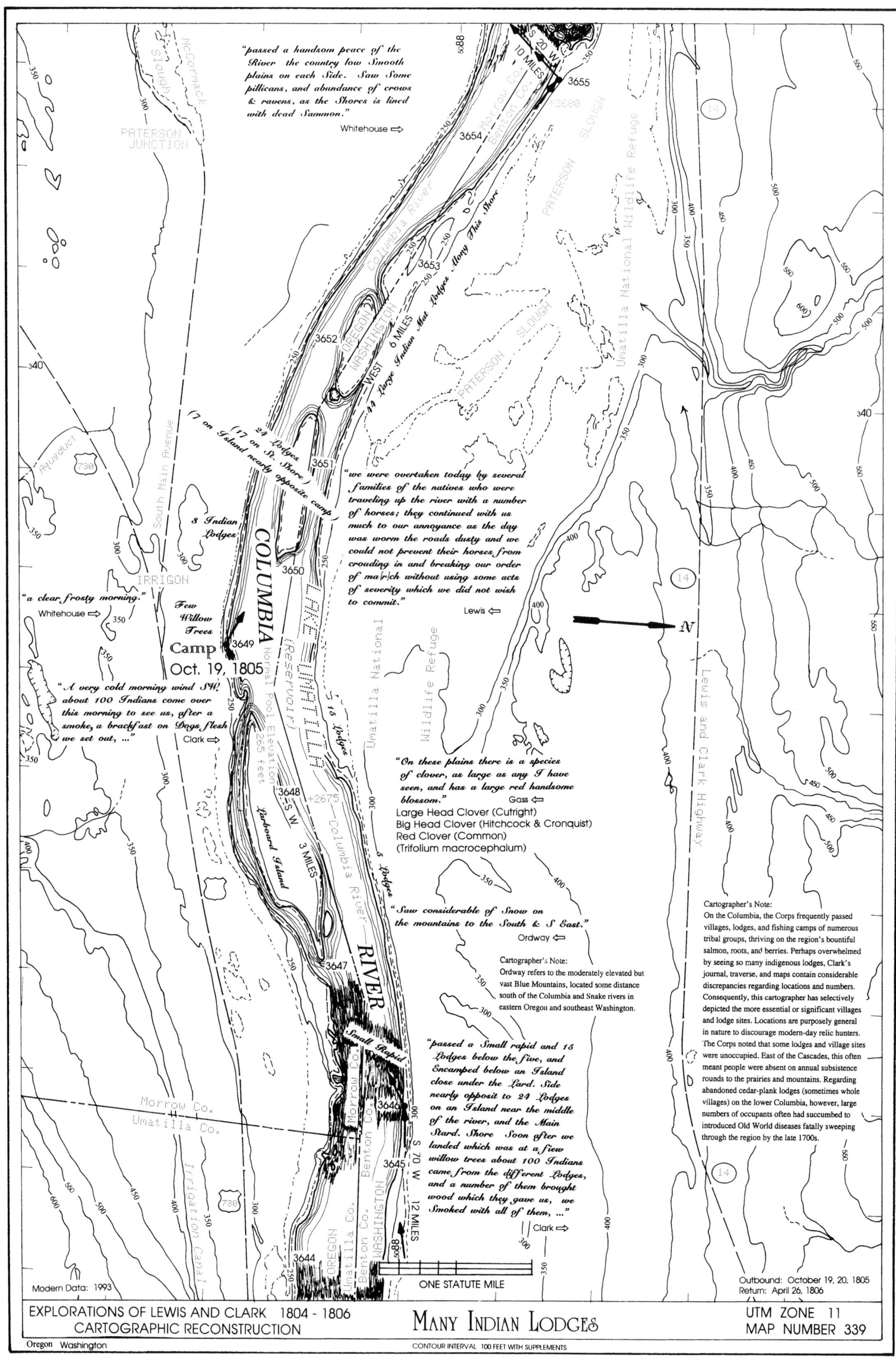
"passed a handsom peace of the River the country low Smooth plains on each Side. Saw Some pillicans, and abundance of crows & ravens, as the Shores is lined with dead Sammon."
Whitehouse ⇨
"we were overtaken today by several families of the natives who were traveling up the river with a number of horses; they continued with us much to our annoyance as the day was worm the roads dusty and we could not prevent their horses from crouding in and breaking our order of ma[r]ch without using some acts of severity which we did not wish to commit."
Lewis ⇦
"a clear frosty morning."
Whitehouse ⇨
Camp
Oct. 19, 1805
"A very cold morning wind SW, about 100 Indians come over this morning to see us, after a smoke, a brackfast on Dogs flesh we set out, ..."
Clark ⇨
"On these plains there is a species of clover, as large as any I have seen, and has a large red handsome blossom."
Gass ⇦
Large Head Clover (Cutright)
Big Head Clover (Hitchcock & Cronquist)
Red Clover (Common)
(Trifolium macrocephalum)
"Saw considerable of Snow on the mountains to the South & S East."
Ordway ⇦
Cartographer's Note:
Ordway refers to the moderately elevated but vast Blue Mountains, located some distance south of the Columbia and Snake rivers in eastern Oregon and southeast Washington.
"passed a Small rapid and 15 Lodges below the five, and Encamped below an Island close under the Lard. Side nearly opposit to 24 Lodges on an Island near the middle of the river, and the Main Stard. Shore Soon after we landed which was at a fiew willow trees about 100 Indians came from the different Lodges, and a number of them brought wood which they gave us, we Smoked with all of them, ..."
Clark ⇨
Cartographer's Note:
On the Columbia, the Corps frequently passed villages, lodges, and fishing camps of numerous tribal groups, thriving on the region's bountiful salmon, roots, and berries. Perhaps overwhelmed by seeing so many indigenous lodges, Clark's journal, traverse, and maps contain considerable discrepancies regarding locations and numbers. Consequently, this cartographer has selectively depicted the more essential or significant villages and lodge sites. Locations are purposely general in nature to discourage modern-day relic hunters. The Corps noted that some lodges and village sites were unoccupied. East of the Cascades, this often meant people were absent on annual subsistence rounds to the prairies and mountains. Regarding abandoned cedar-plank lodges (sometimes whole villages) on the lower Columbia, however, large numbers of occupants often had succumbed to introduced Old World diseases fatally sweeping through the region by the late 1700s.
COLUMBIA
RIVER
LAKE UMATILLA
(Reservoir)
Normal Pool Elevation 265 feet
24 Lodges (17 on St. Shore) (7 on Island nearly opposite camp)
3 Indian Lodges
Few Willow Trees
44 Large Indian Mat Lodges Along This Shore
15 Lodges
5 Lodges
Small Rapid
Larboard Island
S 20 W 10 MILES
WEST 6 MILES
S W 3 MILES
S 70 W 12 MILES
PATERSON JUNCTION
PATERSON SLOUGH
McCormack Slough
IRRIGON
South Main Avenue
Aqueduct
Irrigation Canal
Umatilla National Wildlife Refuge
Lewis and Clark Highway
Morrow Co.
Umatilla Co.
Benton Co.
OREGON
WASHINGTON
N
ONE STATUTE MILE
Modern Data: 1993
Outbound: October 19, 20, 1805
Return: April 26, 1806
EXPLORATIONS OF LEWIS AND CLARK 1804 - 1806
CARTOGRAPHIC RECONSTRUCTION
MANY INDIAN LODGES
UTM ZONE 11
MAP NUMBER 339
Oregon Washington
CONTOUR INTERVAL 100 FEET WITH SUPPLEMENTS

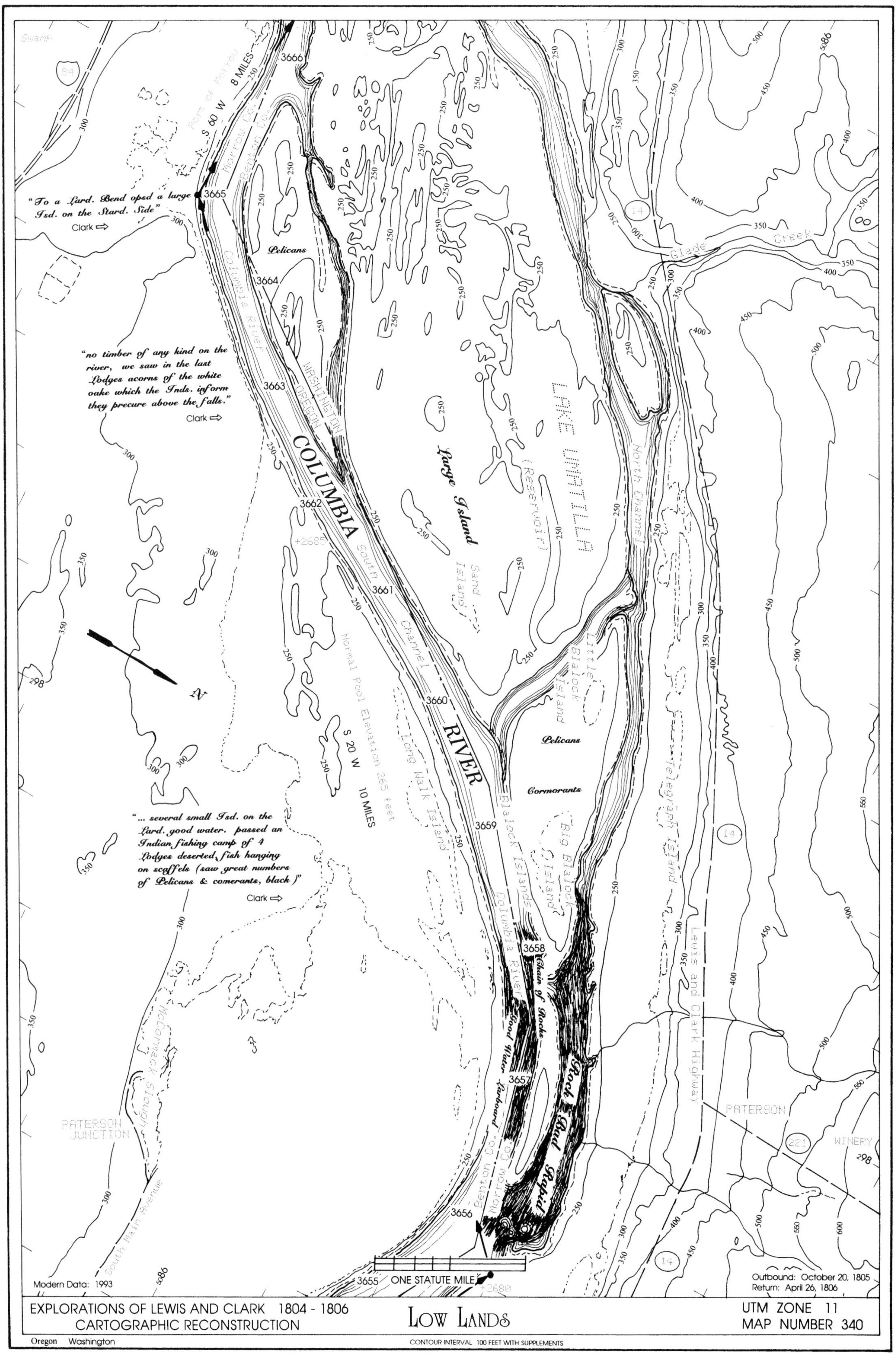
"To a Lard. Bend opsd a large Isd. on the Stard. Side"
Clark ⇨
"no timber of any kind on the river, we saw in the last Lodges acorns of the white oake which the Inds. inform they precure above the falls."
Clark ⇨
"... several small Isd. on the Lard. good water. passed an Indian fishing camp of 4 Lodges deserted fish hanging on scaffels (saw great numbers of Pelicans & comerants, black)"
Clark ⇨
COLUMBIA RIVER
Large Island
Pelicans
Cormorants
Chain of Rocks
Good Water Larboard
Rock Bad Rapid
S 60 W 8 MILES
S 20 W 10 MILES
ONE STATUTE MILE
Modern Data: 1993
Outbound: October 20, 1805
Return: April 26, 1806
EXPLORATIONS OF LEWIS AND CLARK 1804 - 1806
CARTOGRAPHIC RECONSTRUCTION
LOW LANDS
UTM ZONE 11
MAP NUMBER 340
Oregon Washington
CONTOUR INTERVAL 100 FEET WITH SUPPLEMENTS

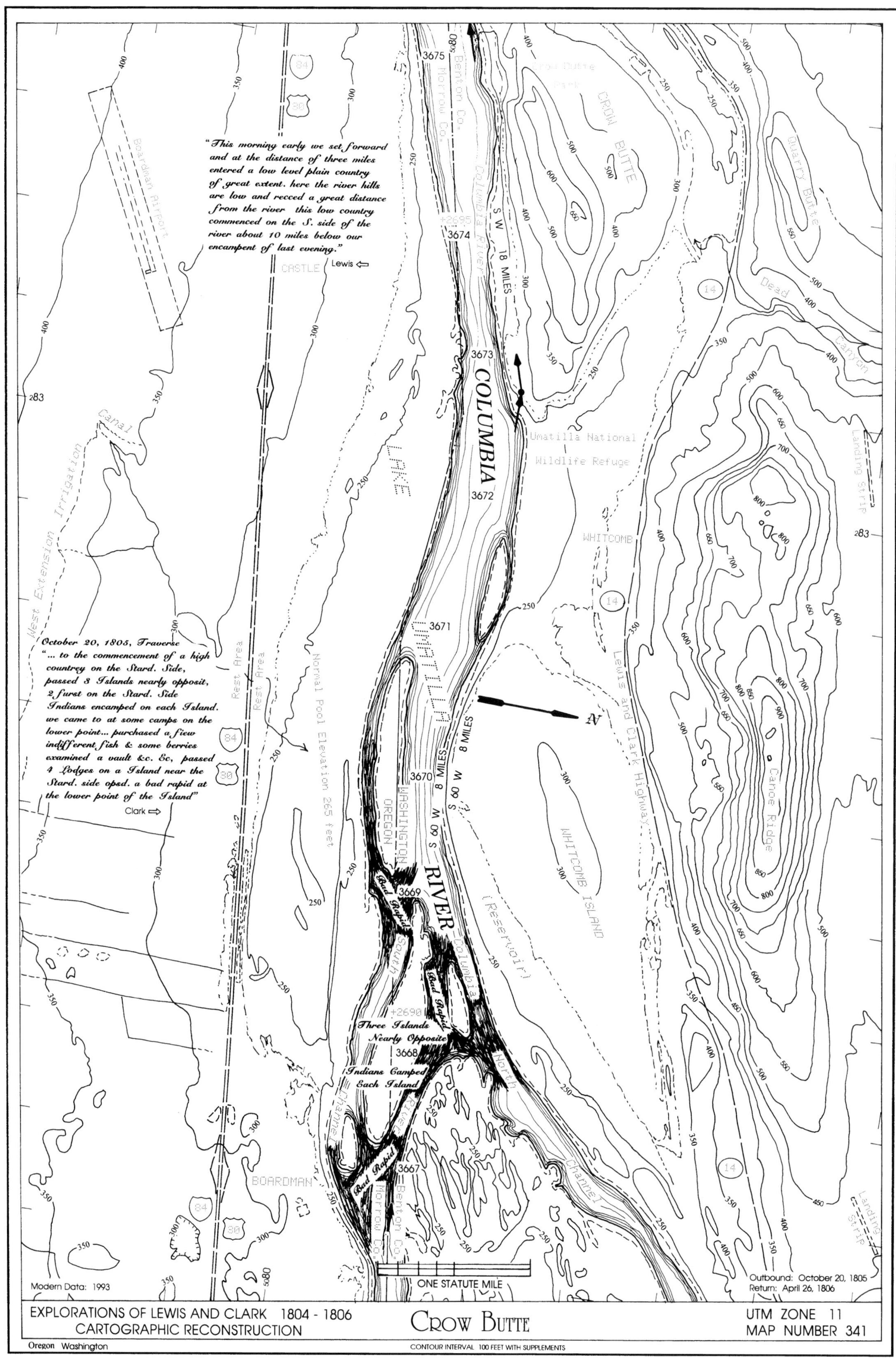
"This morning early we set forward and at the distance of three miles entered a low level plain country of great extent. here the river hills are low and recced a great distance from the river this low country commenced on the S. side of the river about 10 miles below our encampent of last evening."
Lewis ⇦
October 20, 1805, Traverse
"... to the commencement of a high countrey on the Stard. Side, passed 3 Islands nearly opposit, 2 furst on the Stard. Side Indians encamped on each Island. we came to at some camps on the lower point... purchased a fiew indifferent fish & some berries examined a vault &c. Ec, passed 4 Lodges on a Island near the Stard. side opsd. a bad rapid at the lower point of the Island"
Clark ⇨
COLUMBIA
RIVER
LAKE
UMATILLA
Crow Butte Park
CROW BUTTE
Quarry Butte
Dead Canyon
Umatilla National Wildlife Refuge
WHITCOMB
WHITCOMB ISLAND
(Reservoir)
Lewis and Clark Highway
Canoe Ridge
Landing Strip
Boardman Airport
West Extension Irrigation Canal
CASTLE
Rest Area
Normal Pool Elevation 265 feet
WASHINGTON
OREGON
Benton Co.
Morrow Co.
Bad Rapid
South
Columbia
Three Islands Nearly Opposite
Indians Camped Each Island
North
Channel
BOARDMAN
S W 18 MILES
S 60 W 8 MILES
N
ONE STATUTE MILE
Modern Data: 1993
Outbound: October 20, 1805
Return: April 26, 1806
EXPLORATIONS OF LEWIS AND CLARK 1804 - 1806
CARTOGRAPHIC RECONSTRUCTION
CROW BUTTE
UTM ZONE 11
MAP NUMBER 341
Oregon Washington
CONTOUR INTERVAL 100 FEET WITH SUPPLEMENTS

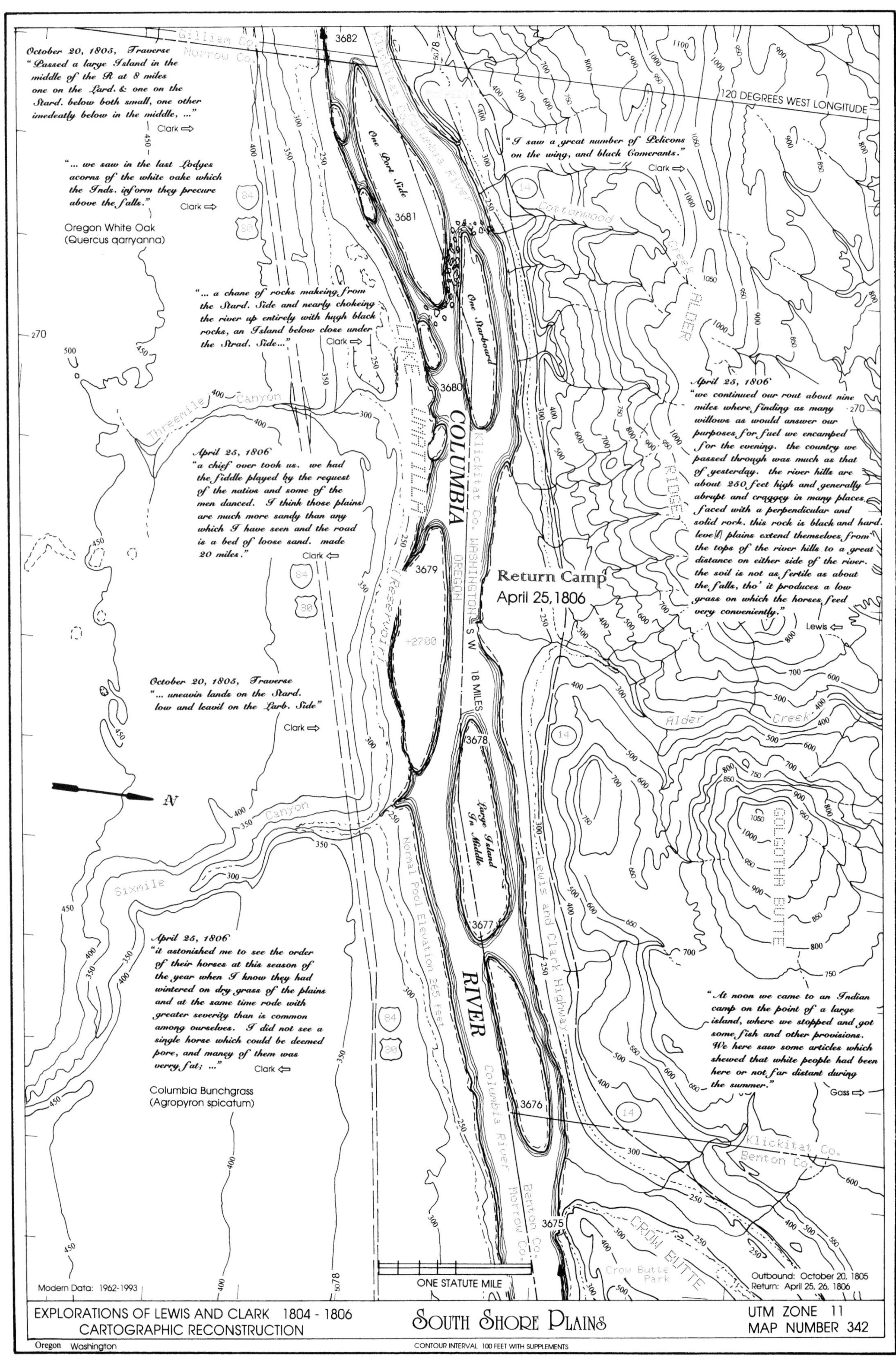

October 20, 1805, Traverse
"Passed a large Island in the middle of the R at 8 miles one on the Lard. & one on the Stard. below both small, one other imedeatly below in the middle, ..."
Clark ⇨
"... we saw in the last Lodges acorns of the white oake which the Inds. inform they precure above the falls."
Clark ⇨
Oregon White Oak (Quercus qarryanna)
"... a chane of rocks makeing from the Stard. Side and nearly chokeing the river up entirely with hugh black rocks, an Island below close under the Strad. Side..."
Clark ⇨
April 25, 1806
"a chief over took us. we had the fiddle played by the request of the natives and some of the men danced. I think those plains are much more sandy than any which I have seen and the road is a bed of loose sand. made 20 miles."
Clark ⇦
October 20, 1805, Traverse
"... uneavin lands on the Stard. low and leavil on the Larb. Side"
Clark ⇨
April 25, 1806
"it astonished me to see the order of their horses at this season of the year when I know they had wintered on dry grass of the plains and at the same time rode with greater severity than is common among ourselves. I did not see a single horse which could be deemed pore, and maney of them was verry fat; ..."
Clark ⇦
Columbia Bunchgrass (Agropyron spicatum)
"I saw a great number of Pelicons on the wing, and black Comerants."
Clark ⇨
120 DEGREES WEST LONGITUDE
April 25, 1806
"we continued our rout about nine miles where finding as many willows as would answer our purposes for fuel we encamped for the evening. the country we passed through was much as that of yesterday. the river hills are about 250 feet high and generally abrupt and craggey in many places faced with a perpendicular and solid rock. this rock is black and hard. leve[l] plains extend themselves from the tops of the river hills to a great distance on either side of the river. the soil is not as fertile as about the falls, tho' it produces a low grass on which the horses feed very conveniently."
Lewis ⇦
Return Camp
April 25, 1806
"At noon we came to an Indian camp on the point of a large island, where we stopped and got some fish and other provisions. We here saw some articles which shewed that white people had been here or not far distant during the summer."
Gass ⇨
Gilliam Co.
Morrow Co.
Klickitat Co.
Benton Co.
One Port Side
One Starboard
Large Island In Middle
LAKE UMATILLA (Reservoir)
COLUMBIA RIVER
Columbia River
Klickitat Co. WASHINGTON
OREGON
18 MILES
Threemile Canyon
Sixmile Canyon
Cottonwood Creek
ALDER RIDGE
Alder Creek
GOLGOTHA BUTTE
CROW BUTTE
Crow Butte Park
Lewis and Clark Highway
Normal Pool Elevation 265 feet
N
ONE STATUTE MILE
Modern Data: 1962-1993
Outbound: October 20, 1805
Return: April 25, 26, 1806
EXPLORATIONS OF LEWIS AND CLARK 1804 - 1806
CARTOGRAPHIC RECONSTRUCTION
SOUTH SHORE PLAINS
UTM ZONE 11
MAP NUMBER 342
Oregon Washington
CONTOUR INTERVAL 100 FEET WITH SUPPLEMENTS

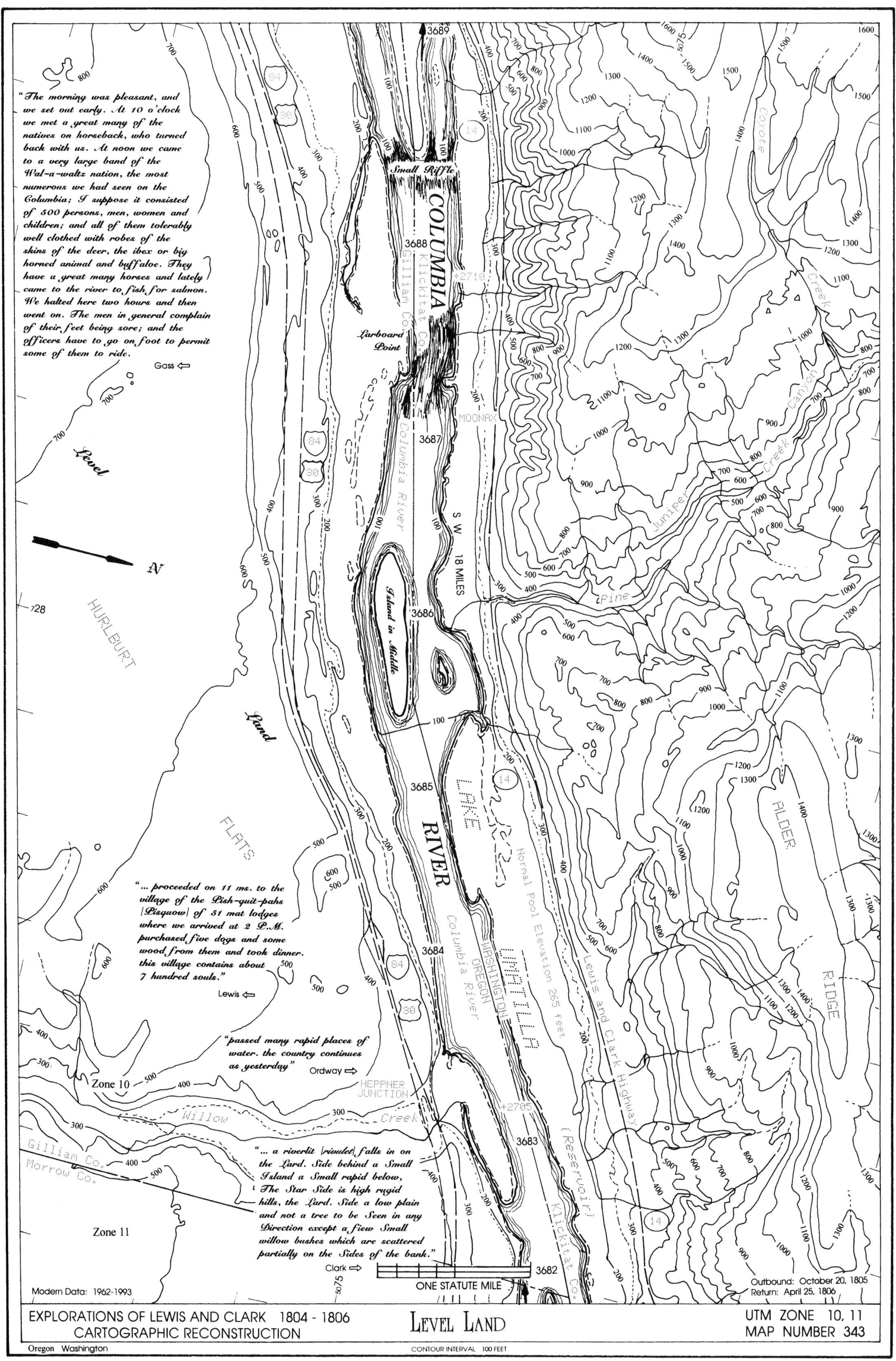
"The morning was pleasant, and we set out early. At 10 o'clock we met a great many of the natives on horseback, who turned back with us. At noon we came to a very large band of the Wal-a-waltz nation, the most numerous we had seen on the Columbia; I suppose it consisted of 500 persons, men, women and children; and all of them tolerably well clothed with robes of the skins of the deer, the ibex or big horned animal and buffaloe. They have a great many horses and lately came to the river to fish for salmon. We halted here two hours and then went on. The men in general complain of their feet being sore; and the officers have to go on foot to permit some of them to ride.
Gass ⇦
"... proceeded on 11 ms. to the village of the Pish-quit-pahs [Pisquow] of 51 mat lodges where we arrived at 2 P.M. purchased five dogs and some wood from them and took dinner. this village contains about 7 hundred souls."
Lewis ⇦
"passed many rapid places of water. the country continues as yesterday"
Ordway ⇨
"... a riverlit [rivolet] falls in on the Lard. Side behind a Small Island a Small rapid below, The Star Side is high rugid hills, the Lard. Side a low plain and not a tree to be Seen in any Direction except a fiew Small willow bushes which are scattered partially on the Sides of the bank."
Clark ⇨
COLUMBIA
RIVER
Small Riffle
Larboard Point
Island in Middle
Level
Land
HURLBURT
FLATS
LAKE
UMATILLA
ALDER
RIDGE
Klickitat Co.
Gilliam Co.
Morrow Co.
Columbia River
WASHINGTON
OREGON
MOONAX
HEPPNER JUNCTION
Willow Creek
Pine
Juniper Creek
Canyon Creek
Coyote
Normal Pool Elevation 265 feet
Lewis and Clark Highway
(Reservoir)
SW 18 MILES
N
Zone 10
Zone 11
Modern Data: 1962-1993
ONE STATUTE MILE
Outbound: October 20, 1805
Return: April 25, 1806
EXPLORATIONS OF LEWIS AND CLARK 1804 - 1806
CARTOGRAPHIC RECONSTRUCTION
LEVEL LAND
UTM ZONE 10, 11
MAP NUMBER 343
Oregon Washington
CONTOUR INTERVAL 100 FEET

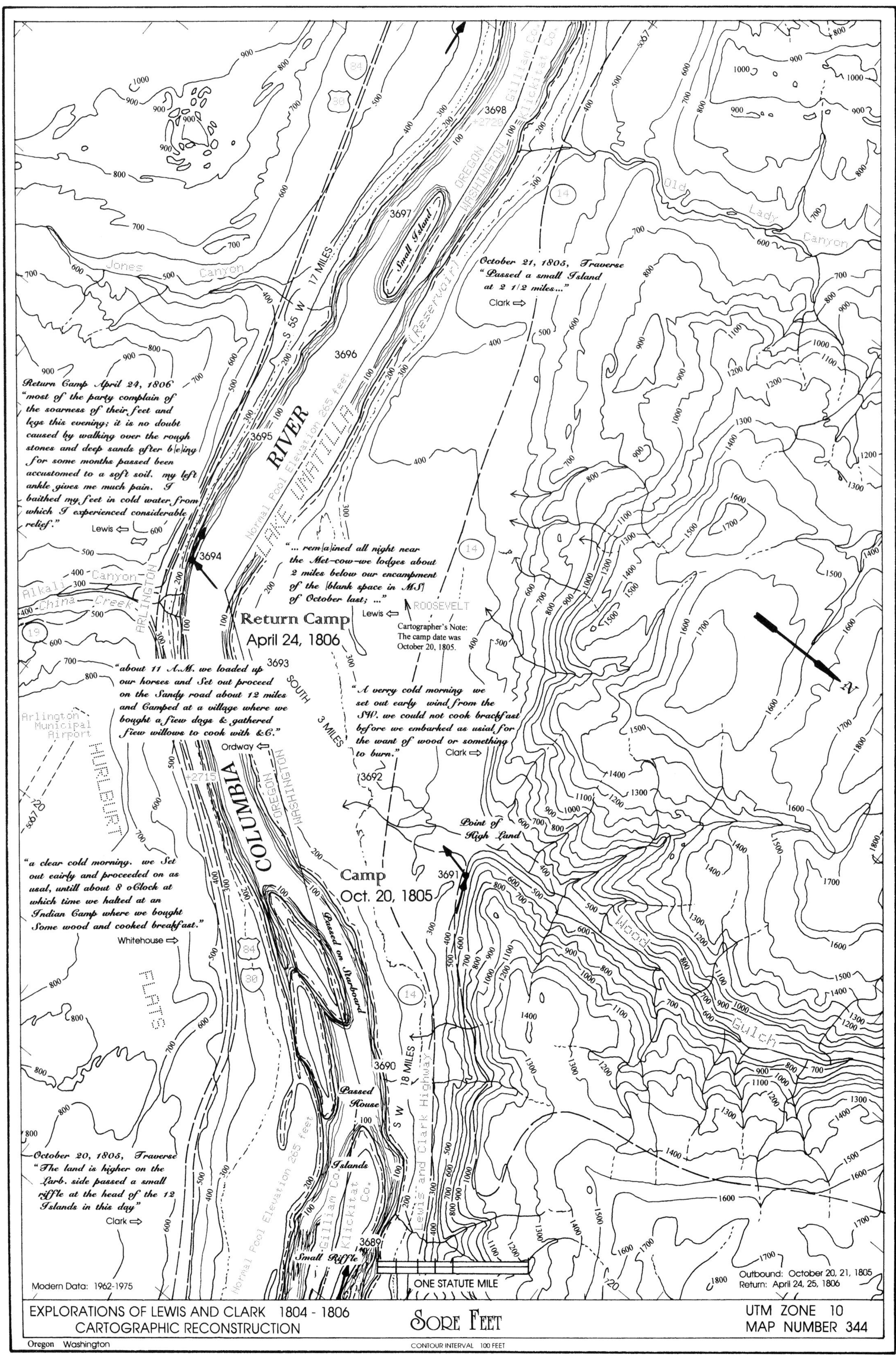
Return Camp April 24, 1806
"most of the party complain of the soarness of their feet and legs this evening; it is no doubt caused by walking over the rough stones and deep sands after b[e]ing for some months passed been accustomed to a soft soil. my left ankle gives me much pain. I baithed my feet in cold water from which I experienced considerable relief."
Lewis ⇦
October 21, 1805, Traverse
"Passed a small Island at 2 1/2 miles..."
Clark ⇨
"... rem[a]ined all night near the Met-cow-we lodges about 2 miles below our encampment of the [blank space in MS] of October last; ..."
Lewis ⇦
Cartographer's Note: The camp date was October 20, 1805.
Return Camp
April 24, 1806
"about 11 A.M. we loaded up our horses and Set out proceed on the Sandy road about 12 miles and Camped at a village where we bought a fiew dogs & gathered fiew willows to cook with &C."
Ordway ⇦
"A verry cold morning we set out early wind from the SW. we could not cook brackfast before we embarked as usial for the want of wood or something to burn."
Clark ⇨
"a clear cold morning. we Set out eairly and proceeded on as usal, untill about 8 oClock at which time we halted at an Indian Camp where we bought Some wood and cooked breakfast."
Whitehouse ⇨
Camp
Oct. 20, 1805
Point of High Land
Passed on Starboard
Passed House
Islands
Small Riffle
October 20, 1805, Traverse
"The land is higher on the Larb. side passed a small riffle at the head of the 12 Islands in this day"
Clark ⇨
RIVER
LAKE UMATILLA
COLUMBIA
Small Island
Jones Canyon
Alkali Canyon
China Creek
ARLINGTON
Arlington Municipal Airport
HURLBURT
FLATS
ROOSEVELT
Old Lady Canyon
Wood Gulch
Lewis and Clark Highway
Normal Pool Elevation 265 feet
OREGON
WASHINGTON
Gilliam Co.
Klickitat Co.
S 55 W 17 MILES
SOUTH 3 MILES
S W 18 MILES
N
ONE STATUTE MILE
Modern Data: 1962-1975
Outbound: October 20, 21, 1805
Return: April 24, 25, 1806
EXPLORATIONS OF LEWIS AND CLARK 1804 - 1806
CARTOGRAPHIC RECONSTRUCTION
SORE FEET
UTM ZONE 10
MAP NUMBER 344
Oregon Washington
CONTOUR INTERVAL 100 FEET

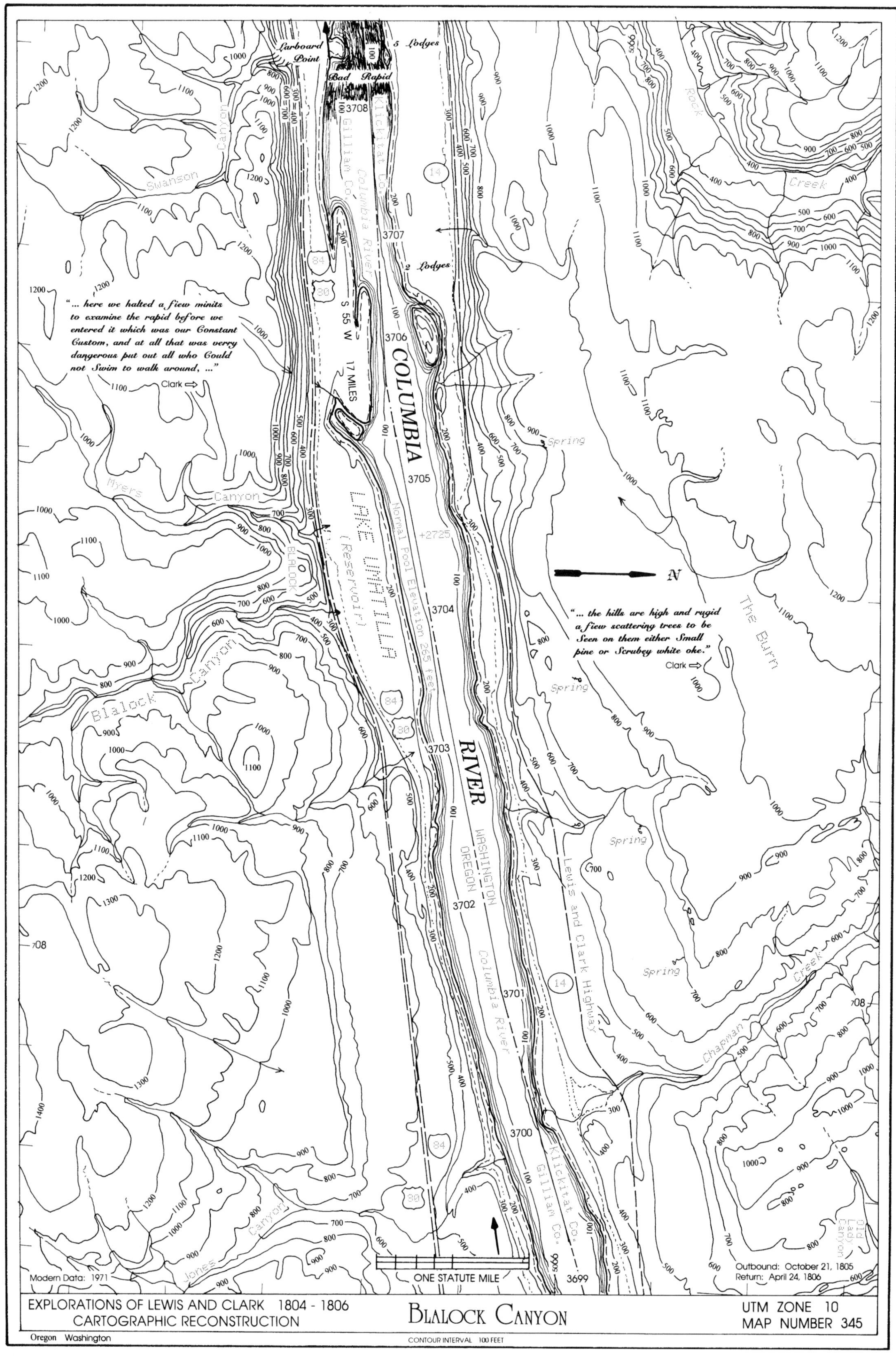
Larboard Point
5 Lodges
Bad Rapid
2 Lodges
S 55 W
17 MILES
COLUMBIA RIVER
LAKE UMATILLA (Reservoir)
Normal Pool Elevation 265 feet
"... here we halted a fiew minits to examine the rapid before we entered it which was our Constant Custom, and at all that was very dangerous put out all who Could not Swim to walk around, ..."
Clark
"... the hills are high and rugid a fiew scattering trees to be Seen on them either Small pine or Scrubey white oke."
Clark
N
Swanson Canyon
Myers Canyon
Blalock Canyon
BLALOCK
Jones Canyon
Rock Creek
The Burn
Chapman Creek
Old Lady Canyon
Spring
Lewis and Clark Highway
Klickitat Co.
Gilliam Co.
WASHINGTON
OREGON
Modern Data: 1971
ONE STATUTE MILE
Outbound: October 21, 1805
Return: April 24, 1806
EXPLORATIONS OF LEWIS AND CLARK 1804 - 1806
CARTOGRAPHIC RECONSTRUCTION
BLALOCK CANYON
UTM ZONE 10
MAP NUMBER 345
Oregon Washington
CONTOUR INTERVAL 100 FEET

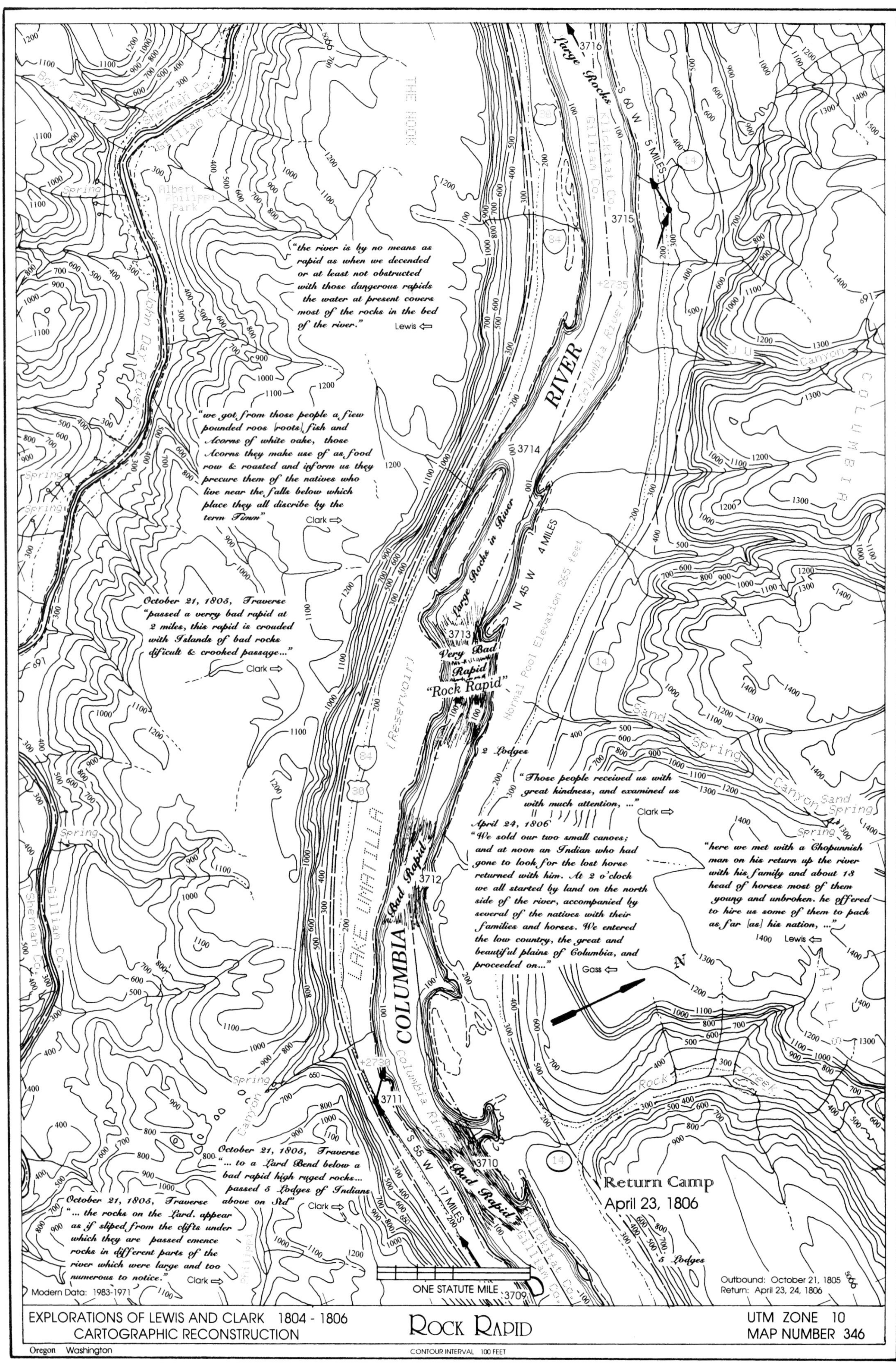
"the river is by no means as rapid as when we decended or at least not obstructed with those dangerous rapids the water at present covers most of the rocks in the bed of the river."
Lewis ⇦
"we got from those people a few pounded roos [roots] fish and Acorns of white oake, those Acorns they make use of as food row & roasted and inform us they precure them of the natives who live near the falls below which place they all discribe by the term Timm"
Clark ⇨
October 21, 1805, Traverse
"passed a verry bad rapid at 2 miles, this rapid is crouded with Islands of bad rocks dificult & crooked passage..."
Clark ⇨
"Those people received us with great kindness, and examined us with much attention, ..."
Clark ⇨
April 24, 1806
"We sold our two small canoes; and at noon an Indian who had gone to look for the lost horse returned with him. At 2 o'clock we all started by land on the north side of the river, accompanied by several of the natives with their families and horses. We entered the low country, the great and beautiful plains of Columbia, and proceeded on..."
Gass ⇦
"here we met with a Chopunnish man on his return up the river with his family and about 18 head of horses most of them young and unbroken. he offered to hire us some of them to pack as far [as] his nation, ..."
Lewis ⇦
October 21, 1805, Traverse
"... to a Lard Bend below a bad rapid high ruged rocks... passed 5 Lodges of Indians above on Std"
Clark ⇨
October 21, 1805, Traverse
"... the rocks on the Lard. appear as if sliped from the clifts under which they are passed emence rocks in different parts of the river which were large and too numerous to notice."
Clark ⇨
Large Rocks
S 60 W 5 MILES
Large Rocks in River
N 45 W 4 MILES
S 55 W 17 MILES
Very Bad Rapid
"Rock Rapid"
Bad Rapid
2 Lodges
5 Lodges
COLUMBIA RIVER
LAKE UMATILLA (Reservoir)
Columbia River
Normal Pool Elevation 265 feet
THE HOOK
John Day River
Albert Philippi Park
Box Canyon
Sand Spring Canyon
Sand Spring
JU Canyon
Rock Creek
COLUMBIA HILLS
Sherman Co.
Gilliam Co.
Klickitat Co.
3716
3715
3714
3713
3712
3711
3710
3709
+2735
+2730
Return Camp
April 23, 1806
ONE STATUTE MILE
Modern Data: 1983-1971
Outbound: October 21, 1805
Return: April 23, 24, 1806
EXPLORATIONS OF LEWIS AND CLARK 1804 - 1806
CARTOGRAPHIC RECONSTRUCTION
ROCK RAPID
UTM ZONE 10
MAP NUMBER 346
Oregon Washington
CONTOUR INTERVAL 100 FEET

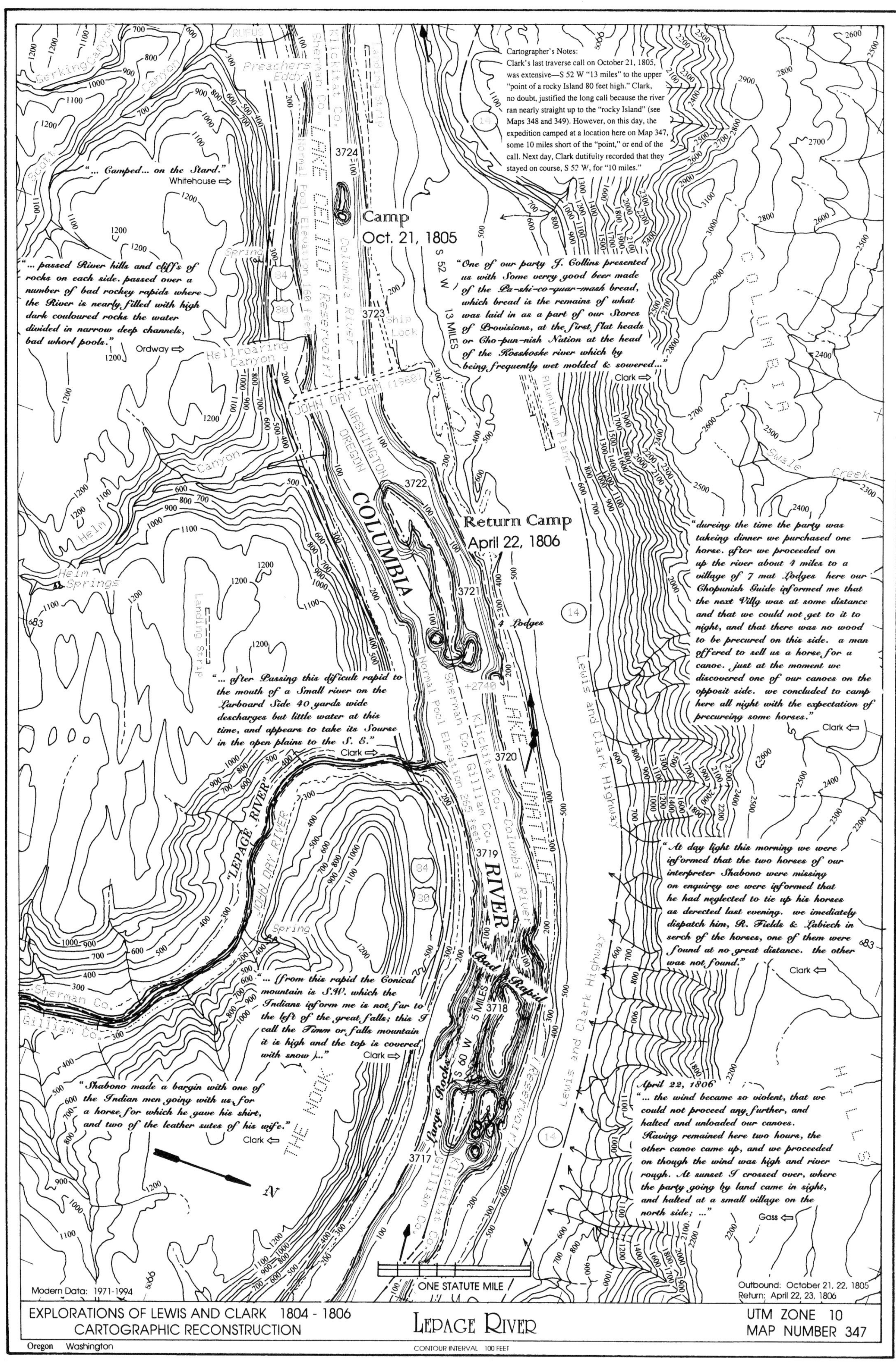

Cartographer's Notes:
Clark's last traverse call on October 21, 1805, was extensive—S 52 W "13 miles" to the upper "point of a rocky Island 80 feet high." Clark, no doubt, justified the long call because the river ran nearly straight up to the "rocky Island" (see Maps 348 and 349). However, on this day, the expedition camped at a location here on Map 347, some 10 miles short of the "point," or end of the call. Next day, Clark dutifully recorded that they stayed on course, S 52 W, for "10 miles."
"... Camped... on the Stard."
Whitehouse ⇨
Camp
Oct. 21, 1805
"... passed River hills and cliffs of rocks on each side. passed over a number of bad rockey rapids where the River is nearly filled with high dark couloured rocks the water divided in narrow deep channels, bad whorl pools."
Ordway ⇨
"One of our party J. Collins presented us with Some verry good beer made of the Pa-shi-co-quar-mash bread, which bread is the remains of what was laid in as a part of our Stores of Provisions, at the first flat heads or Cho-pun-nish Nation at the head of the Kosskoske river which by being frequently wet molded & sowered..."
Clark ⇨
S 52 W
13 MILES
LAKE CELILO (Reservoir)
JOHN DAY DAM (1968)
Return Camp
April 22, 1806
COLUMBIA RIVER
4 Lodges
"dureing the time the party was takeing dinner we purchased one horse. after we proceeded on up the river about 4 miles to a village of 7 mat Lodges here our Chopunish Guide informed me that the next Villg was at some distance and that we could not get to it to night, and that there was no wood to be precured on this side. a man offered to sell us a horse for a canoe. just at the moment we discovered one of our canoes on the opposit side. we concluded to camp here all night with the expectation of precureing some horses."
Clark ⇦
"... after Passing this dificult rapid to the mouth of a Small river on the Larboard Side 40 yards wide descharges but little water at this time, and appears to take its Sourse in the open plains to the S. E."
Clark ⇨
"LEPAGE RIVER"
JOHN DAY RIVER
LAKE UMATILLA (Reservoir)
"At day light this morning we were informed that the two horses of our interpreter Shabono were missing on enquirey we were informed that he had neglected to tie up his horses as derected last evening. we imediately dispatch him, R. Fields & Labiech in serch of the horses, one of them were found at no great distance. the other was not found."
Clark ⇦
Bad Rapid
S 60 W
5 MILES
"... (from this rapid the Conical mountain is S.W. which the Indians inform me is not far to the left of the great falls; this I call the Timm or falls mountain it is high and the top is covered with snow)..."
Clark ⇨
Large Rocks
"Shabono made a bargin with one of the Indian men going with us, for a horse for which he gave his shirt, and two of the leather sutes of his wife."
Clark ⇨
THE HOOK
April 22, 1806
"... the wind became so violent, that we could not proceed any further, and halted and unloaded our canoes. Having remained here two hours, the other canoe came up, and we proceeded on though the wind was high and river rough. At sunset I crossed over, where the party going by land came in sight, and halted at a small village on the north side; ..."
Gass ⇦
COLUMBIA HILLS
Lewis and Clark Highway
Sherman Co.
Gilliam Co.
Klickitat Co.
Helm Springs
Hellroaring Canyon
Preachers Eddy
Ship Lock
Aluminum Plant
Swale Creek
N
ONE STATUTE MILE
Modern Data: 1971-1994
Outbound: October 21, 22, 1805
Return: April 22, 23, 1806
EXPLORATIONS OF LEWIS AND CLARK 1804 - 1806
CARTOGRAPHIC RECONSTRUCTION
LEPAGE RIVER
UTM ZONE 10
MAP NUMBER 347
Oregon Washington
CONTOUR INTERVAL 100 FEET

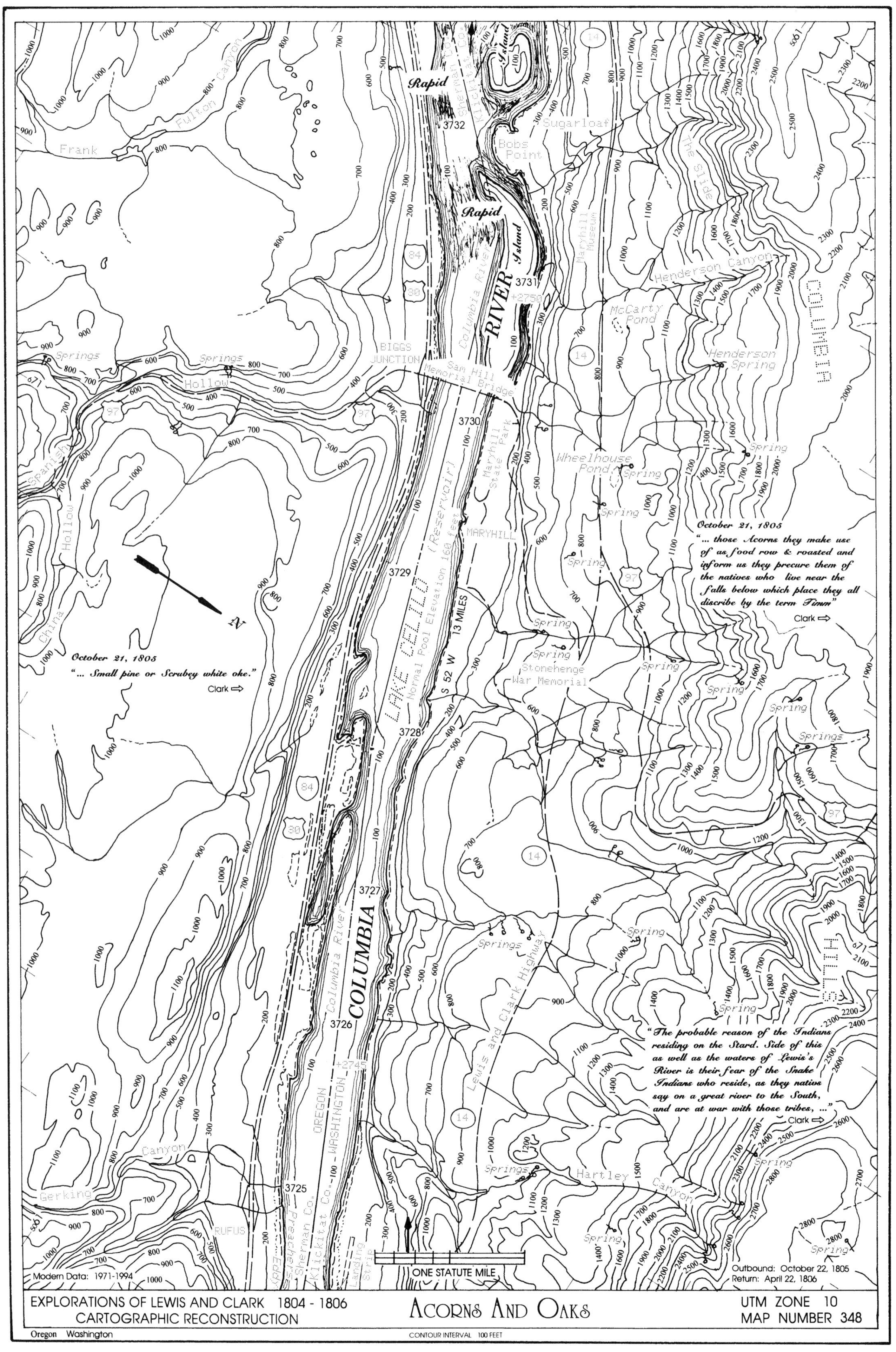
EXPLORATIONS OF LEWIS AND CLARK 1804 - 1806
CARTOGRAPHIC RECONSTRUCTION
ACORNS AND OAKS
UTM ZONE 10
MAP NUMBER 348
Oregon Washington
CONTOUR INTERVAL 100 FEET
Modern Data: 1971-1994
ONE STATUTE MILE
Outbound: October 22, 1805
Return: April 22, 1806
October 21, 1805
"... those Acorns they make use of as food row & roasted and inform us they precure them of the natives who live near the falls below which place they all discribe by the term Timm"
Clark ⇨
October 21, 1805
"... Small pine or Scrubey white oke."
Clark ⇨
"The probable reason of the Indians residing on the Stard. Side of this as well as the waters of Lewis's River is their fear of the Snake Indians who reside, as they nativs say on a great river to the South, and are at war with those tribes, ..."
Clark ⇨
COLUMBIA RIVER
Columbia River
LAKE CELILO (Reservoir)
Normal Pool Elevation 160 feet
S 52 W 13 MILES
Rapid
Rapid
Island
Island
Fulton Canyon
Frank
Springs
Springs
Hollow
Spanish Hollow
China Hollow
BIGGS JUNCTION
Sam Hill Memorial Bridge
Sugarloaf
Bobs Point
Maryhill Museum
The Slide
Henderson Canyon
McCarty Pond
Henderson Spring
COLUMBIA
Wheelhouse Pond
Maryhill State Park
MARYHILL
Stonehenge War Memorial
Spring
Springs
Lewis and Clark Highway
COLUMBIA HILLS
Hartley Canyon
Canyon
Gerking
RUFUS
OREGON
WASHINGTON
Sherman Co.
Klickitat Co.
Landing Strip
N

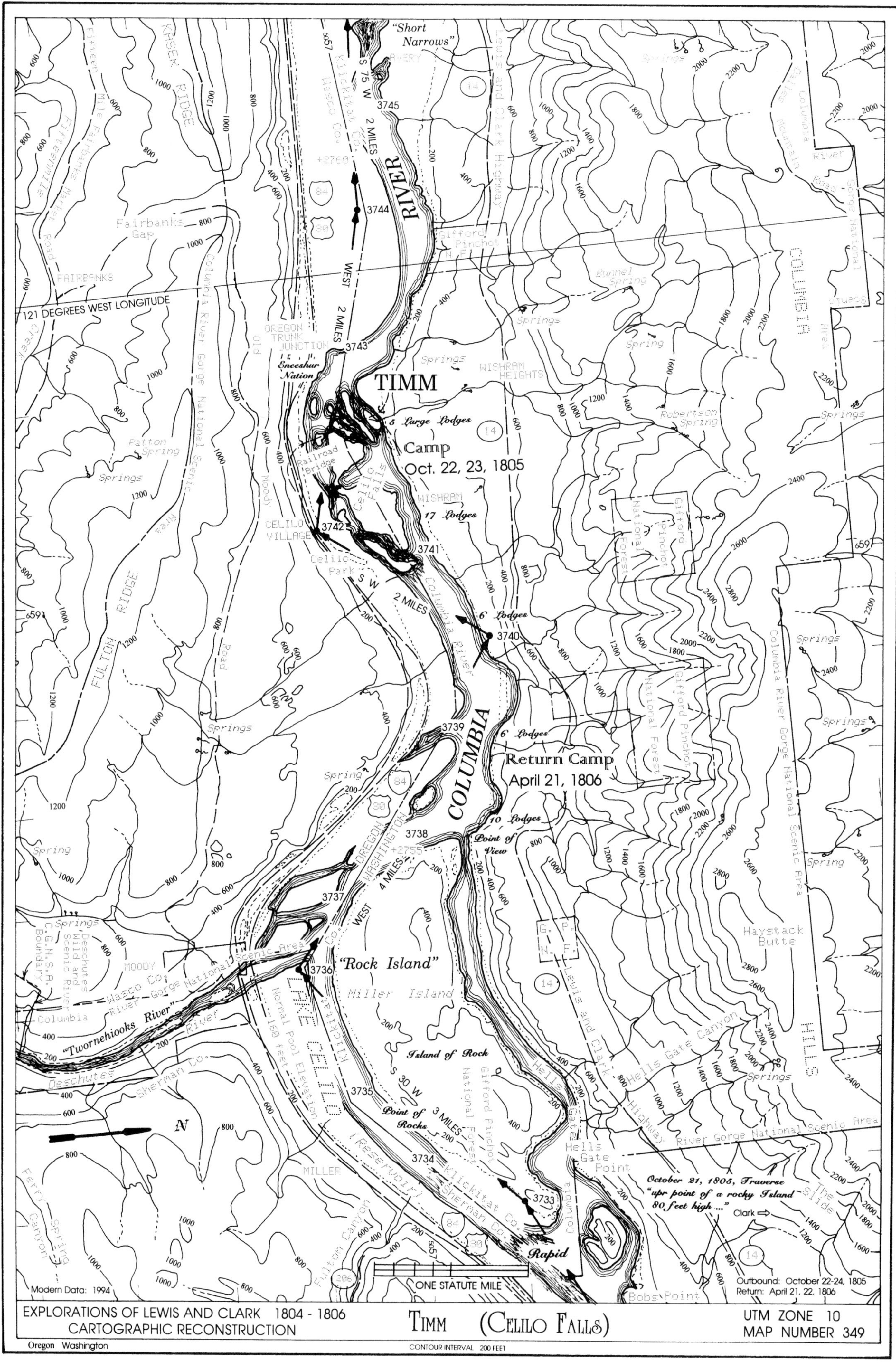
"Short Narrows"
S 75 W
2 MILES
3745
3744
RIVER
WEST
2 MILES
3743
Eneeshur Nation
TIMM
5 Large Lodges
Camp
Oct. 22, 23, 1805
17 Lodges
3742
3741
S W
2 MILES
6 Lodges
3740
3739
6 Lodges
COLUMBIA
Return Camp
April 21, 1806
10 Lodges
3738
Point of View
4 MILES
WEST
3737
"Rock Island"
3736
"Twornehiooks River"
Island of Rock
S 30 W
3735
Point of Rocks
3 MILES
3734
3733
Rapid
N
121 DEGREES WEST LONGITUDE
October 21, 1805, Traverse
"upr point of a rocky Island 80 feet high ..."
Clark ⇨
ONE STATUTE MILE
Modern Data: 1994
Outbound: October 22-24, 1805
Return: April 21, 22, 1806
EXPLORATIONS OF LEWIS AND CLARK 1804 - 1806
CARTOGRAPHIC RECONSTRUCTION
TIMM (CELILO FALLS)
UTM ZONE 10
MAP NUMBER 349
Oregon Washington
CONTOUR INTERVAL 200 FEET

TEXT SUPPLEMENT TO MAP 349, TIMM (CELILO FALLS)

Towahnahiooks River (Deschutes) River

October 22, 1805

"A fine morning calm and fare we set out at 9 oClock passed a verry bad rapid at the head of an Island close under the Stard. side... at 9 mls. passed a bad rapid at the head of a large Island of high & uneaven [rocks], jutting over the water, a Small Island in a Stard. Bend opposit the upper point, on which I counted 20 parcels of dryed and pounded fish; on the main Stard. Shore opposit to this Island five Lodges of Indians are Situated, ... opposit the center of this Island of rocks which is about 4 miles long we discovered the enterence of a large river on the Larb. Side which appeared to come from the S.E. we landed at some distance above the mouth of this river and Capt. Lewis and my Self set out to view this river above its mouth, as our rout was intersepted by a deep narrow Chanel which runs out of this river into the Columbia a little below the place we landed, leaveing a high dry rich Island of about 400 yards wide and 800 yards long here we Seperated, I proceeded on to the river and Struck it at the foot of a verry Considerable rapid, here I beheld an emence body of water compressd in a narrow chanel of about 200 yds in width, fomeing over rocks maney of which presented their tops above the water, when at this place Capt. Lewis joined me haveing delayed on the way to examine a root of which the nativs had been digging great quantities in the bottoms of this River. at about two miles above this river appears to be confined between two high hils below which it [is] divided by numbers of large rocks, and Small Islands covered with a low groth of timber, and has a rapid as far as the narrows, three Small Islands in the mouth of this River, this River haveing no Indian name that we could find out, except "the River on which the Snake Indians live", we think it best to leave the nameing of it untill our return.

we proceeded on pass[ed] the mouth of this river at which place it appears to discharge 1/4 as much water as runs down the Columbia. at two miles below this River passed Eight Lodges on the Lower point of the Rock Island aforesaid at those Lodges we saw large logs of wood which must have been rafted down the To wor-ne hi ooks River, below this Island on the main Stard. Shore is 16 Lodges of nativs, here we landed a fiew minits to Smoke, the lower point of one Island opposit which heads in the mouth of Towornehiooks River which I did not observe untill after pasing these lodges... 6 miles below the upper mouth of Towornehiooks River the comencement of the pitch of the great falls, ..."

Clark ⇨

Portage of the Baggage

October 22, 1805

"we landed and walked down accompanied by an old man to view the falls, and the best rout for to make a portage which we Soon discovered was much nearest on the Stard. Side, and the distance 1200 yards one third of the way on a rock, about 200 yards over a loose Sand collected in a hollar blown by the winds from the bottoms below which was disagreeable to pass, as it was steep and loose. at the lower part of those rapids we arrived at 5 Large Lod[g]es of nativs drying and prepareing fish for market, they gave us Philburts, and berries to eate. we returned droped down to the head of the rapids and took every article except the Canoes across the portag[e] where I formed a camp on [an] ellegable Situation for the protection of our Stores from thieft, which we were more fearfull of, than their arrows."

Clark ⇨

October 22, 1805

"Took our Baggage & formed a camp below the rapids in a cove on the Stard. Side the distance 1200 yards haveing passed at the upper end of the portage 17 Lodges of Indians below the rapids & above camp 5 large Lodges of Indians, great numbers of baskets of Pounded fish on the rocks Islands & near their Lodges those are neetly pounded & put in verry new baskets of about 90 or 100 pounds w[e]ight. hire Indians to take our heavy articles across the portage"

Clark ⇨

October 22, 1805

"So we got all the baggage below the falls this evening and Camped close to a high range of clifts of rocks, where the body of the River beat against it and formed a large Eddy. ... these natives Sign to us that Some white people had been here but were gone four or 5 days journey further down."

Whitehouse ⇨

Canoes around the Falls

October 22, 1805

"we despatched two men to examine the river on the opposit Side, and [they] reported that the canoes could be taken down a narrow Chanel on the opposit Side after a Short portage at the head of the falls, at which place the Indians take over their Canoes. Indians assisted us over the portage with our heavy articles on their horses, the waters is divided into Several narrow chanels which pass through a hard black rock forming Islands of rocks at this Stage of the water, on those Islands of rocks as well as at and about their Lodges I observe great numbers of Stacks of pounded Salmon neetly preserved... their common custom is to Set 7 as close as they can Stand and 5 on the top of them, and secure them with mats which is raped around them and made fast with cords and covered also with mats, those 12 baskets of from 90 to 100 lbs. each form a Stack. thus preserved those fish may be kept Sound and sweet Several years, as those people inform me, Great quantities as they inform us are sold to the whites people who visit the mouth of this river as well as the nativs below.

on one of those Island[s] I saw Several tooms but did not visit them"

Clark ⇨

October 23, 1805

"a fine morning, I with the greater part of the men crossed in the canoes to opposit side above the falls and hauled them across the portage of 457 yards which is on the Lard. Side and certainly the best side to pass the canoes, I then decended through a narrow chanel of about 150 yards wide forming a kind of half circle in it[s] course of a mile to a pitch of 8 feet in which the chanel is divided by 2 large rocks, at this place we were obliged to let the Canoes down by strong ropes of Elk Skin which we had for the purpose, one Canoe in passing this place got loose by the cords breaking, and was cought by the Indians below. I accomplished this necessary business and landed Safe with all the canoes at our Camp below the falls by 3 oClock P. M. nearly covered with flees which were so thick amongst the Straw and fish Skins at the upper part of the portage at which place the nativs had been Camped not long since; that every man of the party was obliged to Strip naked dureing the time of takeing over the canoes, that they might have an oppertunity of brushing the flees of[f] their legs and bodies. ... one of the old Chiefs who had accompanied us from the head of the river, informed us that he herd the Indians Say that the nation below intended to kill us. we examined all the arms &c. complete the amunition to 100 rounds. ...

I obsererved on the beach near the Indian Lodges two butifull canoes of different Shape & Size to what we had Seen above wide in the midd[l]e and tapering to each end, on the bow curious figures were cut in the wood &c. Capt. Lewis went up to the Lodges to See those Canoes and exchanged our Smallest canoe for one of them by giveing a Hatchet & few trinkets to the owner..."

Clark ⇨

October 24, 1805

"a fine morning the Indians approached us with caution, our 2 old Chiefs deturmine to return home, saying they were at war with Indians below and they would kill them we purswaded them to stay 2 nights longer with us, with a view to make a peace with those Indians... I set out with the party at 9 oClock a m at 2 1/2 miles passed a rock which makes from the Stard Side... and confined the river in a narrow channel of about 45 yards this continued for about 1/4 of a mile & widened to about 200 yards, in those narrows the water was agitated in a most shocking manner boils swells & whorpools, we passed with great risque It being impossible to make a portage of the canoes, ..."

Clark ⇨

The Return

April 21, 1806

"... determined to remain no longer with these villains. they stole another tomahawk from us this morning I surched many of them but could not find it. I ordered all spare poles, paddles and the balance of our canoe put on the fire as the morning was cold and also that not a particle should be left for the benefit of the indians. I detected a fellow in stealing an iron socket of a canoe pole and gave him Several severe blows and mad[e] the men kick him out of camp. I now informed the indians that I would shoot the first of them that attempted to steal an article from us."

Lewis ⇦

April 21, 1806

"we soon made the portage with our canoes and baggage and halted about 1/2 mile above the Village where we graized our horses and took dinner on some dogs which we purchased of those people. after dinner we proceeded on about four miles to a village of 9 mat lodges of the Enesher a little below the entrance of Clark's river (Towanahiooks) [Des Chutes] and encamped; ..."

Lewis ⇦

April 21, 1806

"we loaded our horses 9 in number. 4 men took 2 small canoes by water. we set out about 8 oClock and proceed on about noon arived at the village below the big falls. joined Capt Clark who had not purchased any horses. ... the Indians returned us a horse in lieu of one of those we lost &C. we carryed the canoes past the portage and mooved all above the portage and dined and proced on the N. Side to a village opposite the mouth of Clarks River where we Camped as the road leaves the river at this place."

Ordway ⇦

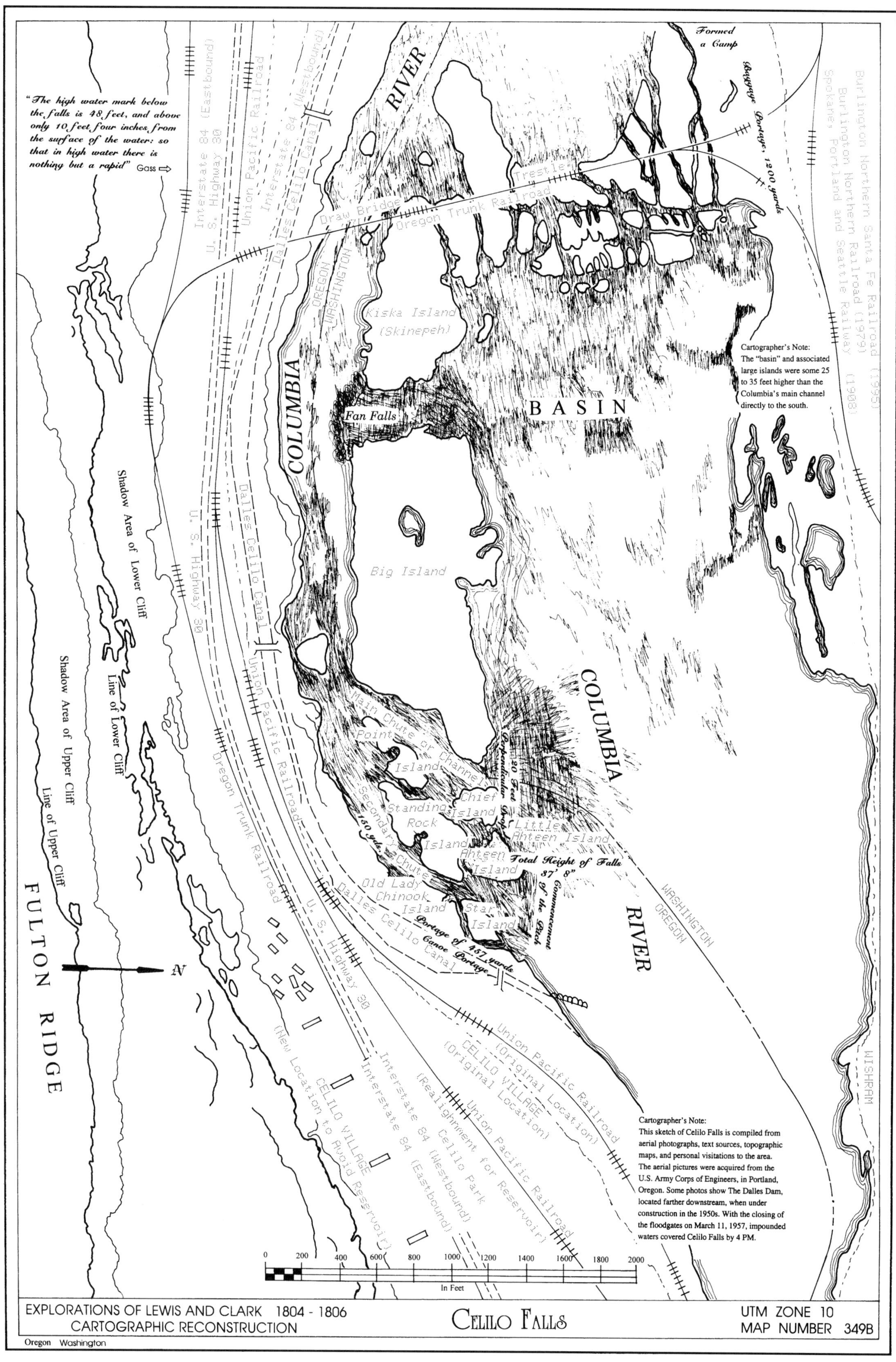
"The high water mark below the falls is 48 feet, and above only 10 feet four inches from the surface of the water: so that in high water there is nothing but a rapid" Gass
Cartographer's Note:
The "basin" and associated large islands were some 25 to 35 feet higher than the Columbia's main channel directly to the south.
Cartographer's Note:
This sketch of Celilo Falls is compiled from aerial photographs, text sources, topographic maps, and personal visitations to the area. The aerial pictures were acquired from the U.S. Army Corps of Engineers, in Portland, Oregon. Some photos show The Dalles Dam, located farther downstream, when under construction in the 1950s. With the closing of the floodgates on March 11, 1957, impounded waters covered Celilo Falls by 4 PM.
Formed a Camp
Baggage Portage, 1200 yards
COLUMBIA RIVER
BASIN
Fan Falls
Kiska Island (Skinepeh)
Big Island
Total Height of Falls 37' 8"
Portage of 457 yards
Canoe Portage
FULTON RIDGE
EXPLORATIONS OF LEWIS AND CLARK 1804 - 1806
CARTOGRAPHIC RECONSTRUCTION
CELILO FALLS
UTM ZONE 10
MAP NUMBER 349B
Oregon Washington
In Feet

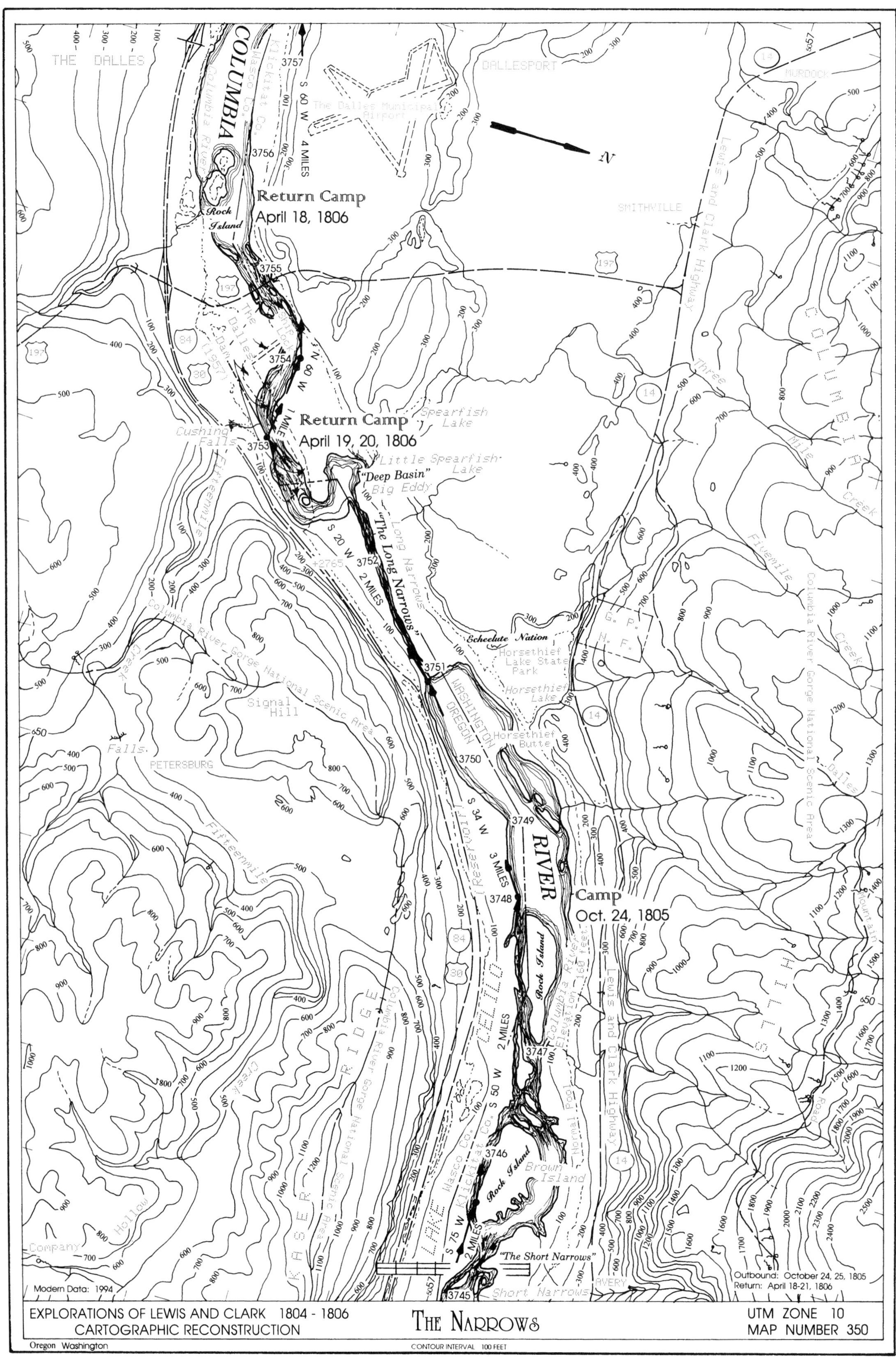
Return Camp
April 18, 1806
Return Camp
April 19, 20, 1806
"Deep Basin"
"The Long Narrows"
Camp
Oct. 24, 1805
"The Short Narrows"
Modern Data: 1994
Outbound: October 24, 25, 1805
Return: April 18-21, 1806
EXPLORATIONS OF LEWIS AND CLARK 1804 - 1806
CARTOGRAPHIC RECONSTRUCTION
THE NARROWS
UTM ZONE 10
MAP NUMBER 350
Oregon Washington
CONTOUR INTERVAL 100 FEET

TEXT SUPPLEMENT TO MAP NUMBER 350, THE NARROWS

THE SHORT NARROWS

October 24, 1805
"... the natives went with me to the top of this rock which makes from the Stard. Side from the top of which I could See the dificuelties we had to pass for Several miles below; at this place the water of this great river is compressed into a chanel between two rocks not exceeding forty five yards wide and continues for a 1/4 of a mile when it again widens to 200 yards and continues this width for about 2 miles when it is again intersepted by rocks. This obstruction in the river accounts for the water in high floods riseing to Such a hite at the last falls. The whole of the Current of this great river must at all Stages pass thro' this narrow chanel of 45 yards wide. as the portage of our canoes over this high rock would be impossible with our Strength, and the only danger in passing thro those narrows was the whorls and swills [swells] arriseing from the Compression of the water, and which I thought (as also our principal waterman Peter Crusat) by good Stearing we could pass down Safe, accordingly I deturmined to pass through this place notwithstanding the horrid appearance of this agitated gut swelling, boiling & whorling in every direction, (which from the top of the rock did not appear as bad as when I was in it:"

Clark ⇨

October 24, 1805, Traverse
"... a rock Island in the middle of the river at a bad rapid & who[r]ls, passed thro a narrow bad chanel 45 yds wide for 1/4 of a mile."

Clark ⇨

October 24, 1805
"... we passed Safe to the astonishment of all the Inds. of the last Lodges who viewed us from the top of the rock."

Clark ⇨

October 24, 1805
"... the houses of those Indians are 20 feet square and sunk 8 feet under ground & covered with bark with a small door round at top rose about 18 inches above ground, to keep out the snow..."

Clark ⇨

THE LONG NARROWS

October 25, 1805
"Capt. Lewis and my Self walked down to See the place the Indians pointed out as the worst place in passing through the gut, which we found difficuelt of passing without great danger, but as the portage was imprat[c]able with our large canoes, we concluded to Make a portage of our most valuable articles and run the canoes thro. accordingly on our return divided the party Some to take over the Canoes, and others to take our Stores across a portage of a mile to a place on the channel below this bad whorl & Suck, with Some others I had fixed on the Chanel with roapes to throw out to any who Should unfortunately meet with difficuelty in passing through: great number of Indians viewing us from the high rocks under which we had to pass, ..."

Clark ⇨

October 24, 1805
"... I counted 107 stacks of dried pounded fish in different places on those rocks which must have contained 10.000 lb of neet fish, ..."

Clark ⇨

October 25, 1805
"... this Chanel is through a hard rough black rock from 50 to 100 yards wide, swelling and boiling in a most tremendious maner..."

Clark ⇨

October 25, 1805, Traverse
"3 miles thro a narrow swift bad chanel from 50 to 100 yards wide, of swels whorls & bad places a verry bad place a 1 mile, a rock in the middle at 2 miles to a rock above a Deep bason to the Stad. Side above the rock"

Clark ⇨

October 25, 1805
"... the 3 fir[s]t canoes passed thro very well, the 4th nearly filled with water, the last passed through by takeing in a little water, thus Safely below what I conceved to be the worst part of this chanel felt my self extreamly gratified and pleased."

Clark ⇨

April 19, 1806
"this morning early had some rain had the small canoes hauled out to dry every man capable of carrying a load comenced the portage and by 5 P. M had every part of our baggage and canoes across the portage."

Clark ⇦

April 19, 1806
"The long narrows are much more formidable then they were when we decended them last fall there would be no possibility of passing either up or down them in any vessle at this time."

Clark ⇦

THE DEEP BASIN

October 25, 1805
"... we passed through a deep bason to the Stard. Side of 1 mile below which the River narrows and [is] divided by a rock the curent we found quit[e] gentle, ..."

Clark ⇨

April 18, 1806
"... here we found it necessary to unload the perogues and canoes and make a portage of 70 paces over a rock; we then drew our vessels up by a cord and the assistance of settingpoles. from hence we proceeded to the bason below the long narrows 5 ms. further..."

Lewis ⇦

THE DALLES (FLAT STEPPING STONES)

October 25, 1805
"... we proceeded on down the water fine, rocks in every derection for a fiew miles when the river widens and becomes a butifull jentle Stream of about half a mile wide, Great numbers of the Sea orter [or Seals] about those narrows..."
Possibly Harbor Seal
(Phoca vitulina

Clark ⇨

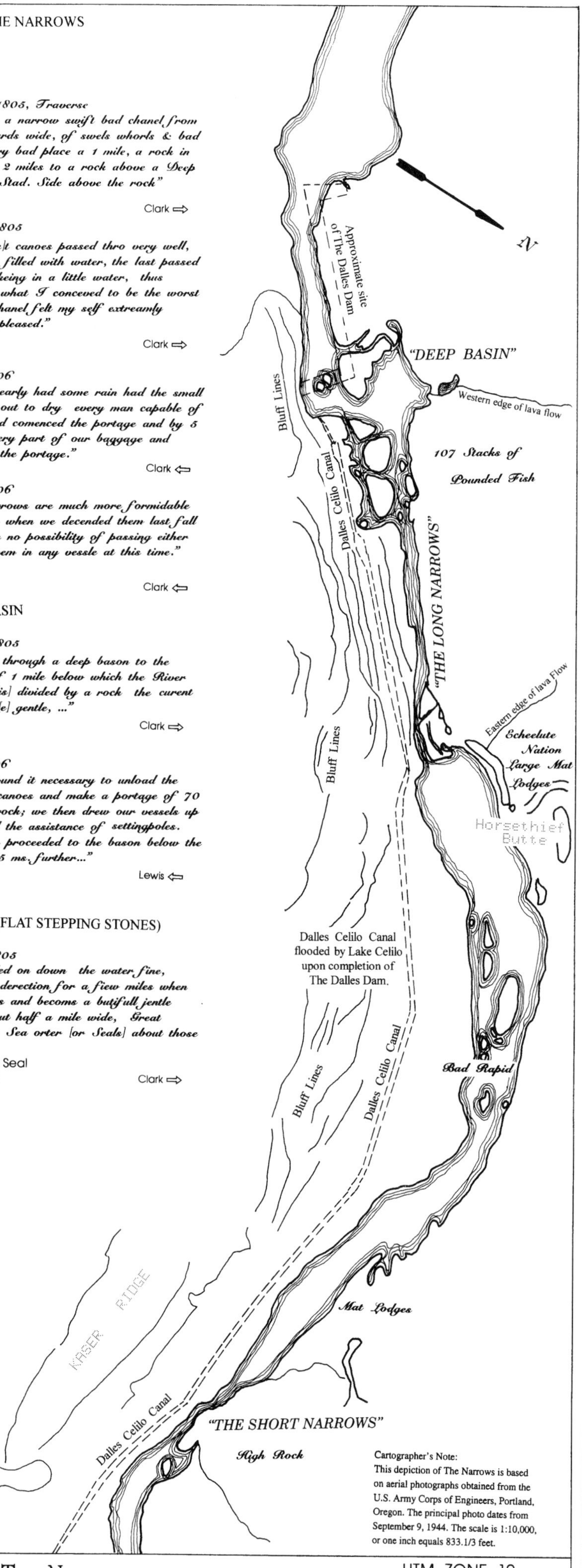

N
Approximate site of The Dalles Dam
"DEEP BASIN"
Western edge of lava flow
Bluff Lines
107 Stacks of Pounded Fish
Dalles Celilo Canal
"THE LONG NARROWS"
Eastern edge of lava Flow
Echeelute Nation Large Mat Lodges
Bluff Lines
Horsethief Butte
Dalles Celilo Canal flooded by Lake Celilo upon completion of The Dalles Dam.
Bluff Lines
Dalles Celilo Canal
Bad Rapid
KASER RIDGE
Mat Lodges
Dalles Celilo Canal
"THE SHORT NARROWS"
High Rock

Cartographer's Note:
This depiction of The Narrows is based on aerial photographs obtained from the U.S. Army Corps of Engineers, Portland, Oregon. The principal photo dates from September 9, 1944. The scale is 1:10,000, or one inch equals 833.1/3 feet.

EXPLORATIONS OF LEWIS AND CLARK 1804 - 1806
CARTOGRAPHIC RECONSTRUCTION
Oregon Washington

THE NARROWS (CONTINUED)

UTM ZONE 10
MAP NUMBER 350A

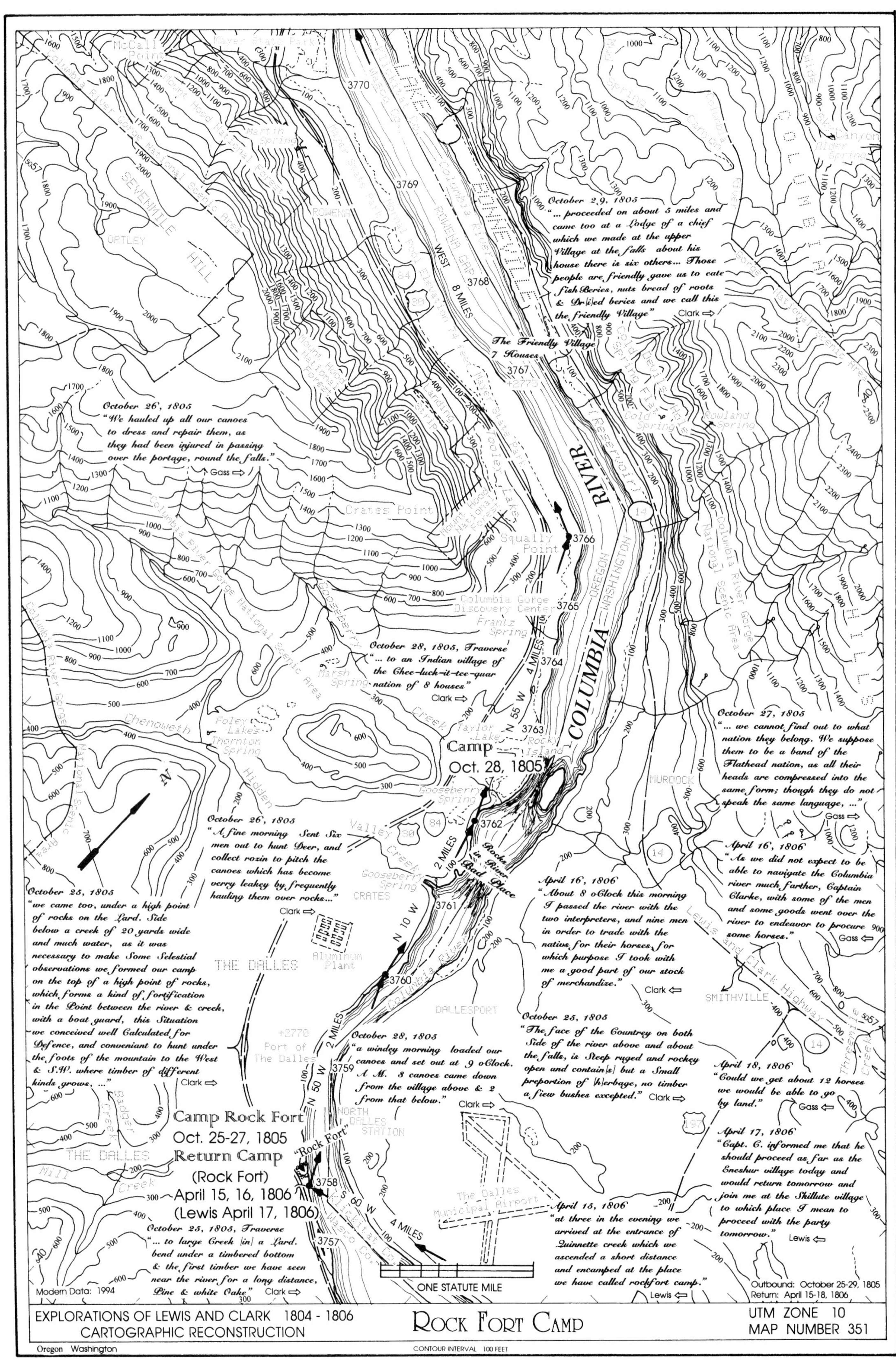
October 29, 1805
"... proceeded on about 5 miles and came too at a Lodge of a chief which we made at the upper Village at the falls about his house there is six others... Those people are friendly gave us to eate fish Beries, nuts bread of roots & Dr[i]ed beries and we call this the friendly Village"
Clark ⇨
The Friendly Village
7 Houses
October 26, 1805
"We hauled up all our canoes to dress and repair them, as they had been injured in passing over the portage, round the falls."
Gass ⇨
October 28, 1805, Traverse
"... to an Indian village of the Chee-luck-it-tee-quar nation of 8 houses"
Clark ⇨
Camp
Oct. 28, 1805
October 27, 1805
"... we cannot find out to what nation they belong. We suppose them to be a band of the Flathead nation, as all their heads are compressed into the same form; though they do not speak the same language, ..."
Gass ⇨
October 26, 1805
"A fine morning Sent Six men out to hunt Deer, and collect rozin to pitch the canoes which has become verry leakey by frequently hauling them over rocks..."
Clark ⇨
Rocks in River Bad Place
April 16, 1806
"As we did not expect to be able to navigate the Columbia river much farther, Captain Clarke, with some of the men and some goods went over the river to endeavor to procure some horses."
Gass ⇦
October 25, 1805
"we came too, under a high point of rocks on the Lard. Side below a creek of 20 yards wide and much water, as it was necessary to make Some Selestial observations we formed our camp on the top of a high point of rocks, which forms a kind of fortification in the Point between the river & creek, with a boat guard, this Situation we conceived well Calculated for Defence, and conveniant to timber under the foots of the mountain to the West & S.W. where timber of different kinds grows, ..."
Clark ⇨
April 16, 1806
"About 8 oClock this morning I passed the river with the two interpreters, and nine men in order to trade with the natives for their horses, for which purpose I took with me a good part of our stock of merchandize."
Clark ⇦
October 25, 1805
"The face of the Countrey on both Side of the river above and about the falls, is Steep raged and rockey open and contain[s] but a Small preportion of [h]erbage, no timber a fiew bushes excepted."
Clark ⇨
October 28, 1805
"a windey morning loaded our canoes and set out at 9 oClock. A. M. 8 canoes came down from the village above & 2 from that below."
Clark ⇨
April 18, 1806
"Could we get about 12 horses we would be able to go by land."
Gass ⇦
Camp Rock Fort
Oct. 25-27, 1805
Return Camp
(Rock Fort)
April 15, 16, 1806
(Lewis April 17, 1806)
"Rock Fort"
April 17, 1806
"Capt. C. informed me that he should proceed as far as the Eneshur village today and would return tomorrow and join me at the Skillute village to which place I mean to proceed with the party tomorrow."
Lewis ⇦
October 25, 1805, Traverse
"... to large Creek [in] a Lard. bend under a timbered bottom & the first timber we have seen near the river for a long distance, Pine & white Oake"
Clark ⇨
April 15, 1806
"at three in the evening we arrived at the entrance of Quinnette creek which we ascended a short distance and encamped at the place we have called rockfort camp."
Lewis ⇦
COLUMBIA RIVER
OREGON
WASHINGTON
THE DALLES
DALLESPORT
MURDOCK
SMITHVILLE
ROWENA
LAKE CELILO
Crates Point
Squally Point
Columbia Gorge Discovery Center
Port of The Dalles
The Dalles Municipal Airport
Aluminum Plant
Lewis and Clark Highway
Mill Creek
Chenoweth Creek
ONE STATUTE MILE
Modern Data: 1994
Outbound: October 25-29, 1805
Return: April 15-18, 1806
EXPLORATIONS OF LEWIS AND CLARK 1804 - 1806
CARTOGRAPHIC RECONSTRUCTION
ROCK FORT CAMP
UTM ZONE 10
MAP NUMBER 351
Oregon Washington
CONTOUR INTERVAL 100 FEET

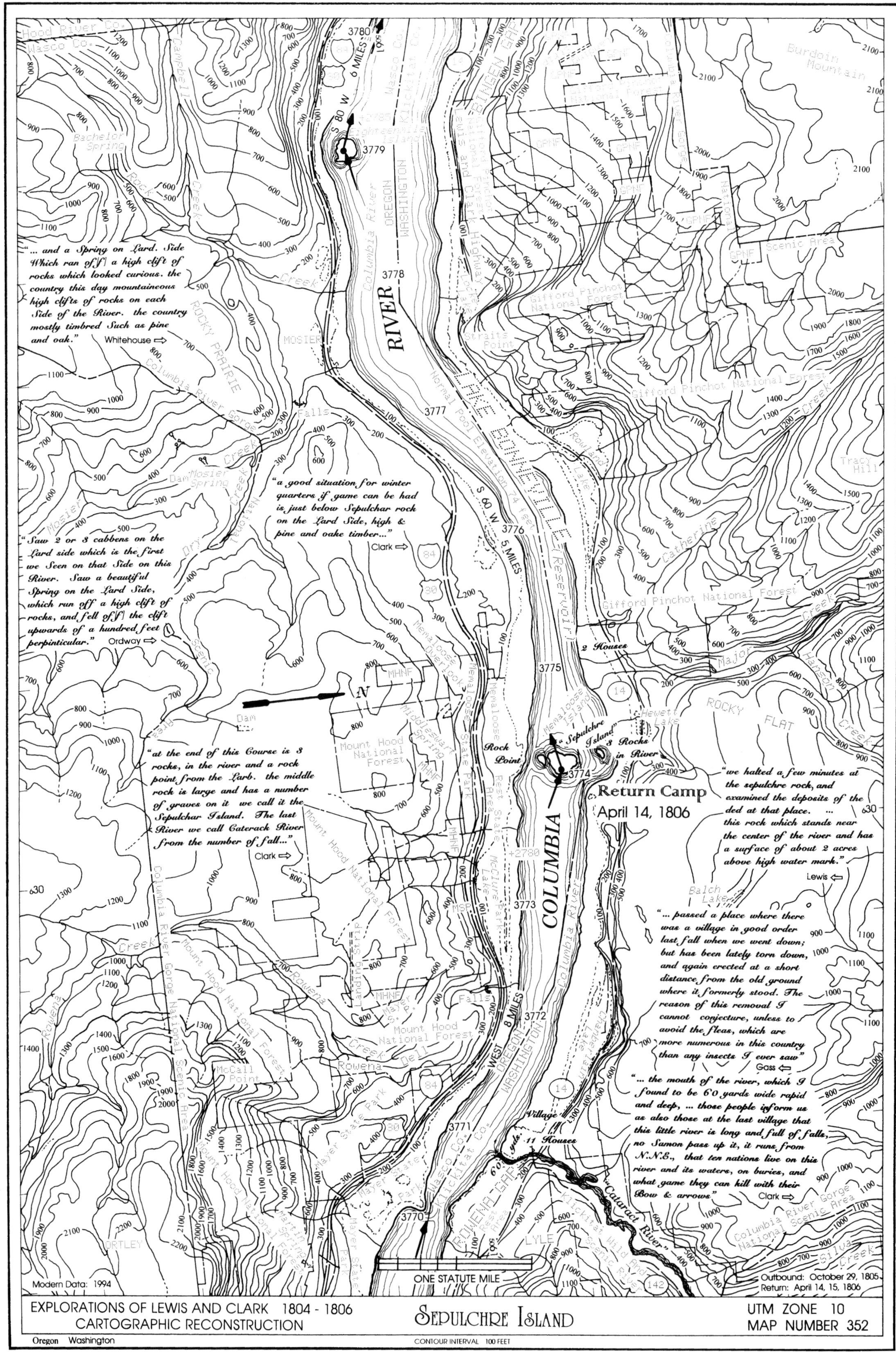
"... and a Spring on Lard. Side Which ran of[f] a high clift of rocks which looked curious. the country this day mountaineous high clifts of rocks on each Side of the River. the country mostly timbred Such as pine and oak." Whitehouse ⇨
"Saw 2 or 3 cabbens on the Lard side which is the first we Seen on that Side on this River. Saw a beautiful Spring on the Lard Side, which run off a high clift of rocks, and fell of[f] the clift upwards of a hundred feet perpinticular." Ordway ⇨
"a good situation for winter quarters if game can be had is just below Sepulchar rock on the Lard Side, high & pine and oake timber..." Clark ⇨
"at the end of this Course is 3 rocks, in the river and a rock point from the Larb. the middle rock is large and has a number of graves on it we call it the Sepulchar Island. The last River we call Caterack River from the number of fall..." Clark ⇨
"we halted a few minutes at the sepulchre rock, and examined the deposits of the ded at that place. ... this rock which stands near the center of the river and has a surface of about 2 acres above high water mark." Lewis ⇦
"... passed a place where there was a village in good order last fall when we went down; but has been lately torn down, and again erected at a short distance from the old ground where it formerly stood. The reason of this removal I cannot conjecture, unless to avoid the fleas, which are more numerous in this country than any insects I ever saw" Gass ⇦
"... the mouth of the river, which I found to be 60 yards wide rapid and deep, ... those people inform us as also those at the last village that this little river is long and full of falls, no Samon pass up it, it runs from N.N.E., that ten nations live on this river and its waters, on buries, and what game they can kill with their Bow & arrows" Clark ⇨
Return Camp
April 14, 1806
Sepulchre Island
3 Rocks in River
Rock Point
2 Houses
Village
11 Houses
Cataract River
COLUMBIA
RIVER
LAKE BONNEVILLE (Reservoir)
Normal Pool Elevation 74 feet
Columbia River
OREGON
WASHINGTON
S 80 W 6 MILES
S 60 W 5 MILES
WEST 8 MILES
ROCKY PRAIRIE
ROCKY FLAT
MOSIER
LYLE
Gifford Pinchot National Forest
Mount Hood National Forest
Columbia River Gorge National Scenic Area
Memaloose Island
Memaloose State Park
Mayer State Park
Burdoin Mountain
Tracy Hill
Balch Lake
Hewett Lake
Rowland Lake
Klickitat Wild And Scenic River
Modern Data: 1994
ONE STATUTE MILE
Outbound: October 29, 1805
Return: April 14, 15, 1806
EXPLORATIONS OF LEWIS AND CLARK 1804 - 1806
CARTOGRAPHIC RECONSTRUCTION
SEPULCHRE ISLAND
UTM ZONE 10
MAP NUMBER 352
Oregon Washington
CONTOUR INTERVAL 100 FEET

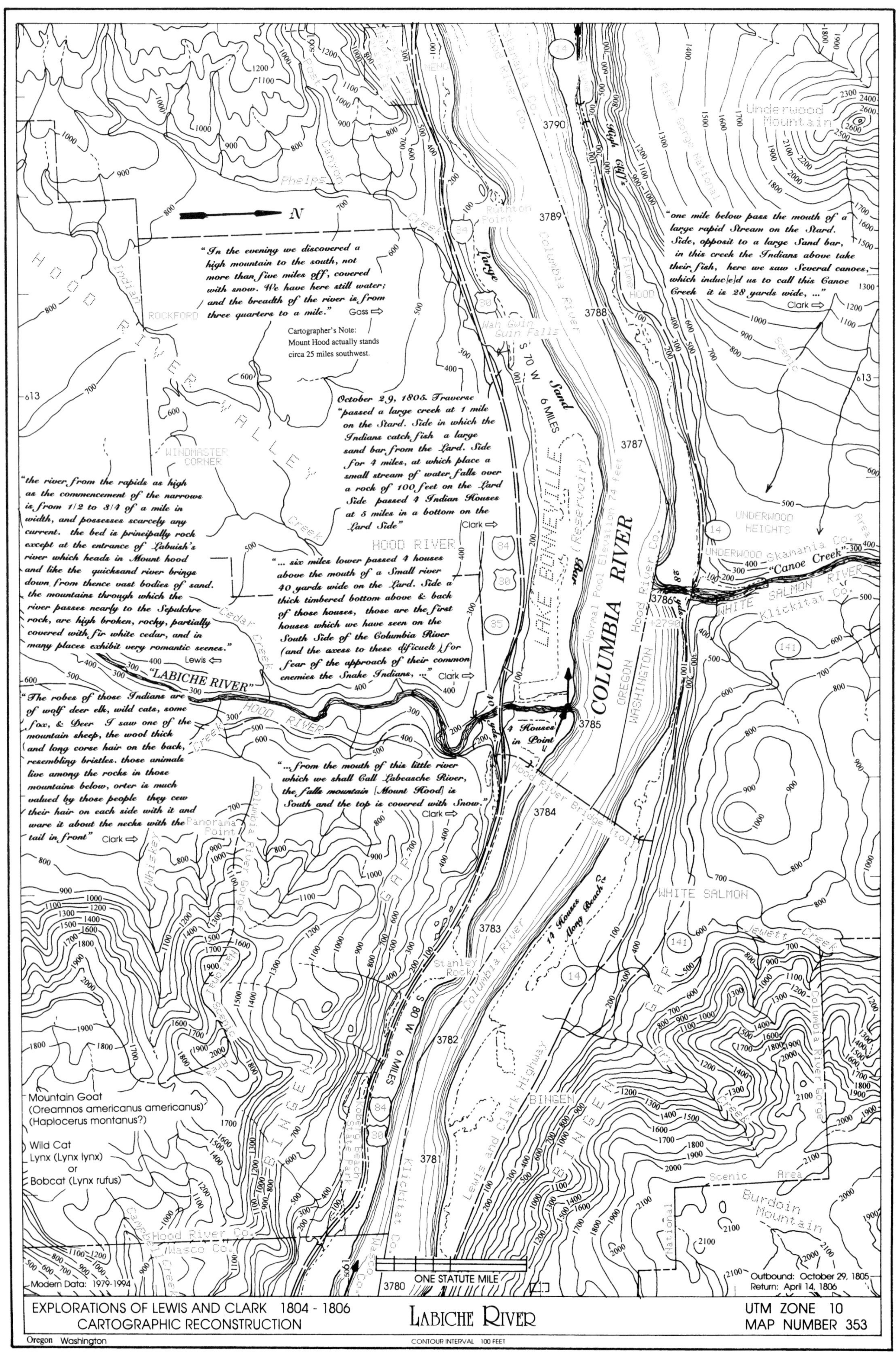
"In the evening we discovered a high mountain to the south, not more than five miles off, covered with snow. We have here still water; and the breadth of the river is from three quarters to a mile." Gass ⇨
Cartographer's Note: Mount Hood actually stands circa 25 miles southwest.
"one mile below pass the mouth of a large rapid Stream on the Stard. Side, opposit to a large Sand bar, in this creek the Indians above take their fish, here we saw Several canoes, which induc[e]d us to call this Canoe Creek it is 28 yards wide, ..." Clark ⇨
October 29, 1805. Traverse "passed a large creek at 1 mile on the Stard. Side in which the Indians catch fish a large sand bar from the Lard. Side for 4 miles, at which place a small stream of water falls over a rock of 100 feet on the Lard Side passed 4 Indian Houses at 5 miles in a bottom on the Lard Side" Clark ⇨
"the river from the rapids as high as the commencement of the narrows is from 1/2 to 3/4 of a mile in width, and possesses scarcely any current. the bed is principally rock except at the entrance of Labuish's river which heads in Mount hood and like the quicksand river brings down from thence vast bodies of sand. the mountains through which the river passes nearly to the Sepulchre rock, are high broken, rocky, partially covered with fir white cedar, and in many places exhibit very romantic scenes." Lewis ⇦
"... six miles lower passed 4 houses above the mouth of a Small river 40 yards wide on the Lard. Side a thick timbered bottom above & back of those houses, those are the first houses which we have seen on the South Side of the Columbia River (and the axess to these dificuelt) for fear of the approach of their common enemies the Snake Indians, ..." Clark ⇨
"The robes of those Indians are of wolf deer elk, wild cats, some fox, & Deer I saw one of the mountain sheep, the wool thick and long corse hair on the back, resembling bristles. those animals live among the rocks in those mountains below, orter is much valued by those people they cew their hair on each side with it and ware it about the necks with the tail in front" Clark ⇨
"...from the mouth of this little river which we shall Call Labeasche River, the falls mountain [Mount Hood] is South and the top is covered with Snow." Clark ⇨
N
HOOD RIVER VALLEY
LAKE BONNEVILLE (Reservoir)
Normal Pool Elevation 74 feet
COLUMBIA RIVER
"LABICHE RIVER"
"Canoe Creek"
WHITE SALMON RIVER
HOOD RIVER
WHITE SALMON
BINGEN
UNDERWOOD HEIGHTS
Underwood Mountain
Burdoin Mountain
Stanley Rock
Panorama Point
Ruthton Point
Wah Gwin Gwin Falls
WINDMASTER CORNER
ROCKFORD
Hood River Bridge (toll)
Lewis and Clark Highway
S 70 W 6 MILES
S 80 W 6 MILES
4 Houses in Point
14 Houses Along Beach
Mountain Goat (Oreamnos americanus americanus) (Haplocerus montanus?)
Wild Cat Lynx (Lynx lynx) or Bobcat (Lynx rufus)
Modern Data: 1979-1994
ONE STATUTE MILE
Outbound: October 29, 1805
Return: April 14, 1806
EXPLORATIONS OF LEWIS AND CLARK 1804 - 1806
CARTOGRAPHIC RECONSTRUCTION
LABICHE RIVER
UTM ZONE 10
MAP NUMBER 353
Oregon Washington
CONTOUR INTERVAL 100 FEET

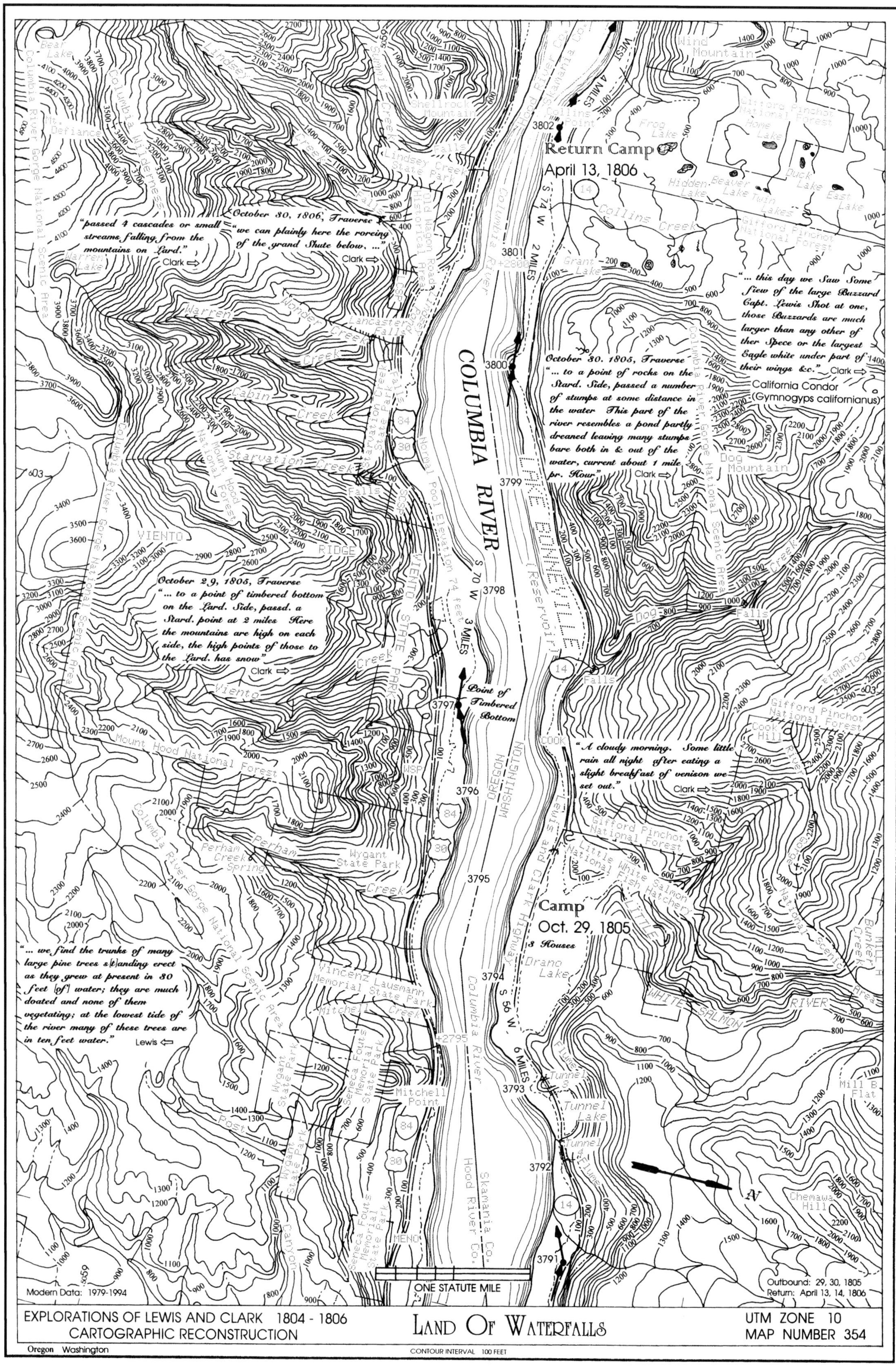
"passed 4 cascades or small streams falling from the mountains on Lard."
Clark ⇨
October 30, 1806, Traverse "we can plainly here the roreing of the grand Shute below. ..."
Clark ⇨
Return Camp
April 13, 1806
"... this day we Saw Some fiew of the large Buzzard Capt. Lewis Shot at one, those Buzzards are much larger than any other of ther Spece or the largest Eagle white under part of their wings &c." Clark ⇨
California Condor
(Gymnogyps californianus)
October 30, 1805, Traverse "... to a point of rocks on the Stard. Side, passed a number of stumps at some distance in the water This part of the river resembles a pond partly dreaned leaving many stumps bare both in & out of the water, current about 1 mile pr. Hour" Clark ⇨
COLUMBIA RIVER
LAKE BONNEVILLE
(Reservoir)
October 29, 1805, Traverse "... to a point of timbered bottom on the Lard. Side, passd. a Stard. point at 2 miles Here the mountains are high on each side, the high points of those to the Lard. has snow"
Clark ⇨
Point of Timbered Bottom
"A cloudy morning. Some little rain all night after eating a slight breakfast of venison we set out."
Clark ⇨
Camp
Oct. 29, 1805
3 Houses
"... we find the trunks of many large pine trees s[t]anding erect as they grew at present in 30 feet [of] water; they are much doated and none of them vegetating; at the lowest tide of the river many of these trees are in ten feet water."
Lewis ⇦
Wind Mountain
Shellrock Mountain
Mt. Defiance
Columbia Wilderness
Lindsey Creek State Park
Starvation Creek State Park
Viento State Park
Wygant State Park
Vincenz Lausmann Memorial State Park
Seneca Fouts Memorial State Park
Mitchell Point
Dog Mountain
Collins Point
Drano Lake
Tunnel Lake
Chemawa Hill
Cook Hill
Gifford Pinchot National Forest
Mount Hood National Forest
Columbia River Gorge National Scenic Area
Little White Salmon National Fish Hatchery
WHITE SALMON RIVER
Lewis and Clark Highway
Modern Data: 1979-1994
ONE STATUTE MILE
Outbound: 29, 30, 1805
Return: April 13, 14, 1806
EXPLORATIONS OF LEWIS AND CLARK 1804 - 1806
CARTOGRAPHIC RECONSTRUCTION
LAND OF WATERFALLS
UTM ZONE 10
MAP NUMBER 354
Oregon Washington
CONTOUR INTERVAL 100 FEET

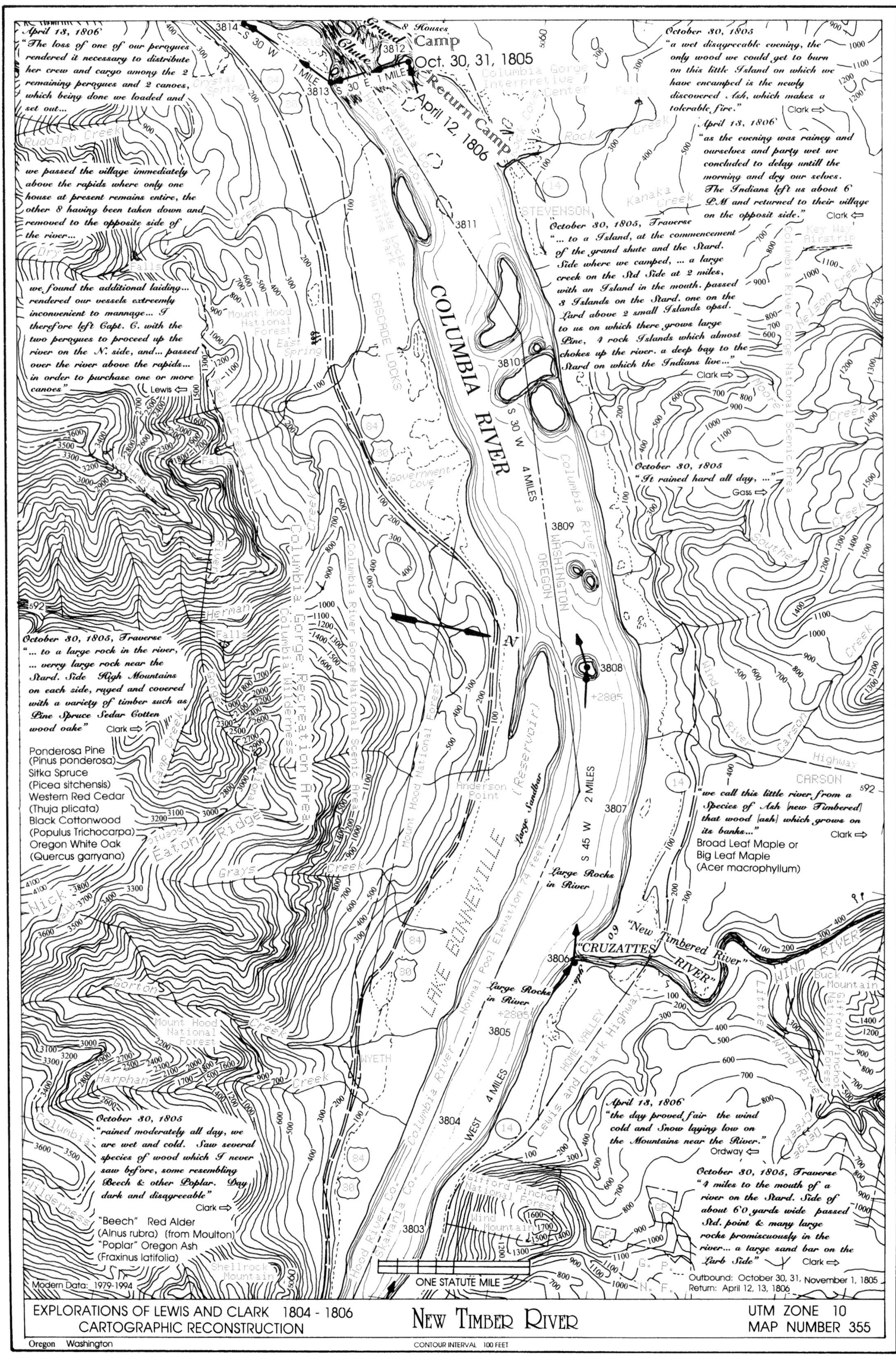
April 13, 1806
"The loss of one of our perogues rendered it necessary to distribute her crew and cargo among the 2 remaining perogues and 2 canoes, which being done we loaded and set out...
we passed the village immediately above the rapids where only one house at present remains entire, the other 8 having been taken down and removed to the opposite side of the river...
we found the additional laiding... rendered our vessels extreemly inconvenient to mannage... I therefore left Capt. C. with the two perogues to proceed up the river on the N. side, and... passed over the river above the rapids... in order to purchase one or more canoes"
Lewis
Grand Chute
Camp Oct. 30, 31, 1805
Return Camp April 12, 1806
October 30, 1805
"a wet disagreeable evening, the only wood we could get to burn on this little Island on which we have encamped is the newly discovered Ash, which makes a tolerable fire."
Clark
April 13, 1806
"as the evening was rainey and ourselves and party wet we concluded to delay untill the morning and dry our selves. The Indians left us about 6 P.M and returned to their village on the opposit side."
Clark
October 30, 1805, Traverse
"... to a Island, at the commencement of the grand shute and the Stard. Side where we camped, ... a large creek on the Std Side at 2 miles, with an Island in the mouth. passed 3 Islands on the Stard. one on the Lard above 2 small Islands opsd. to us on which there grows large Pine, 4 rock Islands which almost chokes up the river. a deep bay to the Stard on which the Indians live..."
Clark
October 30, 1805
"It rained hard all day, ..."
Gass
COLUMBIA RIVER
CASCADE LOCKS
STEVENSON
WASHINGTON
OREGON
October 30, 1805, Traverse
"... to a large rock in the river, ... very large rock near the Stard. Side High Mountains on each side, ruged and covered with a variety of timber such as Pine Spruce Sedar Cotten wood oake"
Clark
Ponderosa Pine
(Pinus ponderosa)
Sitka Spruce
(Picea sitchensis)
Western Red Cedar
(Thuja plicata)
Black Cottonwood
(Populus Trichocarpa)
Oregon White Oak
(Quercus garryana)
"we call this little river from a Species of Ash [new Timbered] that wood [ash] which grows on its banks..."
Clark
Broad Leaf Maple or
Big Leaf Maple
(Acer macrophyllum)
CARSON
LAKE BONNEVILLE
Large Sandbar
Large Rocks in River
"CRUZATTES RIVER"
"New Timbered River"
WIND RIVER
HOME VALLEY
Lewis and Clark Highway
WYETH
April 13, 1806
"the day proved fair the wind cold and Snow laying low on the Mountains near the River."
Ordway
October 30, 1805
"rained moderately all day, we are wet and cold. Saw several species of wood which I never saw before, some resembling Beech & other Poplar. Day dark and disagreeable"
Clark
"Beech" Red Alder
(Alnus rubra) (from Moulton)
"Poplar" Oregon Ash
(Fraxinus latifolia)
October 30, 1805, Traverse
"4 miles to the mouth of a river on the Stard. Side of about 60 yards wide passed Std. point & many large rocks promiscuously in the river... a large sand bar on the Larb Side"
Clark
ONE STATUTE MILE
Modern Data: 1979-1994
Outbound: October 30, 31, November 1, 1805
Return: April 12, 13, 1806
EXPLORATIONS OF LEWIS AND CLARK 1804 - 1806
CARTOGRAPHIC RECONSTRUCTION
NEW TIMBER RIVER
UTM ZONE 10
MAP NUMBER 355
Oregon Washington
CONTOUR INTERVAL 100 FEET

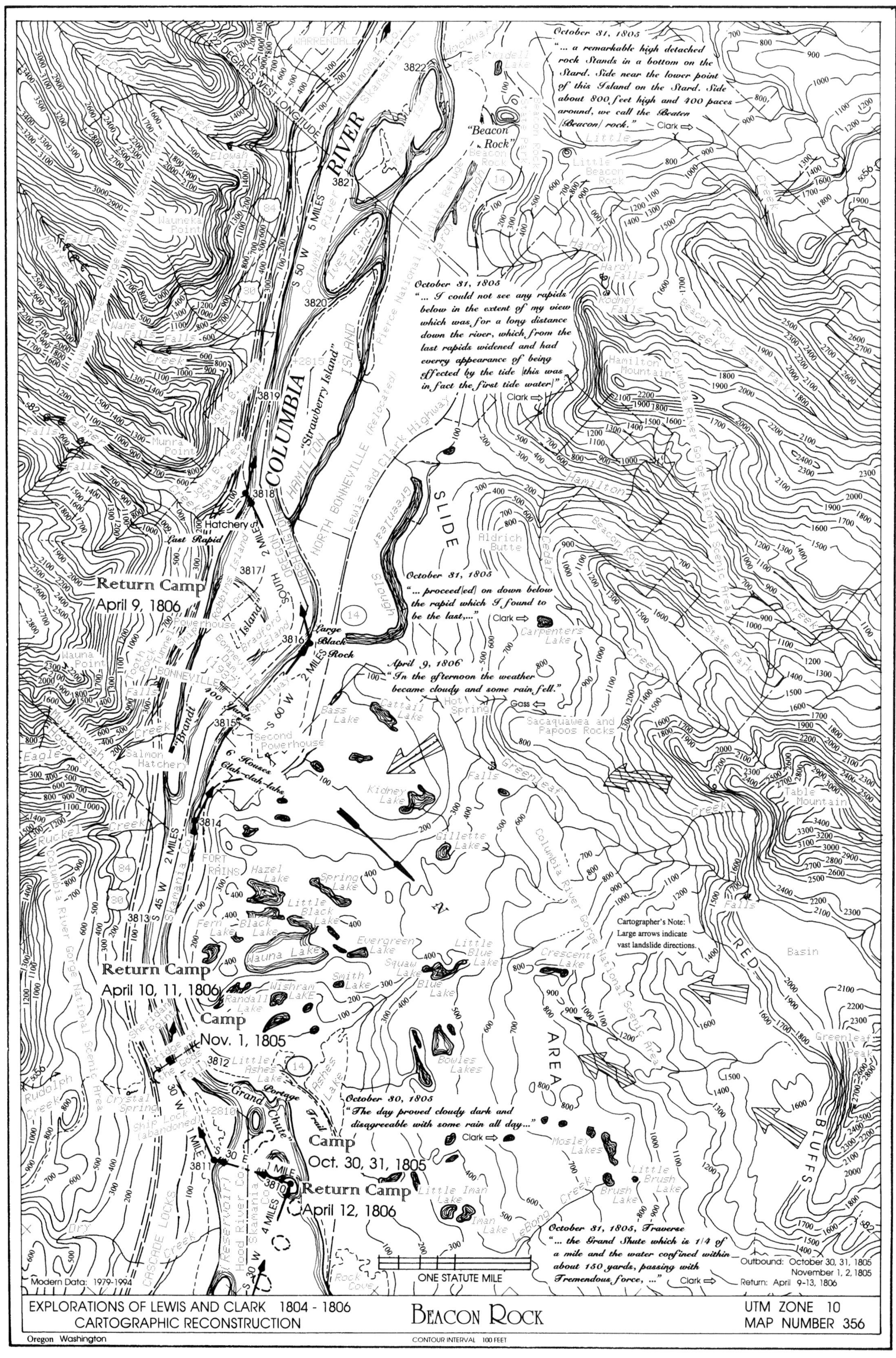
October 31, 1805
"... a remarkable high detached rock Stands in a bottom on the Stard. Side near the lower point of this Island on the Stard. Side about 800 feet high and 400 paces around, we call the Beaten [Beacon] rock."
Clark ⇨
"Beacon Rock"
October 31, 1805
"... I could not see any rapids below in the extent of my view which was for a long distance down the river, which from the last rapids widened and had every appearance of being effected by the tide [this was in fact the first tide water]"
Clark ⇨
October 31, 1805
"... proceed[ed] on down below the rapid which I found to be the last,..."
Clark ⇨
April 9, 1806
"In the afternoon the weather became cloudy and some rain fell."
Gass ⇦
October 30, 1805
"The day proved cloudy dark and disagreeable with some rain all day..."
Clark ⇨
October 31, 1805, Traverse
"... the Grand Shute which is 1/4 of a mile and the water confined within about 150 yards, passing with Tremendous force, ..."
Clark ⇨
Return Camp
April 9, 1806
Return Camp
April 10, 11, 1806
Camp
Nov. 1, 1805
Camp
Oct. 30, 31, 1805
Return Camp
April 12, 1806
Last Rapid
Large Black Rock
6 Houses
Clah-clah-lahs
"Grand Chute"
Portage Trail
"Strawberry Island"
COLUMBIA RIVER
SLIDE AREA
RED BLUFFS
Cartographer's Note: Large arrows indicate vast landslide directions.
Outbound: October 30, 31, 1805
November 1, 2, 1805
Return: April 9-13, 1806
Modern Data: 1979-1994
ONE STATUTE MILE
EXPLORATIONS OF LEWIS AND CLARK 1804 - 1806
CARTOGRAPHIC RECONSTRUCTION
BEACON ROCK
UTM ZONE 10
MAP NUMBER 356
Oregon Washington
CONTOUR INTERVAL 100 FEET

Selected quotes for the outbound, 1805

October 30, 1805
"... landed on an Island close under the Stard. Side at the head of the great Shute, and a little below a village of 8 large houses on a Deep bend on the Stard. Side, and opposit 2 Small Islands imediately in the head of the Shute, which Islands are covered with Pine, maney large rocks also, ..." Clark ⇨

October 30, 1805
"I with 2 men proceeded down the river 2 miles on an old Indian parth to view the rapids, which I found inpassible for our canoes without a portage, the roade bad at 1 mile I saw a Town of Houses entirly abandoned, ..." Clark ⇨

October 31, 1805,
Traverse [30 W 1 Mile]
"... passed several rocks in the river & a rapid at 3/4 of a mile the water being confined between large rocks, maney of which is under water." Clark ⇨

October 31, 1805,
Traverse [S 60 W 2 Miles]
"... to a large black rock in a Stard. bend at the commencement of a rapid opsd. lower point of an Island Lard passed a rapid at 1/2 ml not bad." Clark ⇨

October 31, 1805
"... here is the head of a large Island in high water, at this time no water passes on the Stard. Side I walked thro this Island which I found to be verry rich, open & covered with Strawberry vines, ... The natives has dug roots in some parts of this Isl. which is about 3 miles long & 1 wide, a small Island covered with timber opposit the lower point... opposit the Strawberry Island on its Stard Side a creek falls in which has no running water at present, it has the appearance of throwing out emence torrents." Clark ⇨

October 31, 1805
"This Great Shute or falls is about 1/2 a mile, with the water of this great river compressed within the space of 150 paces in which there is great numbers of both large and Small rocks, water passing with great velocity forming [foaming] & boiling in a most horriable manner, with a fall of about 20 feet, ... the high Stones which are continually braking loose from the mountain on the Stard. Side and roleing down into the Shute aded to those which brake loose from these Islands above and lodge in the Shute, must be the cause of the rivers daming up to such a distance above, ..." Clark ⇨

Selected quotes for the return, 1806

April 9, 1806
"as we could not ascend the rapid by the North side of the river with our large canoes, we passed to the opposite side and entered the narrow channel which separates brant Island from the South shore; ..." Lewis ⇦

April 9, 1806
"evening wet & disagreeable" Clark ⇦

April 10, 1806
"We set out early and droped down the channel to the lower end of brant Island from whence we drew them up the rapid by a cord about a quarter of a mile..." Lewis ⇦

April 10, 1806
"... they [Collins, Gibson, and Pryor] *were to join us on the oposite side at a small village of six houses of the Clah-clah-lahs..."* Lewis ⇦

April 10, 1806
"we set out and continued our rout up the N. side of the river with great difficulty in consequence of the rapidity of the current and the large rocks which form this shore; the South side of the river is impassable." Lewis ⇦

April 10, 1806
"by evening we arived at the portage on the N. Side where we landed and conveyed our baggage to the top of the hill about 200 paces distant where we formed a camp." Clark ⇦

April 11, 1806
"... we concluded to take our canoes first to the head of the rapids, hoping that by evening the rain would cease and afford us a fair afternoon to take our baggage over the portage. this portage is two thousand eight hundred yards along a narrow rough and slipery road." Lewis ⇦

April 11, 1806
"by the evening Capt. C. took 4 of our canoes above the rapids tho' with much difficulty and labour. the canoes were much damaged by being driven against the rocks... the men complained of being so much fatiegued in the evening that we posponed taking up our 5th canoe untill tomorrow. these rapids are much worse than they were [in the] fall when we passed them, ..." Lewis ⇦

April 11, 1806
"three of this same tribe of villains the Wah-clel-lars, stole my dog this evening, and took him towards their village; I was shortly afterwards informed... sent three men in pursuit... the indians discovering the party in pursuit of them left the dog and fled." Lewis ⇦

April 12, 1806
"rained the greater part of the last night and this morning... we employed all hands in attempting to take up the last canoe. in attempting to pass by a rock against which the current run with emence force, the bow unfortunately took the current at too great a distance from the rock, she turned broad side to the stream, ... the men were compelled to let go the rope and both the canoe and rope went with the stream." Clark ⇦

Cartographer's Note:
Both Lewis and Clark speculated that slides or rockfalls affecting the river's flow probably happened here often, but the captains, perhaps, were unknowing of its massive scale. To the north, the slopes of Greenleaf Peak, Table Mountain, and possibly much of eastern and southern Hamilton Mountain are inherently unstable (see Map 356). Great earth slides coming off the gorge's north rim have dammed the Columbia, backing water upstream almost to the Umatilla River (a cataclysm that later gave rise to a Northwest legend, based on Indian storytelling, about a Bridge of the Gods). The current eventually eroded away much of this natural dam, reducing the backwater to an area somewhat smaller than today's Bonneville Dam reservoir. The backwater, of course, had flooded Indian villages and shoreline forests. When the captains observed long-dead trees standing in 10 or 30 ft. of calm water, they pondered a cataclysmic-like event, estimating its occurrence to a couple of decades earlier. Modern-day geologists variously dated this slide (or slides) from 800 years to a few thousands of years ago. Some scientists are taking a new look at the data, however, and it may be that Lewis and Clark's shorter time estimate might be closer to the actual mark. Some local residents claim that on quiet nights one can hear scrapping and cracking sounds of rocks and earth slowly moving toward the river.

Cartographer's Note:
This map of the Columbia River from the Cascades Rapids down to "Strawberry" or Hamilton Island is based on a U.S. Army Corp of Engineers aerial photograph. Probably dating from the early 1930s, the photo shows Bonneville Dam under construction, though work on the spillway had yet to begin. In regard to the small islands at the upper Cascades, there are variations regarding their size and location between Clark's depictions, twentieth-century USGS quadrangle maps, and the aerial photo.

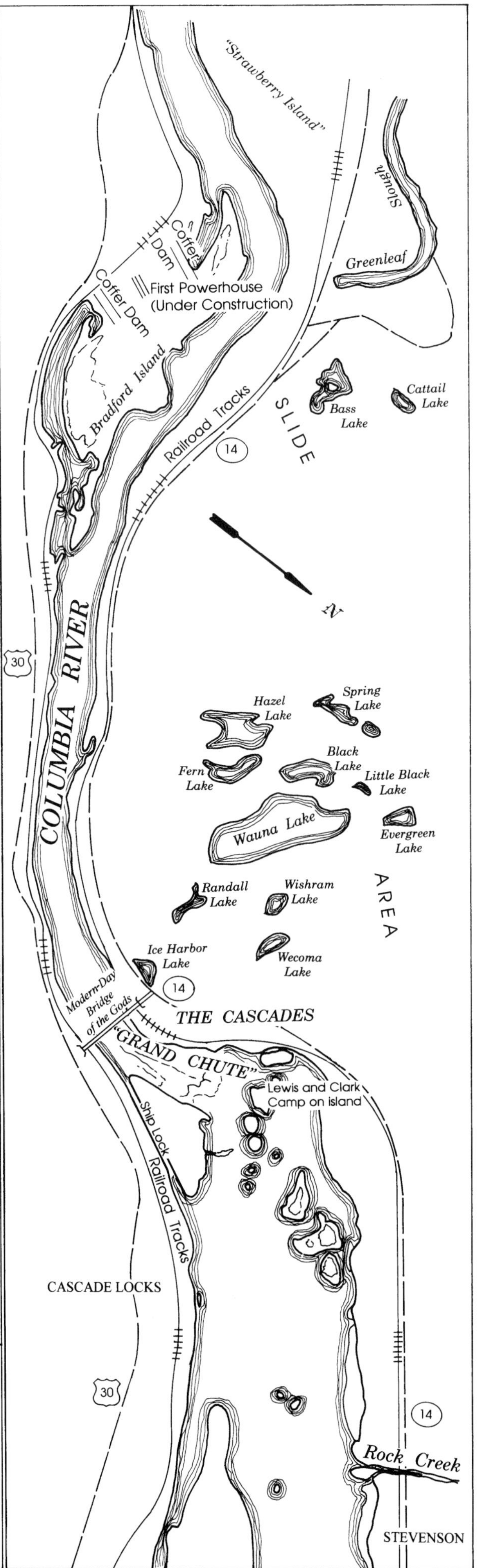

"Strawberry Island"
Slough
Greenleaf
Coffer Dam
Coffer Dam
First Powerhouse (Under Construction)
Bradford Island
Railroad Tracks
SLIDE
Bass Lake
Cattail Lake
14
N
30
COLUMBIA RIVER
Hazel Lake
Spring Lake
Fern Lake
Black Lake
Little Black Lake
Wauna Lake
Evergreen Lake
Randall Lake
Wishram Lake
AREA
Ice Harbor Lake
Wecoma Lake
14
Modern-Day Bridge of the Gods
THE CASCADES
"GRAND CHUTE"
Lewis and Clark Camp on island
Ship Lock
Railroad Tracks
CASCADE LOCKS
30
14
Rock Creek
STEVENSON

EXPLORATIONS OF LEWIS AND CLARK 1804 - 1806
CARTOGRAPHIC RECONSTRUCTION

BEACON ROCK (CONTINUED)

UTM ZONE 10
MAP NUMBER 356A

Oregon Washington

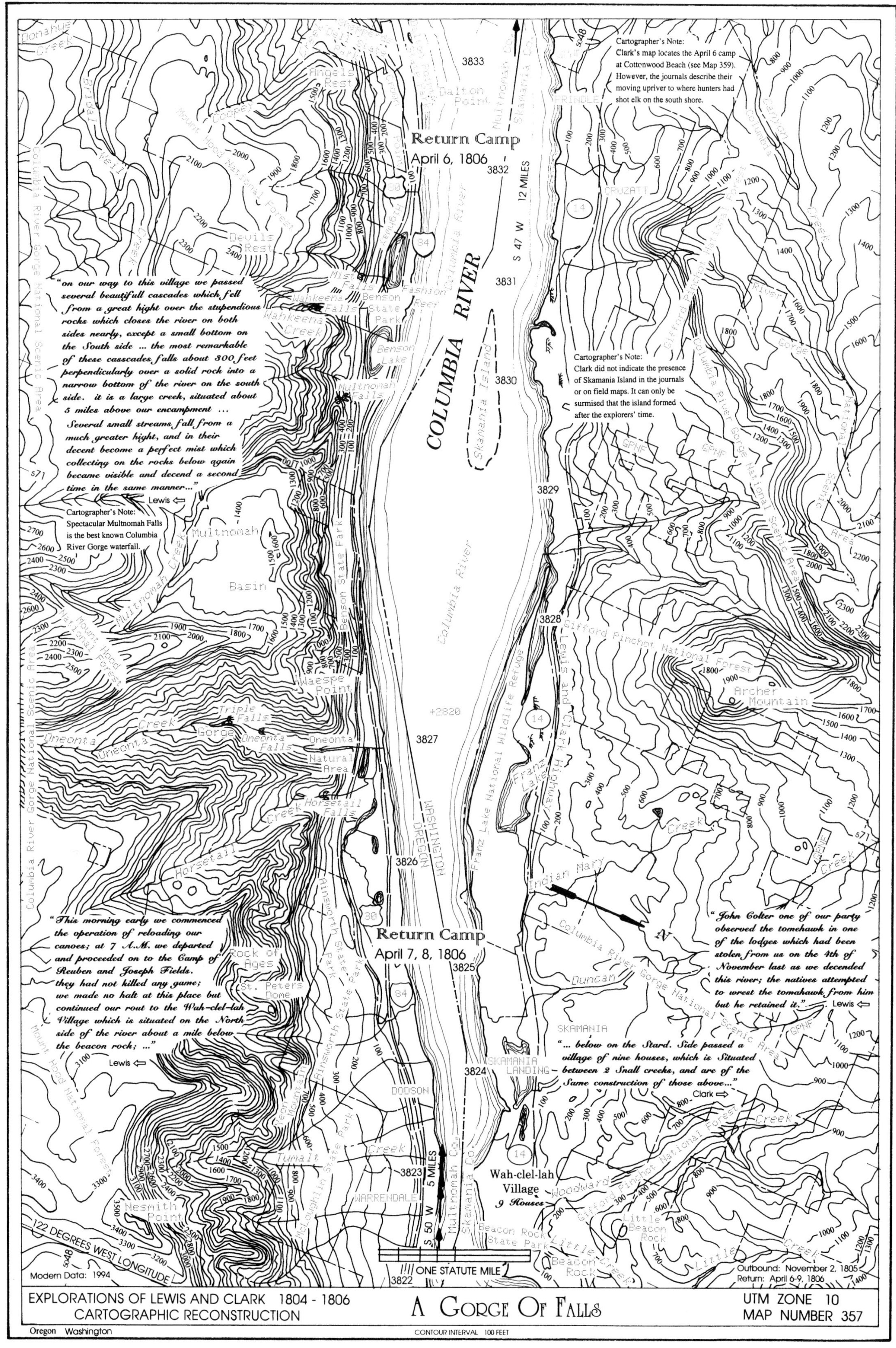

Return Camp
April 6, 1806
Return Camp
April 7, 8, 1806
COLUMBIA RIVER
S 47 W 12 MILES
S 50 W 5 MILES
Cartographer's Note:
Clark's map locates the April 6 camp at Cottonwood Beach (see Map 359). However, the journals describe their moving upriver to where hunters had shot elk on the south shore.
"on our way to this village we passed several beautifull cascades which fell from a great hight over the stupendious rocks which closes the river on both sides nearly, except a small bottom on the South side ... the most remarkable of these casscades falls about 300 feet perpendicularly over a solid rock into a narrow bottom of the river on the south side. it is a large creek, situated about 5 miles above our encampment ... Several small streams fall from a much greater hight, and in their decent become a perfect mist which collecting on the rocks below again became visible and decend a second time in the same manner..."
Lewis
Cartographer's Note:
Spectacular Multnomah Falls is the best known Columbia River Gorge waterfall.
Cartographer's Note:
Clark did not indicate the presence of Skamania Island in the journals or on field maps. It can only be surmised that the island formed after the explorers' time.
"This morning early we commenced the operation of reloading our canoes; at 7 A.M. we departed and proceeded on to the Camp of Reuben and Joseph Fields. they had not killed any game; we made no halt at this place but continued our rout to the Wah-clel-lah Village which is situated on the North side of the river about a mile below the beacon rock; ..."
Lewis
"John Colter one of our party observed the tomehawk in one of the lodges which had been stolen from us on the 4th of November last as we decended this river; the natives attempted to wrest the tomahawk from him but he retained it."
Lewis
"... below on the Stard. Side passed a village of nine houses, which is Situated between 2 Snall creeks, and are of the Same construction of those above..."
Clark
Wah-clel-lah
Village
9 Houses
Multnomah Falls
Wahkeena Falls
Mist Falls
Benson State Park
Benson Lake
Fashion Reef
Angels Rest
Devils Rest
Dalton Point
Skamania Island
Columbia Island
Multnomah Basin
Waespe Point
Triple Falls
Oneonta Gorge
Oneonta Falls
Oneonta Natural Area
Horsetail Falls
Rock of Ages
St. Peters Dome
Ainsworth State Park
Nesmith Point
Tumalt Creek
DODSON
WARRENDALE
McLoughlin State Park
Franz Lake
Franz Lake National Wildlife Refuge
Lewis and Clark Highway
Archer Mountain
Gifford Pinchot National Forest
Mount Hood National Forest
Columbia River Gorge National Scenic Area
Indian Mary
Duncan
SKAMANIA
SKAMANIA LANDING
PRINDLE
CRUZATT
Woodward
Little Beacon Rock
Beacon Rock
Beacon Rock State Park
Little Creek
Multnomah Co.
Skamania Co.
OREGON
WASHINGTON
122 DEGREES WEST LONGITUDE
ONE STATUTE MILE
Modern Data: 1994
Outbound: November 2, 1805
Return: April 6-9, 1806
EXPLORATIONS OF LEWIS AND CLARK 1804 - 1806
CARTOGRAPHIC RECONSTRUCTION
A Gorge Of Falls
UTM ZONE 10
MAP NUMBER 357
Oregon Washington
CONTOUR INTERVAL 100 FEET

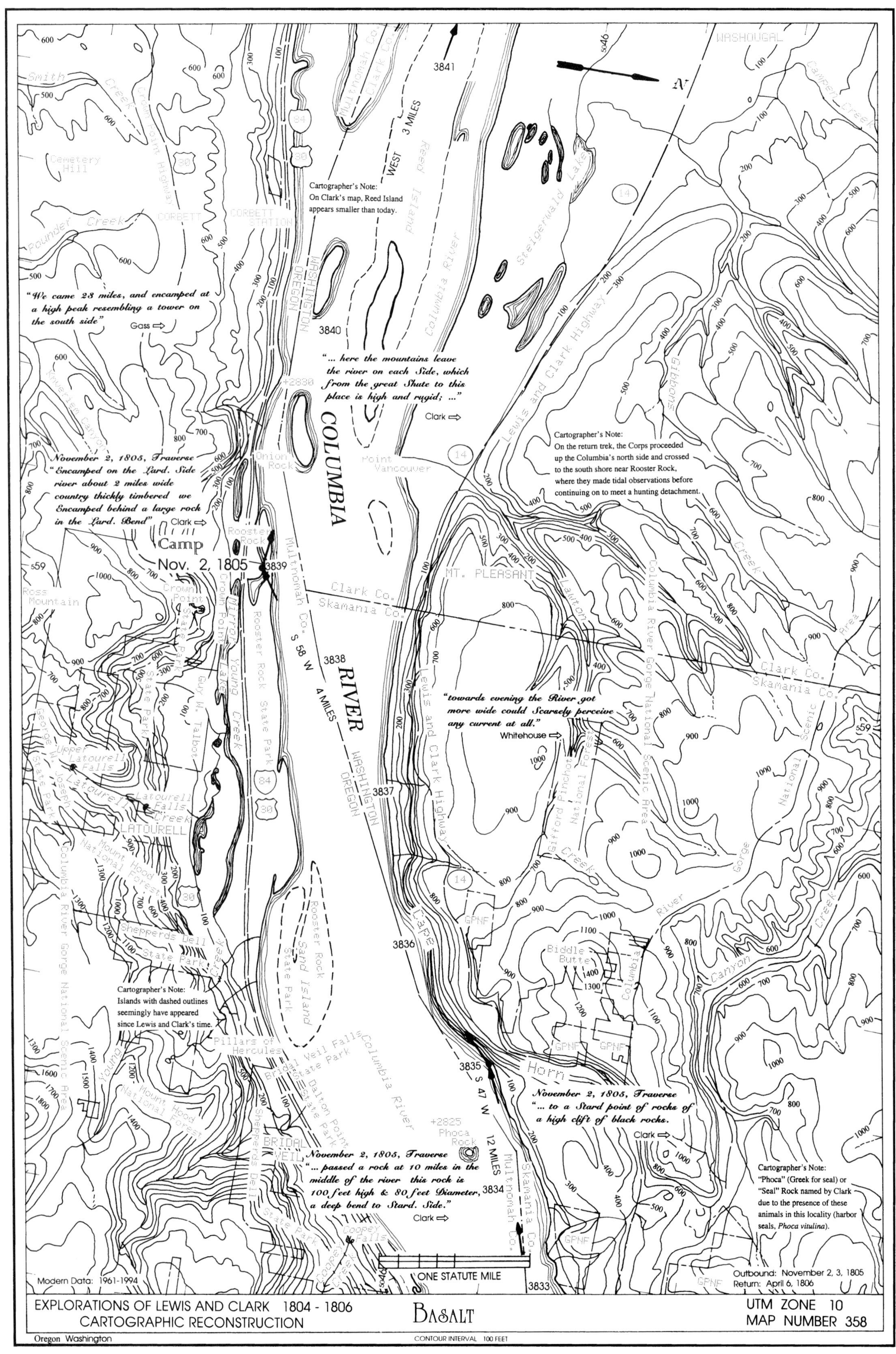
"We came 28 miles, and encamped at a high peak resembling a tower on the south side"
Gass ⇨
Cartographer's Note:
On Clark's map, Reed Island appears smaller than today.
"... here the mountains leave the river on each Side, which from the great Shute to this place is high and rugid; ..."
Clark ⇨
November 2, 1805, Traverse
"Encamped on the Lard. Side river about 2 miles wide country thickly timbered we Encamped behind a large rock in the Lard. Bend"
Clark ⇨
Camp
Nov. 2, 1805
Cartographer's Note:
On the return trek, the Corps proceeded up the Columbia's north side and crossed to the south shore near Rooster Rock, where they made tidal observations before continuing on to meet a hunting detachment.
"towards evening the River got more wide could Scarsely perceive any current at all."
Whitehouse ⇨
COLUMBIA
RIVER
Cartographer's Note:
Islands with dashed outlines seemingly have appeared since Lewis and Clark's time.
November 2, 1805, Traverse
"... to a Stard point of rocks of a high clift of black rocks.
Clark ⇨
November 2, 1805, Traverse
"... passed a rock at 10 miles in the middle of the river this rock is 100 feet high & 80 feet Diameter, a deep bend to Stard. Side."
Clark ⇨
Cartographer's Note:
"Phoca" (Greek for seal) or "Seal" Rock named by Clark due to the presence of these animals in this locality (harbor seals, Phoca vitulina).
Modern Data: 1961-1994
ONE STATUTE MILE
Outbound: November 2, 3, 1805
Return: April 6, 1806
EXPLORATIONS OF LEWIS AND CLARK 1804 - 1806
CARTOGRAPHIC RECONSTRUCTION
BASALT
UTM ZONE 10
MAP NUMBER 358
Oregon Washington
CONTOUR INTERVAL 100 FEET

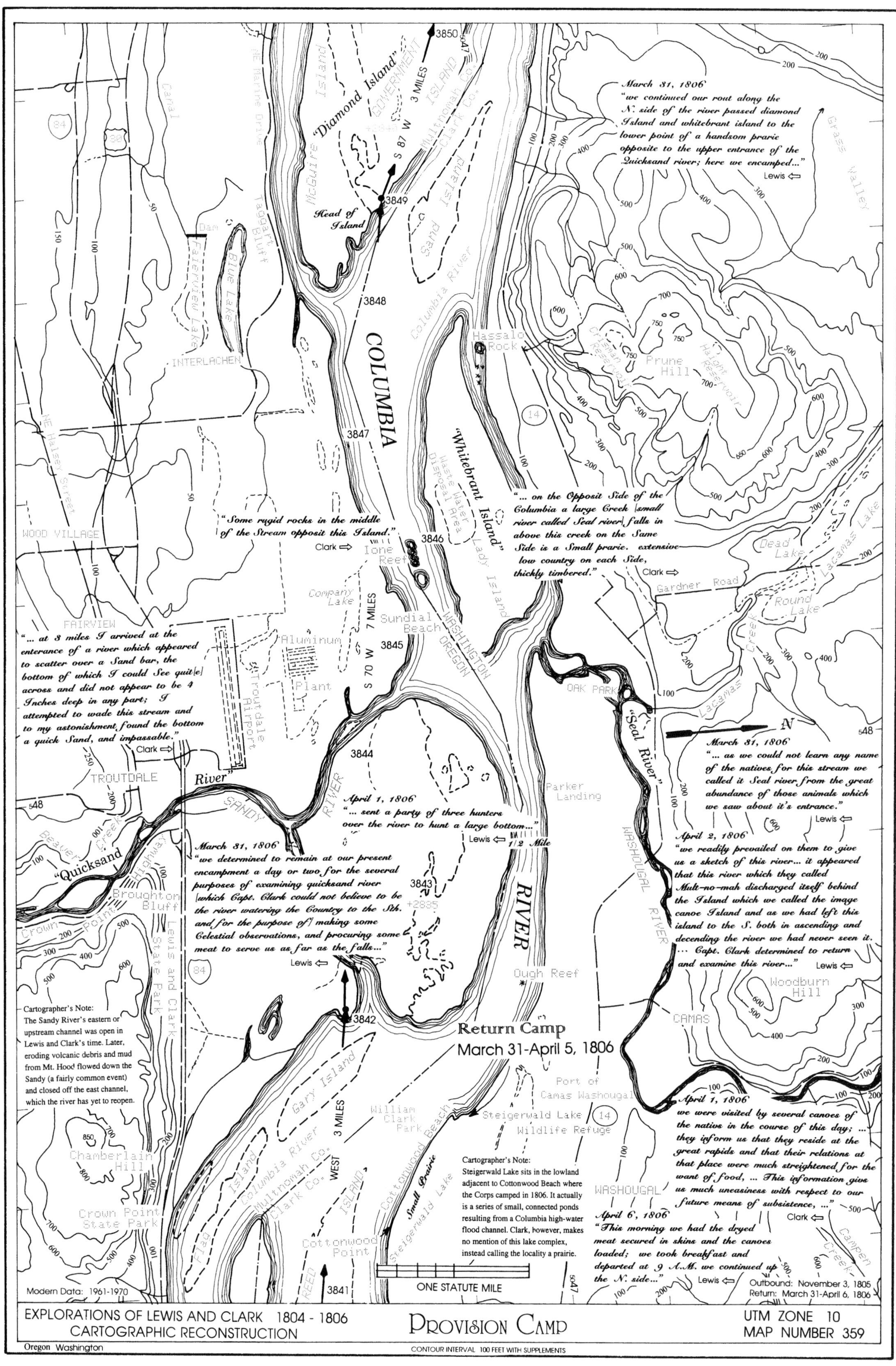

March 31, 1806
"we continued our rout along the N. side of the river passed diamond Island and whitebrant island to the lower point of a handsom prarie opposite to the upper entrance of the Quicksand river; here we encamped..."
Lewis ⇦
"Diamond Island"
Head of Island
COLUMBIA
"Whitebrant Island"
"Some rugid rocks in the middle of the Stream opposit this Island."
Clark ⇨
"... on the Opposit Side of the Columbia a large Creek [small river called Seal river] falls in above this creek on the Same Side is a Small prarie. extensive low country on each Side, thickly timbered."
Clark ⇨
"... at 3 miles I arrived at the enterance of a river which appeared to scatter over a Sand bar, the bottom of which I could See quit[e] across and did not appear to be 4 Inches deep in any part; I attempted to wade this stream and to my astonishment found the bottom a quick Sand, and impassable."
Clark ⇨
"Quicksand River"
"Seal River"
March 31, 1806
"... as we could not learn any name of the natives for this stream we called it Seal river from the great abundance of those animals which we saw about it's entrance."
Lewis ⇦
April 1, 1806
"... sent a party of three hunters over the river to hunt a large bottom..."
Lewis ⇦
1/2 Mile
March 31, 1806
"we determined to remain at our present encampment a day or two for the several purposes of examining quicksand river [which Capt. Clark could not believe to be the river watering the Country to the Sth. and for the purpose of] making some Celestial observations, and procuring some meat to serve us as far as the falls..."
Lewis ⇦
April 2, 1806
"we readily prevailed on them to give us a sketch of this river... it appeared that this river which they called Mult-no-mah discharged itself behind the Island which we called the image canoe Island and as we had left this island to the S. both in ascending and decending the river we had never seen it. ... Capt. Clark determined to return and examine this river..."
Lewis ⇦
RIVER
Return Camp
March 31-April 5, 1806
Cartographer's Note:
The Sandy River's eastern or upstream channel was open in Lewis and Clark's time. Later, eroding volcanic debris and mud from Mt. Hood flowed down the Sandy (a fairly common event) and closed off the east channel, which the river has yet to reopen.
April 1, 1806
we were visited by several canoes of the nativs in the course of this day; ... they inform us that they reside at the great rapids and that their relations at that place were much streightened for the want of food, ... This information givs us much uneasiness with respect to our future means of subsistence, ..."
Clark ⇦
Cartographer's Note:
Steigerwald Lake sits in the lowland adjacent to Cottonwood Beach where the Corps camped in 1806. It actually is a series of small, connected ponds resulting from a Columbia high-water flood channel. Clark, however, makes no mention of this lake complex, instead calling the locality a prairie.
April 6, 1806
"This morning we had the dryed meat secured in skins and the canoes loaded; we took breakfast and departed at 9 A.M. we continued up the N. side..."
Lewis ⇦
Small Prairie
S 87 W
3 MILES
S 70 W
7 MILES
WEST
3 MILES
3850
3849
3848
3847
3846
3845
3844
3843
3842
3841
N
ONE STATUTE MILE
Modern Data: 1961-1970
Outbound: November 3, 1805
Return: March 31-April 6, 1806
EXPLORATIONS OF LEWIS AND CLARK 1804 - 1806
CARTOGRAPHIC RECONSTRUCTION
PROVISION CAMP
UTM ZONE 10
MAP NUMBER 359
Oregon Washington
CONTOUR INTERVAL 100 FEET WITH SUPPLEMENTS

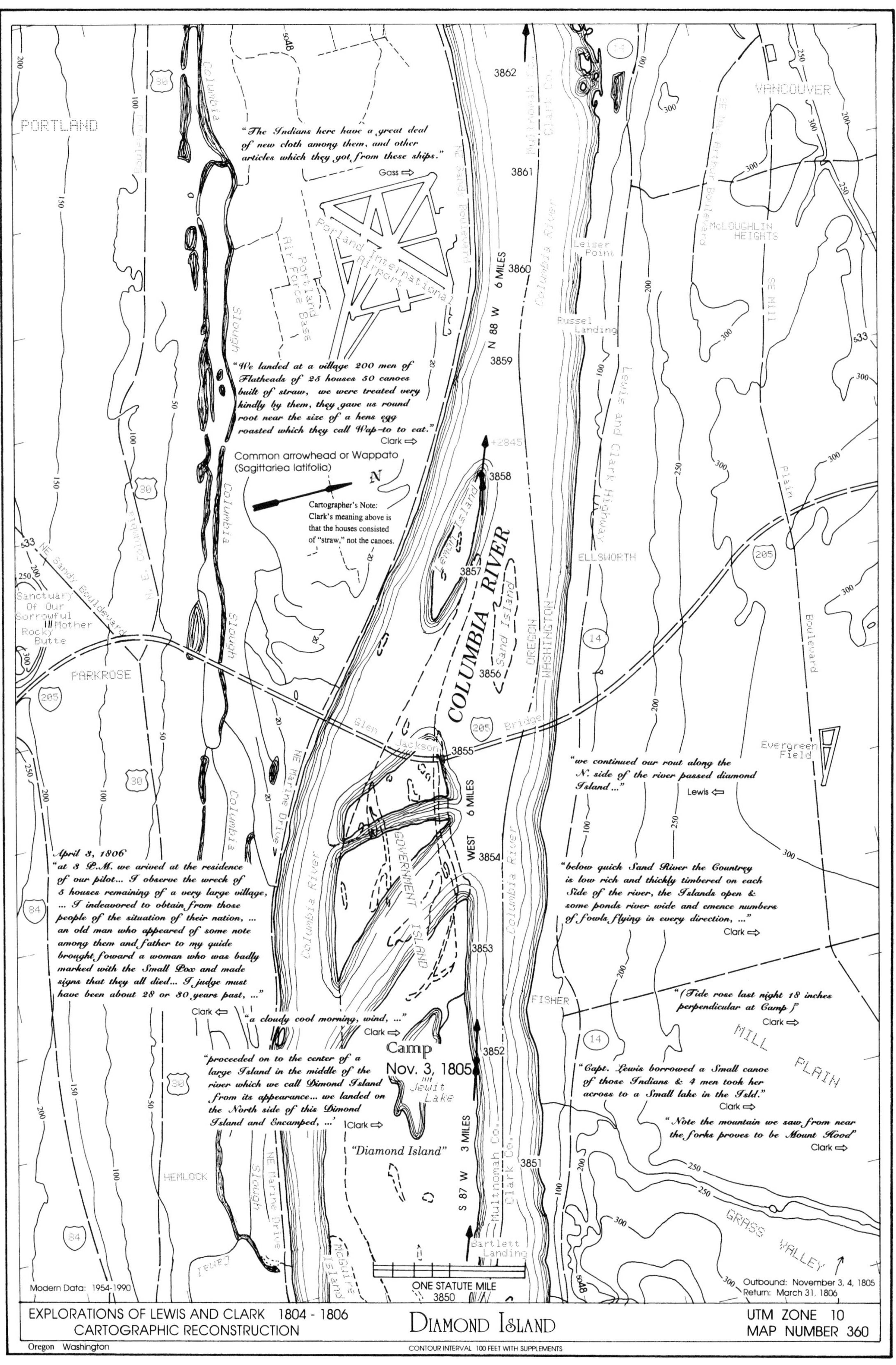
"The Indians here have a great deal of new cloth among them, and other articles which they got from these ships."
Gass ⇨
"We landed at a village 200 men of Flatheads of 25 houses 50 canoes built of straw, we were treated very kindly by them, they gave us round root near the size of a hens egg roasted which they call Wap-to to eat."
Clark ⇨
Common arrowhead or Wappato (Sagittariea latifolia)
Cartographer's Note: Clark's meaning above is that the houses consisted of "straw," not the canoes.
April 3, 1806
"at 3 P.M. we arived at the residence of our pilot... I observe the wreck of 5 houses remaining of a very large village, ... I indeavored to obtain from those people of the situation of their nation, ... an old man who appeared of some note among them and father to my guide brought foward a woman who was badly marked with the Small Pox and made signs that they all died... I judge must have been about 28 or 30 years past, ..."
Clark ⇦
"a cloudy cool morning, wind, ..."
Clark ⇨
"proceeded on to the center of a large Island in the middle of the river which we call Dimond Island from its appearance... we landed on the North side of this Dimond Island and Encamped, ...'
Clark ⇨
Camp
Nov. 3, 1805
"Diamond Island"
"we continued our rout along the N. side of the river passed diamond Island ..."
Lewis ⇦
"below quick Sand River the Countrey is low rich and thickly timbered on each Side of the river, the Islands open & some ponds river wide and emence numbers of fowls flying in every direction, ..."
Clark ⇨
"(Tide rose last night 18 inches perpendicular at Camp)"
Clark ⇨
"Capt. Lewis borrowed a Small canoe of those Indians & 4 men took her across to a Small lake in the Isld."
Clark ⇨
"Note the mountain we saw from near the forks proves to be Mount Hood"
Clark ⇨
PORTLAND
VANCOUVER
Portland International Airport
Portland Air Force Base
COLUMBIA RIVER
Columbia River
Lemon Island
Sand Island
GOVERNMENT ISLAND
McGuire Island
Jewit Lake
Glen Jackson Bridge
PARKROSE
HEMLOCK
ELLSWORTH
FISHER
MILL PLAIN
GRASS VALLEY
McLOUGHLIN HEIGHTS
Leiser Point
Russel Landing
Bartlett Landing
Evergreen Field
Sanctuary Of Our Sorrowful Mother
Rocky Butte
Columbia Slough
NE Marine Drive
NE Sandy Boulevard
Lewis and Clark Highway
SE McArthur Boulevard
SE Mill Plain Boulevard
OREGON
WASHINGTON
Multnomah Co.
Clark Co.
N 88 W 6 MILES
WEST 6 MILES
S 87 W 3 MILES
3862
3861
3860
3859
3858
3857
3856
3855
3854
3853
3852
3851
3850
ONE STATUTE MILE
Modern Data: 1954-1990
Outbound: November 3, 4, 1805
Return: March 31, 1806
EXPLORATIONS OF LEWIS AND CLARK 1804 - 1806
CARTOGRAPHIC RECONSTRUCTION
DIAMOND ISLAND
UTM ZONE 10
MAP NUMBER 360
Oregon Washington
CONTOUR INTERVAL 100 FEET WITH SUPPLEMENTS

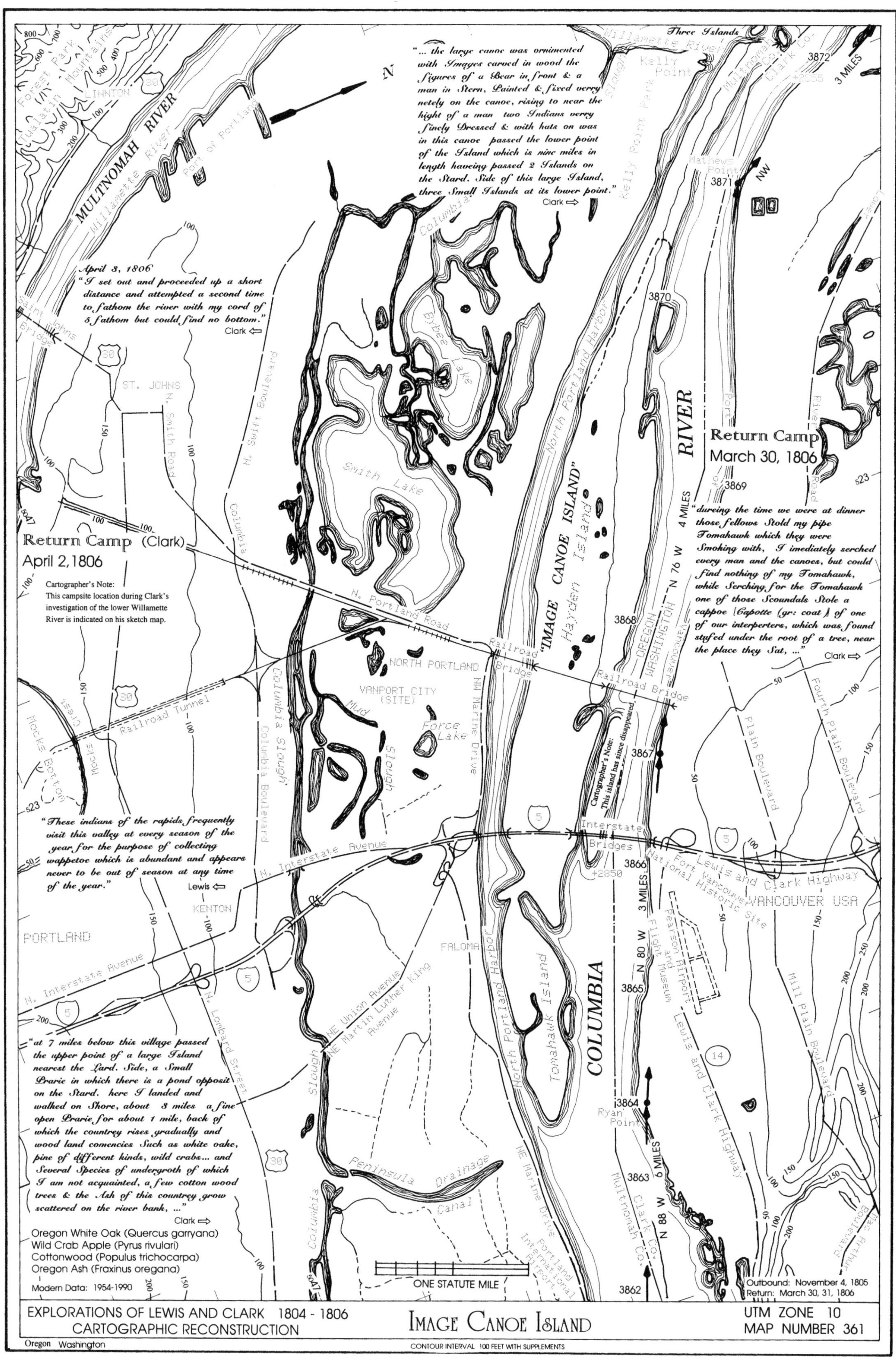

"... the large canoe was ornimented with Images carved in wood the figures of a Bear in front & a man in Stern, Painted & fixed verry netely on the canoe, rising to near the hight of a man two Indians verry finely Dressed & with hats on was in this canoe passed the lower point of the Island which is nine miles in length haveing passed 2 Islands on the Stard. Side of this large Island, three Small Islands at its lower point."
Clark
April 3, 1806
"I set out and proceeded up a short distance and attempted a second time to fathom the river with my cord of 5 fathom but could find no bottom."
Clark
Return Camp (Clark)
April 2, 1806
Cartographer's Note:
This campsite location during Clark's investigation of the lower Willamette River is indicated on his sketch map.
Return Camp
March 30, 1806
"dureing the time we were at dinner those fellows Stold my pipe Tomahawk which they were Smoking with, I imediately serched every man and the canoes, but could find nothing of my Tomahawk, while Serching for the Tomahawk one of those Scoundals Stole a cappoe (Capotte (gr: coat) of one of our interperters, which was found stufed under the root of a tree, near the place they Sat, ..."
Clark
"These indians of the rapids frequently visit this valley at every season of the year for the purpose of collecting wappetoe which is abundant and appears never to be out of season at any time of the year."
Lewis
"at 7 miles below this village passed the upper point of a large Island nearest the Lard. Side, a Small Prarie in which there is a pond opposit on the Stard. here I landed and walked on Shore, about 3 miles a fine open Prarie for about 1 mile, back of which the countrey rises gradually and wood land comencies Such as white oake, pine of different kinds, wild crabs... and Several Species of undergroth of which I am not acquainted, a few cotton wood trees & the Ash of this countrey grow scattered on the river bank, ..."
Clark
Oregon White Oak (Quercus garryana)
Wild Crab Apple (Pyrus rivulari)
Cottonwood (Populus trichocarpa)
Oregon Ash (Fraxinus oregana)
Modern Data: 1954-1990
Cartographer's Note:
This island has since disappeared.
"IMAGE CANOE ISLAND"
Hayden Island
Tomahawk Island
MULTNOMAH RIVER
Willamette River
COLUMBIA
RIVER
OREGON
WASHINGTON
Forest Park
Tualatin Mountains
LINNTON
Port of Portland
Three Islands
Willamette River
Kelly Point
Kelly Point Park
Mathews Point
Multnomah Co.
Clark Co.
Columbia
Bybee Lake
Smith Lake
Saint Johns Bridge
ST. JOHNS
N. Smith Road
N. Swift Boulevard
Columbia
N. Portland Road
North Portland Harbor
Railroad Bridge
Railroad Tunnel
NORTH PORTLAND
VANPORT CITY (SITE)
Mud Slough
Force Lake
NE Marine Drive
Columbia Slough
Columbia Boulevard
Mocks Bottom
Mocks Crest
Interstate Bridges
N. Interstate Avenue
KENTON
PORTLAND
FALOMA
N. Lombard Street
NE Union Avenue
NE Martin Luther King Avenue
Peninsula Drainage Canal
Portland International Airport
Ryan Point
Fort Vancouver National Historic Site
Lewis and Clark Highway
VANCOUVER USA
Pearson Airport and Flight Museum
Port of Vancouver
Plain Boulevard
Fourth Plain Boulevard
Mill Plain Boulevard
Mac Arthur Boulevard
River Road
N 76 W 4 MILES
N 80 W 3 MILES
N 88 W 6 MILES
3 MILES
NW
N
ONE STATUTE MILE
Outbound: November 4, 1805
Return: March 30, 31, 1806
EXPLORATIONS OF LEWIS AND CLARK 1804 - 1806
CARTOGRAPHIC RECONSTRUCTION
IMAGE CANOE ISLAND
UTM ZONE 10
MAP NUMBER 361
Oregon Washington
CONTOUR INTERVAL 100 FEET WITH SUPPLEMENTS

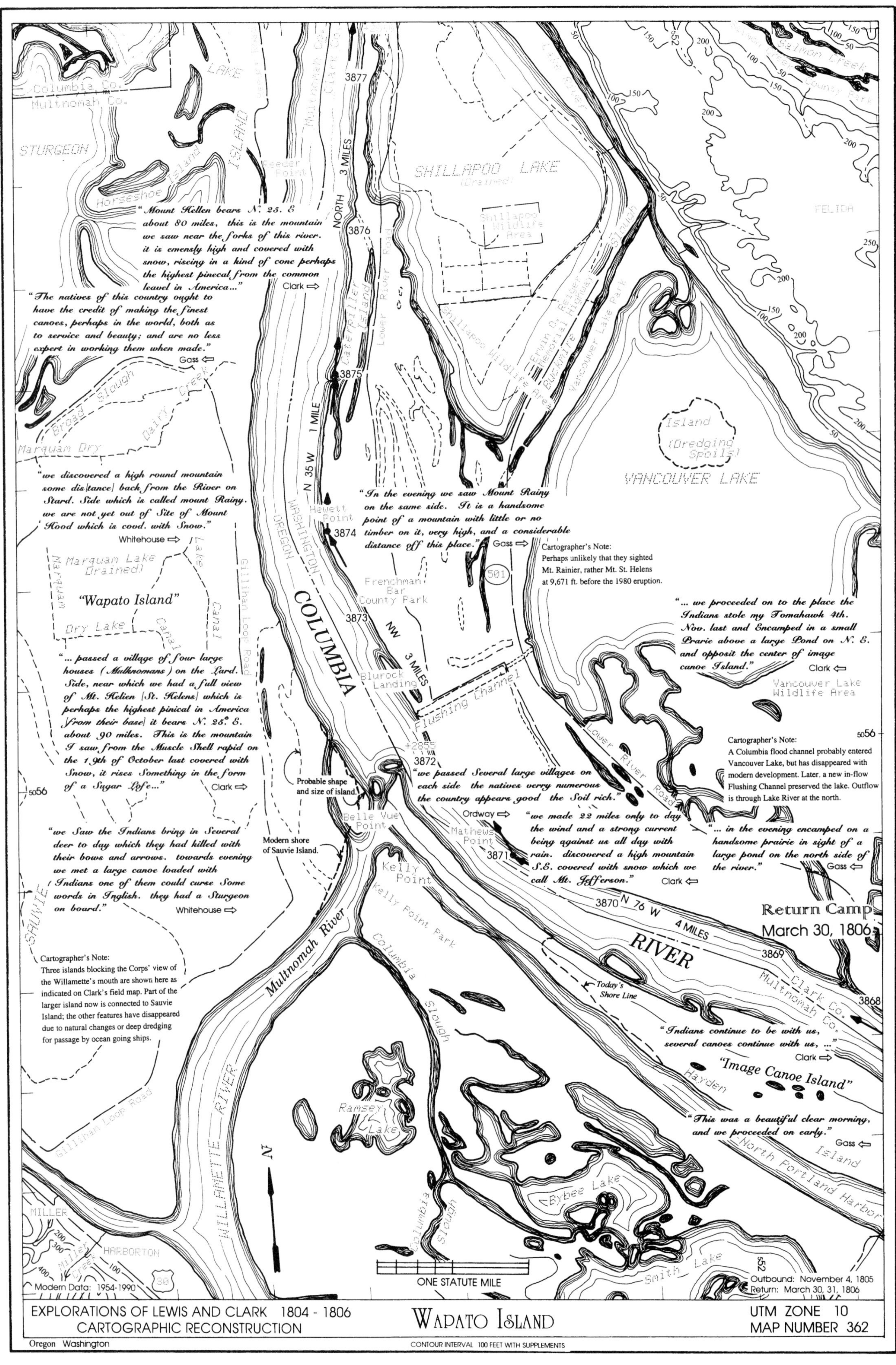
"Mount Hellen bears N. 25. E about 80 miles, this is the mountain we saw near the forks of this river. it is emensly high and covered with snow, riseing in a kind of cone perhaps the highest pinecal from the common leavel in America..."
Clark ⇨
"The natives of this country ought to have the credit of making the finest canoes, perhaps in the world, both as to service and beauty; and are no less expert in working them when made."
Gass ⇦
"we discovered a high round mountain some dis[tance] back from the River on Stard. Side which is called mount Rainy. we are not yet out of Site of Mount Hood which is cood. with Snow."
Whitehouse ⇨
"In the evening we saw Mount Rainy on the same side. It is a handsome point of a mountain with little or no timber on it, very high, and a considerable distance off this place."
Gass ⇨
Cartographer's Note:
Perhaps unlikely that they sighted Mt. Rainier, rather Mt. St. Helens at 9,671 ft. before the 1980 eruption.
"Wapato Island"
"... passed a village of four large houses (Mulknomans) on the Lard. Side, near which we had a full view of Mt. Helien [St. Helens] which is perhaps the highest pinical in America [from their base] it bears N. 25° E. about 90 miles. This is the mountain I saw from the Muscle Shell rapid on the 19th of October last covered with Snow, it rises Something in the form of a Sugar Lofe..."
Clark ⇨
"... we proceeded on to the place the Indians stole my Tomahawk 4th. Nov. last and Encamped in a small Prarie above a large Pond on N. E. and opposit the center of image canoe Island."
Clark ⇦
Cartographer's Note:
A Columbia flood channel probably entered Vancouver Lake, but has disappeared with modern development. Later, a new in-flow Flushing Channel preserved the lake. Outflow is through Lake River at the north.
"we passed Several large villages on each side the natives very numerous the country appears good the Soil rich."
Ordway ⇨
"we made 22 miles only to day the wind and a strong current being against us all day with rain. discovered a high mountain S.E. covered with snow which we call Mt. Jefferson."
Clark ⇦
"... in the evening encamped on a handsome prairie in sight of a large pond on the north side of the river."
Gass ⇦
Probable shape and size of island.
Modern shore of Sauvie Island.
"we Saw the Indians bring in Several deer to day which they had killed with their bows and arrows. towards evening we met a large canoe loaded with Indians one of them could curse Some words in Inglish. they had a Sturgeon on board."
Whitehouse ⇨
Return Camp
March 30, 1806
Cartographer's Note:
Three islands blocking the Corps' view of the Willamette's mouth are shown here as indicated on Clark's field map. Part of the larger island now is connected to Sauvie Island; the other features have disappeared due to natural changes or deep dredging for passage by ocean going ships.
Today's Shore Line
"Indians continue to be with us, several canoes continue with us, ..."
Clark ⇨
"Image Canoe Island"
"This was a beautiful clear morning, and we proceeded on early."
Gass ⇦
COLUMBIA
RIVER
Multnomah River
WILLAMETTE RIVER
VANCOUVER LAKE
SHILLAPOO LAKE (Drained)
STURGEON LAKE
SAUVIE ISLAND
Vancouver Lake Wildlife Area
Kelly Point Park
Ramsey Lake
Bybee Lake
Smith Lake
North Portland Harbor
ONE STATUTE MILE
Modern Data: 1954-1990
Outbound: November 4, 1805
Return: March 30, 31, 1806
EXPLORATIONS OF LEWIS AND CLARK 1804 - 1806
CARTOGRAPHIC RECONSTRUCTION
WAPATO ISLAND
UTM ZONE 10
MAP NUMBER 362
Oregon Washington
CONTOUR INTERVAL 100 FEET WITH SUPPLEMENTS

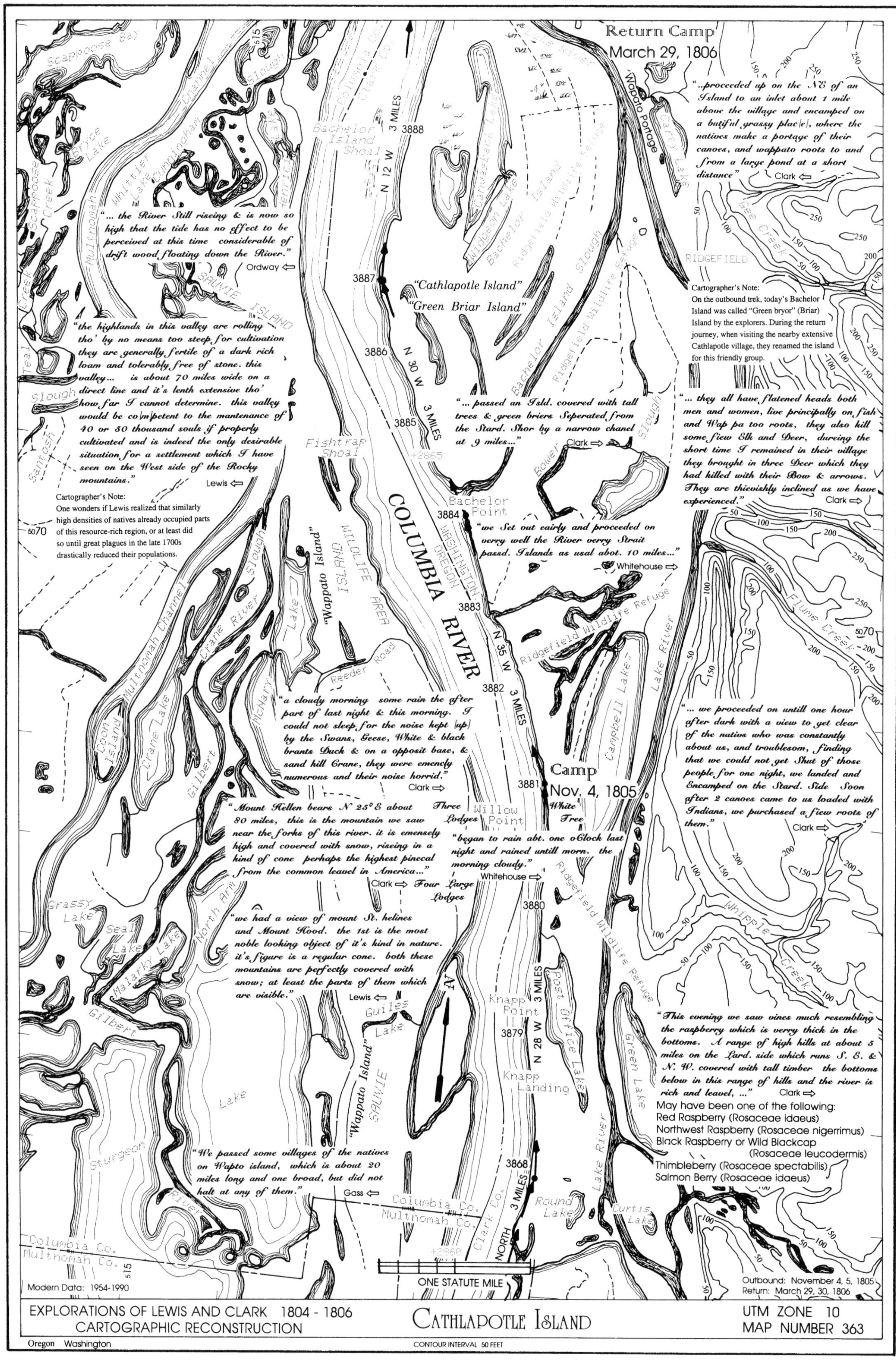

Return Camp
March 29, 1806
"...proceeded up on the N E of an Island to an inlet about 1 mile above the village and encamped on a butiful grassy plac[e], where the natives make a portage of their canoes, and wappato roots to and from a large pond at a short distance"
Clark ⇦
"... the River Still riseing & is now so high that the tide has no effect to be perceived at this time considerable of drift wood floating down the River."
Ordway ⇦
"Cathlapotle Island"
"Green Briar Island"
Cartographer's Note:
On the outbound trek, today's Bachelor Island was called "Green bryor" (Briar) Island by the explorers. During the return journey, when visiting the nearby extensive Cathlapotle village, they renamed the island for this friendly group.
"the highlands in this valley are rolling tho' by no means too steep for cultivation they are generally fertile of a dark rich loam and tolerably free of stone. this valley... is about 70 miles wide on a direct line and it's lenth extensive tho' how far I cannot determine. this valley would be co[m]petent to the mantenance of 40 or 50 thousand souls if properly cultivated and is indeed the only desirable situation for a settlement which I have seen on the West side of the Rocky mountains."
Lewis ⇦
Cartographer's Note:
One wonders if Lewis realized that similarly high densities of natives already occupied parts of this resource-rich region, or at least did so until great plagues in the late 1700s drastically reduced their populations.
"... passed an Isld. covered with tall trees & green briers Seperated from the Stard. Shor by a narrow chanel at 9 miles..."
Clark ⇨
"... they all have flatened heads both men and women, live principally on fish and Wap pa too roots, they also kill some fiew Elk and Deer, dureing the short time I remained in their village they brought in three Deer which they had killed with their Bow & arrows. They are thievishly inclined as we have experienced."
Clark ⇨
"we Set out eairly and proceeded on verry well the River verry Strait passd. Islands as usal abot. 10 miles..."
Whitehouse ⇨
COLUMBIA RIVER
"Wappato Island"
"a cloudy morning some rain the after part of last night & this morning. I could not sleep for the noise kept [up] by the Swans, Geese, White & black brants Duck & on a opposit base, & sand hill Crane, they were emencly numerous and their noise horrid."
Clark ⇨
Camp
Nov. 4, 1805
"... we proceeded on untill one hour after dark with a view to get clear of the natives who was constantly about us, and troublesom, finding that we could not get Shut of those people for one night, we landed and Encamped on the Stard. Side Soon after 2 canoes came to us loaded with Indians, we purchased a fiew roots of them."
Clark ⇨
"Mount Hellen bears N 25° E about 80 miles, this is the mountain we saw near the forks of this river. it is emensely high and covered with snow, riseing in a kind of cone perhaps the highest pinecal from the common leavel in America..."
Clark ⇨
Three Lodges
White Tree
Four Large Lodges
"began to rain abt. one oClock last night and rained untill morn. the morning cloudy."
Whitehouse ⇨
"we had a view of mount St. helines and Mount Hood. the 1st is the most noble looking object of it's kind in nature. it's figure is a regular cone. both these mountains are perfectly covered with snow; at least the parts of them which are visible."
Lewis ⇦
"This evening we saw vines much resembling the raspberry which is verry thick in the bottoms. A range of high hills at about 5 miles on the Lard. side which runs S. E. & N. W. covered with tall timber the bottoms below in this range of hills and the river is rich and leavel, ..."
Clark ⇨
May have been one of the following:
Red Raspberry (Rosaceae idaeus)
Northwest Raspberry (Rosaceae nigerrimus)
Black Raspberry or Wild Blackcap (Rosaceae leucodermis)
Thimbleberry (Rosaceae spectabilis)
Salmon Berry (Rosaceae idaeus)
"Wappato Island"
"We passed some villages of the natives on Wapto island, which is about 20 miles long and one broad, but did not halt at any of them."
Gass ⇦
Scappoose Bay
Scappoose Creek
Multnomah Channel
Cunningham Lake
Whittier Lake
Bryce Lake
Sauvie Island
Bachelor Island Shoal
Lake River
Wapato Portage
Carty Lake
Canvasback Lake
Widgeon Lake
Bachelor Island
Ridgefield Wildlife Refuge
Bachelor Island Slough
Gee Creek
RIDGEFIELD
Fishtrap Shoal
Bachelor Point
Bower Slough
Sauvie Island Wildlife Area
Washington
Oregon
Crane River
Reeder Road
McNary
Coon Island
Crane Lake
Gilbert River
Campbell Lake
Flume Creek
Willow Point
Grassy Lake
Seal Lake
Malarky Lake
North Arm
Whipple Creek
Guiles Lake
Knapp Point
Post Office Lake
Knapp Landing
Green Lake
Sturgeon Lake
Round Lake
Curtis Lake
Columbia Co.
Multnomah Co.
Clark Co.
N 12 W
N 30 W
N 35 W
N 28 W
NORTH
3 MILES
3888
3887
3886
3885
3884
3883
3882
3881
3880
3879
3868
ONE STATUTE MILE
Modern Data: 1954-1990
Outbound: November 4, 5, 1805
Return: March 29, 30, 1806
EXPLORATIONS OF LEWIS AND CLARK 1804 - 1806
CARTOGRAPHIC RECONSTRUCTION
CATHLAPOTLE ISLAND
UTM ZONE 10
MAP NUMBER 363
Oregon Washington
CONTOUR INTERVAL 50 FEET

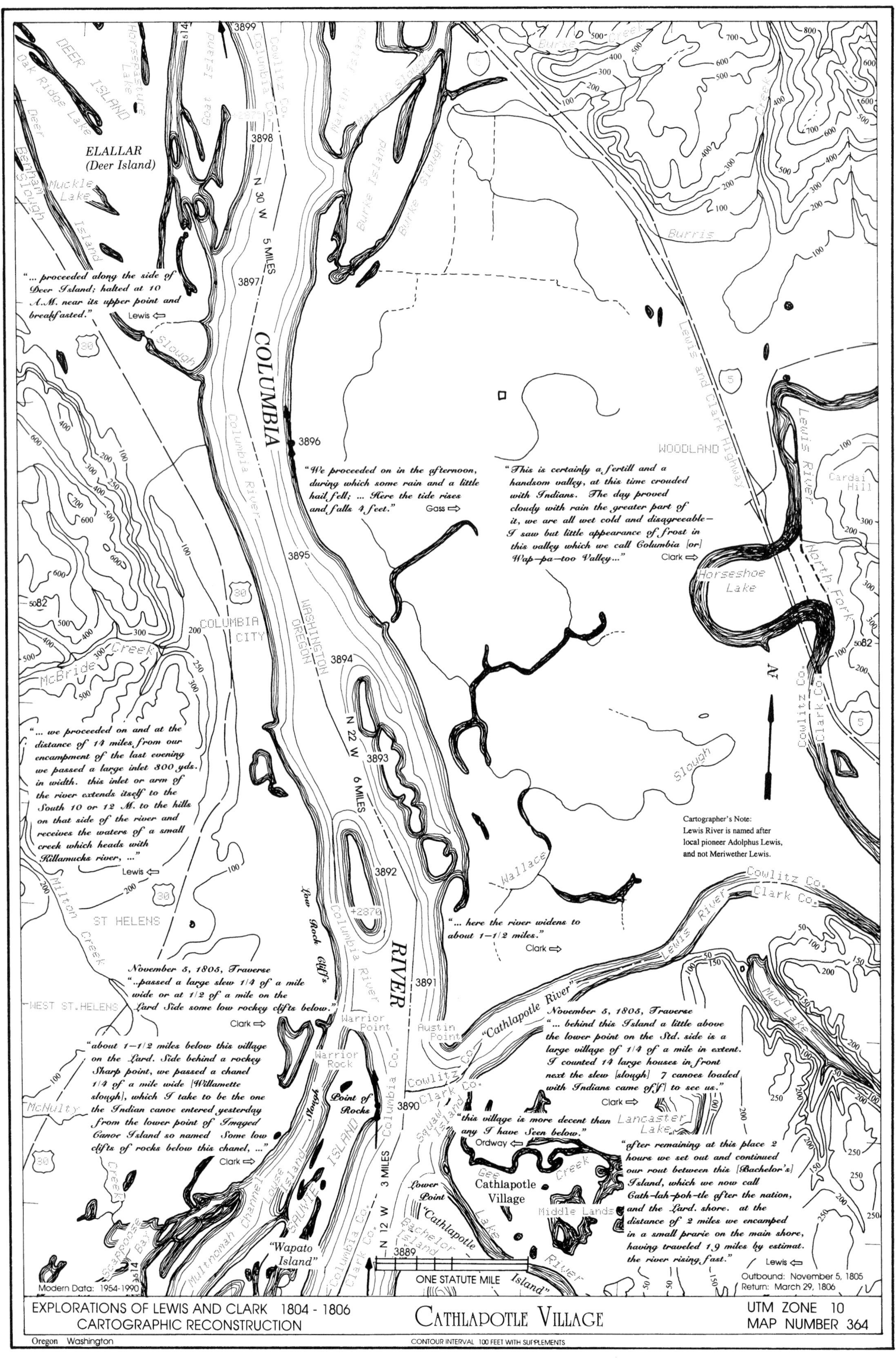
ELALLAR
(Deer Island)
"... proceeded along the side of Deer Island; halted at 10 A.M. near its upper point and breakfasted."
Lewis ⇦
COLUMBIA
RIVER
"We proceeded on in the afternoon, during which some rain and a little hail fell; ... Here the tide rises and falls 4 feet."
Gass ⇨
"This is certainly a fertill and a handsom valley, at this time crouded with Indians. The day proved cloudy with rain the greater part of it, we are all wet cold and disagreeable— I saw but little appearance of frost in this valley which we call Columbia [or] Wap-pa-too Valley..."
Clark ⇨
"... we proceeded on and at the distance of 14 miles from our encampment of the last evening we passed a large inlet 300 yds. in width. this inlet or arm of the river extends itself to the South 10 or 12 M. to the hills on that side of the river and receives the waters of a small creek which heads with Killamucks river, ..."
Lewis ⇦
Cartographer's Note:
Lewis River is named after local pioneer Adolphus Lewis, and not Meriwether Lewis.
"... here the river widens to about 1-1/2 miles."
Clark ⇨
November 5, 1805, Traverse
"...passed a large slew 1/4 of a mile wide or at 1/2 of a mile on the Lard Side some low rockey clifts below."
Clark ⇨
"about 1-1/2 miles below this village on the Lard. Side behind a rockey Sharp point, we passed a chanel 1/4 of a mile wide [Willamette slough], which I take to be the one the Indian canoe entered yesterday from the lower point of Imaged Canoe Island so named Some low clifts of rocks below this chanel, ..."
Clark ⇨
November 5, 1805, Traverse
"... behind this Island a little above the lower point on the Std. side is a large village of 1/4 of a mile in extent. I counted 14 large houses in front next the slew [slough] 7 canoes loaded with Indians came off] to see us."
Clark ⇨
"this village is more decent than any I have Seen below."
Ordway ⇦
"after remaining at this place 2 hours we set out and continued our rout between this [Bachelor's] Island, which we now call Cath-lah-poh-tle after the nation, and the Lard. shore. at the distance of 2 miles we encamped in a small prarie on the main shore, having traveled 19 miles by estimat. the river rising fast."
Lewis ⇨
Low Rock Cliffs
Warrior Point
Austin Point
Warrior Rock
Point of Rocks
"Cathlapotle River"
Lower Point
Cathlapotle Village
"Cathlapotle Island"
"Wapato Island"
SAUVIE ISLAND
COLUMBIA CITY
ST HELENS
WEST ST.HELENS
WOODLAND
Horseshoe Lake
Lancaster Lake
Lewis and Clark Highway
Lewis River
North Fork
Cowlitz Co.
Clark Co.
WASHINGTON
OREGON
N 30 W 5 MILES
N 22 W 6 MILES
N 12 W 3 MILES
3899
3898
3897
3896
3895
3894
3893
3892
3891
3890
3889
ONE STATUTE MILE
Modern Data: 1954-1990
Outbound: November 5, 1805
Return: March 29, 1806
EXPLORATIONS OF LEWIS AND CLARK 1804 - 1806
CARTOGRAPHIC RECONSTRUCTION
CATHLAPOTLE VILLAGE
UTM ZONE 10
MAP NUMBER 364
Oregon Washington
CONTOUR INTERVAL 100 FEET WITH SUPPLEMENTS

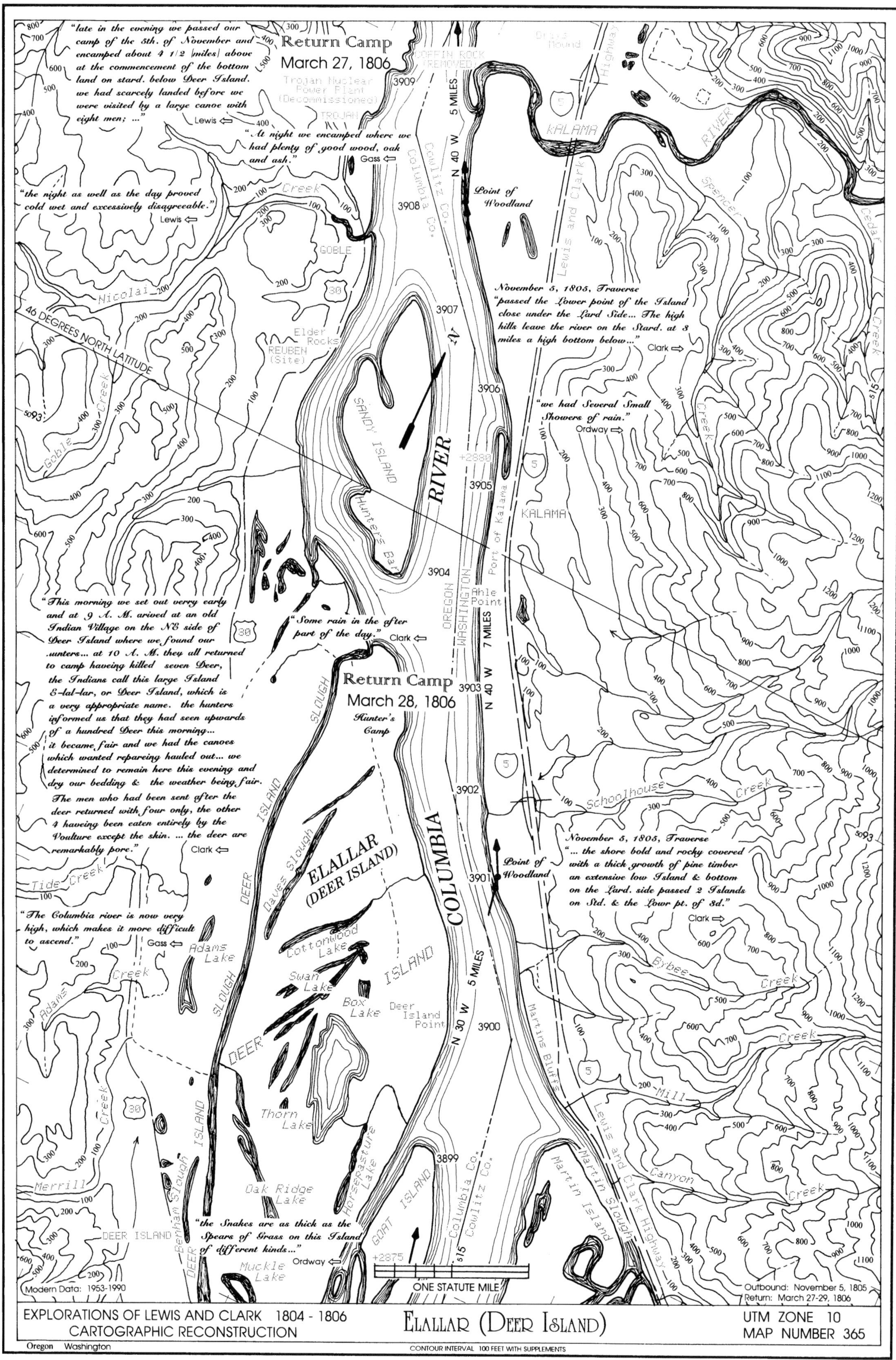
"late in the evening we passed our camp of the 5th. of November and encamped about 4 1/2 [miles] above at the commencement of the bottom land on stard. below Deer Island. we had scarcely landed before we were visited by a large canoe with eight men; ..."
Lewis
Return Camp
March 27, 1806
COFFIN ROCK (REMOVED)
Trojan Nuclear Power Plant (Decommissioned)
TROJAN
"At night we encamped where we had plenty of good wood, oak and ash."
Gass
"the night as well as the day proved cold wet and excessively disagreeable."
Lewis
Creek
GOBLE
Nicolai
46 DEGREES NORTH LATITUDE
Elder Rocks
REUBEN (Site)
Goble Creek
SANDY ISLAND
Hunters Bar
RIVER
Point of Woodland
KALAMA
Lewis and Clark Highway
Spencer Creek
Cedar Creek
"November 5, 1805, Traverse "passed the Lower point of the Island close under the Lard Side... The high hills leave the river on the Stard. at 3 miles a high bottom below..."
Clark
"we had Several Small Showers of rain."
Ordway
Port of Kalama
KALAMA
Ahle Point
OREGON
WASHINGTON
N 40 W 5 MILES
N 40 W 7 MILES
"This morning we set out verry early and at 9 A. M. arived at an old Indian Village on the NE side of Deer Island where we found our .unters... at 10 A. M. they all returned to camp haveing killed seven Deer, the Indians call this large Island E-lal-lar, or Deer Island, which is a very appropriate name. the hunters informed us that they had seen upwards of a hundred Deer this morning... it became fair and we had the canoes which wanted repareing hauled out... we determined to remain here this evening and dry our bedding & the weather being fair. The men who had been sent after the deer returned with four only, the other 4 haveing been eaten entirely by the Voulture except the skin. ... the deer are remarkably pore."
Clark
"Some rain in the after part of the day."
Clark
Return Camp
March 28, 1806
Hunter's Camp
DEER ISLAND SLOUGH
Dawes Slough
ELALLAR (DEER ISLAND)
COLUMBIA
Schoolhouse Creek
"November 5, 1805, Traverse "... the shore bold and rocky covered with a thick growth of pine timber an extensive low Island & bottom on the Lard. side passed 2 Islands on Std. & the Lowr pt. of 3d."
Clark
Point of Woodland
Tide Creek
"The Columbia river is now very high, which makes it more difficult to ascend."
Gass
Adams Lake
Adams Creek
Cottonwood Lake
Swan Lake
Box Lake
DEER ISLAND
Deer Island Point
N 30 W 5 MILES
Bybee Creek
Martins Bluffs
Mill Creek
Canyon Creek
Thorn Lake
Horsepasture Lake
Oak Ridge Lake
GOAT ISLAND
Columbia Co.
Cowlitz Co.
Martin Island
Martin Slough
Lewis and Clark Highway
Merrill Creek
Benham Slough
DEER ISLAND
"the Snakes are as thick as the Spears of Grass on this Island of different kinds..."
Ordway
Muckle Lake
ONE STATUTE MILE
Modern Data: 1953-1990
Outbound: November 5, 1805
Return: March 27-29, 1806
EXPLORATIONS OF LEWIS AND CLARK 1804 - 1806
CARTOGRAPHIC RECONSTRUCTION
ELALLAR (DEER ISLAND)
UTM ZONE 10
MAP NUMBER 365
Oregon Washington
CONTOUR INTERVAL 100 FEET WITH SUPPLEMENTS

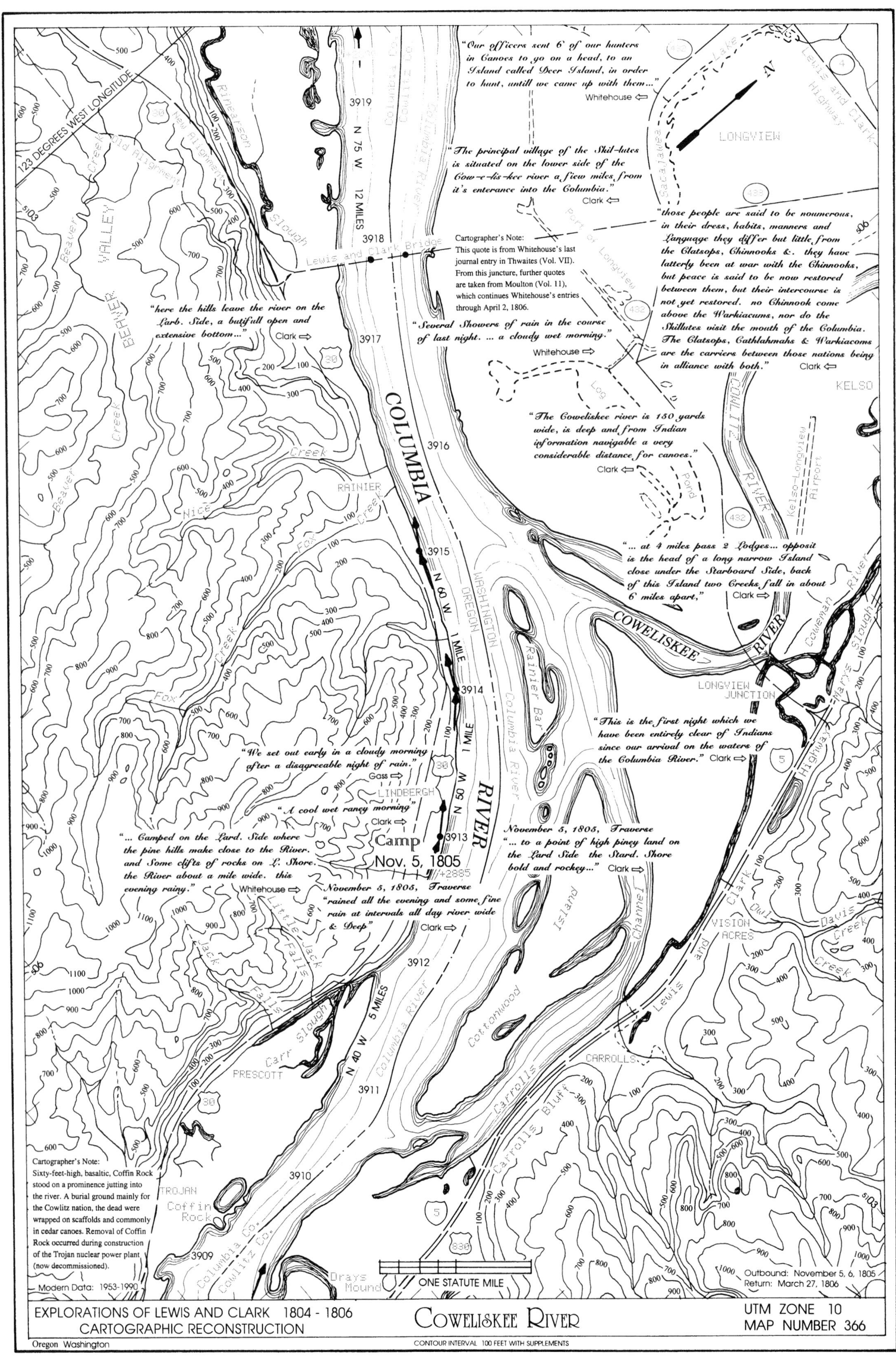
"Our officers sent 6 of our hunters in Canoes to go on a head, to an Island called Deer Island, in order to hunt, untill we came up with them..." Whitehouse
"The principal village of the Skil-lutes is situated on the lower side of the Cow-e-lis-kee river a few miles from it's enterance into the Columbia." Clark
"those people are said to be noumerous, in their dress, habits, manners and Language they differ but little from the Clatsops, Chinnooks &. they have latterly been at war with the Chinnooks, but peace is said to be now restored between them, but their intercourse is not yet restored. no Chinnook come above the Warkiacums, nor do the Skillutes visit the mouth of the Columbia. The Clatsops, Cathlahmahs & Warkiacoms are the carriers between those nations being in alliance with both." Clark
Cartographer's Note: This quote is from Whitehouse's last journal entry in Thwaites (Vol. VII). From this juncture, further quotes are taken from Moulton (Vol. 11), which continues Whitehouse's entries through April 2, 1806.
"here the hills leave the river on the Larb. Side, a butifull open and extensive bottom..." Clark
"Several Showers of rain in the course of last night. ... a cloudy wet morning." Whitehouse
"The Coweliskee river is 150 yards wide, is deep and from Indian information navigable a very considerable distance for canoes." Clark
"... at 4 miles pass 2 Lodges... opposit is the head of a long narrow Island close under the Starboard Side, back of this Island two Creeks fall in about 6 miles apart," Clark
"This is the first night which we have been entirely clear of Indians since our arrival on the waters of the Columbia River." Clark
"We set out early in a cloudy morning after a disagreeable night of rain." Gass
"A cool wet raney morning" Clark
"... Camped on the Lard. Side where the pine hills make close to the River. and Some clifts of rocks on L. Shore. the River about a mile wide. this evening rainy." Whitehouse
Camp
Nov. 5, 1805
November 5, 1805, Traverse "... to a point of high piney land on the Lard Side the Stard. Shore bold and rockey..." Clark
November 5, 1805, Traverse "rained all the evening and some fine rain at intervals all day river wide & Deep" Clark
Cartographer's Note: Sixty-feet-high, basaltic, Coffin Rock stood on a prominence jutting into the river. A burial ground mainly for the Cowlitz nation, the dead were wrapped on scaffolds and commonly in cedar canoes. Removal of Coffin Rock occurred during construction of the Trojan nuclear power plant (now decommissioned).
Modern Data: 1953-1990
COLUMBIA RIVER
COWELISKEE RIVER
LONGVIEW
KELSO
RAINIER
LINDBERGH
LONGVIEW JUNCTION
VISION ACRES
CARROLLS
PRESCOTT
TROJAN
Coffin Rock
Drays Mound
Lewis and Clark Bridge
123 DEGREES WEST LONGITUDE
Beaver Valley
Rainier Bar
Cottonwood Island
Carrolls Bluff
N 75 W 12 MILES
N 60 W 1 MILE
N 50 W 1 MILE
N 40 W 5 MILES
ONE STATUTE MILE
Outbound: November 5, 6, 1805
Return: March 27, 1806
EXPLORATIONS OF LEWIS AND CLARK 1804 - 1806
CARTOGRAPHIC RECONSTRUCTION
COWELISKEE RIVER
UTM ZONE 10
MAP NUMBER 366
Oregon Washington
CONTOUR INTERVAL 100 FEET WITH SUPPLEMENTS

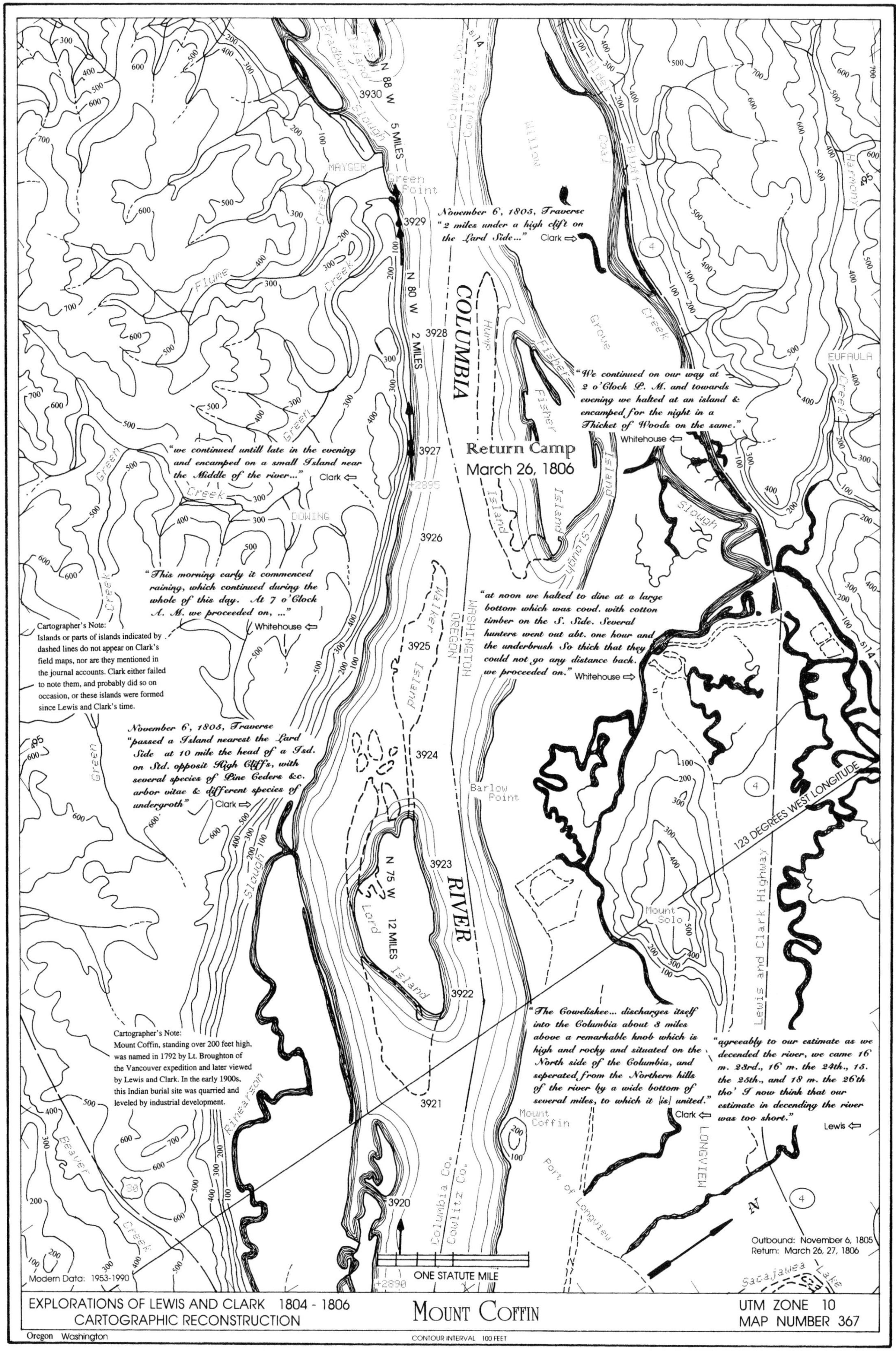
November 6, 1805, Traverse "2 miles under a high clift on the Lard Side..." Clark
"We continued on our way at 2 o'Clock P. M. and towards evening we halted at an island & encamped for the night in a Thicket of Woods on the same." Whitehouse
Return Camp March 26, 1806
"we continued untill late in the evening and encamped on a small Island near the Middle of the river..." Clark
"This morning early it commenced raining, which continued during the whole of this day. At 7 o'Clock A. M. we proceeded on, ..." Whitehouse
"at noon we halted to dine at a large bottom which was covd. with cotton timber on the S. Side. Several hunters went out abt. one hour and the underbrush So thick that they could not go any distance back. we proceeded on." Whitehouse
Cartographer's Note: Islands or parts of islands indicated by dashed lines do not appear on Clark's field maps, nor are they mentioned in the journal accounts. Clark either failed to note them, and probably did so on occasion, or these islands were formed since Lewis and Clark's time.
November 6, 1805, Traverse "passed a Island nearest the Lard Side at 10 mile the head of a Isd. on Std. opposit High Cliffs, with several species of Pine Ceders &c. arbor vitae & different species of undergroth" Clark
Cartographer's Note: Mount Coffin, standing over 200 feet high, was named in 1792 by Lt. Broughton of the Vancouver expedition and later viewed by Lewis and Clark. In the early 1900s, this Indian burial site was quarried and leveled by industrial development.
"The Coweliskee... discharges itself into the Columbia about 3 miles above a remarkable knob which is high and rocky and situated on the North side of the Columbia, and seperated from the Northern hills of the river by a wide bottom of several miles, to which it [is] united." Clark
"agreeably to our estimate as we decended the river, we came 16 m. 23rd., 16 m. the 24th., 13. the 25th., and 18 m. the 26th tho' I now think that our estimate in decending the river was too short." Lewis
COLUMBIA
RIVER
N 88 W
5 MILES
N 80 W
2 MILES
N 75 W
12 MILES
Bradbury
Slough
Crims Island
MAYGER
Green Point
Flume
Creek
Green
DOWING
Hump
Fisher
Island
Slough
Willow
Alder
Bluff
Coal
Grove
Creek
Harmony
EUFAULA
Walker Island
WASHINGTON
OREGON
Barlow Point
Lord Island
Slough
Rinearson
Beaver
Creek
Mount Solo
Mount Coffin
Port of Longview
LONGVIEW
Lewis and Clark Highway
123 DEGREES WEST LONGITUDE
Columbia Co.
Cowlitz Co.
Sacajawea Lake
ONE STATUTE MILE
Outbound: November 6, 1805
Return: March 26, 27, 1806
Modern Data: 1953-1990
EXPLORATIONS OF LEWIS AND CLARK 1804 - 1806
CARTOGRAPHIC RECONSTRUCTION
MOUNT COFFIN
UTM ZONE 10
MAP NUMBER 367
Oregon Washington
CONTOUR INTERVAL 100 FEET

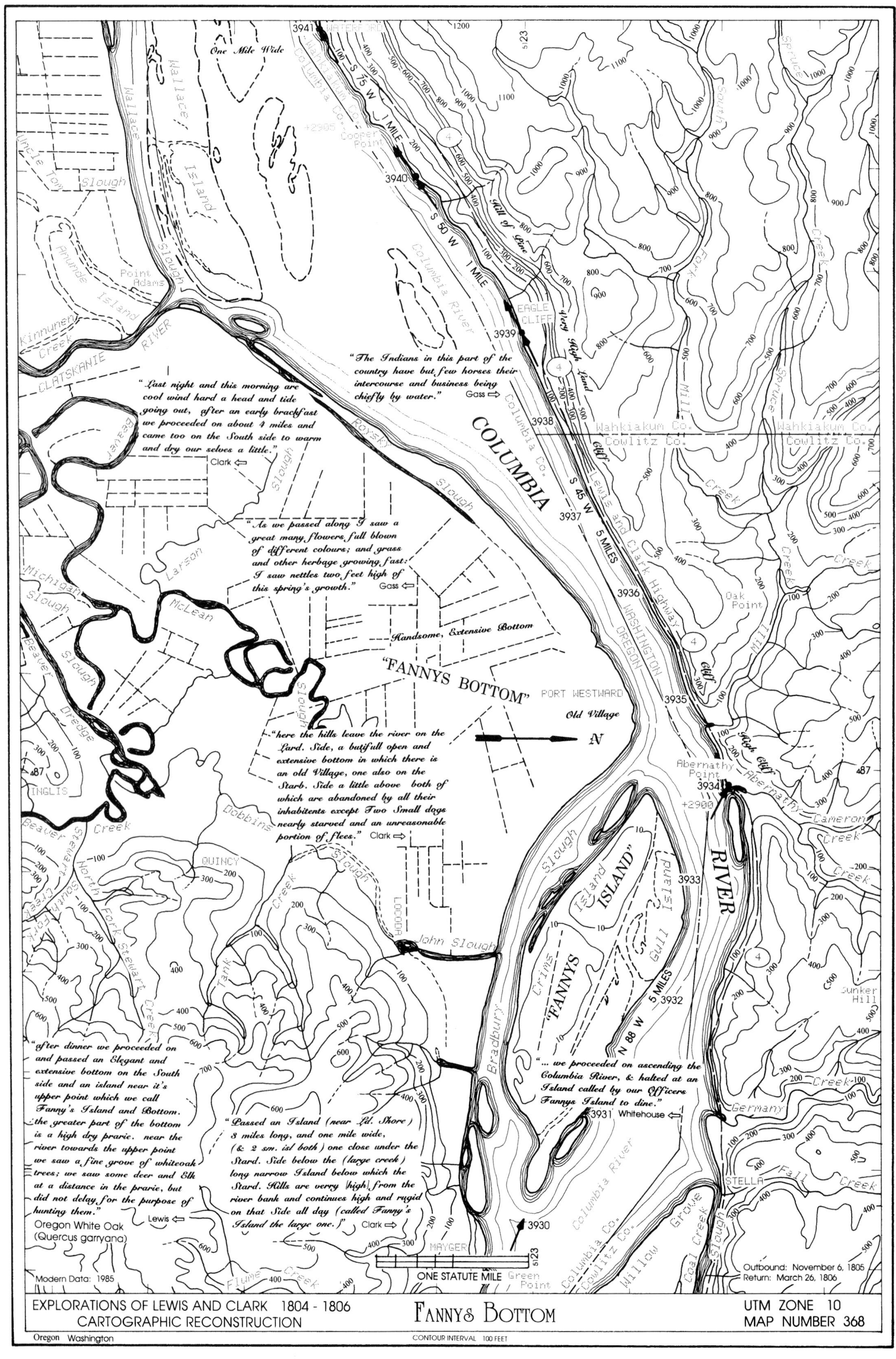

"Last night and this morning are cool wind hard a head and tide going out, after an early brackfast we proceeded on about 4 miles and came too on the South side to warm and dry our selves a little."
Clark
"The Indians in this part of the country have but few horses their intercourse and business being chiefly by water."
Gass
"As we passed along I saw a great many flowers full blown of different colours; and grass and other herbage growing fast: I saw nettles two feet high of this spring's growth."
Gass
Handsome, Extensive Bottom
"FANNYS BOTTOM"
"here the hills leave the river on the Lard. Side, a butifull open and extensive bottom in which there is an old Village, one also on the Starb. Side a little above both of which are abandoned by all their inhabitents except Two Small dogs nearly starved and an unreasonable portion of flees."
Clark
"after dinner we proceeded on and passed an Elegant and extensive bottom on the South side and an island near it's upper point which we call Fanny's Island and Bottom. the greater part of the bottom is a high dry prarie. near the river towards the upper point we saw a fine grove of whiteoak trees; we saw some deer and Elk at a distance in the prarie, but did not delay for the purpose of hunting them."
Lewis
Oregon White Oak (Quercus garryana)
"Passed an Island (near Ld. Shore) 3 miles long, and one mile wide, (& 2 sm. isl both) one close under the Stard. Side below the (large creek) long narrow Island below which the Stard. Side. Hills are verry [high] from the river bank and continues high and rugid on that Side all day (called Fanny's Island the large one.)"
Clark
"... we proceeded on ascending the Columbia River, & halted at an Island called by our Officers Fannys Island to dine."
Whitehouse
One Mile Wide
Hill of Pine
Very High Land
High Cliff
COLUMBIA
RIVER
"FANNYS ISLAND"
Old Village
PORT WESTWARD
Wahkiakum Co.
Cowlitz Co.
Columbia Co.
Lewis and Clark Highway
WASHINGTON
OREGON
S 75 W 1 MILE
S 80 W 1 MILE
S 45 W 5 MILES
N 88 W 5 MILES
CLATSKANIE RIVER
Wallace Island
Crims Island
Gull Island
Abernathy Point
Oak Point
EAGLE CLIFF
STELLA
MAYGER
QUINCY
INGLIS
LOCODA
John Slough
Bradbury Slough
Willow Grove
Coal Creek Slough
Germany Creek
Abernathy Creek
Mill Creek
Fall Creek
Bunker Hill
ONE STATUTE MILE
Modern Data: 1985
Outbound: November 6, 1805
Return: March 26, 1806
EXPLORATIONS OF LEWIS AND CLARK 1804 - 1806
CARTOGRAPHIC RECONSTRUCTION
FANNYS BOTTOM
UTM ZONE 10
MAP NUMBER 368
Oregon Washington
CONTOUR INTERVAL 100 FEET

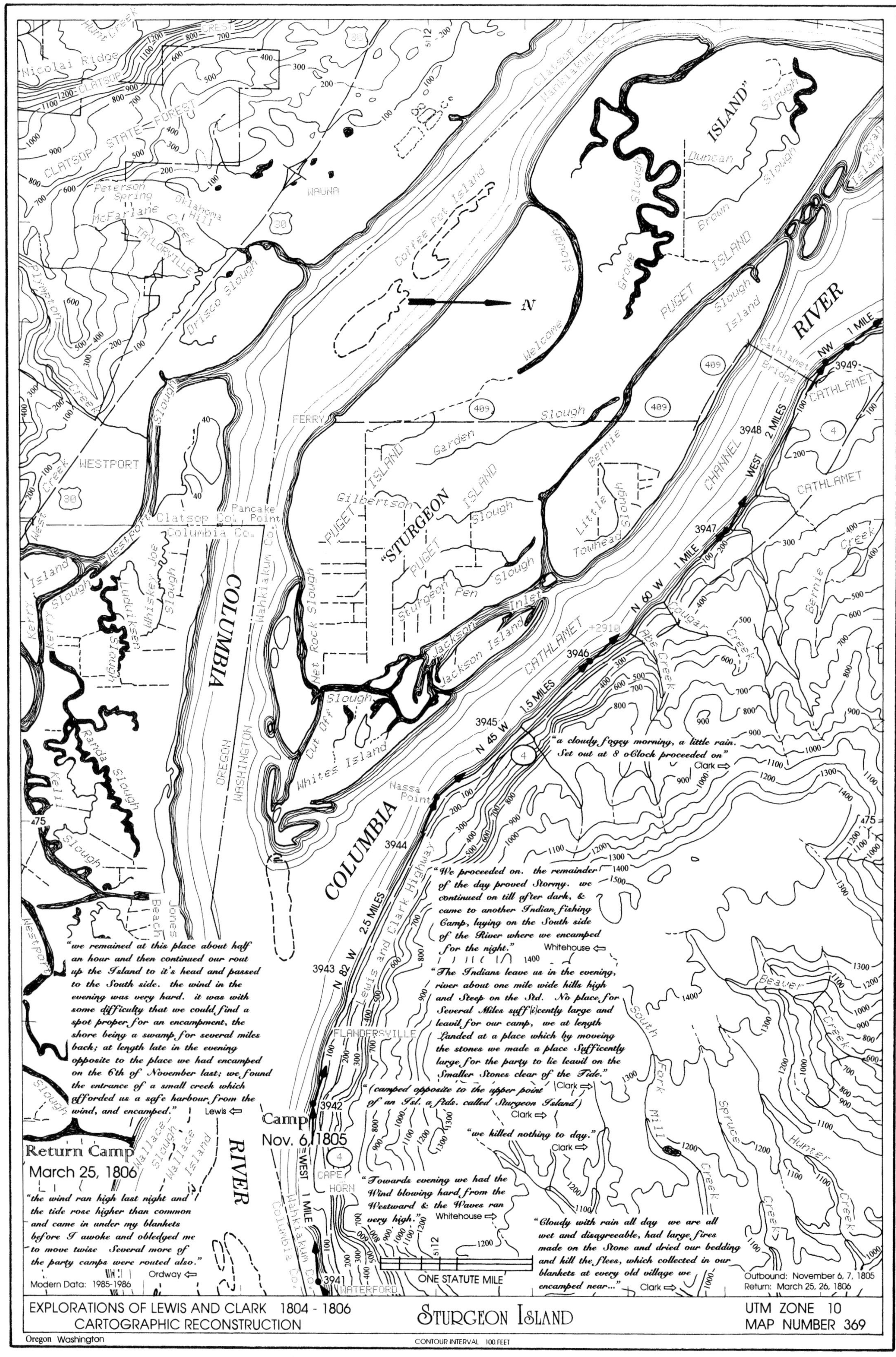

"we remained at this place about half an hour and then continued our rout up the Island to it's head and passed to the South side. the wind in the evening was very hard. it was with some difficulty that we could find a spot proper for an encampment, the shore being a swamp for several miles back; at length late in the evening opposite to the place we had encamped on the 6th of November last; we found the entrance of a small creek which afforded us a safe harbour from the wind, and encamped." Lewis ⇦
Return Camp
March 25, 1806
"the wind ran high last night and the tide rose higher than common and came in under my blankets before I awoke and obledged me to move twise Several more of the party camps were routed also." Ordway ⇦
Camp
Nov. 6, 1805
"a cloudy fogey morning, a little rain. Set out at 8 oClock proceeded on" Clark ⇨
"We proceeded on. the remainder of the day proved Stormy. we continued on till after dark, & came to another Indian fishing Camp, laying on the South side of the River where we encamped for the night." Whitehouse ⇦
"The Indians leave us in the evening, river about one mile wide hills high and Steep on the Std. No place for Several Miles suff[i]cently large and leavil for our camp, we at length Landed at a place which by moveing the stones we made a place Sufficently large for the party to lie leavil on the Smaller Stones clear of the Tide."
"(camped opposite to the upper point of an Isl. a Std. called Sturgeon Island)" Clark ⇨
"we killed nothing to day." Clark ⇨
"Towards evening we had the Wind blowing hard from the Westward & the Waves ran very high." Whitehouse ⇨
"Cloudy with rain all day we are all wet and disagreeable, had large fires made on the Stone and dried our bedding and kill the flees, which collected in our blankets at every old village we encamped near..." Clark ⇨
COLUMBIA RIVER
"STURGEON ISLAND"
PUGET ISLAND
CATHLAMET CHANNEL
OREGON
WASHINGTON
Clatsop Co.
Columbia Co.
Wahkiakum Co.
WESTPORT
WAUNA
CATHLAMET
FLANDERSVILLE
CAPE HORN
WATERFORD
Whites Island
Jackson Island
Coffee Pot Island
Wallace Island
Nassa Point
Pancake Point
Lewis and Clark Highway
N 82 W 2.5 MILES
N 45 W 1.5 MILES
N 60 W 1 MILE
WEST 2 MILES
NW 1 MILE
WEST 1 MILE
ONE STATUTE MILE
Modern Data: 1985-1986
Outbound: November 6, 7, 1805
Return: March 25, 26, 1806
EXPLORATIONS OF LEWIS AND CLARK 1804 - 1806
CARTOGRAPHIC RECONSTRUCTION
STURGEON ISLAND
UTM ZONE 10
MAP NUMBER 369
Oregon Washington
CONTOUR INTERVAL 100 FEET

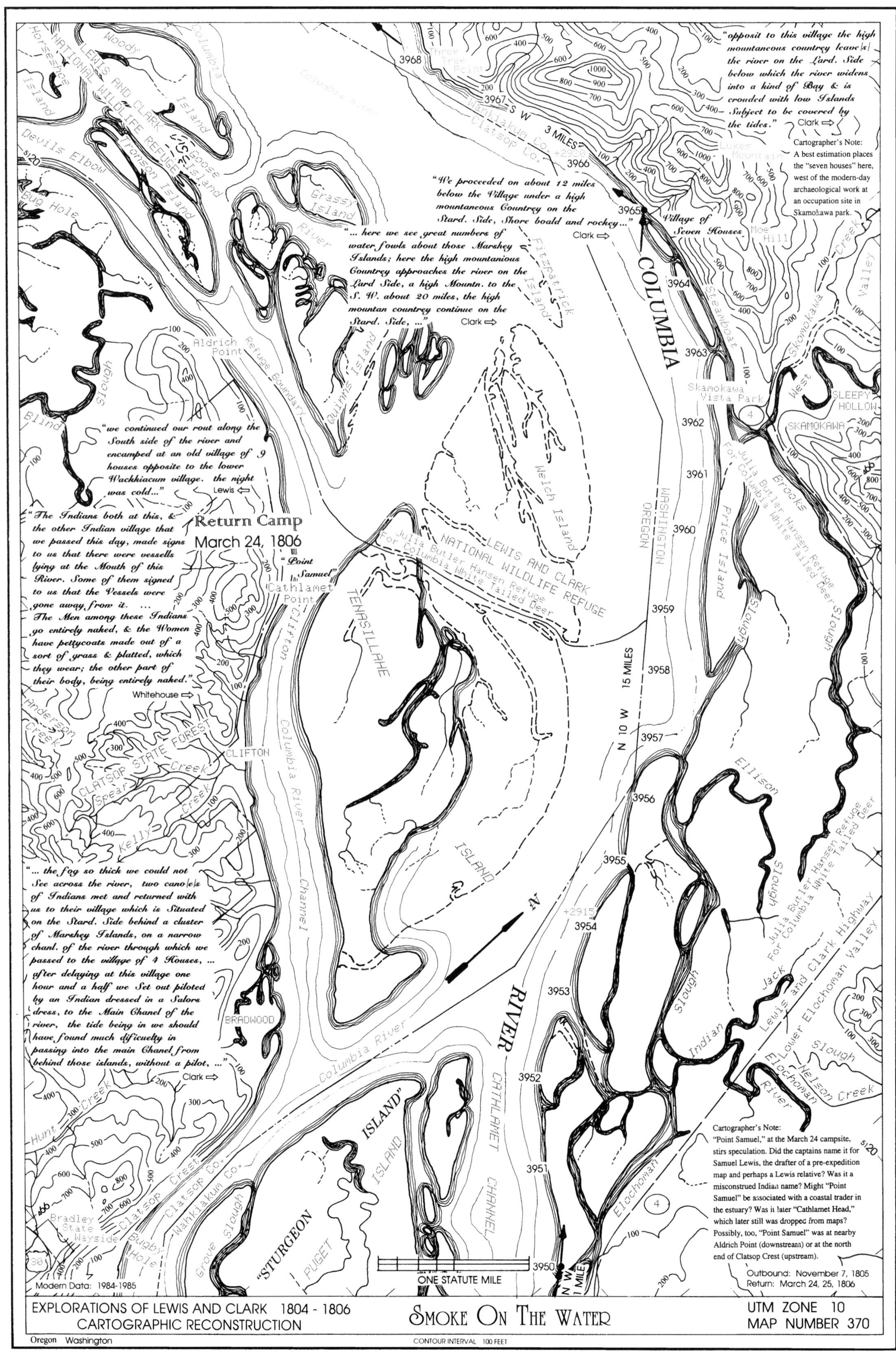

"opposit to this village the high mountanous countrey leave[s] the river on the Lard. Side below which the river widens into a kind of Bay & is crouded with low Islands Subject to be covered by the tides." Clark ⇨
Cartographer's Note: A best estimation places the "seven houses" here, west of the modern-day archaeological work at an occupation site in Skamokawa park.
"We proceeded on about 12 miles below the Village under a high mountaneous Countrey on the Stard. Side, Shore boald and rockey..." Clark ⇨
Village of Seven Houses
"... here we see great numbers of water fowls about those Marshey Islands; here the high mountanious Countrey approaches the river on the Lard Side, a high Mountn. to the S. W. about 20 miles, the high mountan countrey continue on the Stard. Side, ..." Clark ⇨
"we continued our rout along the South side of the river and encamped at an old village of 9 houses opposite to the lower Wackhiacum village. the night was cold..." Lewis ⇦
Return Camp
March 24, 1806
"Point Samuel"
"The Indians both at this, & the other Indian village that we passed this day, made signs to us that there were vessells lying at the Mouth of this River. Some of them signed to us that the Vessels were gone away from it. ... The Men among these Indians go entirely naked, & the Women have pettycoats made out of a sort of grass & platted, which they wear; the other part of their body, being entirely naked." Whitehouse ⇨
"... the fog so thick we could not See across the river, two cano[e]s of Indians met and returned with us to their village which is Situated on the Stard. Side behind a cluster of Marshey Islands, on a narrow chanl. of the river through which we passed to the village of 4 Houses, ... after delaying at this village one hour and a half we Set out piloted by an Indian dressed in a Salors dress, to the Main Chanel of the river, the tide being in we should have found much dificuelty in passing into the main Chanel from behind those islands, without a pilot, ..." Clark ⇨
Cartographer's Note: "Point Samuel," at the March 24 campsite, stirs speculation. Did the captains name it for Samuel Lewis, the drafter of a pre-expedition map and perhaps a Lewis relative? Was it a misconstrued Indian name? Might "Point Samuel" be associated with a coastal trader in the estuary? Was it later "Cathlamet Head," which later still was dropped from maps? Possibly, too, "Point Samuel" was at nearby Aldrich Point (downstream) or at the north end of Clatsop Crest (upstream).
Outbound: November 7, 1805
Return: March 24, 25, 1806
COLUMBIA
RIVER
NATIONAL WILDLIFE REFUGE
LEWIS AND CLARK
LEWIS AND CLARK NATIONAL WILDLIFE REFUGE
Julia Butler Hansen Refuge For Columbia White Tailed Deer
TENASILLAHE
"STURGEON ISLAND"
CLATSOP STATE FOREST
WASHINGTON
OREGON
N 10 W 15 MILES
3 MILES S W
Modern Data: 1984-1985
ONE STATUTE MILE
EXPLORATIONS OF LEWIS AND CLARK 1804 - 1806
CARTOGRAPHIC RECONSTRUCTION
SMOKE ON THE WATER
UTM ZONE 10
MAP NUMBER 370
Oregon Washington
CONTOUR INTERVAL 100 FEET

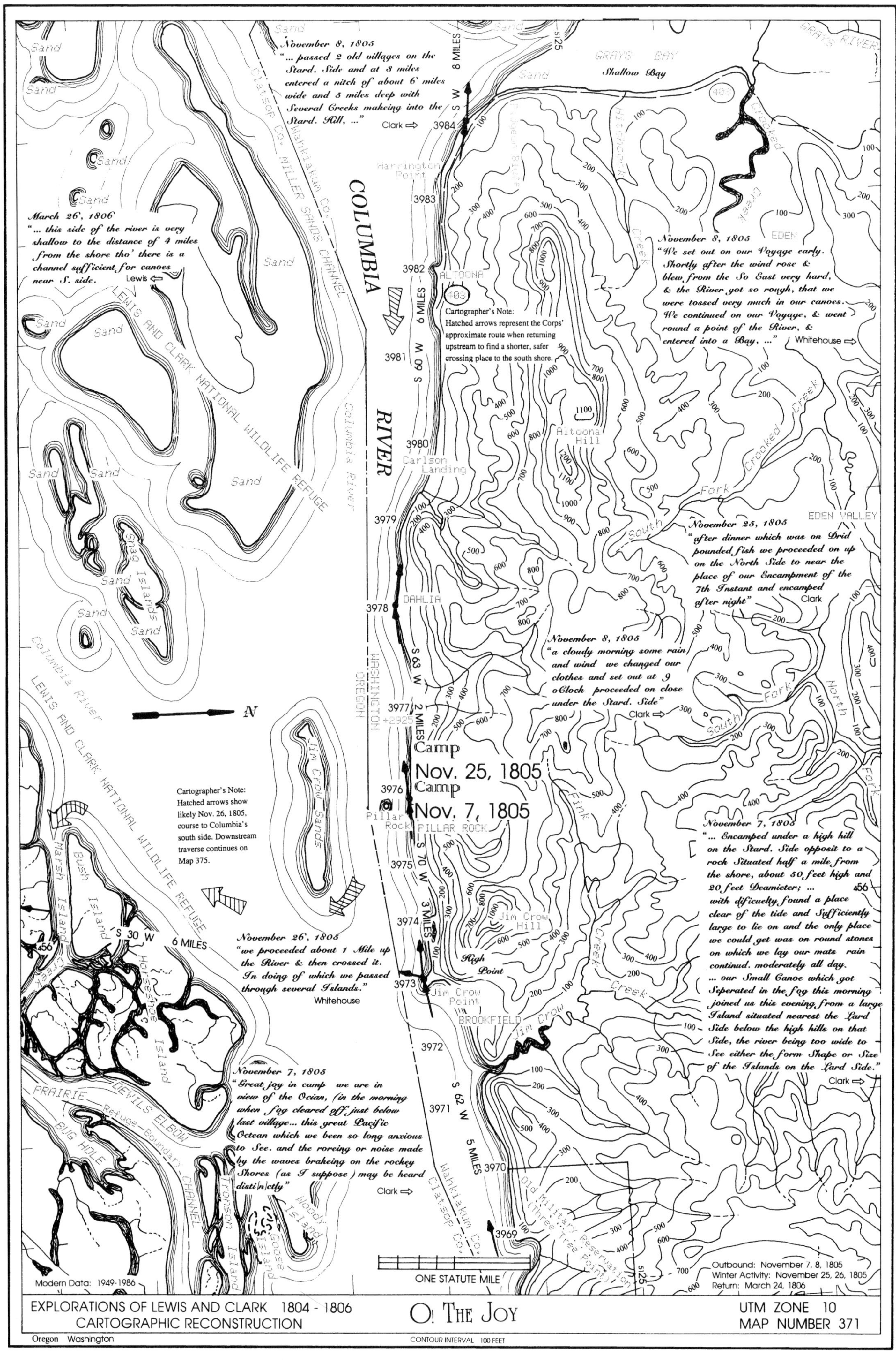
November 8, 1805
"... passed 2 old villages on the Stard. Side and at 8 miles entered a nitch of about 6 miles wide and 5 miles deep with Several Creeks makeing into the Stard. Hill, ..."
Clark
March 26, 1806
"... this side of the river is very shallow to the distance of 4 miles from the shore tho' there is a channel sufficient for canoes near S. side.
Lewis
COLUMBIA RIVER
GRAYS BAY
Shallow Bay
GRAYS RIVER
Harrington Point
ALTOONA
Altoona Hill
Carlson Landing
DAHLIA
Pillar Rock
PILLAR ROCK
Jim Crow Hill
High Point
Jim Crow Point
BROOKFIELD
EDEN
EDEN VALLEY
MILLER SANDS CHANNEL
LEWIS AND CLARK NATIONAL WILDLIFE REFUGE
Snag Islands
Jim Crow Sands
WASHINGTON
OREGON
Clatsop Co.
Wahkiakum Co.
Cartographer's Note: Hatched arrows represent the Corps' approximate route when returning upstream to find a shorter, safer crossing place to the south shore.
November 8, 1805
"We set out on our Voyage early. Shortly after the wind rose & blew from the So East very hard, & the River got so rough, that we were tossed very much in our canoes. We continued on our Voyage, & went round a point of the River, & entered into a Bay, ..."
Whitehouse
November 25, 1805
"after dinner which was on Drid pounded fish we proceeded on up on the North Side to near the place of our Encampment of the 7th Instant and encamped after night"
Clark
November 8, 1805
"a cloudy morning some rain and wind we changed our clothes and set out at 9 oClock proceeded on close under the Stard. Side"
Clark
Camp Nov. 25, 1805
Camp Nov. 7, 1805
Cartographer's Note: Hatched arrows show likely Nov. 26, 1805, course to Columbia's south side. Downstream traverse continues on Map 375.
November 7, 1805
"... Encamped under a high hill on the Stard. Side opposit to a rock Situated half a mile from the shore, about 50 feet high and 20 feet Deamieter; ... with dificulty found a place clear of the tide and Sufficiently large to lie on and the only place we could get was on round stones on which we lay our mats rain continud. moderately all day. ... our Small Canoe which got Seperated in the fog this morning joined us this evening from a large Island situated nearest the Lard Side below the high hills on that Side, the river being too wide to See either the form Shape or Size of the Islands on the Lard Side."
Clark
November 26, 1805
"we proceeded about 1 Mile up the River & then crossed it. In doing of which we passed through several Islands."
Whitehouse
November 7, 1805
"Great joy in camp we are in view of the Ocian, (in the morning when fog cleared off just below last village... this great Pacific Octean which we been so long anxious to See. and the roreing or noise made by the waves brakeing on the rocky Shores (as I suppose) may be heard disti[n]ctly"
Clark
Marsh Island
Bush Island
Horseshoe Island
PRAIRIE CHANNEL
DEVILS ELBOW
BUG HOLE
Tronson Island
Woody Island
Goose Island
Old Military Reservation (Three Tree Point)
ONE STATUTE MILE
Modern Data: 1949-1986
Outbound: November 7, 8, 1805
Winter Activity: November 25, 26, 1805
Return: March 24, 1806
EXPLORATIONS OF LEWIS AND CLARK 1804 - 1806
CARTOGRAPHIC RECONSTRUCTION
O! THE JOY
UTM ZONE 10
MAP NUMBER 371
Oregon Washington
CONTOUR INTERVAL 100 FEET

"Ocian in view!"

"Great joy in camp we are in view of the Ocian, (in the morning when fog cleared off just below last village (first on leaveing this village) of Warkiacum) this great Pacific Octean which we been so long anxious to See. and the roreing or noise made by the waves brakeing on the rockey Shores (as I suppose) may be heard disti[n]ctly"

Captain William Clark, November 7, 1805

No statements by the captains have caused more controversy than Clark's famous, oft quoted "Ocian in view!" and "O! the joy." Normally a reserved, practical observer, Clark's surprising emotional outburst on this day was due to finally accomplishing this key expedition goal. Most major Lewis and Clark scholars, however, have maintained that Clark did not observe the ocean from the Pillar Rock locality. Thwaites (1904–5) typically states: "The ocean could not possibly be seen from this point, although during a storm the breakers might be heard. The explorers probably mistook the great bay of the river, which just below this point widens to fifteen miles, for the expanse of the ocean."

Clark's quote indicates that he first sighted the ocean after paddling from a Skamokawa-area Indian village and rounding Three Tree Point toward Pillar Rock. Perhaps the natives had pointed the way to Clark. Could one of America's finest frontier geographers, who took great pains to accurately maintain (and correct) his journal entries, be mistaken? Not likely!

In recent years, some regional Lewis and Clark investigators have come to the conclusion that Clark, when in the Pillar Rock area, indeed sighted the opening to the sea some 20 miles away between Oregon's Point Adams and the north shore's Point Ellice. Key to this discussion is the fact that modern jetties have dramatically changed the configuration of the Columbia's mouth. Breakwaters were built to improve navigation through one of the world's most dangerous river bars, known for its treacherous, shifting sands. A South Jetty from Point Adams was constructed in several phases, starting in 1885, with final work beginning in 1913 (including a most-westerly underwater section, it is more than 6 miles long). Work began in 1914 on a North Jetty, too, completed by 1917. The North Jetty extends about 3 miles out from Cape Disappointment. Before the breakwater construction, the juxtaposed landmasses of Point Adams and Cape Disappointment forced the river's mouth to make a southwest bend into the sea, which would have been conducive for Clark's sighting from Pillar Rock.

For decades after completion, the jetties accreted thousands of acres of sands and sediments, mostly to seaward. Not until the mid-twentieth century, with many new hydroelectric dams on the upper Columbia restricting sediment flows, has this process reversed. Storms and tides now are removing some of these accreted areas.

Today, however, the South Jetty and its accreted lands and vegetation yet cuts off any sighting of the ocean from Pillar Rock. When Lewis and Clark approached Pillar Rock in 1805, of course, no jetty blocked the view. Looking west, they would have seen Point Adams on the south shore and Point Ellice to the north (Cape Disappointment was hidden behind Point Ellice)—a view of nearly 6 degrees between headlands out to the storm-tossed ocean.

The earth's curvature perhaps is a consideration here. The distance from Pillar Rock to Point Ellice is about 21 miles, with curvature figured at about 8 inches per mile for this area. Thus, the curvature would be approximately 14 feet. However, surface water elevations at Pillar Rock are not constant, but probably vary from about 3 to 6 feet above sea level. Clark sitting or standing in a canoe would add another 3 to 6 feet. Perhaps with his telescope he could see the distant ocean's high surf on the sand bars or possible storm swells up to a score or more feet high. Or, Clark simply noticed a gap of open water between headlands, extending westward to an empty horizon, especially if advised by native informants.

Yet, despite the "Great joy in camp" reported by Clark, it is curious that no other journal writer mentions sighting the ocean on this day, November 7, 1805.

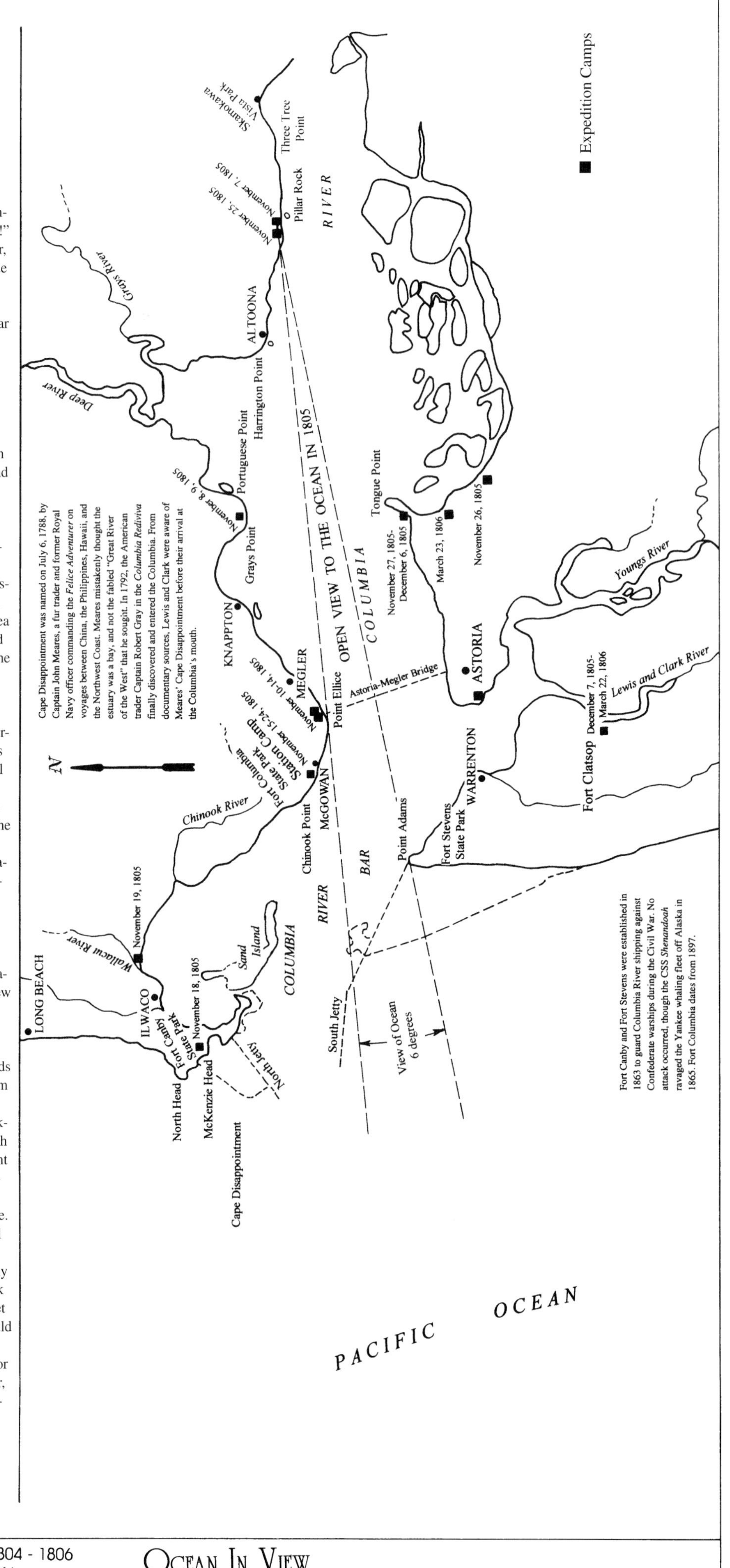

EXPLORATIONS OF LEWIS AND CLARK 1804 - 1806
CARTOGRAPHIC RECONSTRUCTION

OCEAN IN VIEW

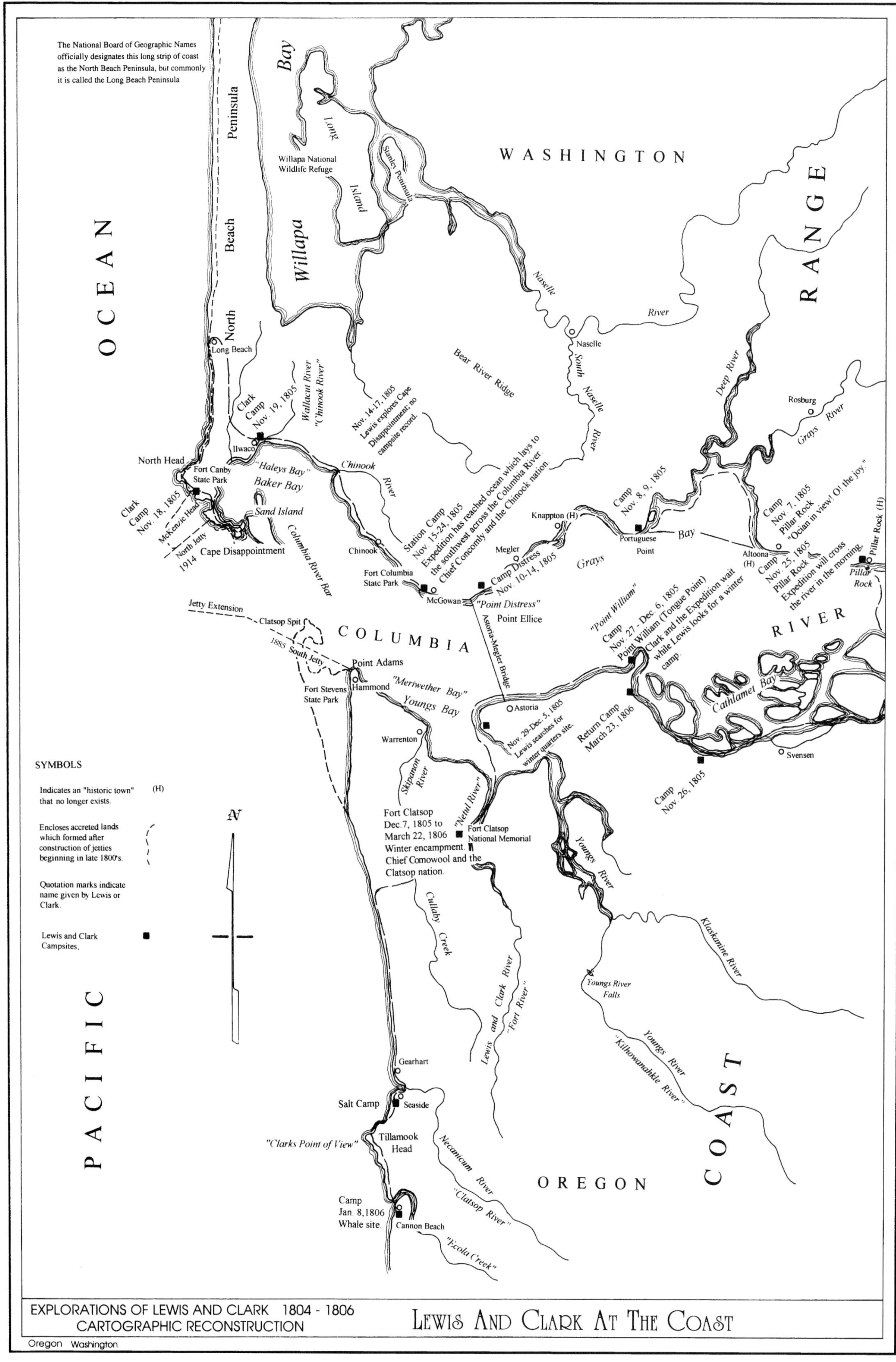
The National Board of Geographic Names officially designates this long strip of coast as the North Beach Peninsula, but commonly it is called the Long Beach Peninsula
North Beach Peninsula
Willapa Bay
Long Island
Willapa National Wildlife Refuge
Stanley Peninsula
WASHINGTON
RANGE
OCEAN
Naselle River
Naselle
South Naselle River
Bear River Ridge
Long Beach
Clark Camp Nov. 19, 1805
Wallacut River "Chinook River"
Nov. 14-17, 1805 Lewis explores Cape Disappointment; no campsite record.
Deep River
Rosburg
Grays River
Ilwaco
North Head
Fort Canby State Park
"Haleys Bay"
Baker Bay
Chinook River
Station Camp Nov. 15-24, 1805 Expedition has reached ocean which lays to the southwest across the Columbia River. Chief Concomly and the Chinook nation.
Camp Nov. 8, 9, 1805
Camp Nov. 7, 1805 Pillar Rock "Ocian in view! O! the joy."
Clark Camp Nov. 18, 1805
McKenzie Head
Sand Island
North Jetty 1914
Cape Disappointment
Columbia River Bar
Chinook
Knappton (H)
Megler
Portuguese Point
Grays Bay
Altoona (H)
Camp Nov. 25, 1805 Pillar Rock Expedition will cross the river in the morning.
Pillar Rock (H)
Pillar Rock
Fort Columbia State Park
McGowan
Camp Distress Nov. 10-14, 1805
"Point Distress"
Point Ellice
Astoria-Megler Bridge
"Point William"
Camp Nov. 27 - Dec. 6, 1805 Point William (Tongue Point) Clark and the Expedition wait while Lewis looks for a winter camp.
Jetty Extension
Clatsop Spit
COLUMBIA
RIVER
1885 South Jetty
Point Adams
Fort Stevens State Park
Hammond
"Meriwether Bay"
Youngs Bay
Astoria
Nov. 29-Dec. 5, 1805 Lewis searches for winter quarters site.
Return Camp March 23, 1806
Cathlamet Bay
Svensen
Camp Nov. 26, 1805
Warrenton
Skipanon River
"Netul River"
SYMBOLS
Indicates an "historic town" that no longer exists. (H)
Encloses accreted lands which formed after construction of jetties beginning in late 1800's.
Quotation marks indicate name given by Lewis or Clark.
Lewis and Clark Campsites.
N
Fort Clatsop Dec. 7, 1805 to March 22, 1806 Winter encampment. Chief Comowool and the Clatsop nation.
Fort Clatsop National Memorial
Youngs River
Cullaby Creek
Klaskanine River
Lewis and Clark River "Fort River"
Youngs River Falls
Youngs River "Kilhowanahkle River"
PACIFIC
COAST
Gearhart
Salt Camp
Seaside
Tillamook Head
"Clarks Point of View"
Necanicum River "Clatsop River"
OREGON
Camp Jan. 8, 1806 Whale site
Cannon Beach
"Ecola Creek"
EXPLORATIONS OF LEWIS AND CLARK 1804 - 1806
CARTOGRAPHIC RECONSTRUCTION
Oregon Washington
LEWIS AND CLARK AT THE COAST

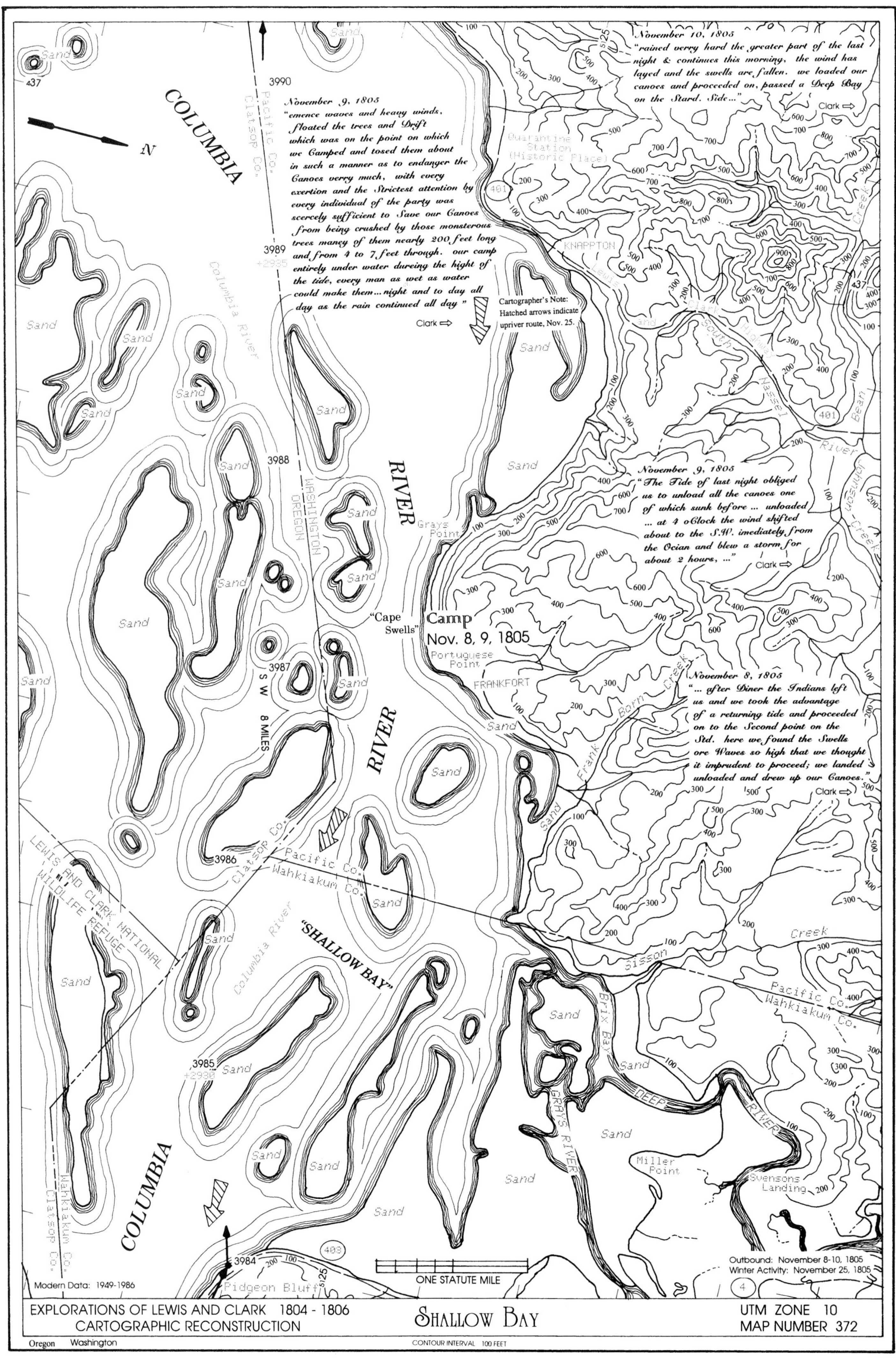
November 10, 1805
"rained verry hard the greater part of the last night & continues this morning, the wind has layed and the swells are fallen. we loaded our canoes and proceeded on, passed a Deep Bay on the Stard. Side..."
November 9, 1805
"emence waves and heavy winds, floated the trees and Drift which was on the point on which we Camped and tosed them about in such a manner as to endanger the Canoes verry much, with every exertion and the Strictest attention by every individual of the party was scercely sufficient to Save our Canoes from being crushed by those monsterous trees maney of them nearly 200 feet long and from 4 to 7 feet through. our camp entirely under water dureing the hight of the tide, every man as wet as water could make them...night and to day all day as the rain continued all day"
Cartographer's Note: Hatched arrows indicate upriver route, Nov. 25.
November 9, 1805
"The Tide of last night obliged us to unload all the canoes one of which sunk before ... unloaded ... at 4 oClock the wind shifted about to the S.W. imediately from the Ocian and blew a storm for about 2 hours, ..."
November 8, 1805
"... after Diner the Indians left us and we took the advantage of a returning tide and proceeded on to the Second point on the Std. here we found the Swells ore Waves so high that we thought it imprudent to proceed; we landed unloaded and drew up our Canoes."
Clark
COLUMBIA
RIVER
COLUMBIA RIVER
N
Pacific Co.
Clatsop Co.
Wahkiakum Co.
WASHINGTON
OREGON
Quarantine Station (Historic Place)
KNAPPTON
Lewis and Clark South Highway
Nassel River
Johnson Creek
Bean Creek
Grays Point
"Cape Swells"
Camp
Nov. 8, 9, 1805
Portuguese Point
FRANKFORT
Frank Born Creek
Sisson Creek
"SHALLOW BAY"
LEWIS AND CLARK NATIONAL WILDLIFE REFUGE
Brix Bay
GRAYS RIVER
DEEP RIVER
Miller Point
Svensons Landing
Pidgeon Bluff
Sand
S W 8 MILES
ONE STATUTE MILE
Modern Data: 1949-1986
Outbound: November 8-10, 1805
Winter Activity: November 25, 1805
EXPLORATIONS OF LEWIS AND CLARK 1804 - 1806
CARTOGRAPHIC RECONSTRUCTION
SHALLOW BAY
UTM ZONE 10
MAP NUMBER 372
Oregon Washington
CONTOUR INTERVAL 100 FEET

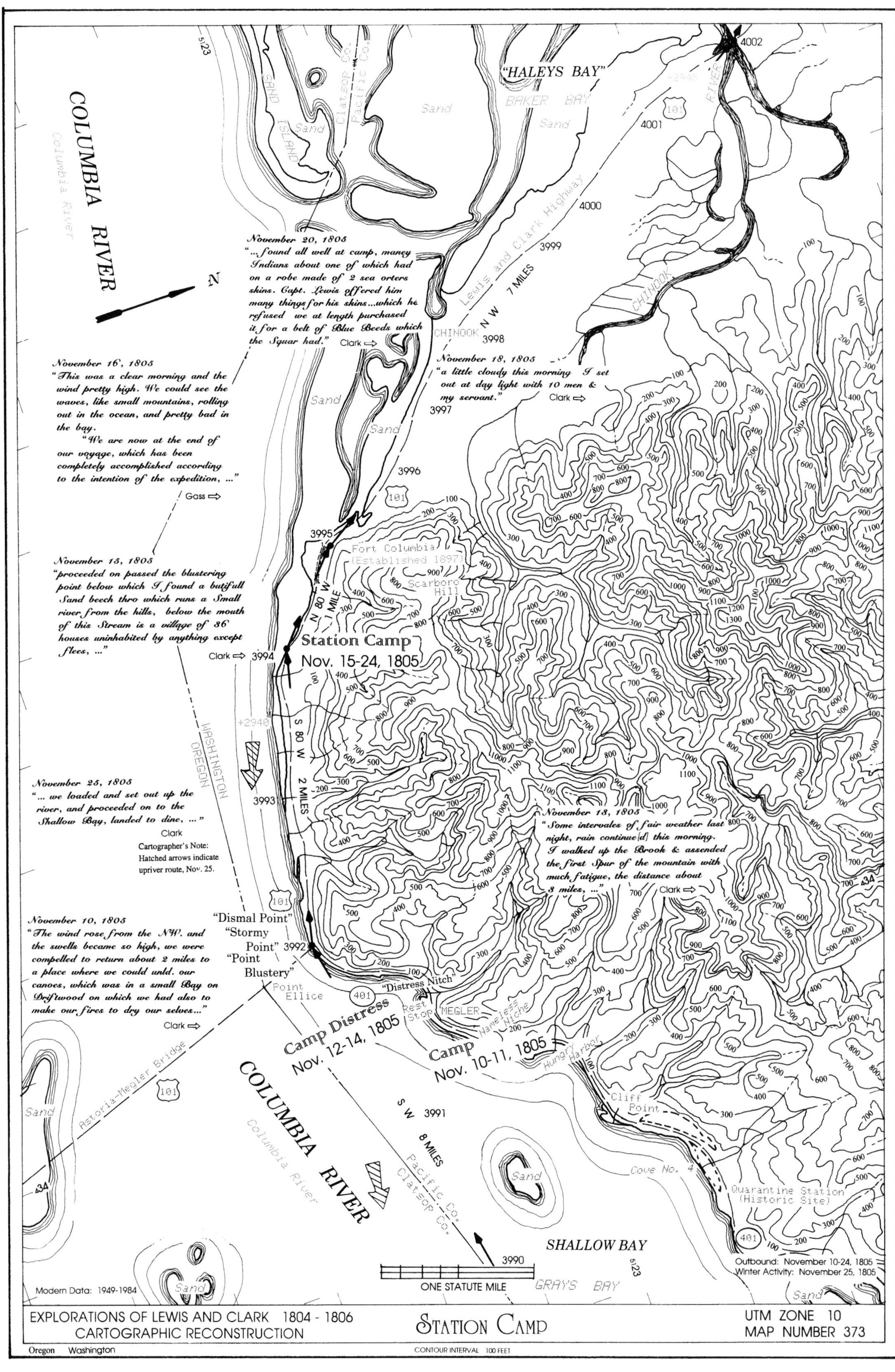

"HALEYS BAY"
BAKER BAY
COLUMBIA RIVER
November 20, 1805
"...found all well at camp, maney Indians about one of which had on a robe made of 2 sea orters skins. Capt. Lewis offered him many things for his skins...which he refused we at length purchased it for a belt of Blue Beeds which the Squar had." Clark ⇨
November 16, 1805
"This was a clear morning and the wind pretty high. We could see the waves, like small mountains, rolling out in the ocean, and pretty bad in the bay.
"We are now at the end of our voyage, which has been completely accomplished according to the intention of the expedition, ..."
Gass ⇨
November 18, 1805
"a little cloudy this morning I set out at day light with 10 men & my servant." Clark ⇨
November 15, 1805
"proceeded on passed the blustering point below which I found a butifull Sand beech thro which runs a Small river from the hills, below the mouth of this Stream is a village of 36 houses uninhabited by anything except flees, ..."
Clark ⇨
Fort Columbia (Established 1897)
Scarboro Hill
Station Camp
Nov. 15-24, 1805
November 25, 1805
"... we loaded and set out up the river, and proceeded on to the Shallow Bay, landed to dine, ..."
Clark
Cartographer's Note: Hatched arrows indicate upriver route, Nov. 25.
November 13, 1805
"Some intervales of fair weather last night, rain continue[d] this morning. I walked up the Brook & assended the first Spur of the mountain with much fatigue, the distance about 3 miles, ..."
Clark ⇨
November 10, 1805
"The wind rose from the NW. and the swells became so high, we were compelled to return about 2 miles to a place where we could unld. our canoes, which was in a small Bay on Driftwood on which we had also to make our fires to dry our selves..."
Clark ⇨
"Dismal Point"
"Stormy Point"
"Point Blustery"
Point Ellice
"Distress Nitch"
Camp Distress
Nov. 12-14, 1805
Camp
Nov. 10-11, 1805
MEGLER
Hungry Harbor
Cliff Point
Cove No. 4
Quarantine Station (Historic Site)
SHALLOW BAY
GRAYS BAY
Astoria-Megler Bridge
Lewis and Clark Highway
CHINOOK
WASHINGTON
OREGON
Pacific Co.
Clatsop Co.
Outbound: November 10-24, 1805
Winter Activity: November 25, 1805
Modern Data: 1949-1984
ONE STATUTE MILE
EXPLORATIONS OF LEWIS AND CLARK 1804 - 1806
CARTOGRAPHIC RECONSTRUCTION
STATION CAMP
UTM ZONE 10
MAP NUMBER 373
Oregon Washington
CONTOUR INTERVAL 100 FEET

Cartographer's Note:

The Corps proceeded haltingly down the Columbia's north shore in the face of waves, wind, and rain before finally establishing "Station Camp," occupied November 15–24 (see Map 373). Looking out over the Pacific Ocean, the men had accomplished their main outbound goal—the "end of our voyage" at the Pacific Ocean. Leaving Station Camp on Nov. 25, the Corps planned to investigate the Columbia's south side for a wintering site. Rather than attempting a dangerous crossing of the Columbia's wide, treacherous estuary, they returned upstream to near Pillar Rock and spent the night. They then crossed the river at a narrower place the following day. Proceeding down the south shore, they established Fort Clatsop as winter quarters, December 7, 1805–March 22, 1806.

November 10, 1805

"we continued on this drift wood untill about 3 oClock when the evening appearing favourable we loaded & set out in hopes to turn the Point below and get into a better harber, but finding the waves & swells continue to rage with great fury below, we got a safe place for our stores & a much beter one for the canoes to lie and formed a campment on Drift logs... The logs on which we lie is all on flote every high tide. The rain continues all day. ... nothing to eate but Pounded fish"

Clark ⇨

November 11, 1805

"a hard rain all the last night we again get wet the rain continue[s] at intervales all day. Wind verry high from S W and blew a storm all day. ... we are truly unfortunate to be compelled to be 4 days nearly in the same place at a time that our day[s] are precious to us, ... the great quantities of rain which has fallen losens the stones on the side of the hill & the small ones fall on us, our situation is truely a disagreeable one our canoes in one place at the mercy of the waves our baggage in another and our selves & party scattered on drift trees of emence size, ..."

Clark ⇨

November 12, 1805

"a tremendous thunder storm abt. 3 oClock this morning accompanied by wind from the S W. and Hail, this Storm of hard clap's of thunder Light[n]ing and hail untill about 6 oClock at intervales It then became light for a short time when the heavens became darkened by a black cloud from the S. W. & a hard rain suckceeded which lasted untill 12 oClock with a hard wind which raised the seas tremendiously high braking with great force and fury against the rocks & trees on which we lie, ... moved our camp around a point a short distance to a small wet bottom at the mouth of a small creek, which we had not observed when we first came to this cove, ... It would be distressing to a feeling person to see our situation at this time all wet and cold with our bedding &c. also wet, in a cove scercely large [e]nough to contain us, our Baggage in a small holler about 1/2 a mile from us, and canoes at the mercy of the waves & drift wood, ... It was clear at 12 for a short time. I observed the mountains on the opposit side was covered with snow. our party has been wet for 8 days and is truly disagreeable, ... we got our selves tolerable comfortable by drying our selves & bedding."

Clark ⇨

November 13, 1805

"on my return we dispatched 3 men Colter, Willard and Shannon in the Indian canoe to get around the point if possible and examine the river, and the Bay below for a go[o]d harber for our canoes to lie in Safty... nothing to eate but pounded fish which we Keep as a reserve and use in Situations of this kind."

Clark ⇨

November 14, 1805 [First Draft]

"The rain continue all day... at 3 oClock Capt. Lewis, Drewyer Jo & R. Fields, & Frasure set out down on the shore to examine if any white men were below within our reach, they took a empty canoe & 5 men to set them around the Point on a Gravelley Beech which Colter informed was at no great distance below."

Clark ⇨

November 14, 1805 [Final Entry]

"rained all the last night without intermition, and this morning. wind blows verry hard, ... one of our canoes is much broken by the waves dashing it against the rocks. ... Colter informed us that "it was but a Short distance from where we lay around the point to a butiful Sand beech, ... that he had proceeded in the canoe as far as he could... Capt. Lewis concluded to proceed on by land & find if possible the white people. ... and examine if a Bay is Situated near the mouth of this river as laid down by Vancouver... having landed Capt. Lewis & his party Safe on the Sand Beech. The rain continues all day. all wet. The rain &c. which has continued without a longer intermition than 2 hours at a time for ten days past has distroyd. the robes and rotted nearly one half of the fiew clothes the party has, perticularley the leather clothes..."

Clark ⇨

November 15, 1805

"Rained all the last night, this morning it became calm and fair, I preposed Setting out, and ordered the canoes Repared and loaded; before we could load our canoes the wind Sudenly Sprung up from the S. E. and blew with Such violence, that we could not proceed in Safety with the loading. ... proceeded on passed the blustering point... I landed and formed a camp on the highest Spot I could find between the hight of the tides, and the Slashers in a small bottom this I could plainly See would be the extent of our journey by water as the waves were too high at any stage for our Canoes to proceed any further down. in full view of the Ocian from Point Adams or Rond (see La Payrouse) to Cape Disapointment, I could not see any Island in the mouth of this river as laid down by Vancouver. the Bay which he laies down in the mouth is immediately below me. This Bay we call Haley's bay from a favourite trader with the Indians."

Clark ⇨

Cartographer's Note:
Haley's Bay is now Baker Bay.

November 16, 1805

"a fine morning cool the latter part of the night, ... The Sea is fomeing and looks truly dismal to day, from the wind which blew to day from the S.W. ... we Encamped just above a Point in a Deep bay to the Stard Side into which falls 2 small rivers Std."

Clark ⇨

November 17, 1805

"The Chief of the nation below us came up to see us the name of the nation is Chin-nook and is noumerous live principally on fish roots a fiew Elk and fowls. they are well armed with good Fusees. I directed all the men who wished to see more of the Ocean to Get ready to set out with me on tomorrow..."

Clark ⇨

Cartographer's Note:
Chief Concomly was the principal Chinook chief. Direct descendents of this great native leader still live in this area today.

Cartographer's Note:
On November 18–20, 1805, Clark was away investigating Cape Disappointment and the northern beaches (see Map 374).

November 21, 1805

"An old woman & Wife to a Cheif of the Chunnooks came and made a Camp near ours. She brought with her 6 young Squars [her daughters & nieces] I believe for the purpose of Gratifying the passions of the men of our party and receving for those indulgiences Such Small [presents] as She (the old woman) thought proper to accept of."

Clark ⇨

November 22, 1805

"Some little rain all the last night with wind, before day the wind increased to a storm from the S.S.E. and blew with violence throwing the water of the river with emence waves out of its banks almost over whelming us in water, O! how horriable is the day. This storm continued all day with equal violence accompanied with rain, ..."

Clark ⇨

November 23, 1805

"in the evening Seven indians of the Clo tsop Nation came over in a Canoe, they brought with them 2 Sea otter Skins for which they asked blue beads... This nation is the remains of a large nation destroyed by the Small pox or Some other [disease] which those people were not aquainted with, ..."

Clark ⇨

Cartographer's Note:
Indians reports about elk herds in the Columbia's lowlands were a significant reason why most of the party favored looking to the river's south side to establish winter quarters. These seven Clatsops, inhabitants of the south shore, probably were a source of this information.

November 24, 1805

"At night the party were consulted by the Commanding Officers, as to the place most proper for winter quarters; and the most of them were of opinion, that it would be best, in the first place, to go over to the south side of the river and ascertain whether good hunting ground could be found there. Should that be the case, it would be a more eligible place than higher up the river, on account of getting salt, as that is a very scarce article with us."

Gass ⇨

Cartographer's Note:
Some have hailed an "election" at Chinook Point—regarding where to seek winter quarters—as a first great incident of democratic enfranchisement in the West, decided not only by the enlisted men, but a woman and black servant as well. This cartographer believes, however, that this was more a consultation—a polling by the captains of their rank-and-file members' views and opinions, and not a mandate that bound the leaders in making their decision. This was common in Lewis and Clark's time, as, in fact, it has been throughout much of our nation's military history. At least four times during the expedition, the captains went to their men for polling, and on one occasion at the Missouri-Marias confluence decided against the wishes of every man in the group. Ideally, military leaders always have been expected to make responsible decisions based on their education, training, experience, and expertise. Military choices never have been dictated by a rank-and-file election. The journals of the Corps enlisted men, too, record on these occasions that the officers "consulted" them for opinions. Never is there an expectation or even a hint that the polling would bind the captains' decision. Surely, however, the important thing here is the level of respect given Ben York and Sacagawea, a black man and an Indian woman.

Cartographer's Note:
The shoreline immediately upstream from Point Ellice is scalloped by small coves or niches backed by high cliffs of exposed rock and soil. It was here that the Corps desperately sought protection, November 10–14, from storms lashing the headlands. Today, State Highway 401 and its embankments have significantly impacted these coves. Four small coves in particular are situated east (upriver) from Point Ellice (see Map 373): (1) "Distress Nitch" just east of Point Ellice and extending to near the Megler rest area, (2) a small nameless niche beyond the highway stop, (3) Hungry Harbor, and (4) an open cove with a gentle rise up from the river to an old quarantine station. In 1805, these coves were filled with massive log drifts in the tide zones. This cartographer believes it likely that on November 10, when failing to pass west around Point Ellice, the party retreated to Hungry Harbor (2 miles east); attempting to paddle by stormy Point Ellice at that time in heavy dugout canoes was too risky. Then, after a second foiled attempt to proceed past the point later in the day, the captains sought refuge in the small "nameless" niche, a slightly less dismal situation. Here, they stayed two nights, before moving a short distance around a small point to a larger cove, "Distress Nitch," remaining here another two nights. (Distress Nitch and Hungry Harbor each have small freshwater streams, whereas the unnamed niche does not. Clark provides no evidence that they retreated as far east as the fourth cove.) On the afternoon of November 15, they finally proceeded around Point Ellice to a much more habitable place, "Station Camp." This location became a key survey station from which Clark determined bearings and distances to Cape Disappointment, Point Adams, and other features at the Columbia's mouth, thus the name "Station Camp."

To avoid confusion with Outbound and Return arrows of the journal quotes, no arrows follow quotes for Winter Activity near the Columbia's mouth, November 25, 1805, to March 22, 1806.

EXPLORATIONS OF LEWIS AND CLARK 1804 - 1806
CARTOGRAPHIC RECONSTRUCTION

STATION CAMP (CONTINUED)

UTM ZONE 10
MAP NUMBER 373A

Washington

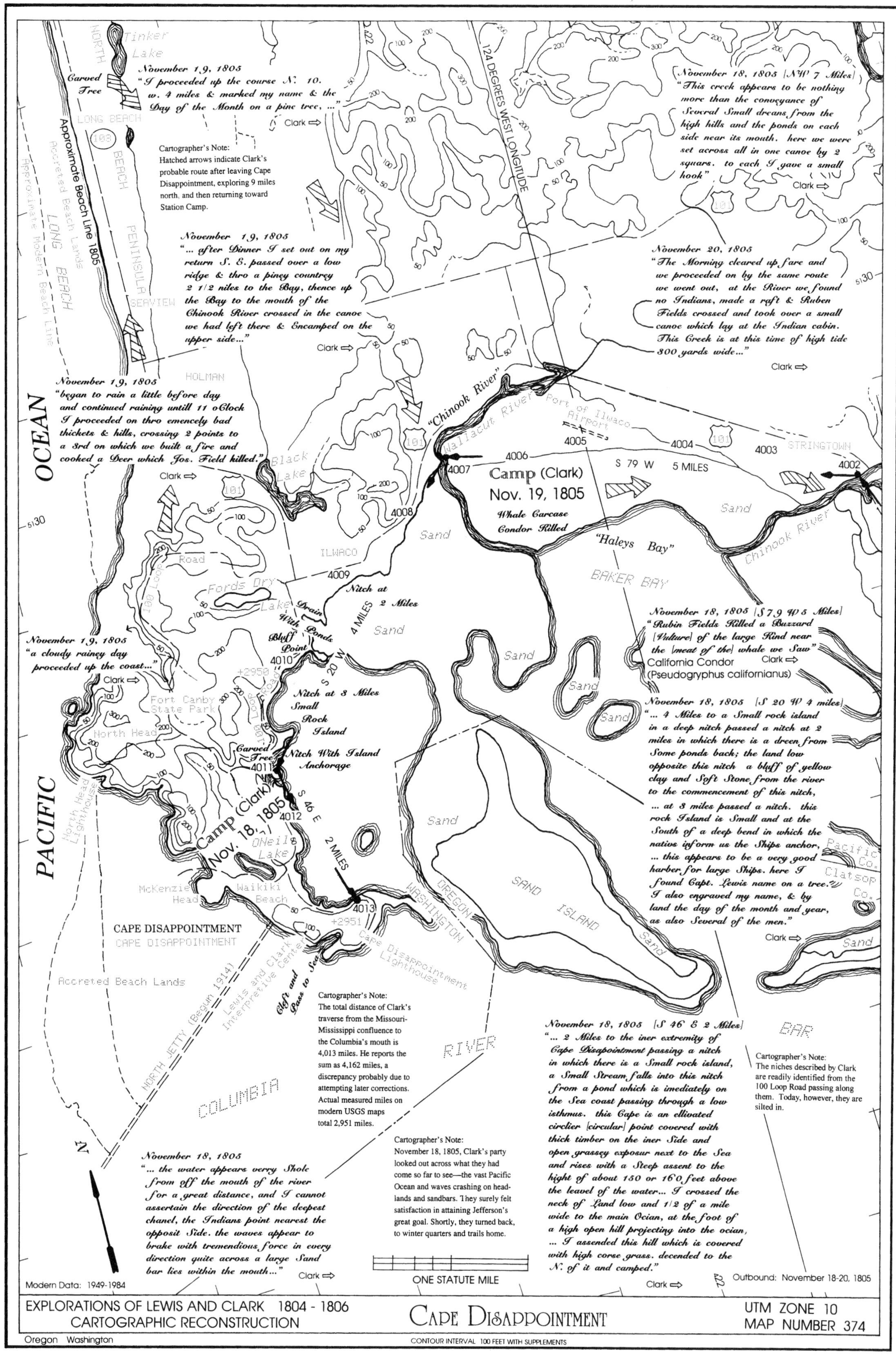

November 1,9, 1805
"I proceeded up the course N: 10. w. 4 miles & marked my name & the Day of the Month on a pine tree, ..."
Carved Tree
Tinker Lake
Clark ⇨
Cartographer's Note:
Hatched arrows indicate Clark's probable route after leaving Cape Disappointment, exploring 9 miles north, and then returning toward Station Camp.
Approximate Beach Line 1805
Accreted Beach Lands
Approximate Modern Beach Line
LONG BEACH
NORTH BEACH PENINSULA
SEAVIEW
124 DEGREES WEST LONGITUDE
[November 18, 1805 [NW 7 Miles]
"This creek appears to be nothing more than the conveyance of Several Small dreans from the high hills and the ponds on each side near its mouth. here we were set across all in one canoe by 2 squars. to each I gave a small hook"
Clark ⇨
November 1,9, 1805
"... after Dinner I set out on my return S. E. passed over a low ridge & thro a piney countrey 2 1/2 niles to the Bay, thence up the Bay to the mouth of the Chinook River crossed in the canoe we had left there & Encamped on the upper side..."
Clark ⇨
November 20, 1805
"The Morning cleared up fare and we proceeded on by the same route we went out, at the River we found no Indians, made a raft & Ruben Fields crossed and took over a small canoe which lay at the Indian cabin. This Creek is at this time of high tide 300 yards wide..."
Clark ⇨
November 1,9, 1805
"began to rain a little before day and continued raining untill 11 oClock I proceeded on thro emencely bad thickets & hills, crossing 2 points to a 3rd on which we built a fire and cooked a Deer which Jos. Field killed."
Clark ⇨
HOLMAN
OCEAN
"Chinook River"
Wallacut River
Port of Ilwaco Airport
Black Lake
4005
4004
4003
STRINGTOWN
4006
4007
4002
Camp (Clark) Nov. 19, 1805
Whale Carcase
Condor Killed
S 79 W 5 MILES
Sand
Chinook River
"Haleys Bay"
BAKER BAY
4008
ILWACO
4009
Road
Fords Dry Lake
100 Loop Road
Drain With Ponds
Nitch at 2 Miles
4 MILES
Bluff Point
4010
S 20 W
November 18, 1805 [S 7,9 W 5 Miles]
"Rubin Fields Killed a Buzzard [Vulture] of the large Kind near the [meat of the] whale we Saw"
California Condor
(Pseudogryphus californianus)
Clark ⇨
November 1,9, 1805
"a cloudy rainey day proceeded up the coast..."
Clark ⇨
+2950
Fort Canby State Park
North Head
Nitch at 3 Miles Small Rock Island
Carved Tree
Nitch With Island Anchorage
4011
November 18, 1805 [S 20 W 4 miles]
"... 4 Miles to a Small rock island in a deep nitch passed a nitch at 2 miles in which there is a dreen from Some ponds back; the land low opposite this nitch a bluff of yellow clay and Soft Stone from the river to the commencement of this nitch, ... at 3 miles passed a nitch. this rock Island is Small and at the South of a deep bend in which the natives inform us the Ships anchor, ... this appears to be a very good harber for large Ships. here I found Capt. Lewis name on a tree. I also engraved my name, & by land the day of the month and year, as also Several of the men."
Clark ⇨
PACIFIC
North Head Lighthouse
Camp (Clark) Nov. 18, 1805
4012
S 46 E
2 MILES
O'Neil Lake
McKenzie Head
Waikiki Beach
4013
+2951
WASHINGTON
OREGON
SAND ISLAND
Pacific Co.
Clatsop Co.
CAPE DISAPPOINTMENT
Cape Disappointment Lighthouse
Lewis and Clark Interpretive Center
Cleft and Pass to Sea
Accreted Beach Lands
NORTH JETTY (Begun 1914)
Cartographer's Note:
The total distance of Clark's traverse from the Missouri-Mississippi confluence to the Columbia's mouth is 4,013 miles. He reports the sum as 4,162 miles, a discrepancy probably due to attempting later corrections. Actual measured miles on modern USGS maps total 2,951 miles.
RIVER
COLUMBIA
BAR
November 18, 1805 [S 46 E 2 Miles]
"... 2 Miles to the iner extremity of Cape Disapointment passing a nitch in which there is a Small rock island, a Small Stream falls into this nitch from a pond which is imediately on the Sea coast passing through a low isthmus. this Cape is an elivated circlier [circular] point covered with thick timber on the iner Side and open grassey exposur next to the Sea and rises with a Steep assent to the hight of about 150 or 160 feet above the leavel of the water... I crossed the neck of Land low and 1/2 of a mile wide to the main Ocian, at the foot of a high open hill projecting into the ocian, ... I assended this hill which is covered with high corse grass. decended to the N. of it and camped."
Clark ⇨
Cartographer's Note:
The niches described by Clark are readily identified from the 100 Loop Road passing along them. Today, however, they are silted in.
N
November 18, 1805
"... the water appears verry Shole from off the mouth of the river for a great distance, and I cannot assertain the direction of the deepest chanel, the Indians point nearest the opposit Side. the waves appear to brake with tremendious force in every direction quite across a large Sand bar lies within the mouth..."
Clark ⇨
Cartographer's Note:
November 18, 1805, Clark's party looked out across what they had come so far to see—the vast Pacific Ocean and waves crashing on headlands and sandbars. They surely felt satisfaction in attaining Jefferson's great goal. Shortly, they turned back, to winter quarters and trails home.
ONE STATUTE MILE
Modern Data: 1949-1984
Outbound: November 18-20, 1805
EXPLORATIONS OF LEWIS AND CLARK 1804 - 1806
CARTOGRAPHIC RECONSTRUCTION
CAPE DISAPPOINTMENT
UTM ZONE 10
MAP NUMBER 374
Oregon Washington
CONTOUR INTERVAL 100 FEET WITH SUPPLEMENTS

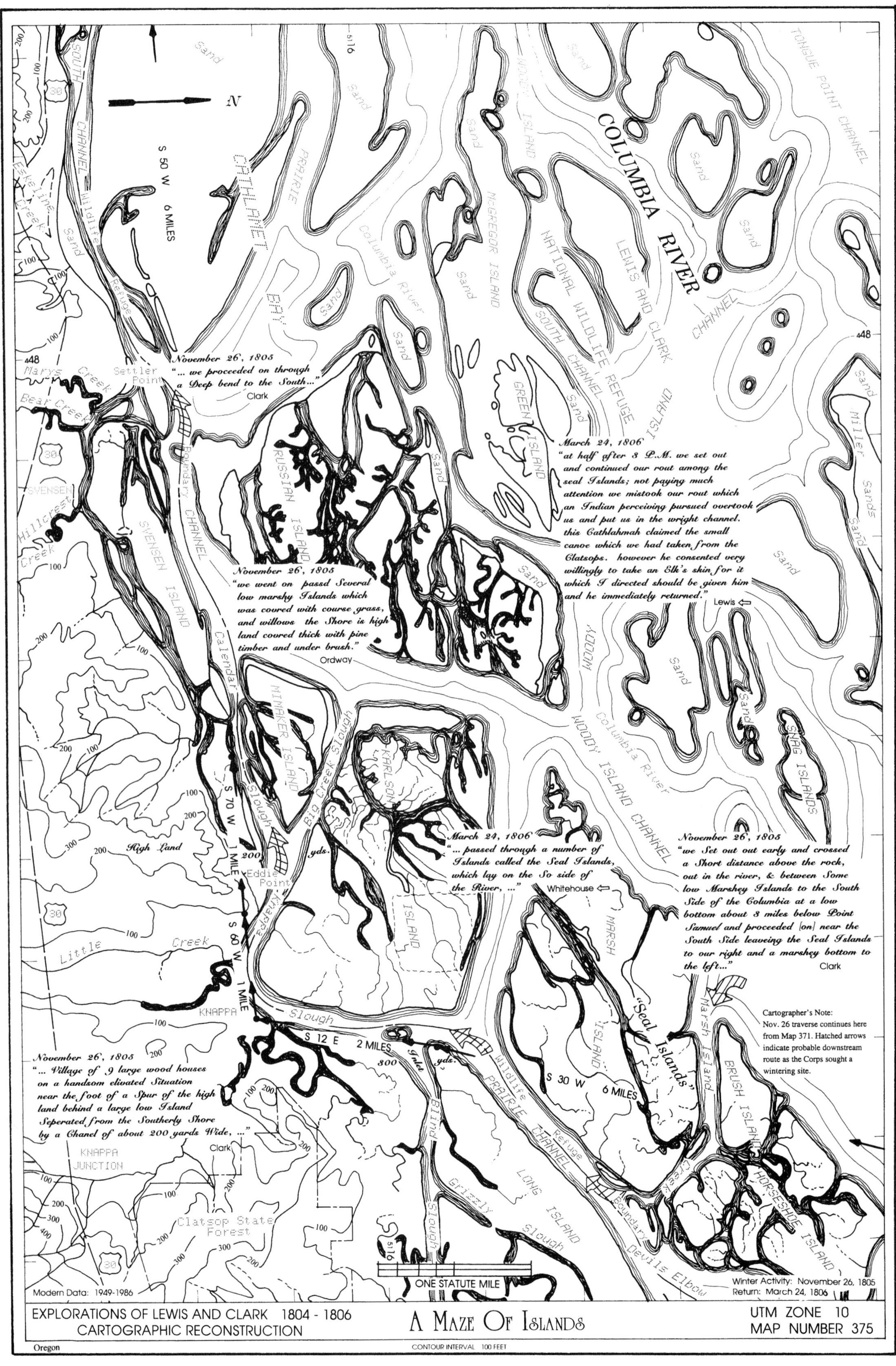

November 26, 1805
"... we proceeded on through a Deep bend to the South..."
Clark
March 24, 1806
"at half after 3 P.M. we set out and continued our rout among the seal Islands; not paying much attention we mistook our rout which an Indian perceiving pursued overtook us and put us in the wright channel. this Cathlahmah claimed the small canoe which we had taken from the Clatsops. however he consented very willingly to take an Elk's skin for it which I directed should be given him and he immediately returned."
Lewis
November 26, 1805
"we went on passd Several low marshy Islands which was coured with course grass, and willows the Shore is high land coured thick with pine timber and under brush."
Ordway
March 24, 1806
"... passed through a number of Islands called the Seal Islands, which lay on the So side of the River, ..."
Whitehouse
November 26, 1805
"we Set out out early and crossed a Short distance above the rock, out in the river, & between Some low Marshey Islands to the South Side of the Columbia at a low bottom about 3 miles below Point Samuel and proceeded [on] near the South Side leaveing the Seal Islands to our right and a marshey bottom to the left..."
Clark
Cartographer's Note:
Nov. 26 traverse continues here from Map 371. Hatched arrows indicate probable downstream route as the Corps sought a wintering site.
November 26, 1805
"... Village of 9 large wood houses on a handsom elivated Situation near the foot of a Spur of the high land behind a large low Island Seperated from the Southerly Shore by a Chanel of about 200 yards Wide, ..."
Clark
COLUMBIA RIVER
"Seal Islands"
N
S 50 W 6 MILES
S 70 W 1 MILE
S 60 W 1 MILE
S 12 E 2 MILES
S 30 W 6 MILES
Modern Data: 1949-1986
ONE STATUTE MILE
Winter Activity: November 26, 1805
Return: March 24, 1806
EXPLORATIONS OF LEWIS AND CLARK 1804 - 1806
CARTOGRAPHIC RECONSTRUCTION
A MAZE OF ISLANDS
UTM ZONE 10
MAP NUMBER 375
Oregon
CONTOUR INTERVAL 100 FEET

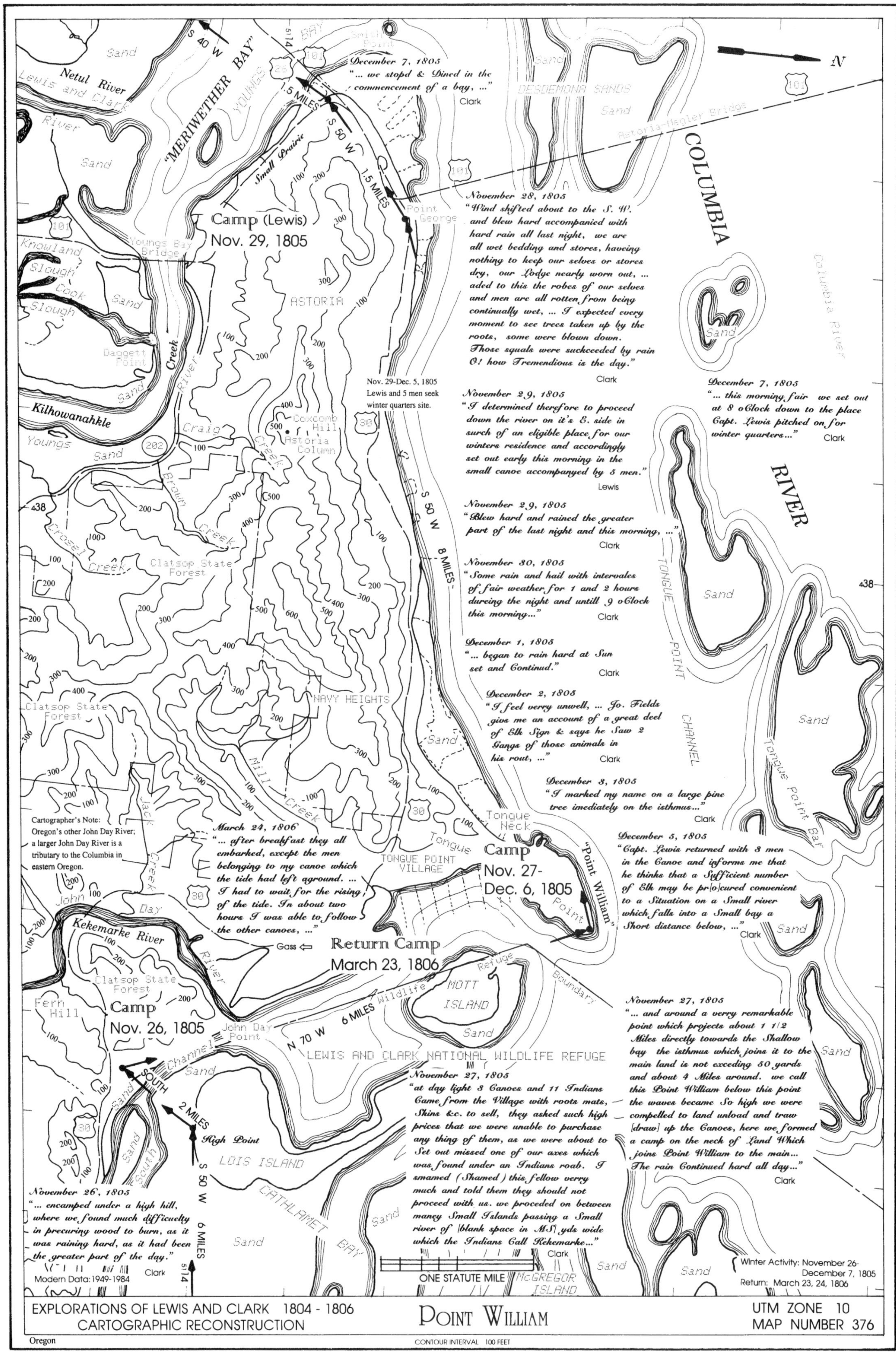
EXPLORATIONS OF LEWIS AND CLARK 1804 - 1806
CARTOGRAPHIC RECONSTRUCTION
POINT WILLIAM
UTM ZONE 10
MAP NUMBER 376
Oregon
CONTOUR INTERVAL 100 FEET
N
COLUMBIA
RIVER
Columbia River
"MERIWETHER BAY"
Netul River
Lewis and Clark River
Youngs
Youngs Bay Bridge
Knowland Slough
Cook Slough
Daggett Point
Kilhowanahkle
Youngs
Craig
Brown Creek
Creek
River
Crossel Creek
Clatsop State Forest
Coxcomb Hill
Astoria Column
ASTORIA
NAVY HEIGHTS
Mill Creek
Jack Creek
John Day River
Kekemarke River
Fern Hill
Clatsop State Forest
John Day Point
Tongue Neck
Tongue Point
TONGUE POINT VILLAGE
"Point William"
Point
TONGUE POINT CHANNEL
Tongue Point Bar
MOTT ISLAND
Refuge
Wildlife
Boundary
LEWIS AND CLARK NATIONAL WILDLIFE REFUGE
LOIS ISLAND
CATHLAMET BAY
McGREGOR ISLAND
High Point
Channel
South Channel
DESDEMONA SANDS
Astoria-Megler Bridge
Smith Point
BAY
YOUNGS
Small Prairie
Point George
Sand
S 40 W
1.5 MILES
S 50 W
1.5 MILES
S 50 W
8 MILES
N 70 W
6 MILES
SOUTH
2 MILES
S 50 W
6 MILES
Camp (Lewis)
Nov. 29, 1805
Camp
Nov. 27-
Dec. 6, 1805
Return Camp
March 23, 1806
Camp
Nov. 26, 1805
Gass
ONE STATUTE MILE
Modern Data: 1949-1984
Winter Activity: November 26-December 7, 1805
Return: March 23, 24, 1806
Nov. 29-Dec. 5, 1805
Lewis and 5 men seek winter quarters site.
Cartographer's Note:
Oregon's other John Day River; a larger John Day River is a tributary to the Columbia in eastern Oregon.
December 7, 1805
"... we stopd & Dined in the commencement of a bay, ..."
Clark
November 28, 1805
"Wind shifted about to the S. W. and blew hard accompanied with hard rain all last night, we are all wet bedding and stores, haveing nothing to keep our selves or stores dry, our Lodge nearly worn out, ... aded to this the robes of our selves and men are all rotten from being continually wet, ... I expected every moment to see trees taken up by the roots, some were blown down. Those squals were suckceeded by rain O! how Tremendious is the day."
Clark
December 7, 1805
"... this morning fair we set out at 8 oClock down to the place Capt. Lewis pitched on for winter quarters..."
Clark
November 29, 1805
"I determined therefore to proceed down the river on it's E. side in surch of an eligible place for our winters residence and accordingly set out early this morning in the small canoe accompanyed by 5 men."
Lewis
November 29, 1805
"Blew hard and rained the greater part of the last night and this morning, ..."
Clark
November 30, 1805
"Some rain and hail with intervales of fair weather for 1 and 2 hours dureing the night and untill 9 oClock this morning..."
Clark
December 1, 1805
"... began to rain hard at Sun set and Continud."
Clark
December 2, 1805
"I feel verry unwell, ... Jo. Fields givs me an account of a great deel of Elk Sign & says he Saw 2 Gangs of those animals in his rout, ..."
Clark
December 3, 1805
"I marked my name on a large pine tree imediately on the isthmus..."
Clark
December 5, 1805
"Capt. Lewis returned with 3 men in the Canoe and informs me that he thinks that a Sufficient number of Elk may be pr[o]cured convenient to a Situation on a Small river which falls into a Small bay a Short distance below, ..."
Clark
March 24, 1806
"... after breakfast they all embarked, except the men belonging to my canoe which the tide had left aground. ... I had to wait for the rising of the tide. In about two hours I was able to follow the other canoes, ..."
November 27, 1805
"... and around a verry remarkable point which projects about 1 1/2 Miles directly towards the Shallow bay the isthmus which joins it to the main land is not exceding 50 yards and about 4 Miles around. we call this Point William below this point the waves became So high we were compelled to land unload and traw [draw] up the Canoes, here we formed a camp on the neck of Land Which joins Point William to the main... The rain Continued hard all day..."
Clark
November 27, 1805
"at day light 3 Canoes and 11 Indians Came from the Village with roots mats, Skins &c. to sell, they asked such high prices that we were unable to purchase any thing of them, as we were about to Set out missed one of our axes which was found under an Indians roab. I smamed (Shamed) this fellow verry much and told them they should not proceed with us. we proceded on between maney Small Islands passing a Small river of [blank space in MS] yds wide which the Indians Call Kekemarke..."
Clark
November 26, 1805
"... encamped under a high hill, where we found much difficuelty in precuring wood to burn, as it was raining hard, as it had been the greater part of the day."
Clark

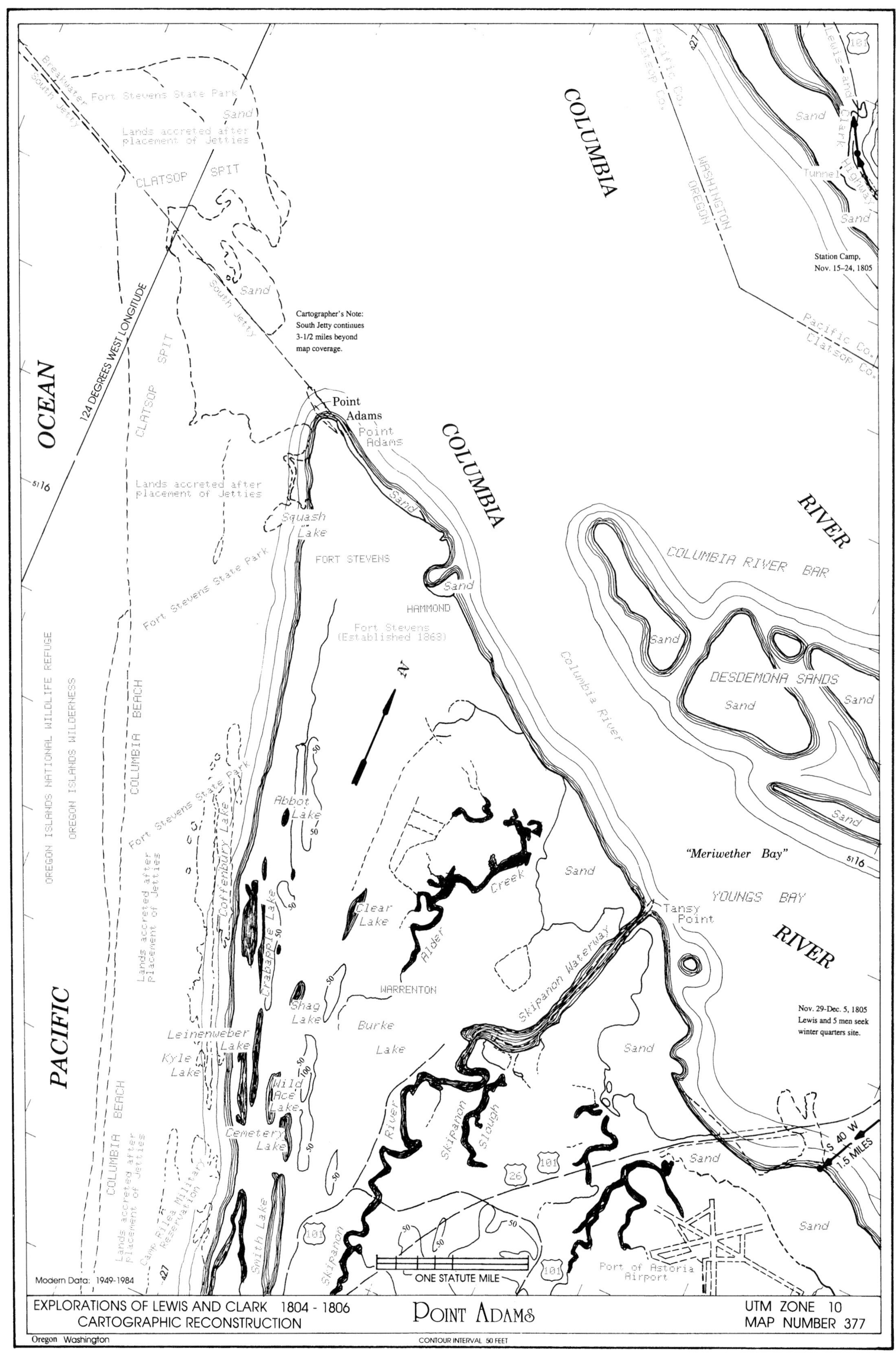
OCEAN
PACIFIC
COLUMBIA
COLUMBIA
RIVER
RIVER
124 DEGREES WEST LONGITUDE
Cartographer's Note: South Jetty continues 3-1/2 miles beyond map coverage.
Point Adams
Station Camp, Nov. 15-24, 1805
Nov. 29-Dec. 5, 1805 Lewis and 5 men seek winter quarters site.
"Meriwether Bay"
S 40 W 1.5 MILES
Modern Data: 1949-1984
ONE STATUTE MILE
EXPLORATIONS OF LEWIS AND CLARK 1804 - 1806
CARTOGRAPHIC RECONSTRUCTION
POINT ADAMS
UTM ZONE 10
MAP NUMBER 377
Oregon Washington
CONTOUR INTERVAL 50 FEET

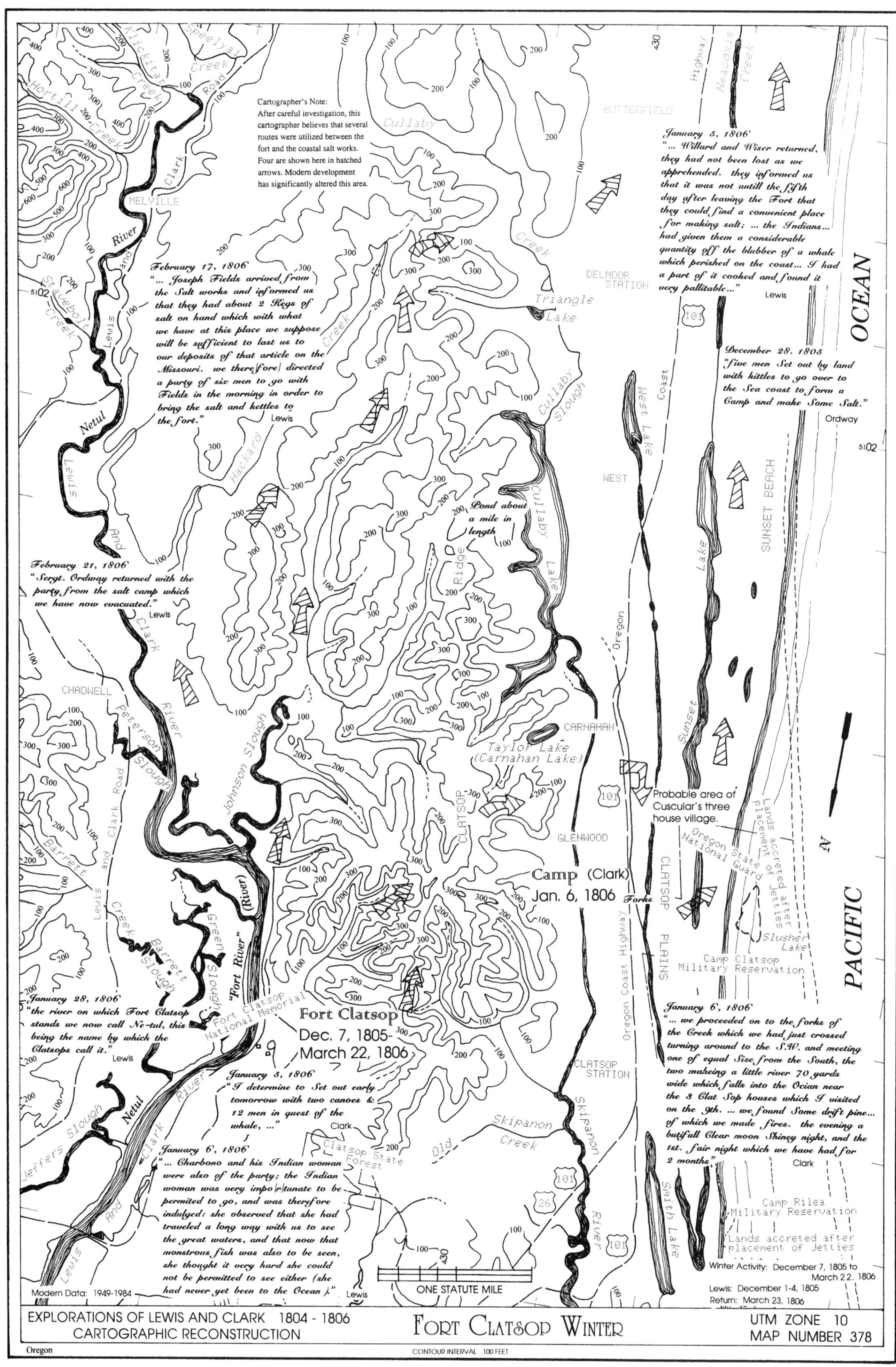

Cartographer's Note: After careful investigation, this cartographer believes that several routes were utilized between the fort and the coastal salt works. Four are shown here in hatched arrows. Modern development has significantly altered this area.
January 5, 1806
"... Willard and Wiser returned, they had not been lost as we apprehended. they informed us that it was not untill the fifth day after leaving the Fort that they could find a convenient place for making salt; ... the Indians... had given them a considerable quantity off the blubber of a whale which perished on the coast... I had a part of it cooked and found it very pallitable..."
Lewis
February 17, 1806
"... Joseph Fields arrived from the Salt works and informed us that they had about 2 Kegs of salt on hand which with what we have at this place we suppose will be sufficient to last us to our deposits of that article on the Missouri. we there[fore] directed a party of six men to go with Fields in the morning in order to bring the salt and kettles to the fort."
Lewis
December 28, 1805
"five men Set out by land with kittles to go over to the Sea coast to form a Camp and make Some Salt."
Ordway
OCEAN
PACIFIC
February 21, 1806
"Sergt. Ordway returned with the party from the salt camp which we have now evacuated."
Lewis
Pond about a mile in length
Probable area of Cuscular's three house village.
Camp (Clark) Jan. 6, 1806
Forks
January 28, 1806
"the river on which Fort Clatsop stands we now call Ne-tul, this being the name by which the Clatsops call it."
Lewis
Fort Clatsop
Dec. 7, 1805- March 22, 1806
"Fort River" (River)
Fort Clatsop National Memorial
January 5, 1806
"I determine to Set out early tomorrow with two canoes & 12 men in quest of the whale, ..."
Clark
January 6, 1806
"... we proceeded on to the forks of the Creek which we had just crossed turning around to the S.W. and meeting one of equal Size from the South, the two makeing a little river 70 yards wide which falls into the Ocian near the 3 Clat Sop houses which I visited on the 9th. ... we found Some drift pine... of which we made fires. the evening a butifull Clear moon Shiney night, and the 1st. fair night which we have had for 2 months"
Clark
January 6, 1806
"... Charbono and his Indian woman were also of the party; the Indian woman was very impo[r]tunate to be permited to go, and was therefore indulged: she observed that she had traveled a long way with us to see the great waters, and that now that monstrous fish was also to be seen, she thought it very hard she could not be permitted to see either (she had never yet been to the Ocean)."
Lewis
Winter Activity: December 7, 1805 to March 22, 1806
Lewis: December 1-4, 1805
Return: March 23, 1806
Modern Data: 1949-1984
ONE STATUTE MILE
MELVILLE
CHADWELL
BUTTERFIELD
DELMOOR STATION
WEST
CARNAHAN
GLENWOOD
CLATSOP STATION
SUNSET BEACH
Taylor Lake (Carnahan Lake)
Triangle Lake
Cullaby Lake
Cullaby Slough
Netul River
Lewis and Clark River
Johnson Slough
Skipanon River
Smith Lake
Slusher Lake
Camp Clatsop Military Reservation
Camp Rilea Military Reservation
Oregon State National Guard
Lands accreted after placement of Jetties
Clatsop State Forest
CLATSOP PLAINS
Oregon Coast Highway
EXPLORATIONS OF LEWIS AND CLARK 1804 - 1806
CARTOGRAPHIC RECONSTRUCTION
FORT CLATSOP WINTER
UTM ZONE 10
MAP NUMBER 378
Oregon
CONTOUR INTERVAL 100 FEET

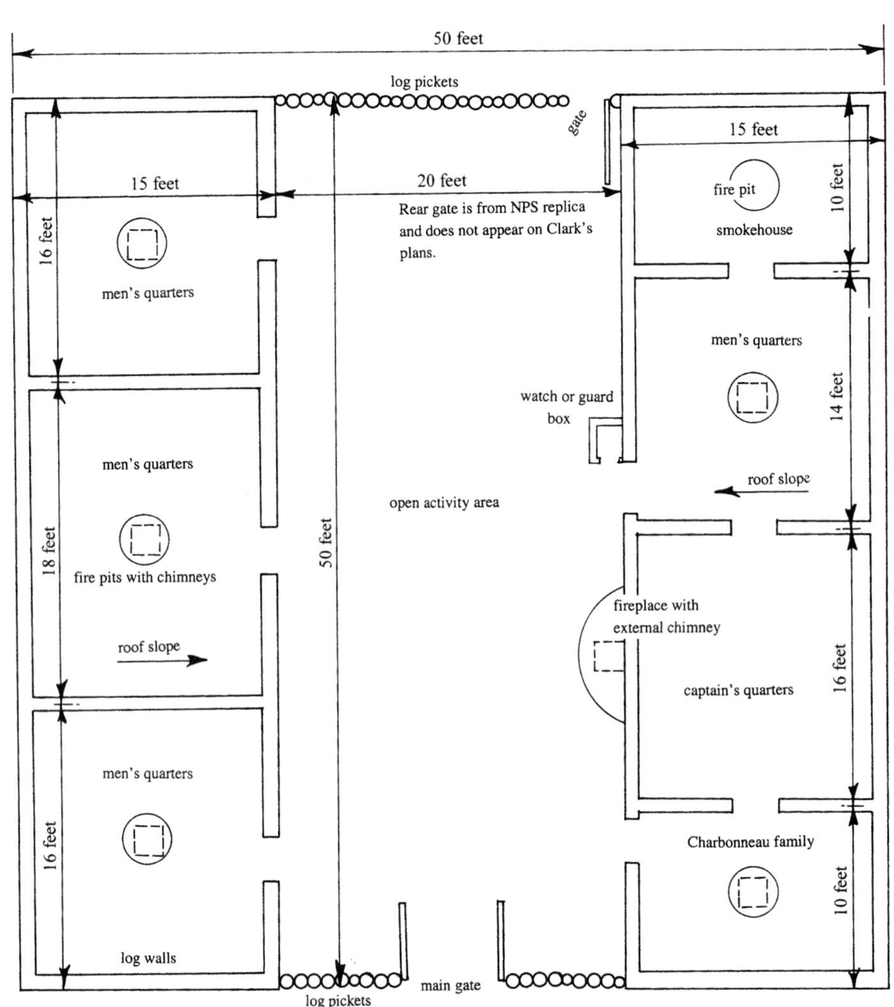

TODAY'S FORT CLATSOP REPLICA

The Oregon Historical Society acquired the Fort Clatsop site in 1901. This drawing depicts the Fort Clatsop replica built on or close to the original location by a citizens coalition during the 1955 Lewis and Clark Sesquicentennial celebration. In 1958, the replica was acquired by the National Park Service as part of the new 125-acre Fort Clatsop National Memorial.

The replica's layout is primarily based on two circa December 7, 1805, planning-sketches made by Clark—one on the inside back-cover of his elk-skin bound field book, and another on a page in the same journal interposed with text entries and a sketch of Youngs Bay. The two fort sketches are very similar, but not identical. Clark did not assign uses to the 7 rooms; thus Clark's intended interior assignments for the expedition's members are speculative.

It is important to note, however, that neither Clark nor any other journalists at the time—Gass, Ordway, and Whitehouse—made any suggestion that Clark's sketch plans actually were used in the building of Fort Clatsop (Lewis was not writing in his diary during this period). Clark's two sketches appear to have been for planning purposes only, and the actual Fort Clatsop erected by the Corps was significantly different in many aspects (see Alternate Version of Fort Clatsop). It appears that the post's actual layout was not decided upon until December 10, 1805.

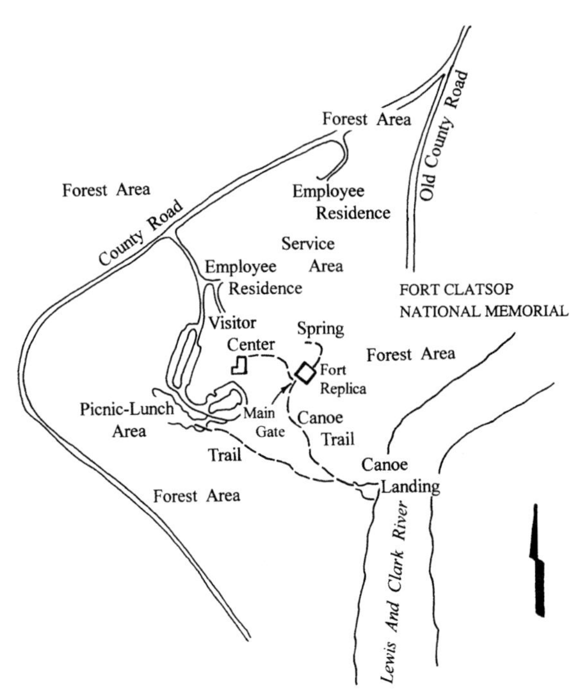

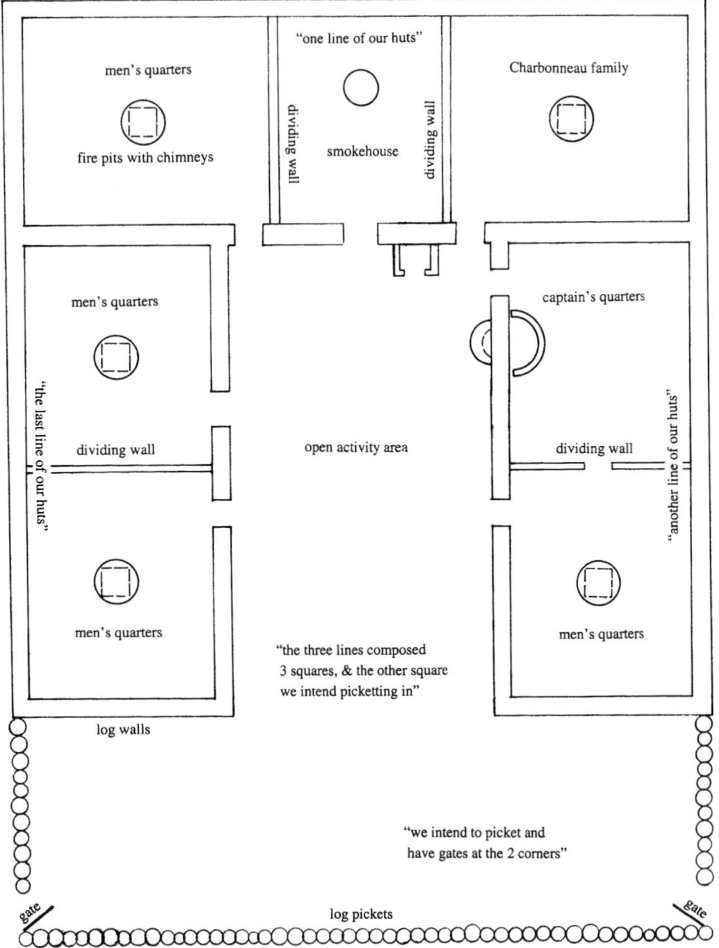

FORT CLATSOP

Construction begins, December 9, 1805
Rooms fully occupied, December 25, 1805
Stockade completed, December 30, 1805
Standing guard instituted, January 1, 1806
Evacuated for return journey, March 23, 1806
Occupied by Comowool and the Clatsop

Represented here is the best understanding of the fort's layout as described in the journals of men who actually participated in its construction— Pvt. Joseph Whitehouse, Sgt. John Ordway, and master carpenter Sgt. Patrick Gass. In regard to the post's features, the enlisted men's journals (particularly Ordway and Whitehouse) are more detailed than Clark's rather short, general remarks, though Clark did mention that the cabins had puncheon floors and bunks, and the smokehouse was the first hut to be roofed.

Fort Clatsop's basic outline is obvious from the enlisted men's accounts. However, evidence is less clear as to actual room uses, who occupied which quarters, and placements of fire pits. These, plus the doors, fireplace, and guard box, are identified here according to this cartographer's best attempted interpretation of the journal entries and U.S. Army practices at the time. The arrangements, however, are open to interpretation.

The actual positions of the two gates along the north and south stockade walls are somewhat uncertain, though the gates probably were not large since only people, and not horses, passed through them. (Lewis's January 1, 1806, detachment order indicates a "main gate" and a "water-gate.")

However, the post's "defensive" nature—an enclosed stockade-front facing the river landing 200 yards away—is clearly evident from the enlisted men's descriptions.

Pvt. Whitehouse uniquely recorded the only real description, albeit briefly, of the "completed" Fort Clatsop—he did so on March 23, 1806, as the Corps left the post for the return trip to St. Louis. (Note: the 1955 replica builders did not have access to Whitehouse's November 7, 1805–April 2, 1806 journal, rediscovered in 1966.)

December 9, 1805—
"*clearing a place for huts and a small fort*." Gass

December 10, 1805
"*Our officers concluded on to build our huts of logs, & to picket them in from the Corners*." Whitehouse

December 11, 1805
"*raised one line of our huts today*." Whitehouse

December 12, 1805—
"*finished raiseing one line of our huts*." Ordway

"*finished 3 rooms of our cabins, all but the covering*" [i.e., roofs]. Gass

December 13, 1805—
"*raised another line of our huts and began the last line of our huts forming three [sides of a] Square and 7 rooms 16 by 18 feet large. the other Square we intend to picket and have gates at the 2 corners, so as to have it a defensive fort*." Ordway

"*raised another line of our Huts. they had 2 Rooms in each hut, & were 16 feet in the clear. We finished raising the huts, & began the foundation of another line of them in the same Manner…the three lines composed 3 Squares, & the other square we intend picketting in, & to have 2 Gates at the two Corners*." Whitehouse

December 14, 1805—
"*completed the building of our huts, 7 in number*." Gass

"*finished raising the line of huts, & began to cover one of them*" [apparently the smokehouse]. Whitehouse

December 24, 1805—
"*got all our huts covered and daubed*." Gass

December 30, 1805—
"*finished puting up our pickets and gates*." Ordway

"*our fortification is Completed*." Clark

March 23, 1806—
"*Fort Clatsop is situated on the South side of Columbia River, and about 1 1/2 Miles up a small River…called by the Natives Ne-tul, and lay a small distance back, from the West bank… The fort was built in the form of an oblong Square, & the front of it facing the River, was picketed in, & had a Gate on the North & one on the South side of it*." Whitehouse

MP & GWL

EXPLORATIONS OF LEWIS AND CLARK 1804 - 1806
CARTOGRAPHIC RECONSTRUCTION

ALTERNATE VERSION OF FORT CLATSOP

TEXT SUPPLEMENT TO MAP 378, FORT CLATSOP WINTER.

December 7, 1805
"... we came to the mouth of a river about 100 yards broad, which we went up about 2 miles to the place fixed upon for winter quarters, unloaded our canoes, and carried our baggage about 200 yards to a spring, where we encamped." Gass

December 8, 1805
"... I prosue'd this gang of Elk through bogs which the wate of a man would Shake for 1/2 an Acre, and maney places I Sunk into the Mud and water up to my hips without finding any bottom on the trale of those Elk. Those bogs are covered with a kind of Moss among' which I observe an ebundance of Cramberries." Clark

December 14, 1805
"all employd in finishing a house to put meat into. all our last Supply of Elk has Spoiled in the repeted rains which has been fallen ever Since our arrival at this place, ..." Clark

December 14, 1805
"We completed the building of our huts, 7 in number, all but the covering, which I now find will not be so difficult as I expected; ..." Gass

December 25, 1805
"We had no ardent spirit of any kind among us; but are mostly in good health, A blessing, which we esteem more, than all the luxuries this life can afford, and the party are all thankful to the Supreme Being, for his goodness towards us.– hoping he will preserve us in the same, & enable us to return to the United States again in safety. We have at present nothing to eat but lean Elk meat & that without Salt, but the whole of our party are content with this fare." Whitehouse

December 26, 27, 1805
"we found that our huts Smoaked by the high winds and hard Storms... we built backs and enside chimneys in our huts which made them much more comfortable..." Ordway

December 28, 1805
"Derected... Jos. Fields, Bratton, Gibson to proceed to the Ocean at some conveneint place form a Camp and Commence makeing Salt with 5 of the largest Kittles, and Willard and Wiser to assist them in carrying the Kittles to the Sea Coast." Clark

December 29, 1805
"we were informed day before yesterday that a whale had foundered on the coast to the S.W. near the Kil a mox N. and that the greater part of the Clatsops were gorn for the oile & blubber, ..." Clark

December 30, 1805
"I saw flies & different kinds of insects in motion to day Snakes are yet to be seen, and Snales without cover is common and large, ..." Clark

January 1, 1806
"This morning I was awoke at an early hour by the discharge of a volley of small arms, which were fired by our party in front of our quarters to usher in the new year; this was the only mark of rispect which we had it in our power to pay this celebrated day." Lewis

January 11, 1806
"... This morning the Sergt. of the guard reported that our Indian Canoe had gone a Drift, ... the tide... had taken her off; ..." Clark

January 12, 1806
"The men who were sent in surch of the canoe returned without being able to find her, we therefore give her over as lost. ..." Lewis

January 13, 1806
"this evening we exhausted the last of our candles, but fortunately had taken the precaution to bring with us moulds and wick, by means of which and some Elk's tallow in our possession we do not yet consider ourslves destitute of this necessary article;" Lewis

January 23, 1806
"The men of the garison are still busily employed in dressing Elk's skins for cloathing, they find great difficulty for the want of branes; we have not soap to supply the deficiency, nor can we procure ashes to make the lye; none of the pines which we use for fuel affords any ashes; extrawdinary as it may seem, the greene wood is consoomed without leaving the residium of a particle of ashes." Lewis

January 25, 1806
"The morning was cloudy and some showers of snow fell in the course of the day; and in the night it fell to the depth of 8 inches. ... This is the first freezing weather of any consequence we have had during the winter." Gass

January 28, 1806
"The party that were sent... for the Elk returned in the evening with three of them only; the Elk had been killed just before the snow fell which had covered them and so altered the apparent face of the country that the hunters could not find the Elk which they had killed." Lewis

February 3, 1806
"About three o'clock Drewyer [and] La Page, returned; Drewyer had killed seven Elk in the point below us, several miles distant but can be approached with in 3/4 of a mile with canoes... we are apprehensive that the Clatsops who know where the meat is will rob us of a part if not the whole of it. ..." Lewis

February 5, 1806
Late this evening one of the hunters fired his gun over the swamp of the Netul opposite to the fort and hooped. I sent sergt. Gass and a party of men over; ... in their way fortunately recovered our Indian Canoe, so long lost and much lamented." Lewis

February 6, 1806
"Late in the evening Sergt. Pryor returned with the flesh of about 2 Elk and 4 skins the Indians having purloined the ballance of seven Elk which Drewyer killed the other day." Lewis

February 7, 1806
"The Small pox had distroyed a great number of the nativs in this quarter. it provailed about 4 or 5 yrs. sin[c]e among the Clatsops, and distroy'd several hundreds of them, ...I think the late ravages of the small pox, may well account for the number of remains of Villages which I saw on my rout to the Kilamox in several places." Clark

February 10, 1806
"Willard arrived late in the evening from the Saltworks, had cut his knee very badly... he informed us that Bratton was very unwell, and that Gibson was so sick that he could not set up or walk alone..." Lewis

February 12, 1806
"This morning we were visited by a Clatsop man who brought with him three dogs as a remuneration for the Elk which himself and nation had stolen from us... however the dogs took the alarm and ran off; ..." Lewis

February 14, 1806
"We are very uneasy with rispect to our sick men at the salt works." Lewis

February 17, 1806
"... Joseph Fields arrived from the Salt works and informed us that they had about 2 Kegs of salt on hand which with what we have at this place we suppose will be sufficient to last us to our deposits of that article on the Missouri. we there[fore] directed a party of six men to go with Fields in the morning in order to bring the salt and kettles to the fort." Lewis

March 1, 1806
"There is a large river that flows into the southeast part of Hailey's Bay; upon which about 20 miles from its mouth, our hunters discovered falls, which had about 60 feet of a perpendicular pitch." Gass

March 3, 1806
"Two of our perogues have been lately injured very much in consequence of the tide leaving them partially on shore. they split by this means with their own weight." Lewis

March 5, 1806
"This morning we were visited by two parties of Clatsops. ... the hunters returned... They had neither killed nor seen any Elk. they informed us that the Elk had all gone off to the mountains a considerable distance from us. this is unwelcome information and reather allarming we have only 2 days provision on hand, and that nearly spoiled. ... if we find that the Elk have left us, we have determined to ascend the river slowly and indeavour to procure subsistence on the way, consuming the Month of March in the woody country." Lewis

March 6, 1806
"Hall had his foot and ankle much injured yesterday by the fall of a large stick of timber; the bones were fortunately not broken... Bratton is now weaker..." Lewis Lewis

March 7, 1806
Bratton is much wo[r]se today, he complains of a violent pain in the small of his back and is unable in consequence to set up." Lewis

March 9, 1806
"Bratton complains of his back being very painfull to him today; I conceive this pain to be something of the rheumatism." Lewis

March 10, 1806
"The hunters who were over the Netull the other day informed us that they measured a pine tree, (or fir...) which at the hight of a man's breast was 42 feet in the girth; about three feet higher, or as high as a tall man could reach, it was 40 feet in the girth which was about the circumpherence for at least 200 feet without a limb, and that it was very lofty above the commencement of the limbs. from the appearance of other trees of this speceis of fir and their account of this tree, I think it may be safely estimated at 300 feet. it had every appearance of being perfectly sound." Lewis

March 12, 1806
"Our party are now furnished with 358 par of Mockersons exclusive of a good portion of Dressed leather, they are also previded with shirts overalls capoes of dressed Elk skins for the homeward journey." Clark

March 16, 1806
"two handkerchiefs would now contain all the small articles of merchandize which we possess; the ballance of the stock consists of 6 blue robes one scarlet do. one uniform artillerist's coat and hat, five robes made of our large flag, and a few old cloaths trimed with ribbon." Lewis

March 18, 1806
"4 men went over to the prarie near the coast to take a canoe which belongd to the Clatsop Indians, as we are in want of it. ... took the canoe near the fort and concealed it, as the chief of the Clatsops is now here." Ordway

March 19, 1806
"It continued to rain and hail today in such manner that nothing further could be done to the canoes." Lewis

March 21, 1806
"we directed Drewyer and the Fieldses to set out tomorrow morning early, and indevour to provide us some provision on the bay beyond point William. ... Bratton is now so much reduced that I am somewhat uneasy with rispect to his recovery; ..." Lewis

March 23, 1806
"... distributed the baggage and directed the canoes to be launched and loaded for our departure. at 1 P.M. we bid a final adieu to Fort Clatsop." Lewis

EXPLORATIONS OF LEWIS AND CLARK 1804 - 1806
CARTOGRAPHIC RECONSTRUCTION

FORT CLATSOP WINTER (CONTINUED)

UTM ZONE 10
MAP NUMBER 378A

Oregon

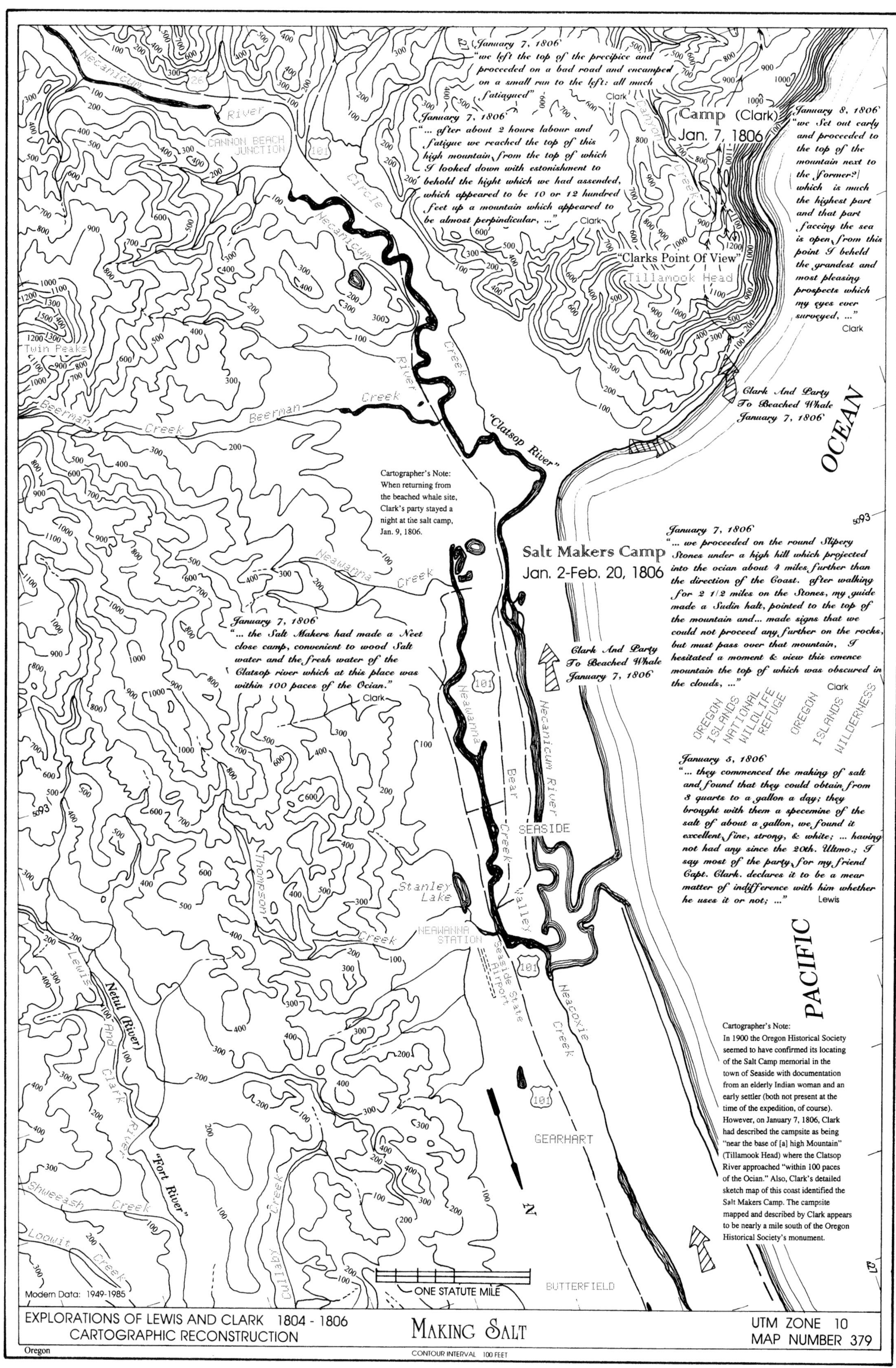

"January 7, 1806"
"we left the top of the precipice and proceeded on a bad road and encamped on a small run to the left: all much fatiagued"
Clark
"January 7, 1806"
"... after about 2 hours labour and fatigue we reached the top of this high mountain, from the top of which I looked down with estonishment to behold the hight which we had assended, which appeared to be 10 or 12 hundred feet up a mountain which appeared to be almost perpindicular, ..."
Clark
Camp (Clark)
Jan. 7, 1806
"January 8, 1806"
"we Set out early and proceeded to the top of the mountain next to the [former?] which is much the highest part and that part faceing the sea is open, from this point I beheld the grandest and most pleasing prospects which my eyes ever surveyed, ..."
Clark
"Clarks Point Of View"
Tillamook Head
Clark And Party To Beached Whale January 7, 1806
OCEAN
PACIFIC
Cartographer's Note:
When returning from the beached whale site, Clark's party stayed a night at the salt camp, Jan. 9, 1806.
Salt Makers Camp
Jan. 2-Feb. 20, 1806
"January 7, 1806"
"... we proceeded on the round Slipery Stones under a high hill which projected into the ocian about 4 miles further than the direction of the Coast. after walking for 2 1/2 miles on the Stones, my guide made a Sudin halt, pointed to the top of the mountain and... made signs that we could not proceed any further on the rocks, but must pass over that mountain, I hesitated a moment & view this emence mountain the top of which was obscured in the clouds, ..."
Clark
"January 7, 1806"
"... the Salt Makers had made a Neet close camp, convenient to wood Salt water and the fresh water of the Clatsop river which at this place was within 100 paces of the Ocian."
Clark
Clark And Party To Beached Whale January 7, 1806
"January 5, 1806"
"... they commenced the making of salt and found that they could obtain from 3 quarts to a gallon a day; they brought with them a specemine of the salt of about a gallon, we found it excellent, fine, strong, & white; ... having not had any since the 20th. Ultmo.; I say most of the party, for my friend Capt. Clark. declares it to be a mear matter of indifference with him whether he uses it or not; ..."
Lewis
OREGON ISLANDS NATIONAL WILDLIFE REFUGE
OREGON ISLANDS WILDERNESS
Cartographer's Note:
In 1900 the Oregon Historical Society seemed to have confirmed its locating of the Salt Camp memorial in the town of Seaside with documentation from an elderly Indian woman and an early settler (both not present at the time of the expedition, of course). However, on January 7, 1806, Clark had described the campsite as being "near the base of [a] high Mountain" (Tillamook Head) where the Clatsop River approached "within 100 paces of the Ocian." Also, Clark's detailed sketch map of this coast identified the Salt Makers Camp. The campsite mapped and described by Clark appears to be nearly a mile south of the Oregon Historical Society's monument.
Necanicum River
Cannon Beach Junction
Circle Creek
Necanicum River
Canyon Creek
Twin Peaks
Beerman Creek
"Clatsop River"
Neawanna Creek
Neawanna Creek
Necanicum River
Bear Creek Valley
Seaside
Stanley Lake
Thompson Creek
Neawanna Station
Seaside State Airport
Neacoxie Creek
Lewis And Clark River
Netul (River)
"Fort River"
Shweeash Creek
Loowit Creek
Cullaby Creek
Gearhart
ONE STATUTE MILE
Butterfield
Modern Data: 1949-1985
EXPLORATIONS OF LEWIS AND CLARK 1804 - 1806
CARTOGRAPHIC RECONSTRUCTION
MAKING SALT
UTM ZONE 10
MAP NUMBER 379
Oregon
CONTOUR INTERVAL 100 FEET

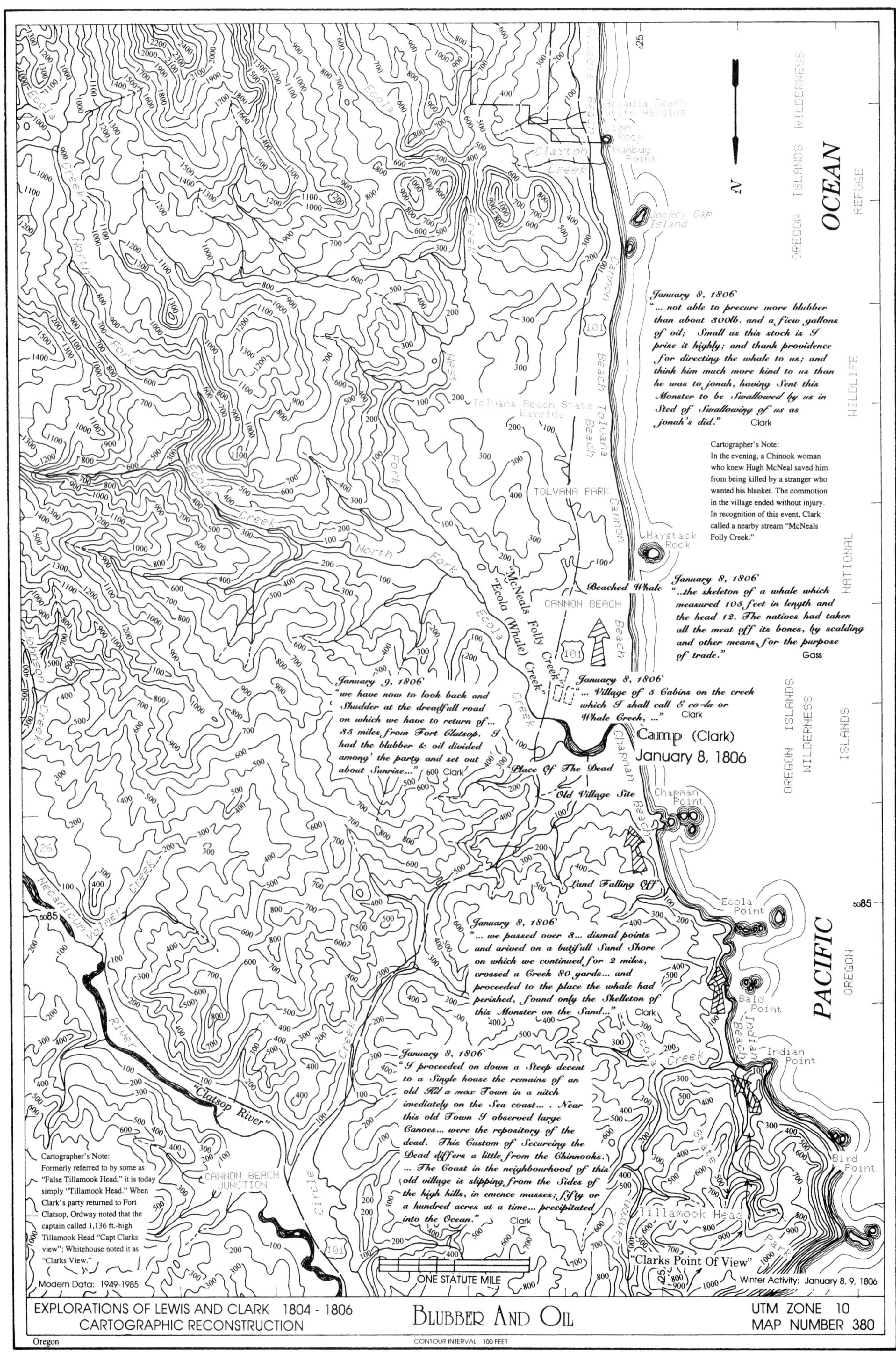

OCEAN
PACIFIC
OREGON ISLANDS WILDERNESS
REFUGE
WILDLIFE
NATIONAL
ISLANDS
OREGON
N
Arcadia Beach State Wayside
Lion Rock
Humbug Point
Clayton Creek
Jockey Cap Island
Cannon Beach
Tolvana Beach
Tolvana Beach State Wayside
TOLVANA PARK
Haystack Rock
CANNON BEACH
Chapman Beach
Chapman Point
Ecola Point
Bald Point
Indian Beach
Indian Point
Bird Point
Tillamook Head
Ecola Creek
North Fork
West Fork
Ecola (Whale) Creek
"McNeals Folly Creek"
Johnson Creek
Necanicum River
Walmer Creek
"Clatsop River"
CANNON BEACH JUNCTION
Circle Creek
Canyon Creek
State Park
January 8, 1806
"... not able to precure more blubber than about 300lb. and a fiew gallons of oil; Small as this stock is I prise it highly; and thank providence for directing the whale to us; and think him much more kind to us than he was to jonah, having Sent this Monster to be Swallowed by us in Sted of Swallowing of us as jonah's did." Clark
Cartographer's Note:
In the evening, a Chinook woman who knew Hugh McNeal saved him from being killed by a stranger who wanted his blanket. The commotion in the village ended without injury. In recognition of this event, Clark called a nearby stream "McNeals Folly Creek."
Beached Whale
January 8, 1806
"..the skeleton of a whale which measured 105 feet in length and the head 12. The natives had taken all the meat off its bones, by scalding and other means, for the purpose of trade." Gass
January 8, 1806
"... Village of 5 Cabins on the creek which I shall call E co-la or Whale Creek, ..." Clark
Camp (Clark)
January 8, 1806
January 9, 1806
"we have now to look back and Shudder at the dreadfull road on which we have to return of... 35 miles from Fort Clatsop. I had the blubber & oil divided among' the party and set out about Sunrise..." Clark
Place Of The Dead
Old Village Site
Land Falling Off
January 8, 1806
"... we passed over 3... dismal points and arived on a butifull Sand Shore on which we continued for 2 miles, crossed a Creek 80 yards... and proceeded to the place the whale had perished, found only the Skelleton of this Monster on the Sand..." Clark
January 8, 1806
"I proceeded on down a Steep decent to a Single house the remains of an old Kil a mox Town in a nitch imediately on the Sea coast... . Near this old Town I observed large Canoes... were the repository of the dead. This Custom of Secureing the Dead differs a little from the Chinnooks. ... The Coast in the neighbourhood of this old village is slipping from the Sides of the high hills, in emence masses; fifty or a hundred acres at a time... precipitated into the Ocean." Clark
Cartographer's Note:
Formerly referred to by some as "False Tillamook Head," it is today simply "Tillamook Head." When Clark's party returned to Fort Clatsop, Ordway noted that the captain called 1,136 ft.-high Tillamook Head "Capt Clarks view"; Whitehouse noted it as "Clarks View."
"Clarks Point Of View"
ONE STATUTE MILE
Modern Data: 1949-1985
Winter Activity: January 8, 9, 1806
EXPLORATIONS OF LEWIS AND CLARK 1804 - 1806
CARTOGRAPHIC RECONSTRUCTION
BLUBBER AND OIL
UTM ZONE 10
MAP NUMBER 380
Oregon
CONTOUR INTERVAL 100 FEET

The Way Home—The Best Route
Alternate Return Routes No. 1–7

THE BEST WAY WEST

"I compleated a map *of the Countrey through which we have been passing from the Mississippi at the Mouth of Missouri to this place [Fort Clatsop]...[T]he most practicable and navigable passage across the Continent of North America...is that which we have traveled with the exception of that part of our rout from the foot of the Falls of the Missouri,...untill we arrive....at the enterance of* Travelers-rest *Creek[.]"*

Captain William Clark, February 14, 1806

While at Fort Clatsop, Clark completed his new cartographic work, including a master map covering the entire westbound journey. By February 14, 1806, after analyzing their notes, sketches, and cartography of the 1804–5 trek, the captains opinioned that they had followed the best possible route across this part of the continent, but with one notable exception. They had not utilized a more northerly overland shortcut via the Sun and Blackfoot rivers between the Missouri's Great Falls and Travelers Rest (near present-day Missoula, Montana). Both plains and mountain tribes had told the explorers about this easier, more direct path. The captains believed this link across the continental divide would be part of the "best and most practicable rout" that they had explored in the West, though it was anything but an all-water route as originally hoped for. The previous year's southwestward diversion to the Three Forks, Beaverhead River, and Lemhi Pass, and then north through the Salmon-Bitterroot country was over 500 miles long and fraught with more obstacles and difficulties than the shorter route now contemplated by the captains.

Consequently, to accomplish several goals on the return journey—including separate investigations of the upper Marias and Yellowstone rivers, as well as taking the shorter trail described by the Indians—the captains decided to divide the Corps upon their summertime return to Travelers Rest in the Bitterroot country. Lewis and 9 men would follow the Indian-recommended shortcut northeast up the Blackfoot River, over the continental divide in today's Lewis and Clark Pass locality, and then across the plains to the Great Falls. Then, as part of the men retrieved cached supplies, equipment, and the white pirogue at the falls, Lewis would lead a small detachment north to probe the upper Marias and determine its relationship to Canadian rivers.

Clark, meanwhile, with two-thirds of the party, including York and the Charbonneau family, would proceed southeast from Travelers Rest, retracing last year's route in the Bitterroot Valley before taking a new, more direct route through the Big Hole valley to retrieve canoes and cached supplies at Camp Fortunate on the Beaverhead River. From there, the combined horse-canoe party would continue down the Beaverhead to Three Forks, where Clark and 12 others, including York and the Charbonneaus, would travel east by horseback to strike the Yellowstone River at today's Livingston, Montana, and explore down to its mouth. Meanwhile, a canoe party of 10 men under Sergeant Ordway would paddle from the Three Forks down the Missouri to the Great Falls, joining Lewis's men in opening caches and portaging canoes and equipment around the falls. When Lewis returned from the upper Marias, the combined Great Falls party of 20 men then would float down the Missouri, intending to meet Clark and his detachment at the mouth of the Yellowstone.

In the spring of 1806, after leaving Fort Clatsop and proceeding up the Columbia into the interior, the Walla Walla tribe would also inform the Corps about another shortcut, in this case across today's southeast Washington. This overland route would bypass a difficult and considerably longer trek up the Snake River. Other short new sections of trail would be taken in Nez Perce country as well.

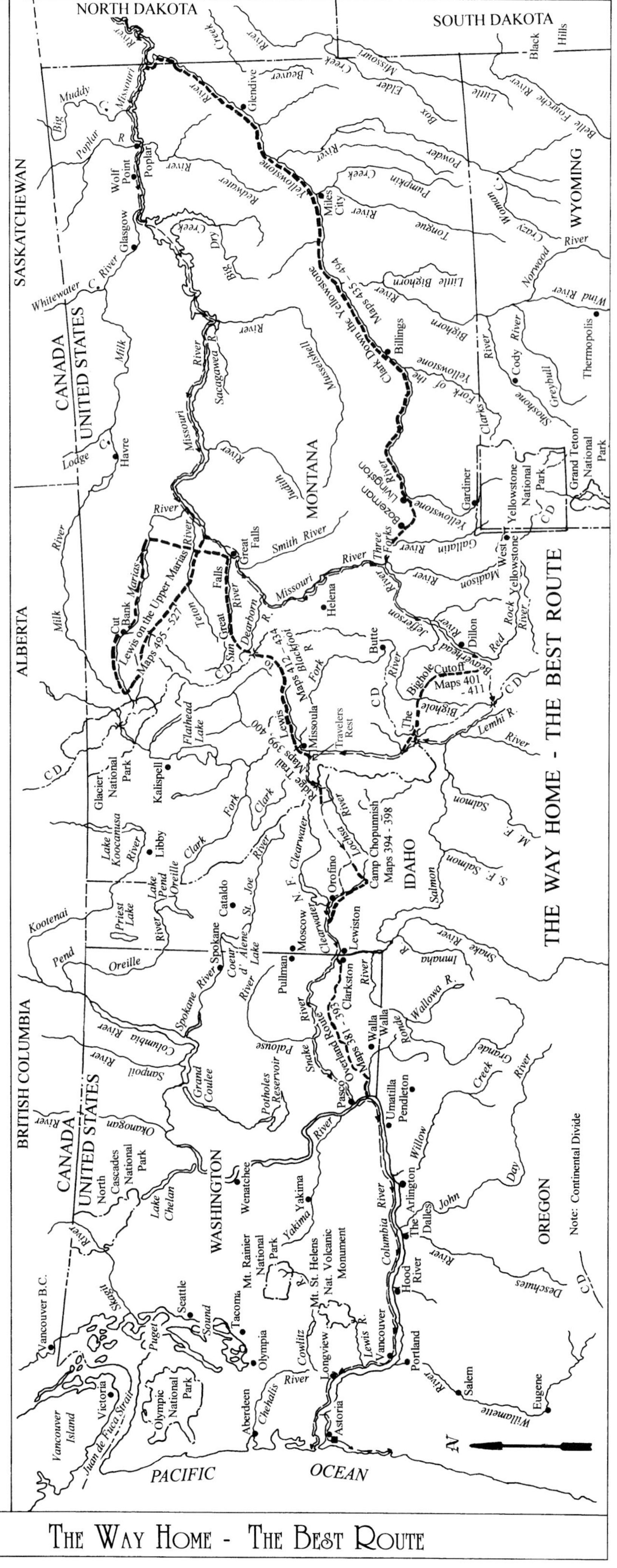

EXPLORATIONS OF LEWIS AND CLARK 1804 - 1806
CARTOGRAPHIC RECONSTRUCTION

THE WAY HOME - THE BEST ROUTE

Alternate Return Route No. 1

MAPS 381 - 393

MAIN PARTY, NEZ PERCE TRAIL,
SOUTHEAST WASHINGTON

Walla Walla River, April 30, 1806—
- Captains Meriwether Lewis and William Clark
- Interpreters George Drouillard and Toussaint Charbonneau
- Sacagawea and Jean Baptiste Charbonneau
- York
- 26 enlisted men
- 2 "Chopunnish" guides (1 with family)
- 23 expedition horses

COLUMBIA RIVER HAZARDS

When the Corps had begun portaging around the Cascades Rapids in the Columbia gorge on April 11, 1806, Lewis observed that the river was "upwards of 20 feet higher than" in the previous fall when they were headed west. The rapids now were one great continuous torrent, 7 miles in length.

Aware of springtime's awesome water hazards, the men looked with dread toward the struggles and portages that might yet lay ahead on the Columbia River.

Consequently, the captains had decided to trade their canoes for horses when it became practicable to do so on the upriver journey. Bartering with the natives, however, initially proved more difficult than expected, but within a few days in The Dalles area they began acquiring the horses they wanted. Canoes no longer needed or not traded were split up for firewood in the treeless region east of the Cascade Range. As the expedition's growing horse party proceeded along Indian trails on the Columbia's north shore, they often saw their companions in the remaining canoes, moving against the current.

On April 24, several miles above the mouth of the John Day River, the last two canoes were broken up and the entire party then proceeded by land with the horses.

THE NEZ PERCE TRAIL CUTOFF

On April 27, when encamped with Chief Yellepit's band at the Columbia-Walla Walla confluence, the explorers were told about a shortcut to Nez Perce country across the prairie lands south of the Snake River. This would avoid the long, circuitous route along the lower Snake followed in the previous autumn, saving about 80 miles of traveling. The natives assured them that it was a fine road, starting from the north bank of the Walla Walla River opposite to Yellepit's village.

Consequently, on April 30, 1806, the expedition set out with 23 horses along the clearly defined trail with two "Chopunnish" guides, one of whom had a wife and children with him.

Traveling fast was a concern for the captains. They hoped to reach the Nez Perces before the latter began crossing the mountains to buffalo country. In particular, the Corps wished to retrieve their horses left for the winter along the Clearwater, and to procure guides before attempting to cross the Bitterroot Range—mountains that caused them so much trouble, discomfort, and fear the year before.

The Corps' journey over the prairie shortcut to the Snake-Clearwater confluence required five days (April 30 to May 4, 1806). Known later by mid-nineteenth-century frontiersmen as the Nez Perce or Overland Trail, this key network of paths across southeast Washington had been well worn by thousands of natives travelers, century after century. The route mostly followed flat or gentle-sloped stream bottoms and ridgelines, but, when intersecting steep hills, the trail usually led straight up and down, without switchbacks.

Today, modern agricultural crops—particularly dry-land wheat farming—have replaced most of the region's original prairie grasses, wild flowers, and low shrubs. Consequently, most traces of the Nez Perce trail have been plowed under, except, of course, in steep or difficult terrain that is inaccessible to tractors and harvesting machines. In these latter places, isolated remnants of the original route can still be seen, worn into the rich loess soil.

A number of local researchers have attempted to retrace the Nez Perce Trail and its variants, using written pioneer accounts, early Government Land Office maps, and other historical sources, and by conducting on-the-ground searches. Four persons, in particular, have achieved success in this endeavor: Walt Gary, a retired WSU Agricultural Extension Agent; Gary Lentz of nearby Lewis and Clark Trail State Park; Steve Plucker, a Walla Walla County farmer; and George Touchette, a former Columbia County commissioner. Through their diligence, virtually every remaining visible section of the route has been located.

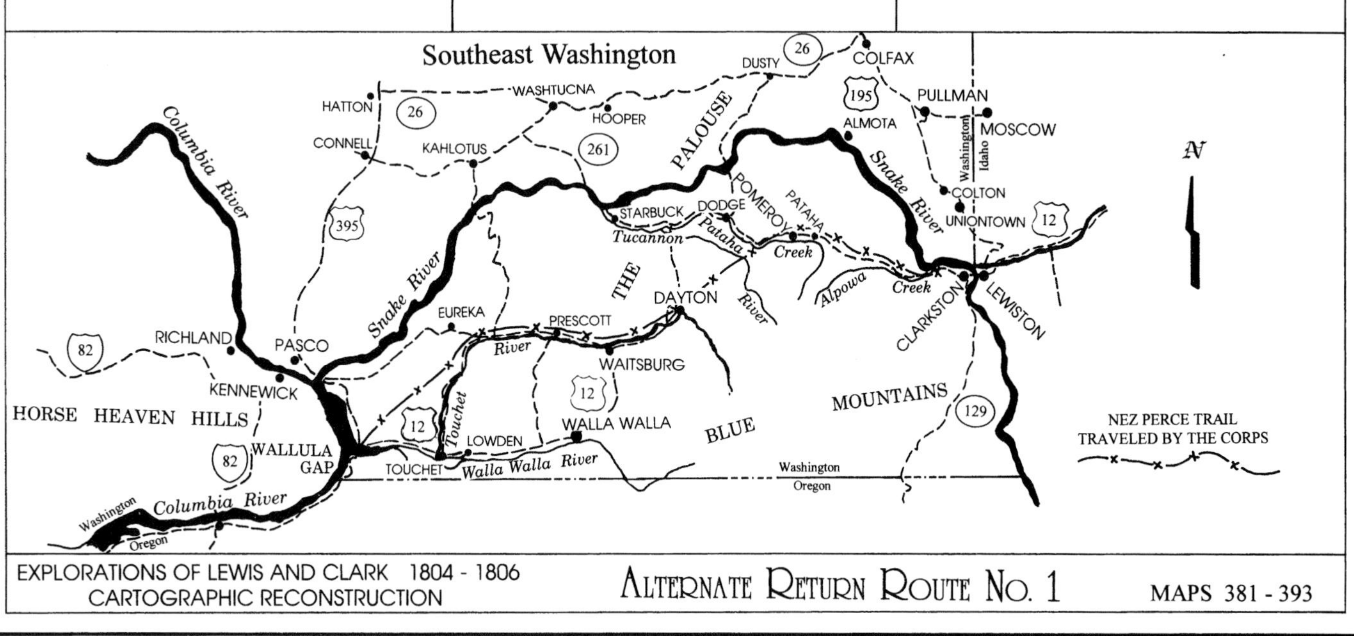

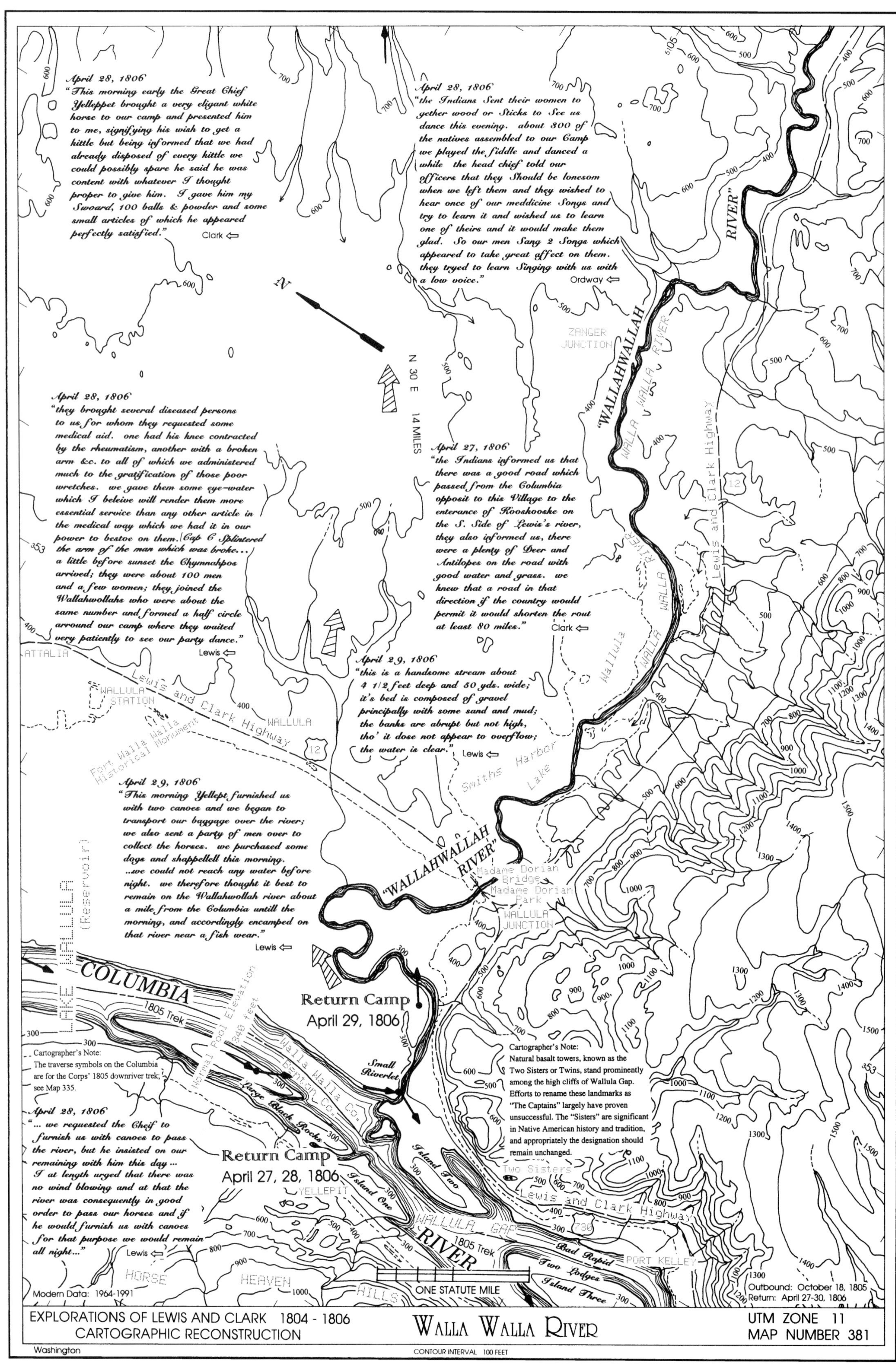
April 28, 1806
"This morning early the Great Chief Yelleppet brought a very eligant white horse to our camp and presented him to me, signifying his wish to get a kittle but being informed that we had already disposed of every kittle we could possibly spare he said he was content with whatever I thought proper to give him. I gave him my Sword, 100 balls & powder and some small articles of which he appeared perfectly satisfied."
Clark ⇦
April 28, 1806
"the Indians Sent their women to gether wood or Sticks to See us dance this evening. about 300 of the natives assembled to our Camp we played the fiddle and danced a while the head chief told our officers that they Should be lonesom when we left them and they wished to hear once of our meddicine Songs and try to learn it and wished us to learn one of theirs and it would make them glad. So our men Sang 2 Songs which appeared to take great affect on them. they tryed to learn Singing with us with a low voice."
Ordway ⇦
N
N 30 E 14 MILES
ZANGER JUNCTION
"WALLAHWALLAH RIVER"
WALLA WALLA RIVER
April 28, 1806
"they brought several diseased persons to us for whom they requested some medical aid. one had his knee contracted by the rheumatism, another with a broken arm &c. to all of which we administered much to the gratification of those poor wretches. we gave them some eye-water which I beleive will render them more essential service than any other article in the medical way which we had it in our power to bestoe on them. [Cap C Splintered the arm of the man which was broke... a little before sunset the Chymnahpos arrived; they were about 100 men and a few women; they joined the Wallahwollahs who were about the same number and formed a half circle arround our camp where they waited very patiently to see our party dance."
Lewis ⇦
April 27, 1806
"the Indians informed us that there was a good road which passed from the Columbia opposit to this Village to the enterance of Kooskooske on the S. Side of Lewis's river, they also informed us, there were a plenty of Deer and Antilopes on the road with good water and grass. we knew that a road in that direction if the country would permit it would shorten the rout at least 80 miles."
Clark ⇦
Lewis and Clark Highway
12
ATTALIA
WALLULA STATION
WALLULA
Fort Walla Walla Historical Monument
April 29, 1806
"this is a handsome stream about 4 1/2 feet deep and 50 yds. wide; it's bed is composed of gravel principally with some sand and mud; the banks are abrupt but not high, tho' it dose not appear to overflow; the water is clear."
Lewis ⇦
Wallula
Smiths Harbor Lake
April 29, 1806
"This morning Yellept furnished us with two canoes and we began to transport our baggage over the river; we also sent a party of men over to collect the horses. we purchased some dogs and shappellell this morning. ...we could not reach any water before night. we therfore thought it best to remain on the Wallahwollah river about a mile from the Columbia untill the morning, and accordingly encamped on that river near a fish wear."
Lewis ⇦
"WALLAHWALLAH RIVER"
Madame Dorian Bridge
Madame Dorian Park
WALLULA JUNCTION
LAKE WALLULA (Reservoir)
COLUMBIA
1805 Trek
Normal Pool Elevation 340 feet
Return Camp
April 29, 1806
Walla Walla Co.
Benton Co.
Small Riverlet
Large Black Rocks
Cartographer's Note:
The traverse symbols on the Columbia are for the Corps' 1805 downriver trek; see Map 335.
April 28, 1806
"... we requested the Cheif to furnish us with canoes to pass the river, but he insisted on our remaining with him this day ... I at length urged that there was no wind blowing and at that the river was consequently in good order to pass our horses and if he would furnish us with canoes for that purpose we would remain all night..."
Lewis ⇦
Return Camp
April 27, 28, 1806
YELLEPIT
Island One
Island Two
Cartographer's Note:
Natural basalt towers, known as the Two Sisters or Twins, stand prominently among the high cliffs of Wallula Gap. Efforts to rename these landmarks as "The Captains" largely have proven unsuccessful. The "Sisters" are significant in Native American history and tradition, and appropriately the designation should remain unchanged.
Two Sisters
Lewis and Clark Highway
730
WALLULA GAP
RIVER
1805 Trek
Bad Rapid
Two Lodges
Island Three
PORT KELLEY
HORSE HEAVEN HILLS
ONE STATUTE MILE
Modern Data: 1964-1991
Outbound: October 18, 1805
Return: April 27-30, 1806
EXPLORATIONS OF LEWIS AND CLARK 1804 - 1806
CARTOGRAPHIC RECONSTRUCTION
Washington
WALLA WALLA RIVER
CONTOUR INTERVAL 100 FEET
UTM ZONE 11
MAP NUMBER 381

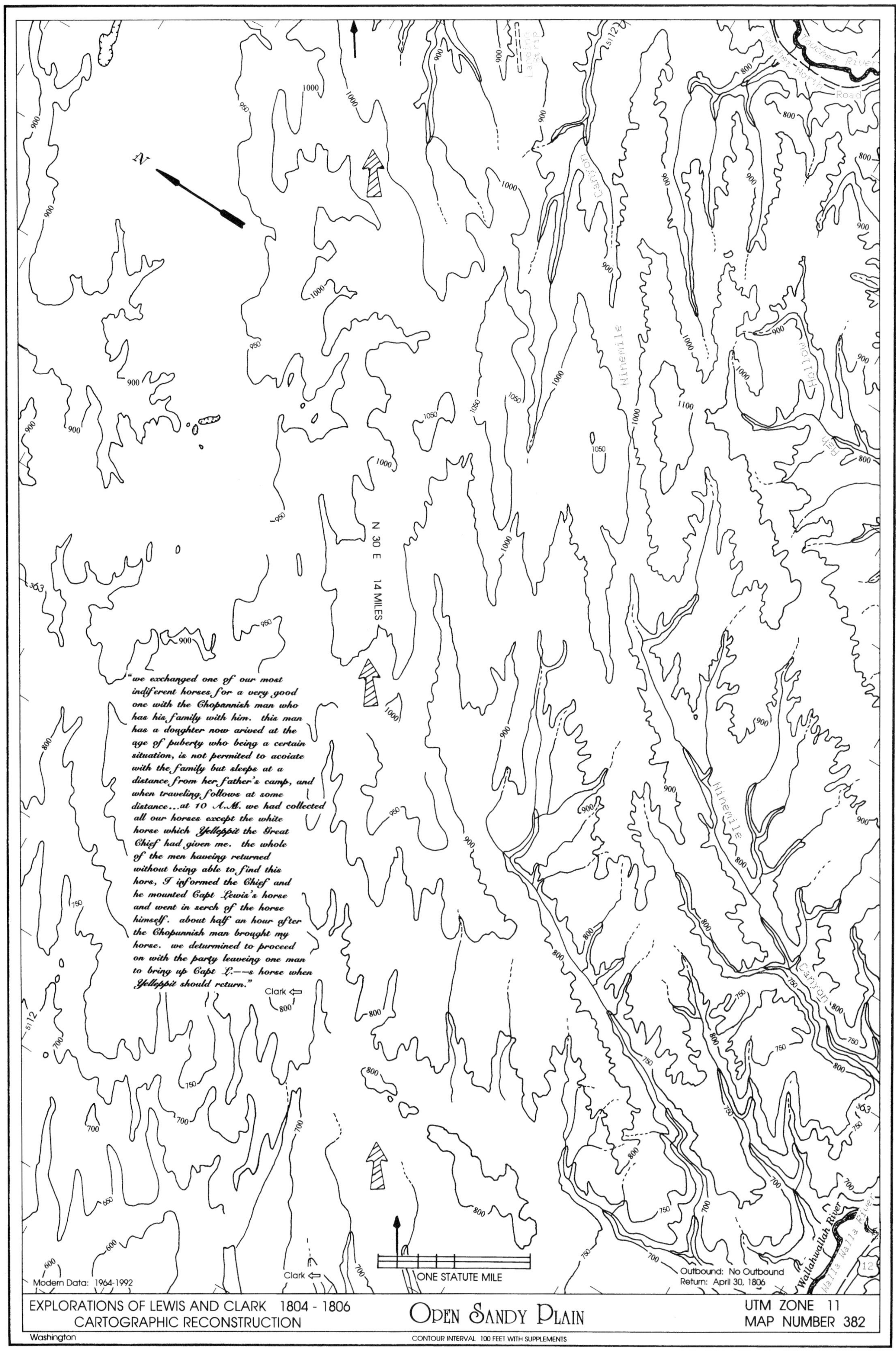
"we exchanged one of our most indiferent horses for a very good one with the Chopannish man who has his family with him. this man has a doughter now arived at the age of puberty who being a certain situation, is not permited to acoiate with the family but sleeps at a distance from her father's camp, and when traveling follows at some distance...at 10 A.M. we had collected all our horses except the white horse which Yelleppit the Great Chief had given me. the whole of the men haveing returned without being able to find this hors, I informed the Chief and he mounted Capt Lewis's horse and went in serch of the horse himself. about half an hour after the Chopunnish man brought my horse. we deturmined to proceed on with the party leaveing one man to bring up Capt L.--s horse when Yellepit should return."
N 30 E 14 MILES
Clark
Touchet River
Touchet-North Road
Landing Strip
Canyon
Ninemile
Ninemile
Hollow
Canyon
Walla Walla River
Wallahwallah River
Outbound: No Outbound
Return: April 30, 1806
ONE STATUTE MILE
Modern Data: 1964-1992
EXPLORATIONS OF LEWIS AND CLARK 1804 - 1806
CARTOGRAPHIC RECONSTRUCTION
OPEN SANDY PLAIN
UTM ZONE 11
MAP NUMBER 382
Washington
CONTOUR INTERVAL 100 FEET WITH SUPPLEMENTS

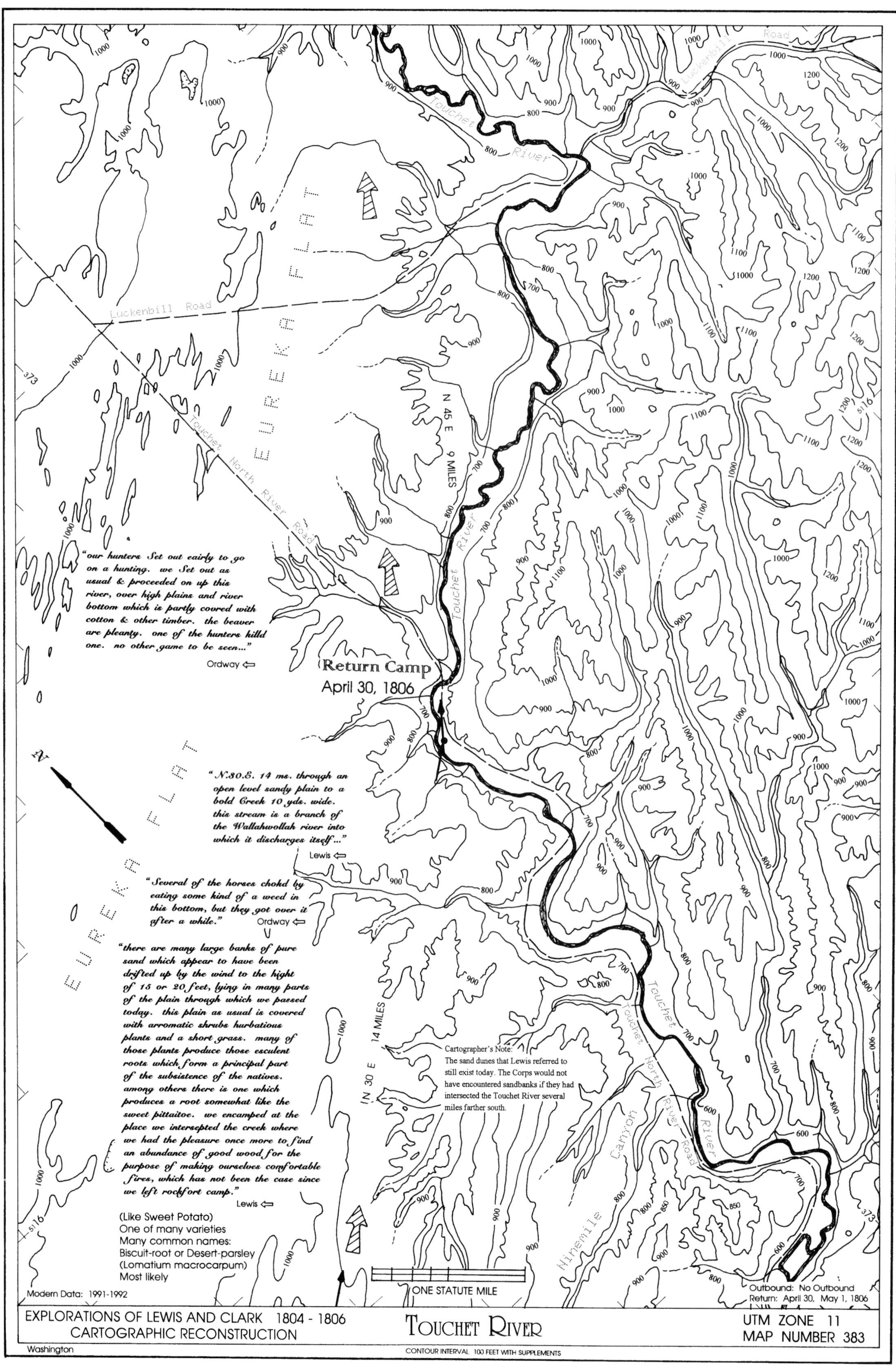

"our hunters Set out eairly to go on a hunting. we Set out as usual & proceeded on up this river, over high plains and river bottom which is partly coured with cotton & other timber. the beaver are pleanty. one of the hunters killd one. no other game to be seen..."
Ordway
Return Camp
April 30, 1806
"N.30.E. 14 ms. through an open level sandy plain to a bold Creek 10 yds. wide. this stream is a branch of the Wallahwollah river into which it discharges itself..."
Lewis
"Several of the horses chokd by eating some kind of a weed in this bottom, but they got over it after a while."
Ordway
"there are many large banks of pure sand which appear to have been drifted up by the wind to the hight of 15 or 20 feet, lying in many parts of the plain through which we passed today. this plain as usual is covered with arromatic shrubs hurbatious plants and a short grass. many of those plants produce those esculent roots which form a principal part of the subsistence of the natives. among others there is one which produces a root somewhat like the sweet pittaitoe. we encamped at the place we intersepted the creek where we had the pleasure once more to find an abundance of good wood for the purpose of making ourselves comfortable fires, which has not been the case since we left rockfort camp."
Lewis
(Like Sweet Potato)
One of many varieties
Many common names:
Biscuit-root or Desert-parsley
(Lomatium macrocarpum)
Most likely
Cartographer's Note:
The sand dunes that Lewis referred to still exist today. The Corps would not have encountered sandbanks if they had intersected the Touchet River several miles farther south.
EUREKA FLAT
Luckenbill Road
Touchet North River Road
Touchet River
Nine Mile Canyon
N 45 E 9 MILES
N 30 E 14 MILES
ONE STATUTE MILE
Modern Data: 1991-1992
Outbound: No Outbound
Return: April 30, May 1, 1806
EXPLORATIONS OF LEWIS AND CLARK 1804 - 1806
CARTOGRAPHIC RECONSTRUCTION
Washington
TOUCHET RIVER
CONTOUR INTERVAL 100 FEET WITH SUPPLEMENTS
UTM ZONE 11
MAP NUMBER 383

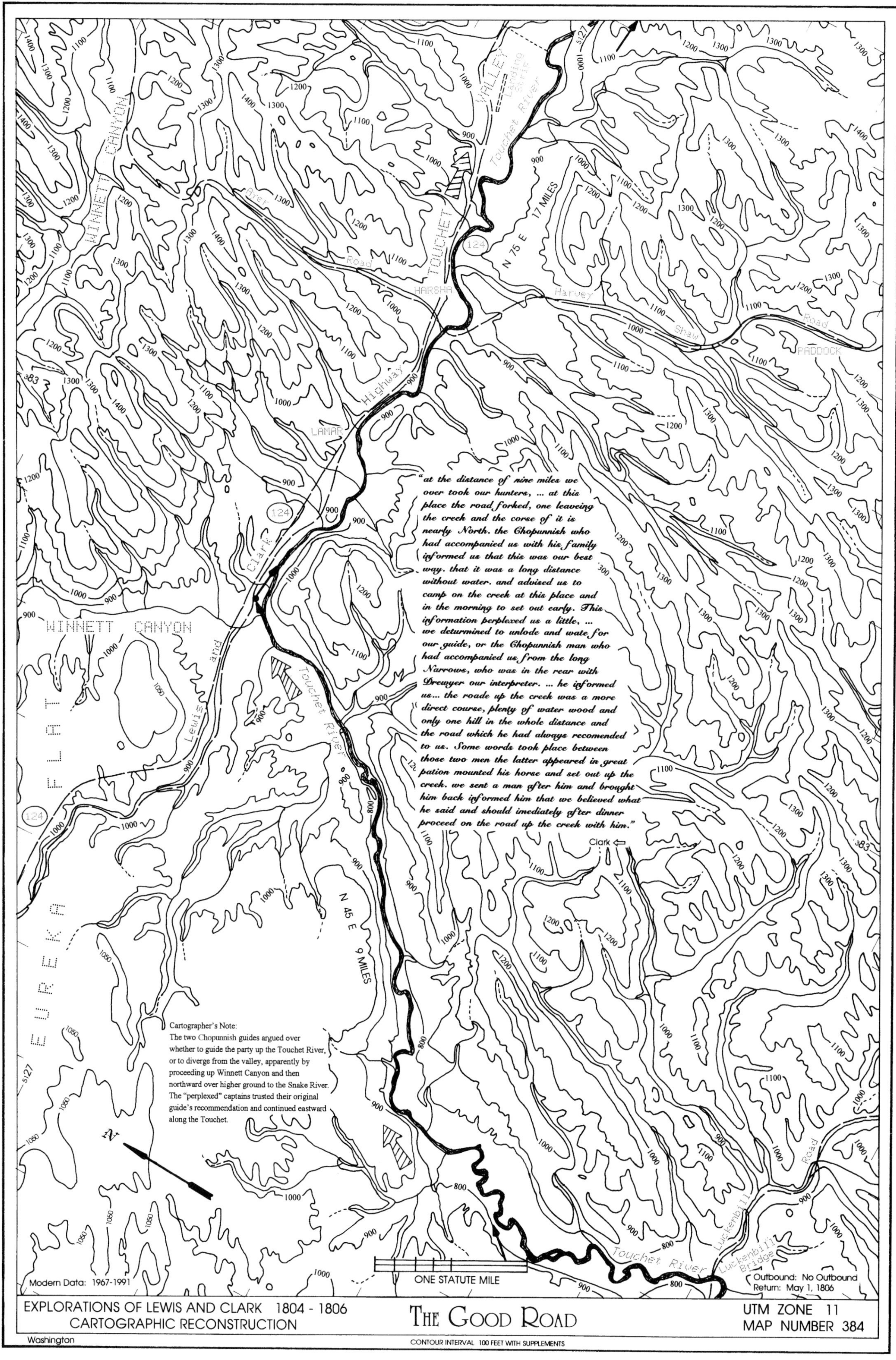
"at the distance of nine miles we over took our hunters, ... at this place the road forked, one leaveing the creek and the corse of it is nearly North. the Chopunnish who had accompanied us with his family informed us that this was our best way. that it was a long distance without water. and advised us to camp on the creek at this place and in the morning to set out early. This information perplexed us a little, ... we deturmined to unlode and wate for our guide, or the Chopunnish man who had accompanied us from the long Narrows, who was in the rear with Drewyer our interpreter. ... he informed us... the roade up the creek was a more direct course, plenty of water wood and only one hill in the whole distance and the road which he had always recomended to us. Some words took place between those two men the latter appeared in great pation mounted his horse and set out up the creek. we sent a man after him and brought him back informed him that we believed what he said and should imediately after dinner proceed on the road up the creek with him."
Clark
Cartographer's Note:
The two Chopunnish guides argued over whether to guide the party up the Touchet River, or to diverge from the valley, apparently by proceeding up Winnett Canyon and then northward over higher ground to the Snake River. The "perplexed" captains trusted their original guide's recommendation and continued eastward along the Touchet.
WINNETT CANYON
EUREKA FLAT
TOUCHET VALLEY
Touchet River
HARSHA
LAMAR
PADDOCK
Harvey Shaw Road
Lewis and Clark Highway
N 75 E 17 MILES
N 45 E 9 MILES
Luckenbill Bridge
ONE STATUTE MILE
Modern Data: 1967-1991
Outbound: No Outbound
Return: May 1, 1806
EXPLORATIONS OF LEWIS AND CLARK 1804 - 1806
CARTOGRAPHIC RECONSTRUCTION
THE GOOD ROAD
UTM ZONE 11
MAP NUMBER 384
Washington
CONTOUR INTERVAL 100 FEET WITH SUPPLEMENTS

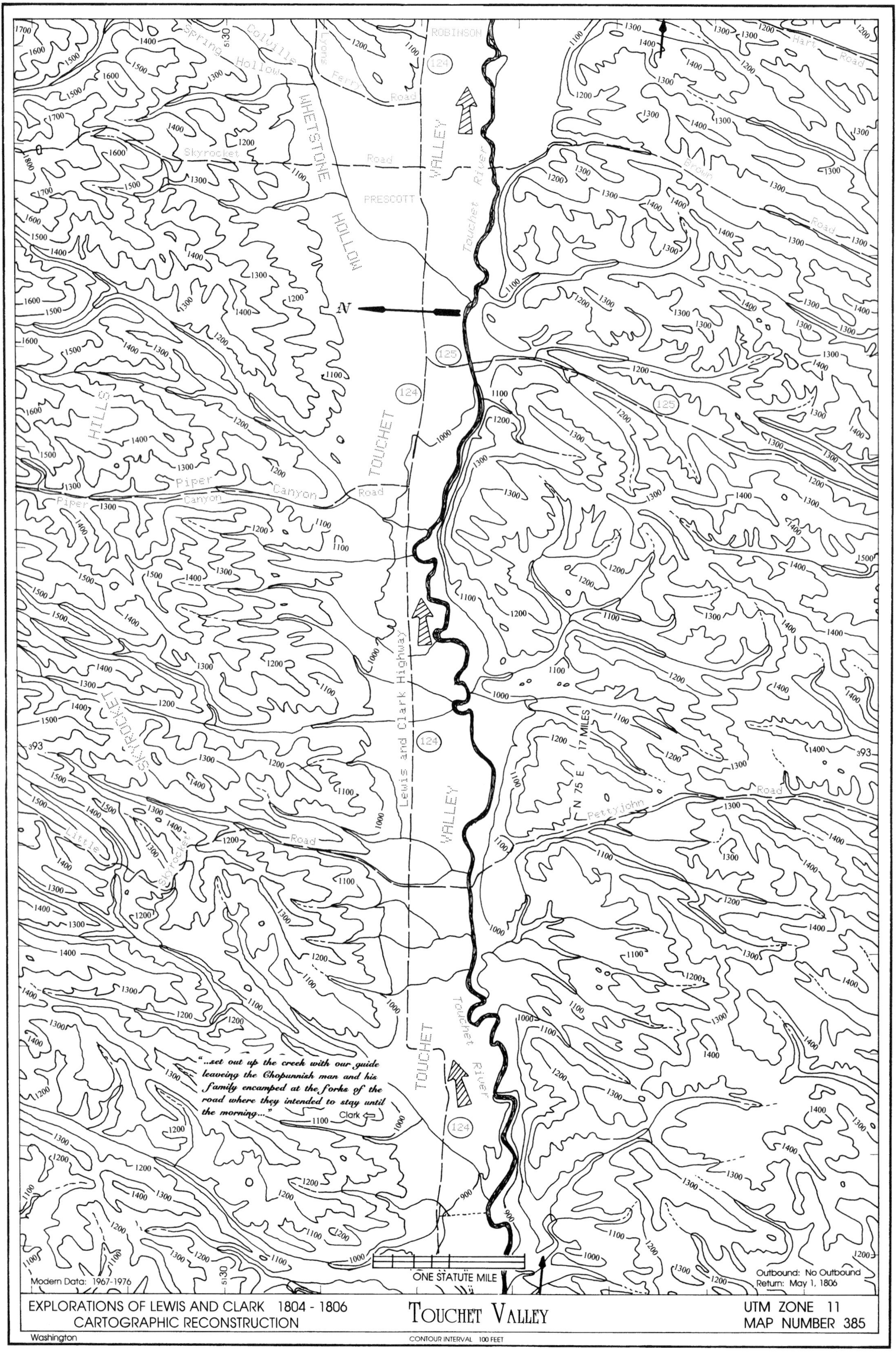
ROBINSON
Spring Hollow
Colville
Lyons Ferry Road
Skyrocket Road
WHETSTONE HOLLOW
PRESCOTT
VALLEY
Touchet River
Hart Road
Brown Road
N
HILLS
TOUCHET
Piper Canyon Road
Piper Canyon
Lewis and Clark Highway
SKYROCKET
N 75 E 17 MILES
Pettyjohn Road
VALLEY
Little Skyrocket Road
TOUCHET
Touchet River
"...set out up the creek with our guide leaveing the Chopunnish man and his family encamped at the forks of the road where they intended to stay until the morning..."
Clark
Modern Data: 1967-1976
ONE STATUTE MILE
Outbound: No Outbound
Return: May 1, 1806
EXPLORATIONS OF LEWIS AND CLARK 1804 - 1806
CARTOGRAPHIC RECONSTRUCTION
TOUCHET VALLEY
UTM ZONE 11
MAP NUMBER 385
Washington
CONTOUR INTERVAL 100 FEET

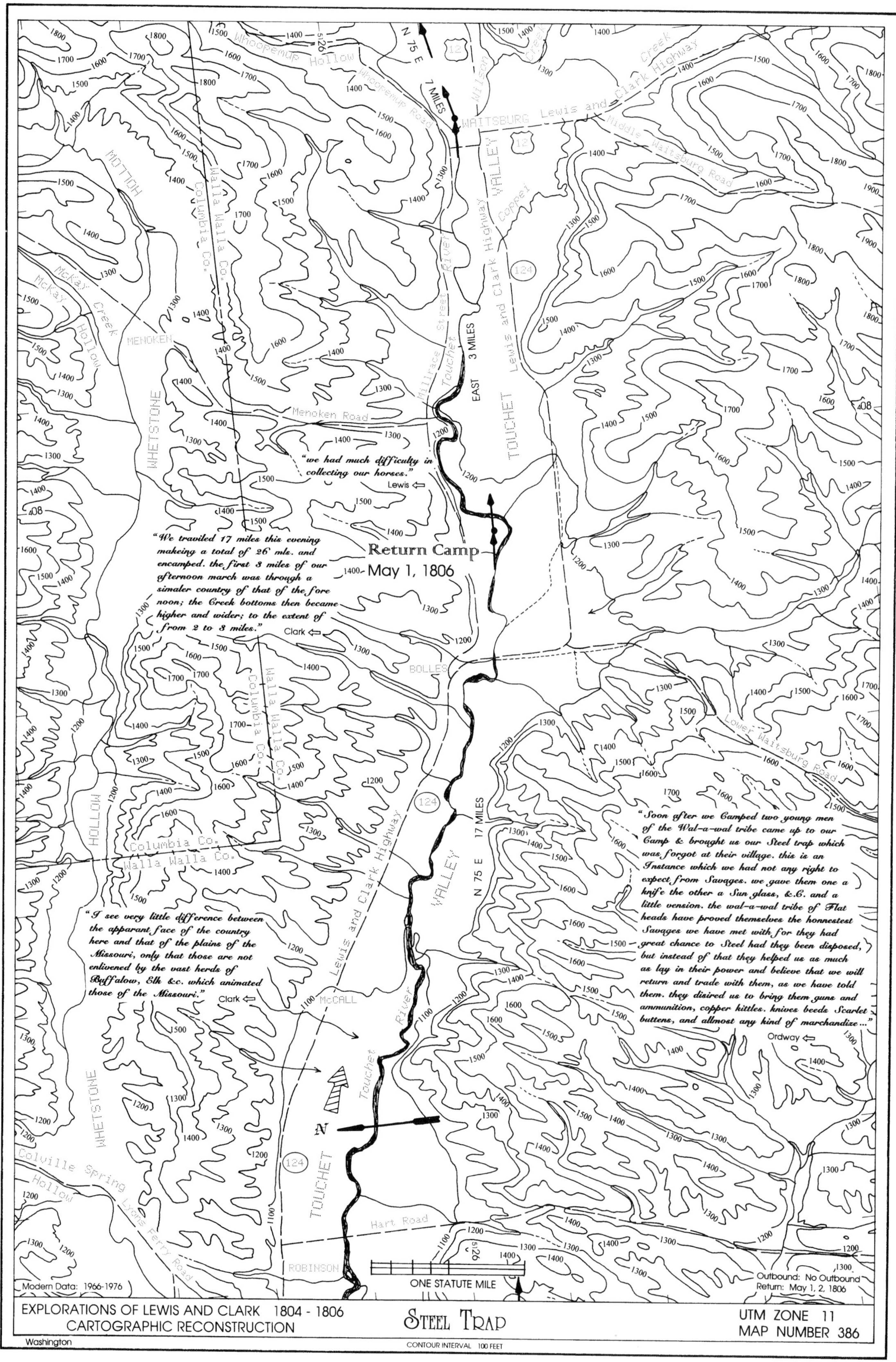

Whoopemup Hollow
Whoopemup Road
N 75 E
7 MILES
WAITSBURG
Lewis and Clark Highway
Wilson Creek
Coppei Creek
Middle Waitsburg Road
VALLEY
TOUCHET
Touchet River
Millrace Street
EAST 3 MILES
McKay Hollow
McKay Creek
MENOKEN
Menoken Road
WHETSTONE
HOLLOW
Columbia Co.
Walla Walla Co.
"we had much difficulty in collecting our horses."
Lewis
"We travild 17 miles this evening makeing a total of 26 mls. and encamped. the first 3 miles of our afternoon march was through a simaler country of that of the fore noon; the Creek bottoms then became higher and wider; to the extent of from 2 to 3 miles."
Clark
Return Camp
May 1, 1806
BOLLES
Lower Waitsburg Road
N 75 E 17 MILES
"Soon after we Camped two young men of the Wal-a-wal tribe came up to our Camp & brought us our Steel trap which was forgot at their village. this is an Instance which we had not any right to expect from Savages. we gave them one a knife the other a Sun glass, &.C. and a little vension. the wal-a-wal tribe of Flat heads have proved themselves the honnestest Savages we have met with for they had great chance to Steel had they been disposed, but instead of that they helped us as much as lay in their power and believe that we will return and trade with them, as we have told them. they disired us to bring them guns and ammunition, copper kittles. knives beeds Scarlet buttens, and allmost any kind of marchandize..."
Ordway
"I see very little difference between the apparant face of the country here and that of the plains of the Missouri, only that those are not enlivened by the vast herds of Buffalow, Elk &c. which animated those of the Missouri."
Clark
McCALL
N
Colville Spring Hollow
Lyons Ferry Road
Hart Road
ROBINSON
Modern Data: 1966-1976
ONE STATUTE MILE
Outbound: No Outbound
Return: May 1, 2, 1806
EXPLORATIONS OF LEWIS AND CLARK 1804 - 1806
CARTOGRAPHIC RECONSTRUCTION
STEEL TRAP
UTM ZONE 11
MAP NUMBER 386
Washington
CONTOUR INTERVAL 100 FEET

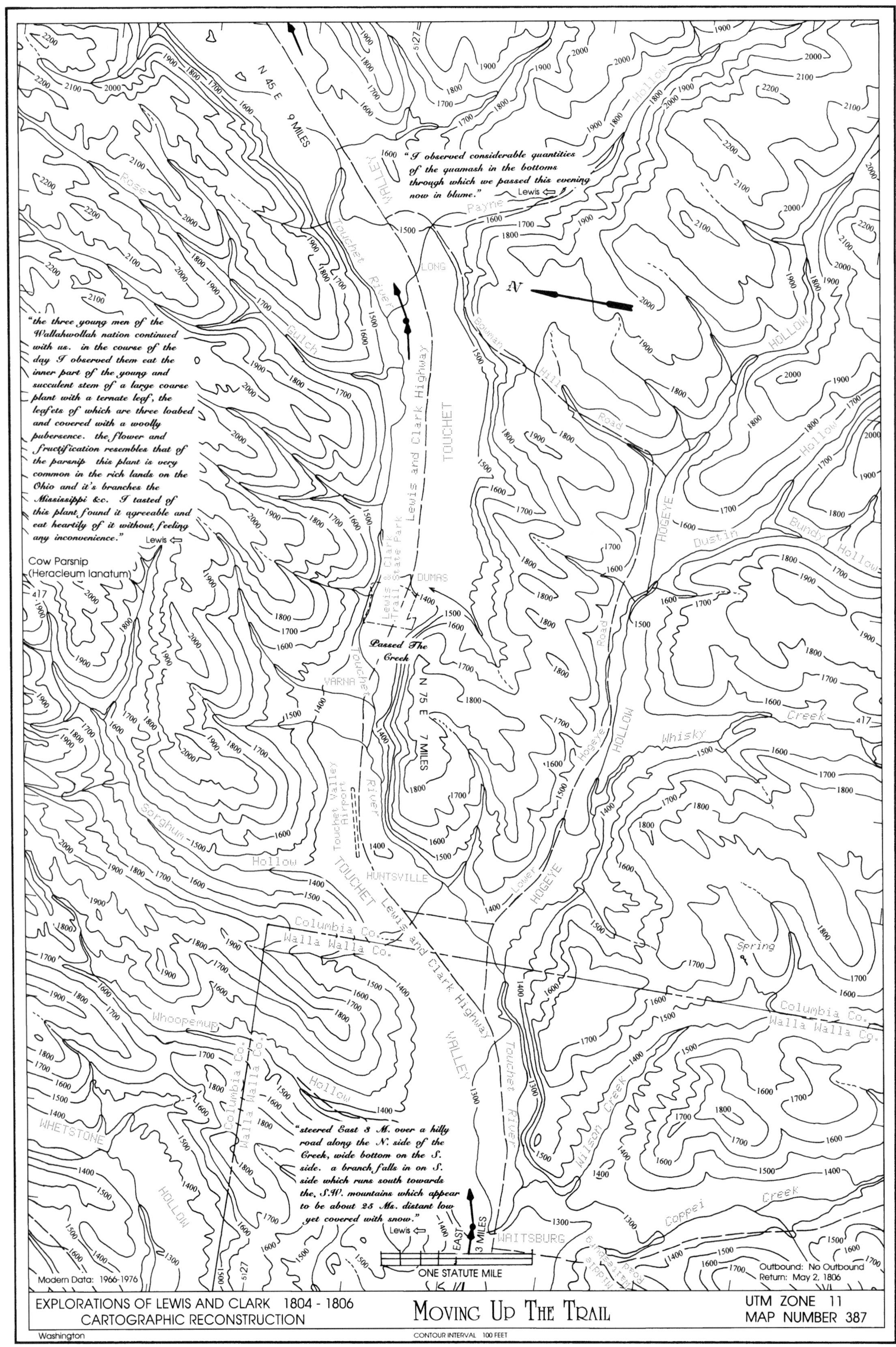
N 45 E 9 MILES
"I observed considerable quantities of the quamash in the bottoms through which we passed this evening now in blume." Lewis
"the three young men of the Wallahwollah nation continued with us. in the course of the day I observed them eat the inner part of the young and succulent stem of a large coarse plant with a ternate leaf, the leafets of which are three loabed and covered with a woolly pubersence. the flower and fructification resembles that of the parsnip this plant is very common in the rich lands on the Ohio and it's branches the Mississippi &c. I tasted of this plant found it agreeable and eat heartily of it without feeling any inconvenience." Lewis
Cow Parsnip
(Heracleum lanatum)
N
Passed The Creek
N 75 E 7 MILES
TOUCHET
Lewis and Clark Highway
Lewis & Clark Trail State Park
Touchet River
DUMAS
VARNA
HUNTSVILLE
Touchet Valley Airport
Columbia Co.
Walla Walla Co.
WAITSBURG
"steered East 3 M. over a hilly road along the N. side of the Creek, wide bottom on the S. side. a branch falls in on S. side which runs south towards the S.W. mountains which appear to be about 25 Ms. distant low yet covered with snow." Lewis
EAST 3 MILES
ONE STATUTE MILE
Modern Data: 1966-1976
Outbound: No Outbound
Return: May 2, 1806
EXPLORATIONS OF LEWIS AND CLARK 1804 - 1806
CARTOGRAPHIC RECONSTRUCTION
MOVING UP THE TRAIL
UTM ZONE 11
MAP NUMBER 387
Washington
CONTOUR INTERVAL 100 FEET

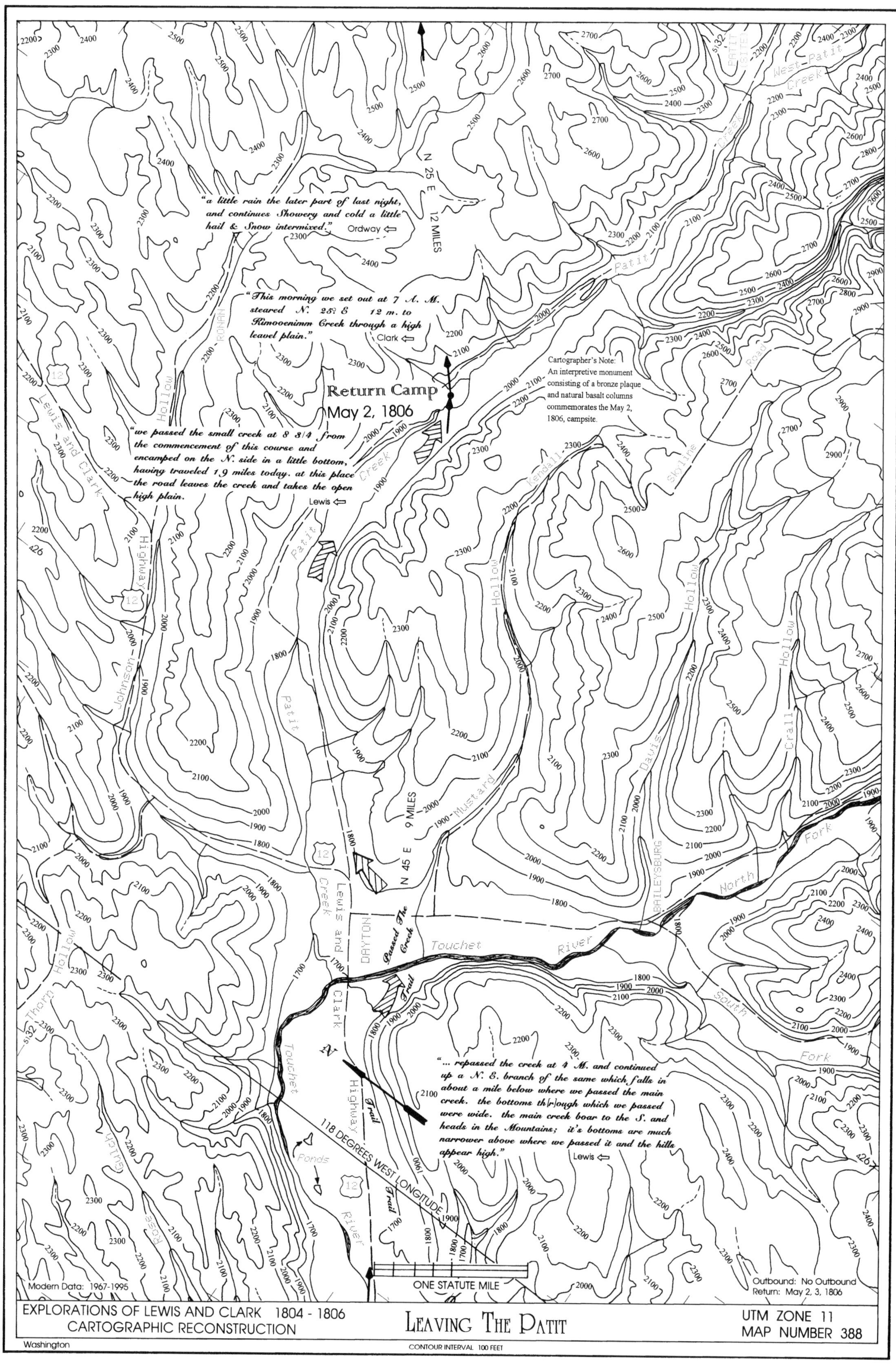

"a little rain the later part of last night, and continues Showery and cold a little hail & Snow intermixed."
Ordway ⇦
N 25 E 12 MILES
"This morning we set out at 7 A. M. steared N. 25° E 12 m. to Kimooenimm Creek through a high leavel plain."
Clark ⇦
Return Camp
May 2, 1806
Cartographer's Note:
An interpretive monument consisting of a bronze plaque and natural basalt columns commemorates the May 2, 1806, campsite.
"we passed the small creek at 8 3/4 from the commencement of this course and encamped on the N. side in a little bottom, having traveled 19 miles today. at this place the road leaves the creek and takes the open high plain."
Lewis ⇦
West Patit Creek
Patit Creek
Kendall
Skyline Road
Lewis and Clark Highway
Johnson Hollow
Hollow
Crall Hollow
Davis Hollow
Mustard
N 45 E 9 MILES
BAILEYSBURG
North Fork
South Fork
Touchet River
DAYTON
Passed The Creek
Trail
Thorn Hollow
Rose Gulch
Ponds
118 DEGREES WEST LONGITUDE
"... repassed the creek at 4 M. and continued up a N. E. branch of the same which falls in about a mile below where we passed the main creek. the bottoms th[r]ough which we passed were wide. the main creek boar to the S. and heads in the Mountains; it's bottoms are much narrower above where we passed it and the hills appear high."
Lewis ⇦
ONE STATUTE MILE
Modern Data: 1967-1995
Outbound: No Outbound
Return: May 2, 3, 1806
EXPLORATIONS OF LEWIS AND CLARK 1804 - 1806
CARTOGRAPHIC RECONSTRUCTION
LEAVING THE PATIT
UTM ZONE 11
MAP NUMBER 388
Washington
CONTOUR INTERVAL 100 FEET

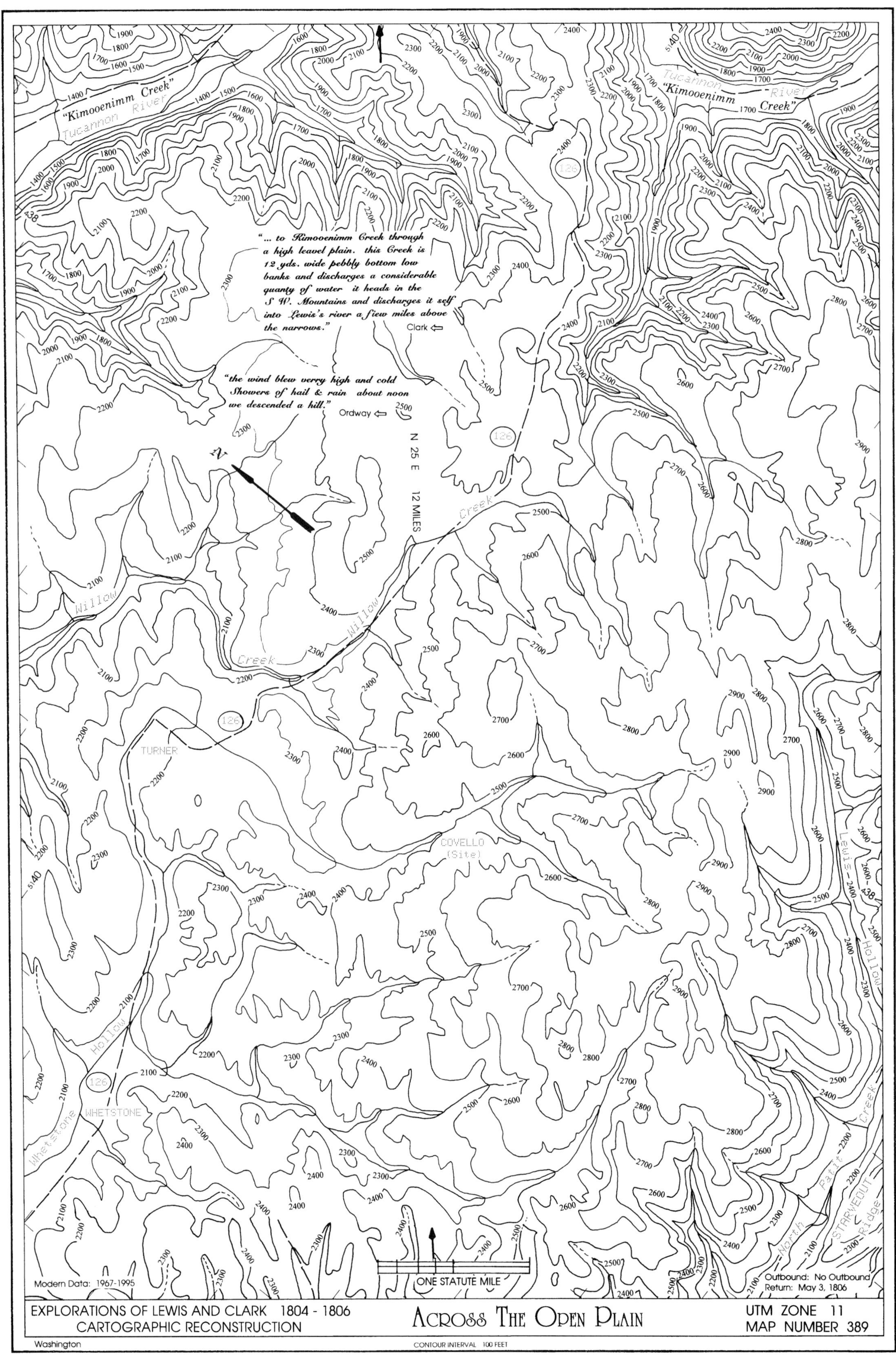
"... to Kimooenimm Creek through a high leavel plain. this Creek is 12 yds. wide pebbly bottom low banks and discharges a considerable quanty of water it heads in the S W. Mountains and discharges it self into Lewis's river a fiew miles above the narrows."
Clark
"the wind blew verry high and cold Showers of hail & rain about noon we descended a hill."
Ordway
"Kimooenimm Creek"
Tucannon River
Willow Creek
N 25 E 12 MILES
TURNER
COVELLO (Site)
WHETSTONE
Whetstone Hollow
Lewis
Hollow
Creek
Patit
North
STARVEDOUT Ridge
ONE STATUTE MILE
Modern Data: 1967-1995
Outbound: No Outbound
Return: May 3, 1806
EXPLORATIONS OF LEWIS AND CLARK 1804 - 1806
CARTOGRAPHIC RECONSTRUCTION
ACROSS THE OPEN PLAIN
UTM ZONE 11
MAP NUMBER 389
Washington
CONTOUR INTERVAL 100 FEET

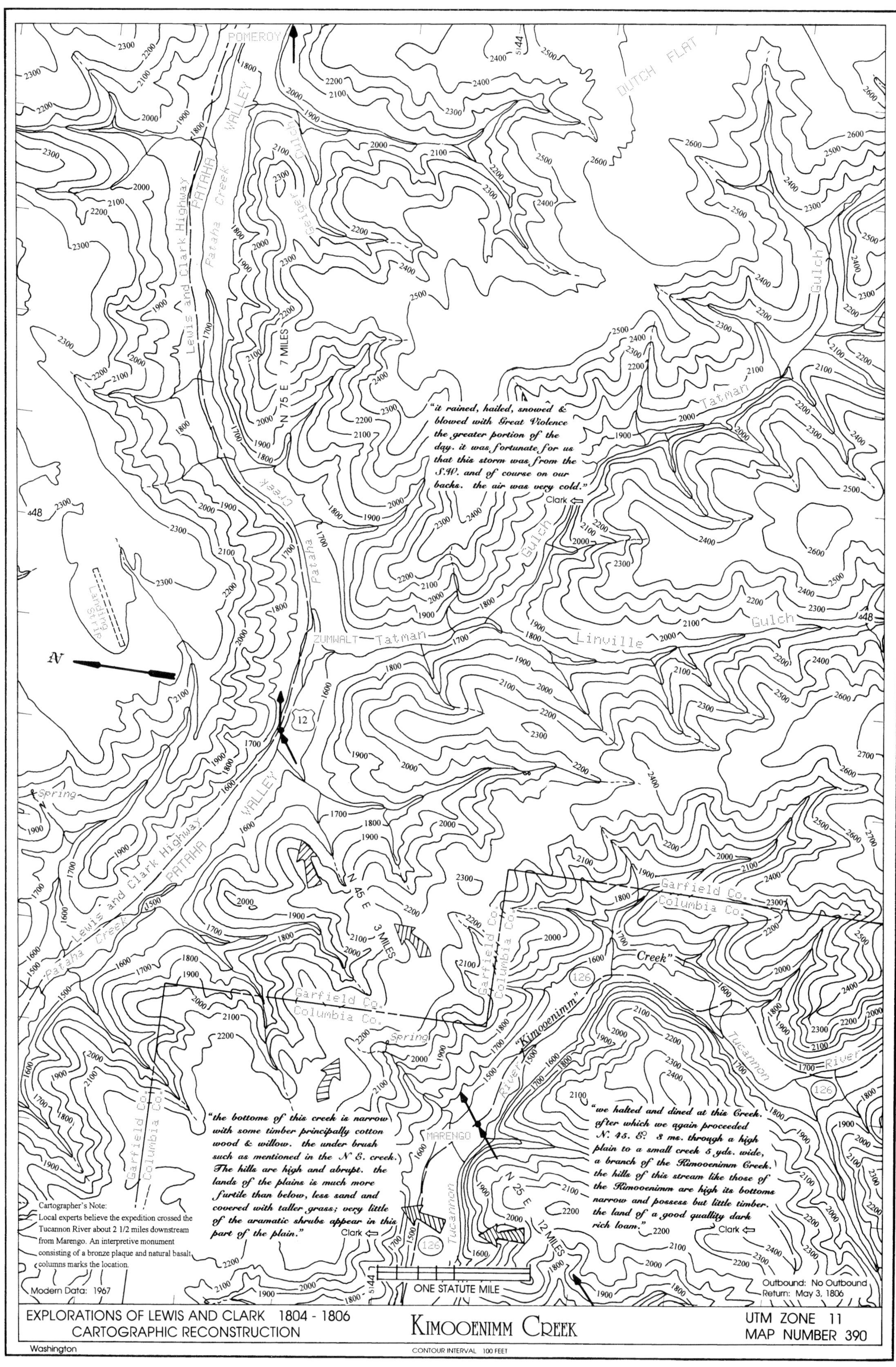
"it rained, hailed, snowed & blowed with Great Violence the greater portion of the day. it was fortunate for us that this storm was from the S.W. and of course on our backs. the air was very cold."
Clark
"the bottoms of this creek is narrow with some timber principally cotton wood & willow. the under brush such as mentioned in the N E. creek. The hills are high and abrupt. the lands of the plains is much more furtile than below, less sand and covered with taller grass; very little of the aramatic shrubs appear in this part of the plain."
Clark
"we halted and dined at this Creek. after which we again proceeded N. 45. E. 3 ms. through a high plain to a small creek 5 yds. wide, a branch of the Kimooenimm Creek. the hills of this stream like those of the Kimooenimm are high its bottoms narrow and possess but little timber. the land of a good quallity dark rich loam."
Clark
Cartographer's Note:
Local experts believe the expedition crossed the Tucannon River about 2 1/2 miles downstream from Marengo. An interpretive monument consisting of a bronze plaque and natural basalt columns marks the location.
POMEROY
DUTCH FLAT
PATAHA VALLEY
Lewis and Clark Highway
Pataha Creek
Geiger Gulch
N 75 E 7 MILES
Tatman Gulch
ZUMWALT
Linville Gulch
Landing Strip
N
Spring
N 45 E 3 MILES
Garfield Co.
Columbia Co.
"Kimooenimm" River
"Creek"
Tucannon River
MARENGO
N 25 E 12 MILES
ONE STATUTE MILE
Modern Data: 1967
Outbound: No Outbound
Return: May 3, 1806
EXPLORATIONS OF LEWIS AND CLARK 1804 - 1806
CARTOGRAPHIC RECONSTRUCTION
KIMOOENIMM CREEK
UTM ZONE 11
MAP NUMBER 390
Washington
CONTOUR INTERVAL 100 FEET

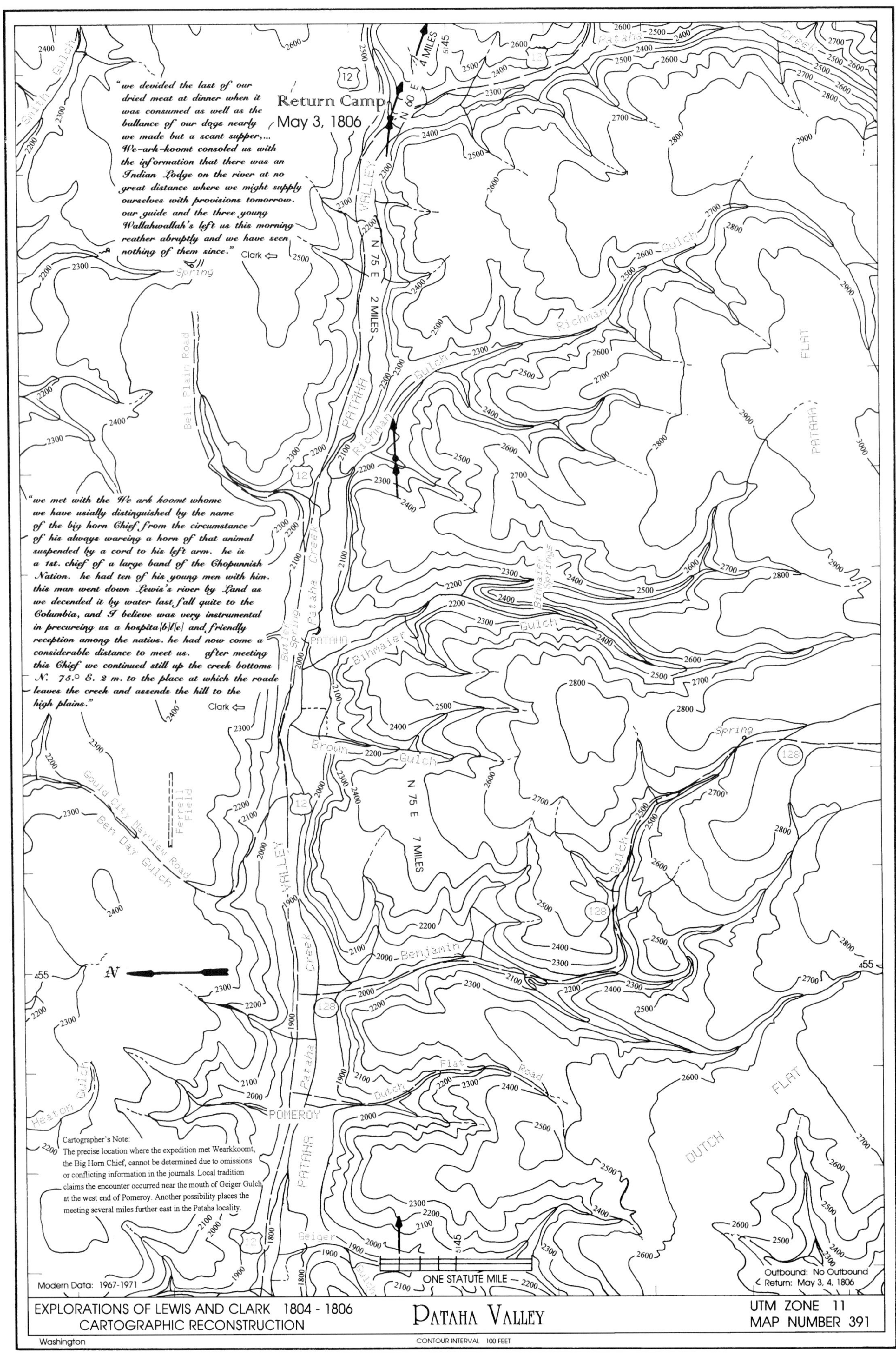
"we devided the last of our dried meat at dinner when it was consumed as well as the ballance of our dogs nearly we made but a scant supper,... We-ark-koomt consoled us with the information that there was an Indian Lodge on the river at no great distance where we might supply ourselves with provisions tomorrow. our guide and the three young Wallahwallah's left us this morning reather abruptly and we have seen nothing of them since."
Clark ⇦
Return Camp
May 3, 1806
N 60 E
4 MILES
N 75 E 2 MILES
"we met with the We ark koomt whome we have usially distinguished by the name of the big horn Chief, from the circumstance of his always wareing a horn of that animal suspended by a cord to his left arm. he is a 1st. chief of a large band of the Chopunnish Nation. he had ten of his young men with him. this man went down Lewis's river by Land as we decended it by water last fall quite to the Columbia, and I believe was very instrumental in precureing us a hospita[b]l[e] and friendly reception among the natives. he had now come a considerable distance to meet us. after meeting this Chief we continued still up the creek bottoms N. 75.° E. 2 m. to the place at which the roade leaves the creek and assends the hill to the high plains."
Clark ⇦
N 75 E 7 MILES
Smith Gulch
Spring
Bell Plain Road
Pataha Creek
Pataha Valley
Richman Gulch
Patahа Flat
Bihmaier Springs
Bihmaier Gulch
Butler Spring
PATAHA
Brown Gulch
Gould City Mayview Road
Ben Day Gulch
Ferrell Field
Benjamin
Heaton Gulch
Flat Road
Dutch
POMEROY
Dutch Flat
Geiger
Spring
Cartographer's Note:
The precise location where the expedition met Wearkkoomt, the Big Horn Chief, cannot be determined due to omissions or conflicting information in the journals. Local tradition claims the encounter occurred near the mouth of Geiger Gulch at the west end of Pomeroy. Another possibility places the meeting several miles further east in the Pataha locality.
Modern Data: 1967-1971
ONE STATUTE MILE
Outbound: No Outbound
Return: May 3, 4, 1806
EXPLORATIONS OF LEWIS AND CLARK 1804 - 1806
CARTOGRAPHIC RECONSTRUCTION
PATAHA VALLEY
UTM ZONE 11
MAP NUMBER 391
Washington
CONTOUR INTERVAL 100 FEET

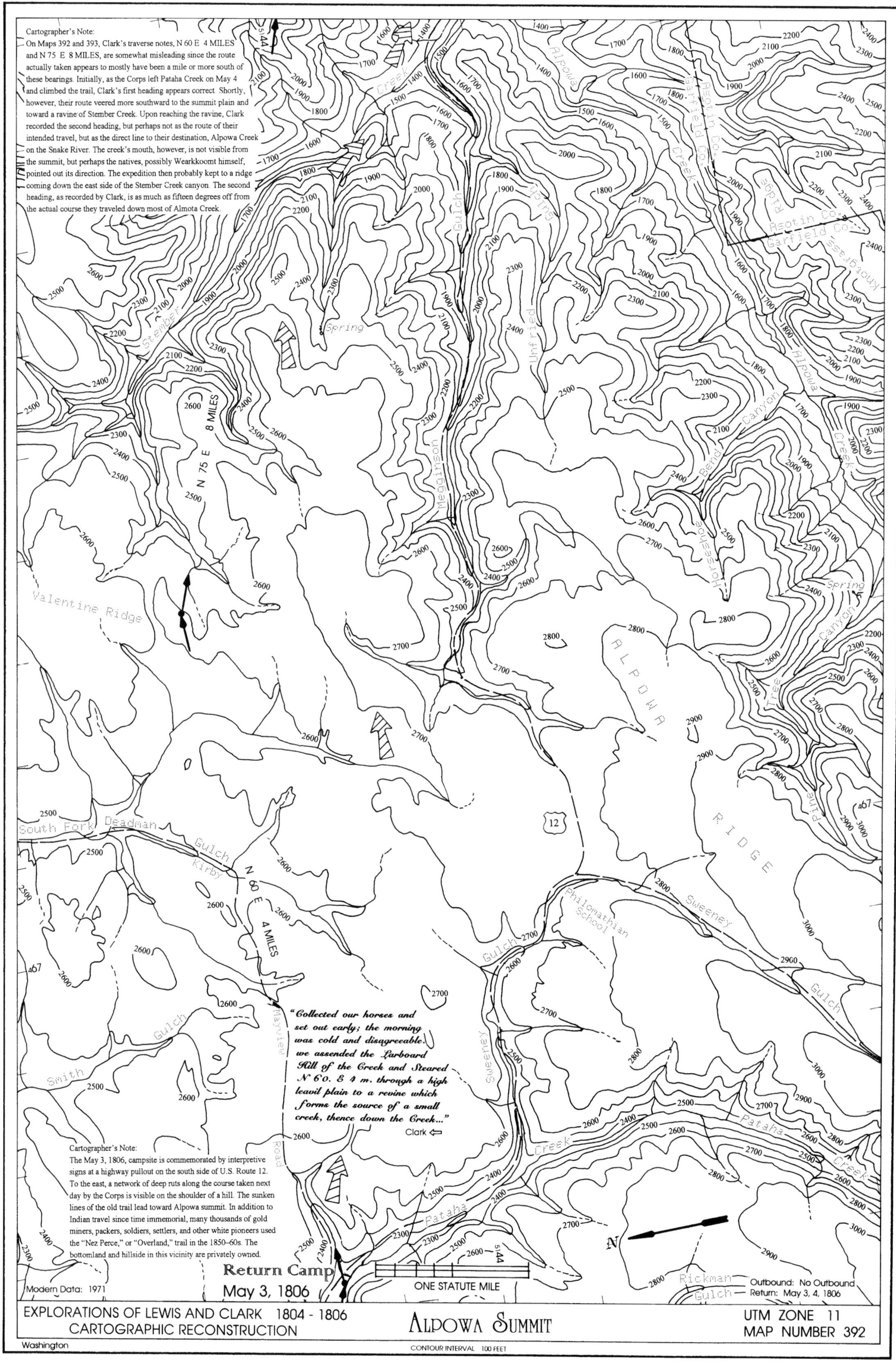
Cartographer's Note:
On Maps 392 and 393, Clark's traverse notes, N 60 E 4 MILES and N 75 E 8 MILES, are somewhat misleading since the route actually taken appears to mostly have been a mile or more south of these bearings. Initially, as the Corps left Pataha Creek on May 4 and climbed the trail, Clark's first heading appears correct. Shortly, however, their route veered more southward to the summit plain and toward a ravine of Stember Creek. Upon reaching the ravine, Clark recorded the second heading, but perhaps not as the route of their intended travel, but as the direct line to their destination, Alpowa Creek on the Snake River. The creek's mouth, however, is not visible from the summit, but perhaps the natives, possibly Wearkkoomt himself, pointed out its direction. The expedition then probably kept to a ridge coming down the east side of the Stember Creek canyon. The second heading, as recorded by Clark, is as much as fifteen degrees off from the actual course they traveled down most of Almota Creek.
N 75 E 8 MILES
Valentine Ridge
Stember
Spring
Meggison Gulch
Unfried Gulch
Alpowa Creek
Asotin Co.
Garfield Co.
Bend Canyon
Horseshoe Bend
Tree Canyon
Pine Tree Canyon
ALPOWA RIDGE
South Fork Deadman Gulch
Kirby
N 60 E 4 MILES
Philomathian School
Sweeney Gulch
Smith Gulch
Mayview Road
Pataha Creek
Rickman Gulch
"Collected our horses and set out early; the morning was cold and disagreeable. we assended the Larboard Hill of the Creek and Steared N 60. E 4 m. through a high leavil plain to a revine which forms the source of a small creek, thence down the Creek..."
Clark ⇦
Cartographer's Note:
The May 3, 1806, campsite is commemorated by interpretive signs at a highway pullout on the south side of U.S. Route 12. To the east, a network of deep ruts along the course taken next day by the Corps is visible on the shoulder of a hill. The sunken lines of the old trail lead toward Alpowa summit. In addition to Indian travel since time immemorial, many thousands of gold miners, packers, soldiers, settlers, and other white pioneers used the "Nez Perce," or "Overland," trail in the 1850–60s. The bottomland and hillside in this vicinity are privately owned.
Return Camp
May 3, 1806
Modern Data: 1971
ONE STATUTE MILE
N
Outbound: No Outbound
Return: May 3, 4, 1806
EXPLORATIONS OF LEWIS AND CLARK 1804 - 1806
CARTOGRAPHIC RECONSTRUCTION
ALPOWA SUMMIT
UTM ZONE 11
MAP NUMBER 392
Washington
CONTOUR INTERVAL 100 FEET

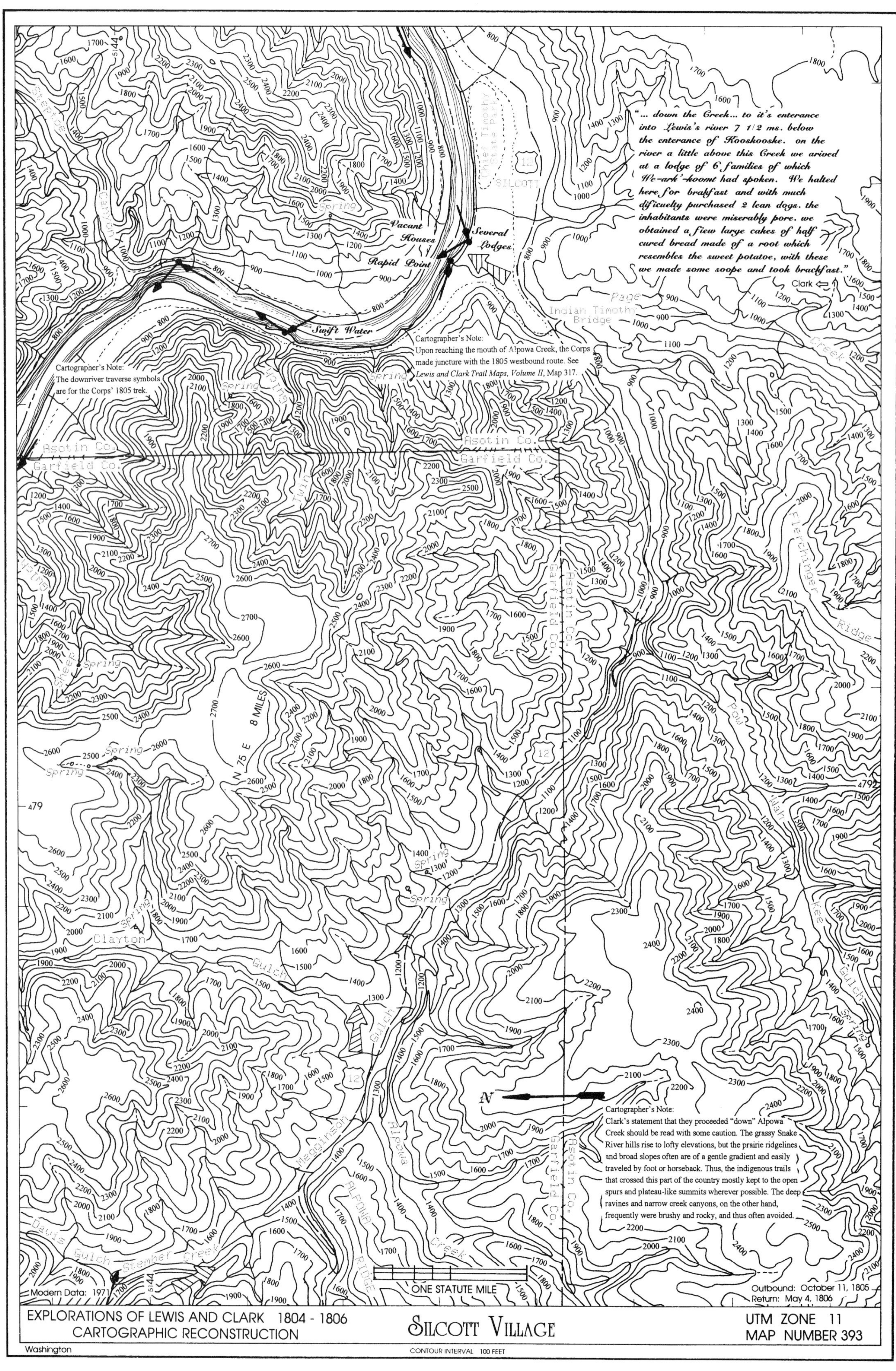
"... down the Creek... to it's enterance into Lewis's river 7 1/2 ms. below the enterance of Kooskooske. on the river a little above this Creek we arived at a lodge of 6 families of which We-ark'-koomt had spoken. We halted here for brakfast and with much dificuelty purchased 2 lean dogs. the inhabitants were miserably pore. we obtained a fiew large cakes of half cured bread made of a root which resembles the sweet potatoe, with these we made some soope and took brackfast."
Clark
Cartographer's Note:
Upon reaching the mouth of Alpowa Creek, the Corps made juncture with the 1805 westbound route. See *Lewis and Clark Trail Maps, Volume II*, Map 317.
Cartographer's Note:
The downriver traverse symbols are for the Corps' 1805 trek.
Cartographer's Note:
Clark's statement that they proceeded "down" Alpowa Creek should be read with some caution. The grassy Snake River hills rise to lofty elevations, but the prairie ridgelines and broad slopes often are of a gentle gradient and easily traveled by foot or horseback. Thus, the indigenous trails that crossed this part of the country mostly kept to the open spurs and plateau-like summits wherever possible. The deep ravines and narrow creek canyons, on the other hand, frequently were brushy and rocky, and thus often avoided.
Several Lodges
Vacant Houses
Rapid Point
Swift Water
Chief Timothy State Park
Silcott
Indian Timothy Bridge
Page Creek
Asotin Co.
Garfield Co.
Twin Springs
Fletchinger Ridge
Pow Wah Kee Gulch
Sheep Spring
Clayton Gulch
Meginson Gulch
Alpowa Creek
Alpowa Ridge
Davis Gulch
Stember Creek
Steptoe Canyon
N 75 E 8 Miles
N
ONE STATUTE MILE
Modern Data: 1971
Outbound: October 11, 1805
Return: May 4, 1806
EXPLORATIONS OF LEWIS AND CLARK 1804 - 1806
CARTOGRAPHIC RECONSTRUCTION
SILCOTT VILLAGE
UTM ZONE 11
MAP NUMBER 393
Washington
CONTOUR INTERVAL 100 FEET

Alternate Return Routes No. 2 & 3

MAPS 394 - 398 AND 399 - 400

MAIN PARTY, CLEARWATER COUNTRY, IDAHO

Clearwater River, May 8, 1806—
Captains Meriwether Lewis and William Clark
Interpreters George Drouillard and Toussaint Charbonneau
Sacagawea and Jean Baptiste Charbonneau
York
26 enlisted men
various Nez Perce

NO. 2, MAPS 394 TO 398—
ROAD TO BROKEN ARM'S VILLAGE

When entering the Snake-Clearwater locality on May 4, 1806, the expedition soon encountered Nez Perce villagers and important tribal leaders, including friendly Chief Tetoharsky, one of the Corps' main guides when traveling down the Snake and much of the Columbia in the previous year. (The Corps already had met up with the Big Horn Chief on the last stages of the cutoff across southeast Washington.)

On May 8, 1806, at a point on the Clearwater River about 9 miles downstream from the previous year's canoe-building camp, the expedition left the 1805 westbound track and proceeded up to the ridges south of the river. The Nez Perce had informed the captains that this was a better route to the main villages located on the Clearwater near Lawyer Creek (today's Kamiah, Idaho).

Sergeant Gass reports the day was "pleasant" when they awoke on the morning of May 8, but, at the campsite and when gaining the high plain, they could see the mountains yet covered in deep snow, confirming the Indians' warning that there would be no crossing of the Bitterroot Range for another "moon and a half."

On this day, they met another very important chief, Cut Nose, for the first time. They also were reacquainted with their old friend, Chief Twisted Hair, who in the autumn of 1805 took responsibility for watching over the Corps' horses for the winter, as the expedition set out in canoes down the Clearwater for the Pacific Ocean. A dispute now arose, however, between Cut Nose and Twisted Hair regarding the latter's and the young men's treatment of the expedition's horses during the winter. With careful negotiation, the dispute soon was resolved and mutual cordiality regained, and the expedition's widely scattered horses were gathered up.

By May 10, the Corps was at Chief Broken Arm's village, located a few miles up Lawyer Creek. In a few days, they moved down to the Clearwater River, soon crossing to the east side. Here, they set up a nearly month-long residence, while waiting for mountain snows to sufficiently melt to allow traveling.

Named Camp Chopunnish or the Long Camp by Lewis and Clark scholars, the Corps spent more days at this campsite than any other, excluding the winter quarters at River Dubois, Fort Mandan, and Fort Clatsop.

On June 10, 1806, the expedition with about 66 horses moved northward to the camas fields of Weippe Prairie, returning to the previous year's westbound course. On June 15, without Nez Perce guides, they began their first assault of the Bitterroots, but deep snow soon brought them to a standstill. On June 17, they hung much of their baggage in trees and retreated to Weippe Prairie.

The second, and successful, attempt to cross the mountains began June 24, this time with Indian guides.

NO. 3, MAPS 399 AND 400—
STAYING ON THE RIDGE TRAIL

On June 28, 1806, the party passed Wendover Ridge where they had climbed up from the Lochsa River in the preceding year. This time, however, the expedition continued along the ridge trail to camp at Papoose Saddle. They proceeded on the next day to Packer Meadows. The expedition's entire complement traveled this alternate return route from Cayuse Junction to Lolo Pass.

The guides informed the captains that when their people traveled across these mountains, some men would descend to a fishery near White Sand Creek. Catching fish for their group, they then proceeded to join the main body at Packer Meadows.

The name "Lolo Trail" originated from a later time period. The route frequently was used by the Nez Perce to visit their allies, the Flathead of western Montana, and to access the Great Plains buffalo ranges. Historically, it never was known as the Nez Perce Trail. In later frontier times, particularly during the 1860s gold rush period, the Nez Perce (or Southern Nez Perce) Trail referred to the Elk City road to Montana in the Salmon River locality (essentially today's Magruder Corridor). It crossed the Bitterroots a considerable distance south of Lolo Pass.

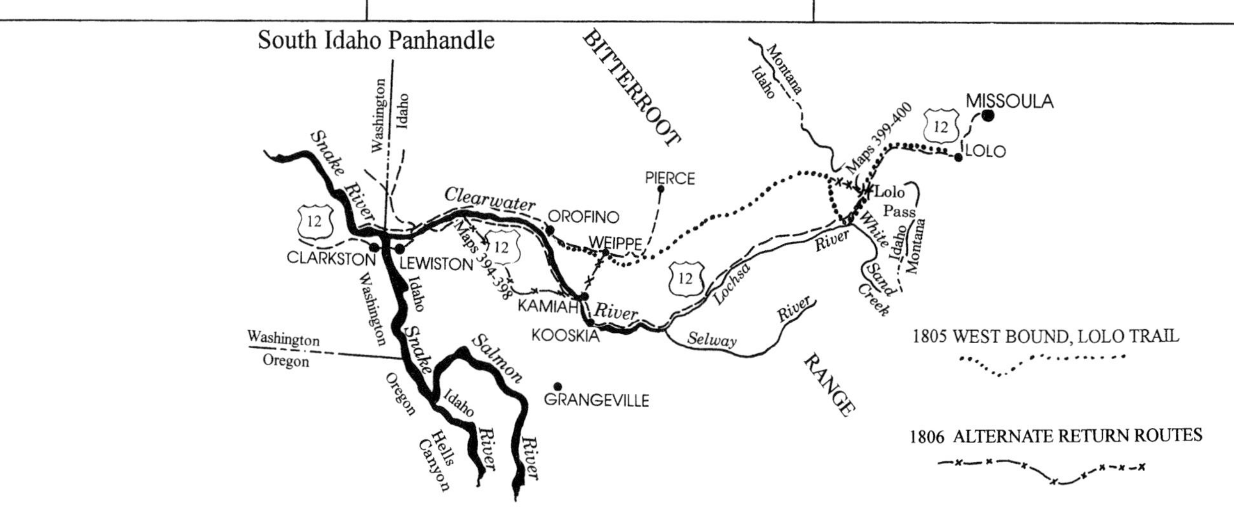

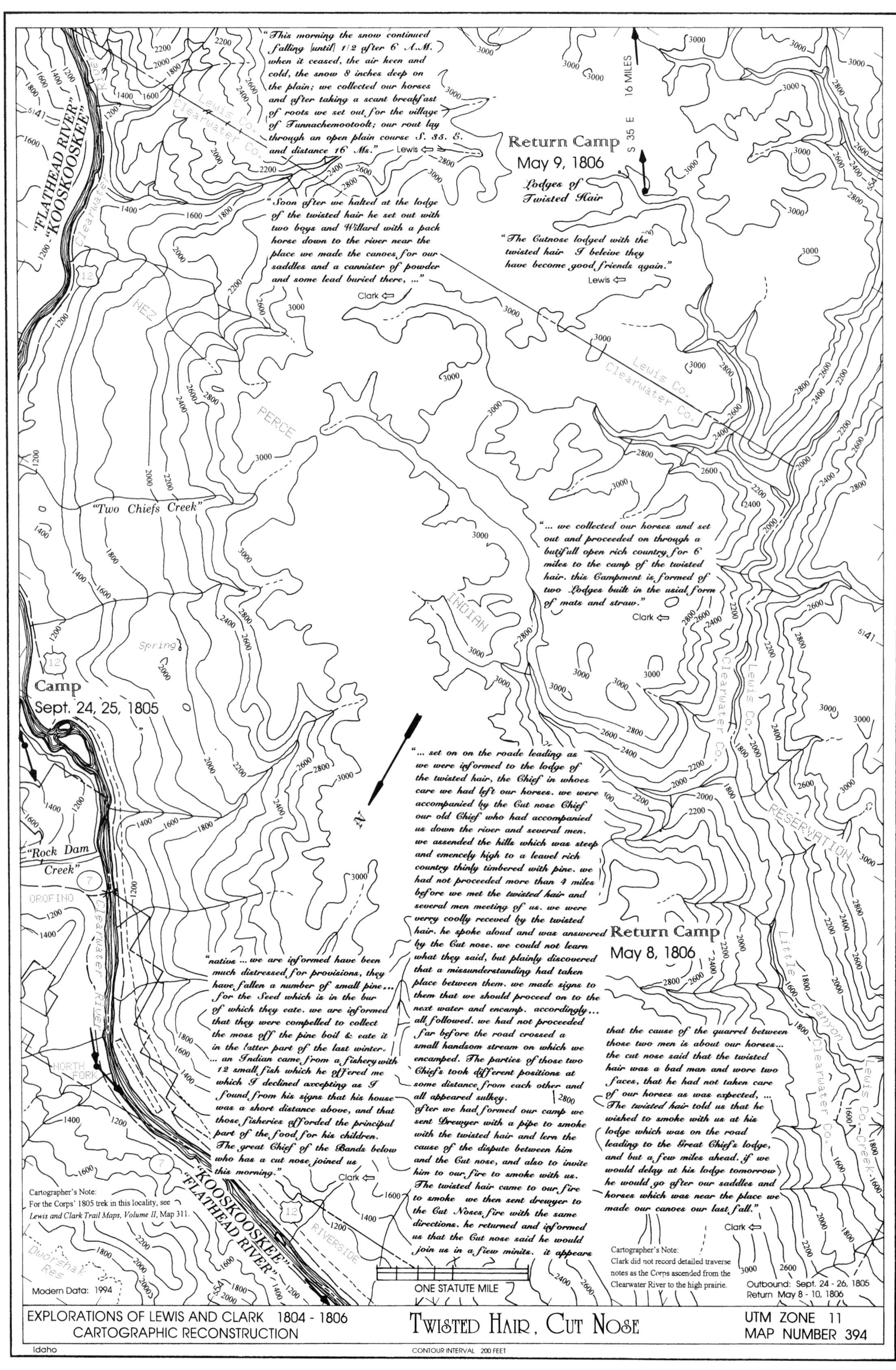
"This morning the snow continued falling [until] 1/2 after 6' A.M. when it ceased, the air keen and cold, the snow 8 inches deep on the plain; we collected our horses and after taking a scant breakfast of roots we set out for the village of Tunnachemootoolt; our rout lay through an open plain course S. 35. E. and distance 16' Ms." Lewis ⇦
"Soon after we halted at the lodge of the twisted hair he set out with two boys and Willard with a pack horse down to the river near the place we made the canoes for our saddles and a cannister of powder and some lead buried there, ..." Clark ⇦
Return Camp
May 9, 1806
Lodges of Twisted Hair
S 35 E 16 MILES
"The Cutnose lodged with the twisted hair I beleive they have become good friends again." Lewis ⇦
"... we collected our horses and set out and proceeded on through a butifull open rich country for 6 miles to the camp of the twisted hair. this Campment is formed of two Lodges built in the usial form of mats and straw." Clark ⇦
"FLATHEAD RIVER"
"KOOSKOOSKEE"
Lewis Co.
Clearwater Co.
NEZ PERCE
INDIAN
RESERVATION
"Two Chiefs Creek"
Spring
Camp
Sept. 24, 25, 1805
"Rock Dam Creek"
OROFINO
Clearwater River
NORTH FORK
Little Canyon Creek
RIVERSIDE
Dworshak Res
N
"... set on on the roade leading as we were informed to the lodge of the twisted hair, the Chief in whoes care we had left our horses. we were accompanied by the Cut nose Chief our old Chief who had accompanied us down the river and several men. we assended the hills which was steep and emencely high to a leavel rich country thinly timbered with pine. we had not proceeded more than 4 miles before we met the twisted hair and several men meeting of us. we were verry coolly receved by the twisted hair. he spoke aloud and was answered by the Cut nose. we could not learn what they said, but plainly discovered that a missunderstanding had taken place between them. we made signs to them that we should proceed on to the next water and encamp. accordingly... all followed. we had not proceeded far before the road crossed a small handsom stream on which we encamped. The parties of those two Chiefs took different positions at some distance from each other and all appeared sulkey. after we had formed our camp we sent Drewyer with a pipe to smoke with the twisted hair and lern the cause of the dispute between him and the Cut nose, and also to invite him to our fire to smoke with us. The twisted hair came to our fire to smoke we then sent drewyer to the Cut Noses fire with the same directions. he returned and informed us that the Cut nose said he would join us in a fiew minits. it appears
Return Camp
May 8, 1806
that the cause of the quarrel between those two men is about our horses... the cut nose said that the twisted hair was a bad man and wore two faces, that he had not taken care of our horses as was expected, ... The twisted hair told us that he wished to smoke with us at his lodge which was on the road leading to the Great Chiefs lodge, and but a few miles ahead. if we would delay at his lodge tomorrow he would go after our saddles and horses which was near the place we made our canoes our last fall." Clark ⇦
"natives ... we are informed have been much distressed for provisions, they have fallen a number of small pine... for the Seed which is in the bur of which they eate. we are informed that they were compelled to collect the moss off the pine boil & eate it in the latter part of the last winter. ... an Indian came from a fishery with 12 small fish which he offered me which I declined accepting as I found from his signs that his house was a short distance above, and that those fisheries afforded the principal part of the food for his children. The great Chief of the Bands below who has a cut nose joined us this morning." Clark ⇦
Cartographer's Note:
For the Corps' 1805 trek in this locality, see Lewis and Clark Trail Maps, Volume II, Map 311.
Cartographer's Note:
Clark did not record detailed traverse notes as the Corps ascended from the Clearwater River to the high prairie.
Outbound: Sept. 24 - 26, 1805
Return May 8 - 10, 1806
Modern Data: 1994
ONE STATUTE MILE
EXPLORATIONS OF LEWIS AND CLARK 1804 - 1806
CARTOGRAPHIC RECONSTRUCTION
TWISTED HAIR, CUT NOSE
UTM ZONE 11
MAP NUMBER 394
Idaho
CONTOUR INTERVAL 200 FEET

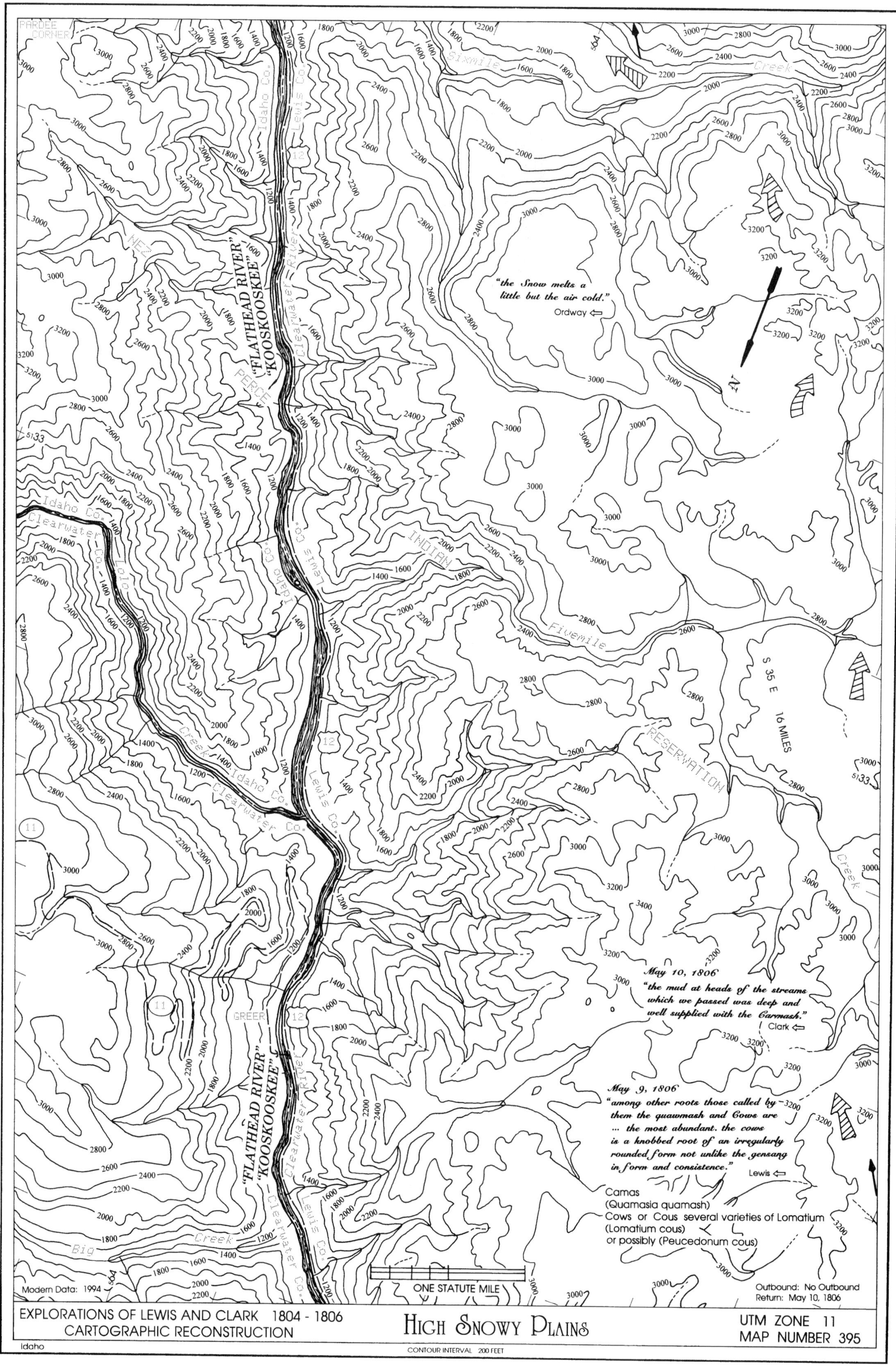
"the Snow melts a little but the air cold."
Ordway
May 10, 1806
"the mud at heads of the streams which we passed was deep and well supplied with the Carmash."
Clark
May 9, 1806
"among other roots those called by them the quawmash and Cows are ... the most abundant. the cows is a knobbed root of an irregularly rounded form not unlike the gensang in form and consistence."
Lewis
Camas
(Quamasia quamash)
Cows or Cous several varieties of Lomatium
(Lomatium cous)
or possibly (Peucedonum cous)
"FLATHEAD RIVER"
"KOOSKOOSKEE"
Clearwater River
NEZ PERCE INDIAN RESERVATION
Sixmile Creek
Fivemile
Lolo Creek
Big Creek
Idaho Co.
Lewis Co.
Clearwater Co.
PARDEE CORNER
GREER
S 35 E
16 MILES
Modern Data: 1994
ONE STATUTE MILE
Outbound: No Outbound
Return: May 10, 1806
EXPLORATIONS OF LEWIS AND CLARK 1804 - 1806
CARTOGRAPHIC RECONSTRUCTION
HIGH SNOWY PLAINS
UTM ZONE 11
MAP NUMBER 395
Idaho
CONTOUR INTERVAL 200 FEET

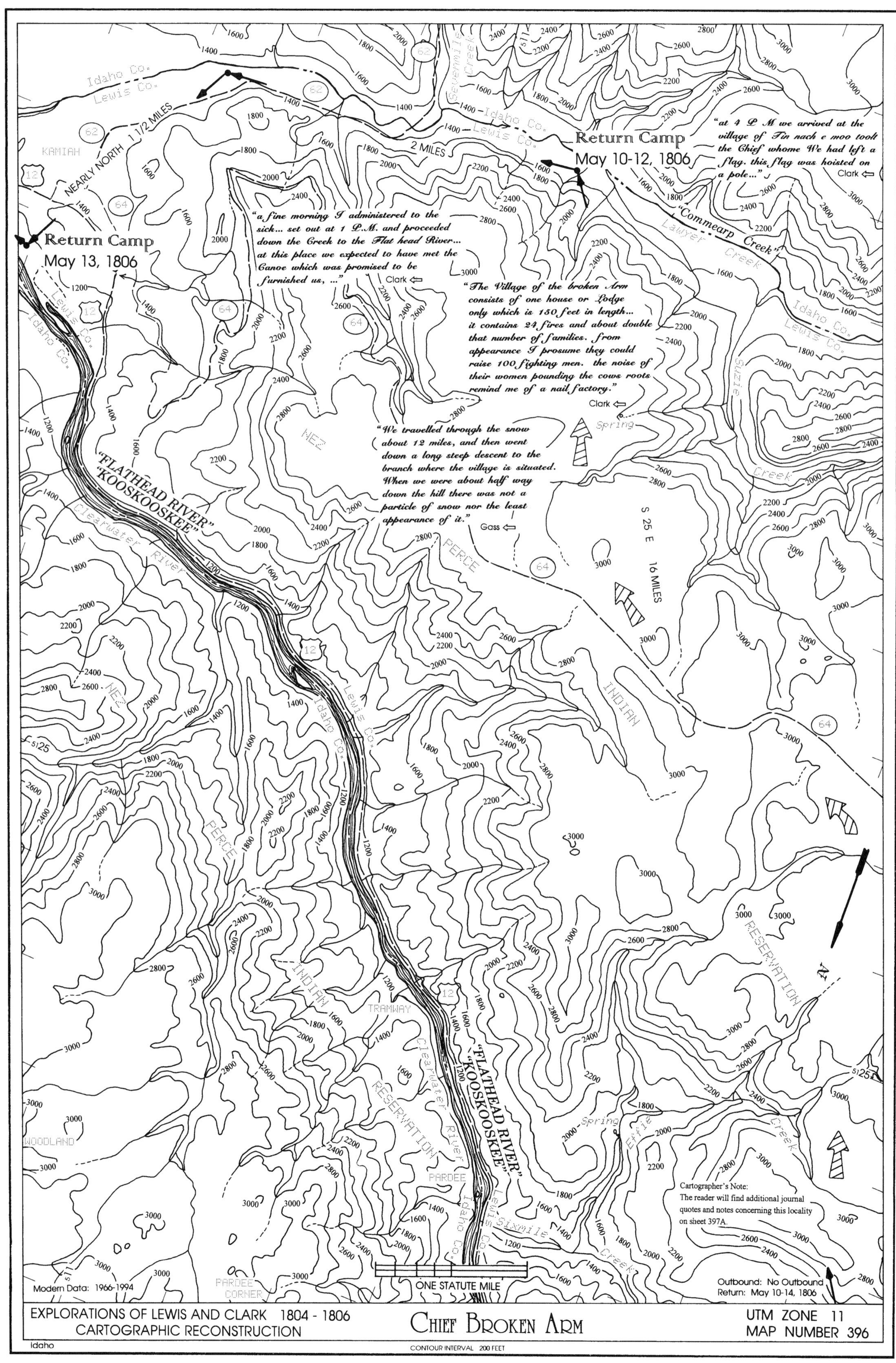
Return Camp
May 10-12, 1806
Return Camp
May 13, 1806
"at 4 P. M we arrived at the village of Tin nach e moo toolt the Chief whome We had left a flag. this flag was hoisted on a pole..."
Clark
"a fine morning I administered to the sick... set out at 1 P.M. and proceeded down the Creek to the Flat head River... at this place we expected to have met the Canoe which was promised to be furnished us, ..."
Clark
"The Village of the broken Arm consists of one house or Lodge only which is 150 feet in length... it contains 24 fires and about double that number of families. from appearance I prosume they could raise 100 fighting men. the noise of their women pounding the cows roots remind me of a nail factory."
Clark
"We travelled through the snow about 12 miles, and then went down a long steep descent to the branch where the village is situated. When we were about half way down the hill there was not a particle of snow nor the least appearance of it."
Gass
"FLATHEAD RIVER"
"KOOSKOOSKEE"
"Commearp Creek"
NEARLY NORTH 1 1/2 MILES
2 MILES
S 25 E 16 MILES
Cartographer's Note:
The reader will find additional journal quotes and notes concerning this locality on sheet 397A.
Outbound: No Outbound
Return: May 10-14, 1806
Modern Data: 1966-1994
ONE STATUTE MILE
EXPLORATIONS OF LEWIS AND CLARK 1804 - 1806
CARTOGRAPHIC RECONSTRUCTION
CHIEF BROKEN ARM
UTM ZONE 11
MAP NUMBER 396
Idaho
CONTOUR INTERVAL 200 FEET

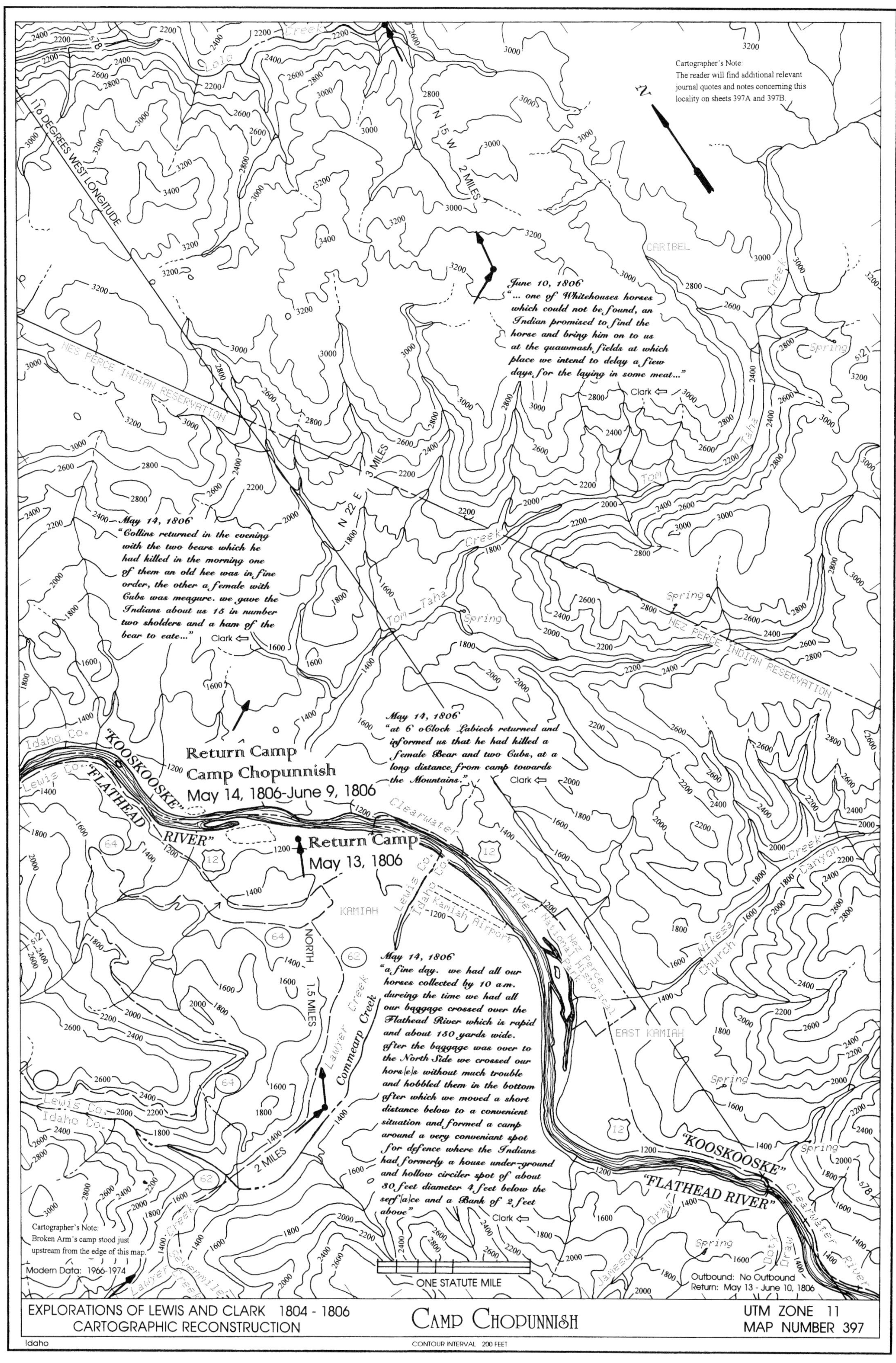

Cartographer's Note:
The reader will find additional relevant journal quotes and notes concerning this locality on sheets 397A and 397B.
N
116 DEGREES WEST LONGITUDE
N 15 W 2 MILES
N 22 E 3 MILES
NORTH 1.5 MILES
2 MILES
NEZ PERCE INDIAN RESERVATION
CARIBEL
Lolo Creek
Tom Taha Creek
Spring
June 10, 1806
"... one of Whitehouses horses which could not be found, an Indian promised to find the horse and bring him on to us at the quawmash fields at which place we intend to delay a fiew days for the laying in some meat..."
Clark ⇦
May 14, 1806
"Collins returned in the evening with the two bears which he had killed in the morning one of them an old hee was in fine order, the other a female with Cubs was meagure. we gave the Indians about us 15 in number two sholders and a ham of the bear to eate..."
Clark ⇦
May 14, 1806
"at 6 oClock Labiech returned and informed us that he had killed a female Bear and two Cubs, at a long distance from camp towards the Mountains."
Clark ⇦
Return Camp
Camp Chopunnish
May 14, 1806-June 9, 1806
Return Camp
May 13, 1806
"KOOSKOOSKE"
"FLATHEAD RIVER"
Idaho Co.
Lewis Co.
Clearwater River
KAMIAH
EAST KAMIAH
Kamiah Airport
Nez Perce National Historical Park
Lawyer Creek
Commearp Creek
Canyon Creek
Nikesa Church
Jameson Draw
Doty Draw
Sevenmile Creek
May 14, 1806
"a fine day. we had all our horses collected by 10 a.m. dureing the time we had all our baggage crossed over the Flathead River which is rapid and about 150 yards wide. after the baggage was over to the North Side we crossed our hors[e]s without much trouble and hobbled them in the bottom after which we moved a short distance below to a convenient situation and formed a camp around a very conveniant spot for defence where the Indians had formerly a house under-ground and hollow circiler spot of about 30 feet diameter 4 feet below the serf[a]ce and a Bank of 2 feet above"
Clark ⇦
Cartographer's Note:
Broken Arm's camp stood just upstream from the edge of this map.
Modern Data: 1966-1974
ONE STATUTE MILE
Outbound: No Outbound
Return: May 13 - June 10, 1806
EXPLORATIONS OF LEWIS AND CLARK 1804 - 1806
CARTOGRAPHIC RECONSTRUCTION
CAMP CHOPUNNISH
UTM ZONE 11
MAP NUMBER 397
Idaho
CONTOUR INTERVAL 200 FEET

Additional Journal Quotes For Map 396

May 10, 1806
"... told this Chief our situation in respect to provisions. they brought forward about 2 bushels of quawmash 4 cakes of bread made of roots and a dried fish. we informed the Chief that our party was not accustomed to eate roots without flesh & proposed to exchange some of our oald horses for young ones to eate. they said that they would not exchange horses, but would furnish us with such as we wished, and produced 2 one of which we killed and informd. them that we did not wish to kill the other at this time. ...
a large Lodge of Leather was pitched... a parcel of wood was collected and laid at the dore and a fire made in this conic lodge before we entered it. the Chief requested that we might make the Lodge our homes while we remained with him. ...
as those people had been liberal I directed the men not to crowd their Lodges in serch of food the manner hunger has compelled them to do, at most lodges we have passed, and which the Twisted Hair had informed us was disagreeable to the nativs."
Clark ⇦

Cartographer's Note:
It is noteworthy that the captains allowed the men, apparently individually and in small groups, to approach native people for the purpose of requesting or bartering for provisions.

May 10, 1806
"The Village of the broken Arm consists of one house or Lodge only which is 150 feet in length... it contains 24 fires and about double that number of families. ... the noise of their women pounding the cows roots remind me of a nail factory. ...
Those people has shewn much greater acts of hospitallity than we have witnessed from any nation or tribe since we have passed the rocky Mountains. in short be it spoken to their immortal honor it is the only act which diserves the appelation of hospitallity which we have witnessed in this quarter."
Clark ⇦

Cartographer's Note:
Elsewhere in the journals, Corps members expressed similar tributes to other Columbia Plateau peoples as well, such as the Walla Walla.

May 11, 1806
"The last evening we were much crouded with the indians in our lodge, the whole floor of which was covered with their sleeping carcases."
Lewis ⇦

May 11, 1806
"we were crouded in the Lodge with Indians who continued all night..."
Clark ⇦

Cartographer's Note:
It is interesting to note the differing tones in the way that the two captains referred to their generous hosts.

May 11, 1806
"We are now pretty well informed that Tunnachemootoolt, Hohastillpilp, Neshneparkkeeook, and Yoomparkkartim were the principal Chiefs of the Chopunnish Nation and ranked in the order here mentioned; ..."
Clark ⇦

Cartographer's Note:
The names of the four chiefs, as written in English in the journals, are Broken Arm, Red Flute, Cut Nose, and Five Big Hearts, in the order listed above.

May 11, 1806
"... by the assistance of the Snake boy and our interpreters were enabled to make ourselves understood by them altho' it had to pass through French, Minnetare, Shoshone and Chopunnish languages. the interpretation being tegious it occupied the greater part of the day, before we had communicated to them what we wished. ...
In the evening a man was brought in a robe by four Indians and laid down near me. they informed me that this man was a Chief of considerable note who has been in the situation I see him for 5 years. this man is incapable of moveing a single limb... he eats hartily, dejests his food perfectly, enjoys his understanding."
Clark ⇦

TEXT SUPPLEMENT TO MAPS 396 AND 397

May 12, 1806
"The Indians held a council among themselves this morning with respect to the subjects on which we had spoken to them yesterday. the result as we learnt was favourable.... the broken Arm, took the flour of the roots of cows and thickened the soope in the kettles and baskets of all his people, this being ended he made a harangue the purport of which was making known the deliberations of their council... he concluded by inviting all such men as had resolved to abide by the decrees of the council to come and eat and requested such as would not be so bound to shew themselves by not partaking of the feast."
Lewis ⇦

May 13, 1806
"a fine morning I administered to the sick... set out at 1 P.M. and proceeded down the Creek to the Flat head River... at this place we expected to have met the Canoe which was promised to be furnished us, ..."
Clark ⇦

Cartographer's Note:
Broken Arm's village was situated several miles up Commearp Creek (Lawyer Creek). The captains decided to move their camp to the east side of the "Flathead" or Clearwater River, where they possibly expected better hunting. They also may have wished to avoid crossing the Clearwater during the worst of the oncoming spring flooding, as deep mountain snows melted.

Additional Journal Quotes For Map 397

May 14, 1806
"made several attempts to exchange our Stalions for Geldin[g]s or mar[e]s without success... those horses are troublesom and cut each other very much and as we can't exchange them we think it best to castrate them and began the opperation this evening one of the indians present offered his services on this occasion."
Clark ⇦

May 16, 1806
"Sahcargarweah geathered a quantity of the roots of a speceis of fennel which we found very agreeable food, the flavor of this root is not unlike annis seed, and they dispell the wind which the roots called Cows and quawmash are apt to create particularly the latter."
Lewis ⇦

Cartographer's Note:
Here again is reference to roots causing excessive digestive gases, which had afflicted the men the year before. This cartographer is convinced that the men's gastric problems in the fall of 1805 probably were a combination of eating pounded fish as well as consuming cous and camas.

May 16, 1806
"Drewyer and Cruzatte returned having killed one deer only. ... they informed us that the hunting was but bad in the quarter they had been, ...
a little after dark Shannon and Labuish returned with one deer; they informed us that game was wild and scarce, that a large creek (Collins' Creek) ran parallel with the river at the distance of about 5 or 6 miles which they found impracticable to pass with their horses in consequence of the debth and rapidity of it's current."
Lewis ⇦

May 17, 1806
"It rained the greater part of the last night and this morning untill 8 OCk. the water passed through [the] flimzy covering and wet our bed most perfectly in sho[r]t we lay in the water all the latter part of the night. unfortunately my chronometer which for greater security I have woarn in my fob for ten days past, got wet last night; it seemed a little extraordinary that every part of my breechies which were under my head, should have escaped the moisture except the fob where the time peice was. I opened it and founded [it] nearly filled with water which I carefully drained out exposed it to the air and wiped the works as well as I could with dry feathers after which I touched them with a little bears oil. several parts of the iron and steel works were rusted a little which I wiped with all the care in my power. I set her to going and from her apparent motion hope she has sustained no material injury."
Lewis ⇦

Cartographer's Note:
This was typical of the misfortune that befell the timepiece, which was essential for taking longitude sightings. Restarting it was insufficient for operational use; Lewis needed to perform a special series of sightings over a period of three days to reset the instrument with any measure of accuracy.

May 17, 1806
"I am pleased at finding the river rise so rapidly, it now doubt is attributeable to the me[l]ting snows of the mountains; that icy barier which seperates me from my friends and Country, from all which makes life esteemable.-patience, patience"
Lewis ⇦

May 18, 1806
"The Squar wife to Shabono busied her self gathering the roots of the fenel called by the Snake Indians Year-pah for the purpose of drying to eate on the Rocky Mountains."
Clark ⇦

May 20, 1806
"rained the greater part of the last night and this morning untill meridian when it cleared away for an hour and began to rain and rained at intervals untill 4 P. M."
Clark ⇦

May 21, 1806
"we set 5 men at work to build a canoe for the purpose of takeing fish and passing the river and for which we can get a good horse. ...
we divided our store of merchindize amongst our pary for the purpose of precureing some roots &c. of the nativs to each mans part amounted to about an awl knitting pin a little paint and some thread & 2 Needles which is but a scanty dependance for roots to take us over those Great snowey Barriers..."
Clark ⇦

Cartographer's Note:
The captains determined that each person would receive a small trading stock to procure provisions for themselves in preparation for the mountain crossing. This helped encourage personal responsibility for rationing while traveling along the bleak Lolo Trail.

May 22, 1806
"Charbono's Child is very ill this evening; he is cuting teeth, and for several days past has had a violent lax, which having suddonly stoped he was attacked with a high fever and his neck and throat are much swolen this evening. We gave him a doze of creem of tartar and flour of sulpher and ... a poltice of boiled onions to his neck as warm as he could well bear it."
Lewis ⇦

May 24, 1806
"a fine morning the Child was very restless last night its jaw and back of its neck is much more swelled than it was yesterday. I gave it a dost of creme of Tarter and a fresh Poltice of Onions. ... Brattin is yet very low he eats hartily but he is so weak in the small of his back that he can't walk. we have made use of every remedy to restore him without it's haveing the desired effect. one of our party, John Shields observed that he had seen men in similar situations restored by violent swets, and bratten requested that he might be Swetted in the way Shields purposed which we agreed to. Shields dug a round hole 4 feet deep & 3 feet Diamuter in which he made a large fire so as to heet the hole after which the fire was taken out a seet placed in the hole, the pat[i]ent was then set on the seat with a board under his feet and a can of water handed him to throw on the bottom & sides of the hole so as to create as greate a heat as he could bear and the hole covered with blankets supported by hoops. after about 20 minits the patient was taken out and put in cold water a few minits, & returned to the hole in which he was kept about 1 hour. then taken out and covered with several blankets, which was taken off by degrees untill he became cool. this remedy took place yesterday and bratten is walking about to day and is much better than he has been."
Clark ⇦

May 25, 1806
"rained moderately the greater part of last night and this morning... The child is not so well to day as yesterday. I repeted the creem of tarter and the onion poltice."
Clark ⇦

TEXT SUPPLEMENT TO MAP 397

Additional Journal Quotes For Map 397 (Continued)

Cartographer's Note:
Eldon G. Chuinard, MD, author of *Only One Man Died: The Medical Aspects of the Lewis and Clark Expedition* (1979) supports Bernard DeVoto's belief that Pvt. William E. Bratton probably suffered from an inflammation or strain of the sacroiliac joint. In regard to little Baptiste, Chuinard believed the infant "had either an external abscess on the side of the neck, or mastoiditis. There is no information about this boy's future life to indicate that he had chronic drainage or impairment of hearing, which often occur as sequels to an acute mastoiditis." William Clark also was asked by the Nez Perces to attend to a chief, who for five years had been in a paralytic or a stroke-like state (see below). The Indians and the man's father, "a very good looking old man" according to Clark, were extremely attentive to the afflicted man.

May 25, 1806
"I caused a swet to be prepared for the Indn. in the same hole which bratten had been swetted in two days past" Clark ⇐

May 26, 1806
"one of our men saw a salmon in the river to day, and two others eat of salmon at the near village which was brought from Lewis's river. our canoe finished and put into the water. it will carry 12 men. the [river] riseing very fast and snow appear to melt on the mountains." Clark ⇐

May 27, 1806
"Serjt. Ordway and two men are ordered to cross this river and proceed on through the plains to Lewis's [river] and precure some salmon on that river, and return tomorrow if possible …
we sent Rub: Field in serch of the horse which the indians had given us to kill. at 10 A. M. he returned with the horse and he was killed and butchered; he was large and in good order. Hohastillpilp told us that most of the horses which we saw runing in those pains in this neighbourhood at large belonged to himself and his people, and whenever we were in want of meet, he requested that we would kill any of them we wished; this is a piece of liberallity which would do honour to such as bost of civilization. …
Shabono's child is much better to day; …
The Indians were so anxious that the sick Chief (who has lost the use of his limbs) should be sweted under our inspection they requested me to make a 2d. attempt to day; accordingly the hole was enlargened and his father a very good looking old man performed all the drugery &c. we could not make him swet as copously as we wished, being compelled to keep him erect in the hole by means of cords. after the oppiration he complained of considerable pain, I gave him 30 drops of Laudnom which soon composed him and he rested very well." Clark ⇐

May 28, 1806
"The Sick Chief is much better this morning he can use his hands and arms and seems much pleased with the prospects of recovering, he says he feels much better than he has done for a great number of months." Clark ⇐

May 29, 1806
"Bratten is recovering his strength very fast. the Child, and the Indian Cheif are also on the recovery." Clark ⇐

May 30, 1806
"Shannon and Collins were permitted to pass the river in order to trade with the natives and lay in a store of roots and bread for themselves with their proportion of the merchandize as the others had done; on landing on the opposite shore the canoe was driven broad side with the full forse of a very strong current against some Standing trees and instantly filled with water and sunk. Potts who was with them is an indifferent swimer, it was with dificulty he made the land. they lost three blankets a Blanket Cappo and their pittance of merchandize." Lewis ⇐

May 30, 1806
"The loss of these blankets is the greatest which hath happened to any individuals since we began our voyage, as there are only three men in the party, who have more than a blanket a piece. The river is so high that the trees stand some distance in the water." Gass ⇐

May 30, 1806
"we gave the sick Chief a severe Swet to day, shortly after which he could move one of his legs and thy's and work his toes pritty well, the other leg he can move a little; his fingers and arms seem to be almost entirely restored." Clark ⇐

May 30, 1806
"the honey bee is not found here. the bumble bee is." Clark ⇐

May 31, 1806
"The Indians pursued a mule deer to the river opposit to our Camp this evening; the deer swam over and one of our hunters killed it. there being a large party of indians assembled on this occasion on the opposit side with Tin-nach-e-moo-tolt they attempted to rais our canoe which was sunk on that side of the river yesterday; they made the attempt but were unable to effect it." Clark ⇐

June 1, 1806
"This morning Geo: Drewyer accompanied by Hohastillpilp set out in serch of two tomahawks of ours which we have understood were in the possession of certain indians resideing at a distance in the plains on the South Side of Flat Head river; … feel some anxiety with respect to Sergt. Ordway and party who were sent to Lewis's river for salmon; we have receved no intellegence of them since they set out. we desired Drewyer to make some enquiry after the Twisted hair; the old man has not been as good as his word with respect to encamping near us, and we fear we shall be at a loss to procure guides to conduct us by the different routs we wish to pursue from Travillers rest to the waters of the Missouri." Clark ⇐

June 2, 1806
"our horses many of them have become so wild that we cannot take them without the assistance of the Indians who are extreemly dextrous in throwing a rope… had a strong pound formed… to take them at pleasure. Drewyer arrived this morning with Neeshneparkkeeook and Hohastillpilp who had accompanyed him to the lodges of the persons who had our tomahawks. he obtained both the tomahawks principally by the influence of the former of these Cheifs. the one which had been stolen we prized most as it was the private property of the late Sergt. Floyd and Capt. C. was desirous of returning it to his friends. the man who had this tomahawk had purchased it from the Indian that had stolen it, and was himself at the moment of their arrival just expiring. his relations were unwilling to give up the tomehawk as they intended to bury it with the disceased owner, but were at length induced to do so for the consideration of a ha[n]dherchief, two strands of beads, which (Cap C sent by) Drewyer gave them and two horses given by the cheifs to be killed agreeably to their custom at the grave of the disceased…Ordway Frazier and Wizer returned with 17 salmon and some roots of cows; the distance was so great from which they had brought the fish that most of them were nearly spoiled. these fish were as fat as any I ever saw; …" Lewis ⇐

June 2, 1806
"our horses are all recovering & I have no hesitation in declareing that I believe that the Indian method of guilding [is] preferable to that practised by ourselves." Clark ⇐

June 3, 1806
"To day the Indians dispatched an express over the mountains to Travellers rest or to the neighbourhood of that creek on Clark's river in order to learn from a band of Flat-Heads who inhabit that river and who have probably wintered on Clarks river near the enterance of travellers rest Creek, the occurences which have taken place on the East side of the mountains dureing the last winter." Clark ⇐

June 3, 1806
"the mountains being practicable for this express we thought it probable that we could also pass, but the Chief's informs us that several of the Creek's would yet swim our horses, that there was no grass and that the road was extreemly deep and slipery; … we have come to a resolution to remove from hence to the quawmash Grounds beyond Colins Creek on the 10th, to hunt in that neighbourhood a fiew days, if possible lay in a stock of meat, and then attempt the mountains about the middle of this month." Clark ⇐

June 6, 1806
"the young men who we had requested to accompany us to the falls of Missouri, [they] were not yet selected for that purpose nor could they be so untill they had a meeting of the nation in council. that this would happen in the course of ten or 12 days as the whole of the Lodges were about to move to the head of Commeap Creek… on examonation we find our whole party have a sufficient store of bread and roots for our Voyage. a circumstance not unpleasing." Clark ⇐

June 8, 1806
"in the evening several foot races were run by the men of our party and the Indians; after which our party devided and played at prisoners base untill night. after dark the fiddle was played and the party amused themselves in danceing. one of those Indians informed us that we could not cross the mountains untill the full of the next moon; or about the 1st of July." Clark ⇐

June 9, 1806
"The flat head river is still falling fast and [is] nearly as low as it was at the time we arrived at this place. this fall of water is what the natives have informed us was a proper token for us. when this river fell the Snows would be sufficiently melted for us to cross the Mountains. the greater length of time we delayed after that time, the higher the grass would grow on th[e] Mountains."

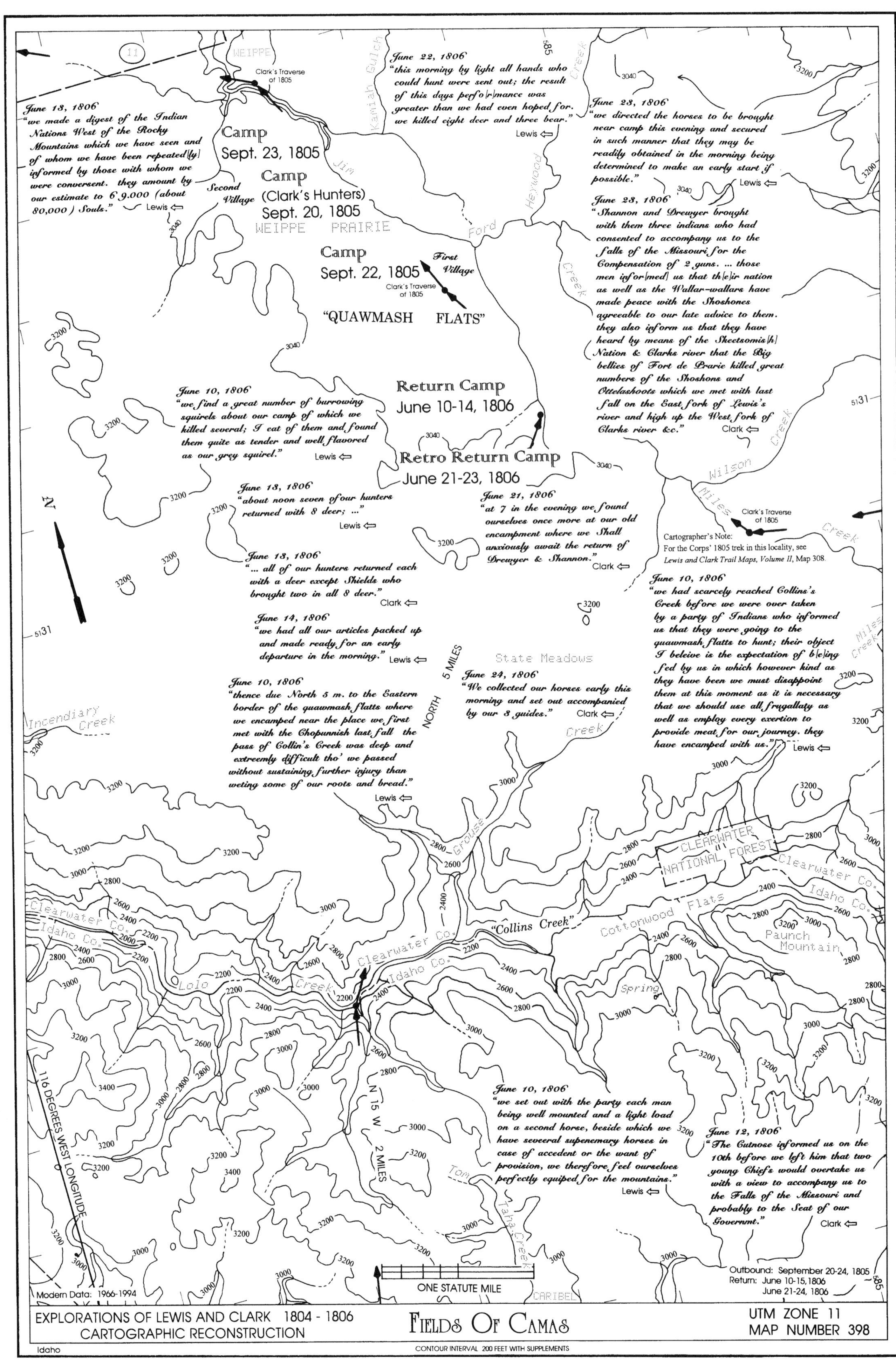
June 22, 1806
"this morning by light all hands who could hunt were sent out; the result of this days perfo[r]mance was greater than we had even hoped for. we killed eight deer and three bear." Lewis
June 13, 1806
"we made a digest of the Indian Nations West of the Rocky Mountains which we have seen and of whom we have been repeated[ly] informed by those with whom we were conversent. they amount by our estimate to 6'9.000 (about 80,000) Souls." Lewis
Camp
Sept. 23, 1805
Camp
(Clark's Hunters)
Sept. 20, 1805
Second Village
WEIPPE PRAIRIE
Camp
Sept. 22, 1805
First Village
Clark's Traverse of 1805
"QUAWMASH FLATS"
June 23, 1806
"we directed the horses to be brought near camp this evening and secured in such manner that they may be readily obtained in the morning being determined to make an early start if possible." Lewis
June 23, 1806
"Shannon and Drewyer brought with them three indians who had consented to accompany us to the falls of the Missouri for the Compensation of 2 guns. ... those men info[r]m[ed] us that th[e]ir nation as well as the Wallar-wallars have made peace with the Shoshones agreeable to our late advice to them. they also inform us that they have heard by means of the Sheetsomis[h] Nation & Clarks river that the Big bellies of Fort de Prarie killed great numbers of the Shoshons and Ottelashoots which we met with last fall on the East fork of Lewis's river and high up the West fork of Clarks river &c." Clark
June 10, 1806
"we find a great number of burrowing squirels about our camp of which we killed several; I eat of them and found them quite as tender and well flavored as our grey squirel." Lewis
Return Camp
June 10-14, 1806
Retro Return Camp
June 21-23, 1806
June 13, 1806
"about noon seven of our hunters returned with 8 deer; ..." Lewis
June 21, 1806
"at 7 in the evening we found ourselves once more at our old encampment where we Shall anxiously await the return of Drewyer & Shannon." Clark
Cartographer's Note:
For the Corps' 1805 trek in this locality, see Lewis and Clark Trail Maps, Volume II, Map 308.
June 13, 1806
"... all of our hunters returned each with a deer except Shields who brought two in all 8 deer." Clark
June 14, 1806
"we had all our articles packed up and made ready for an early departure in the morning." Lewis
June 10, 1806
"we had scarcely reached Collins's Creek before we were over taken by a party of Indians who informed us that they were going to the quawmash flatts to hunt; their object I beleive is the expectation of b[e]ing fed by us in which however kind as they have been we must disappoint them at this moment as it is necessary that we should use all frugallaty as well as employ every exertion to provide meat for our journey. they have encamped with us." Lewis
June 10, 1806
"thence due North 5 m. to the Eastern border of the quawmash flatts where we encamped near the place we first met with the Chopunnish last fall the pass of Collin's Creek was deep and extreemly difficult tho' we passed without sustaining further injury than weting some of our roots and bread." Lewis
June 24, 1806
"We collected our horses early this morning and set out accompanied by our 3 guides." Clark
NORTH 5 MILES
State Meadows
WEIPPE
Kamiah Gulch
Jim Ford Creek
Heywood Creek
Wilson Creek
Miles Creek
Incendiary Creek
Grouse Creek
"Collins Creek"
Lolo Creek
Clearwater Co.
Idaho Co.
CLEARWATER NATIONAL FOREST
Cottonwood Flats
Paunch Mountain
Spring
Tom Taha Creek
N 15 W 2 MILES
116 DEGREES WEST LONGITUDE
June 10, 1806
"we set out with the party each man being well mounted and a light load on a second horse, beside which we have several supenemary horses in case of accedent or the want of provision, we therefore feel ourselves perfectly equiped for the mountains." Lewis
June 12, 1806
"The Cutnose informed us on the 10th before we left him that two young Chiefs would overtake us with a view to accompany us to the Falls of the Missouri and probably to the Seat of our Governmt." Clark
ONE STATUTE MILE
CARIBEL
Outbound: September 20-24, 1805
Return: June 10-15, 1806
June 21-24, 1806
Modern Data: 1966-1994
EXPLORATIONS OF LEWIS AND CLARK 1804 - 1806
CARTOGRAPHIC RECONSTRUCTION
Idaho
FIELDS OF CAMAS
CONTOUR INTERVAL 200 FEET WITH SUPPLEMENTS
UTM ZONE 11
MAP NUMBER 398

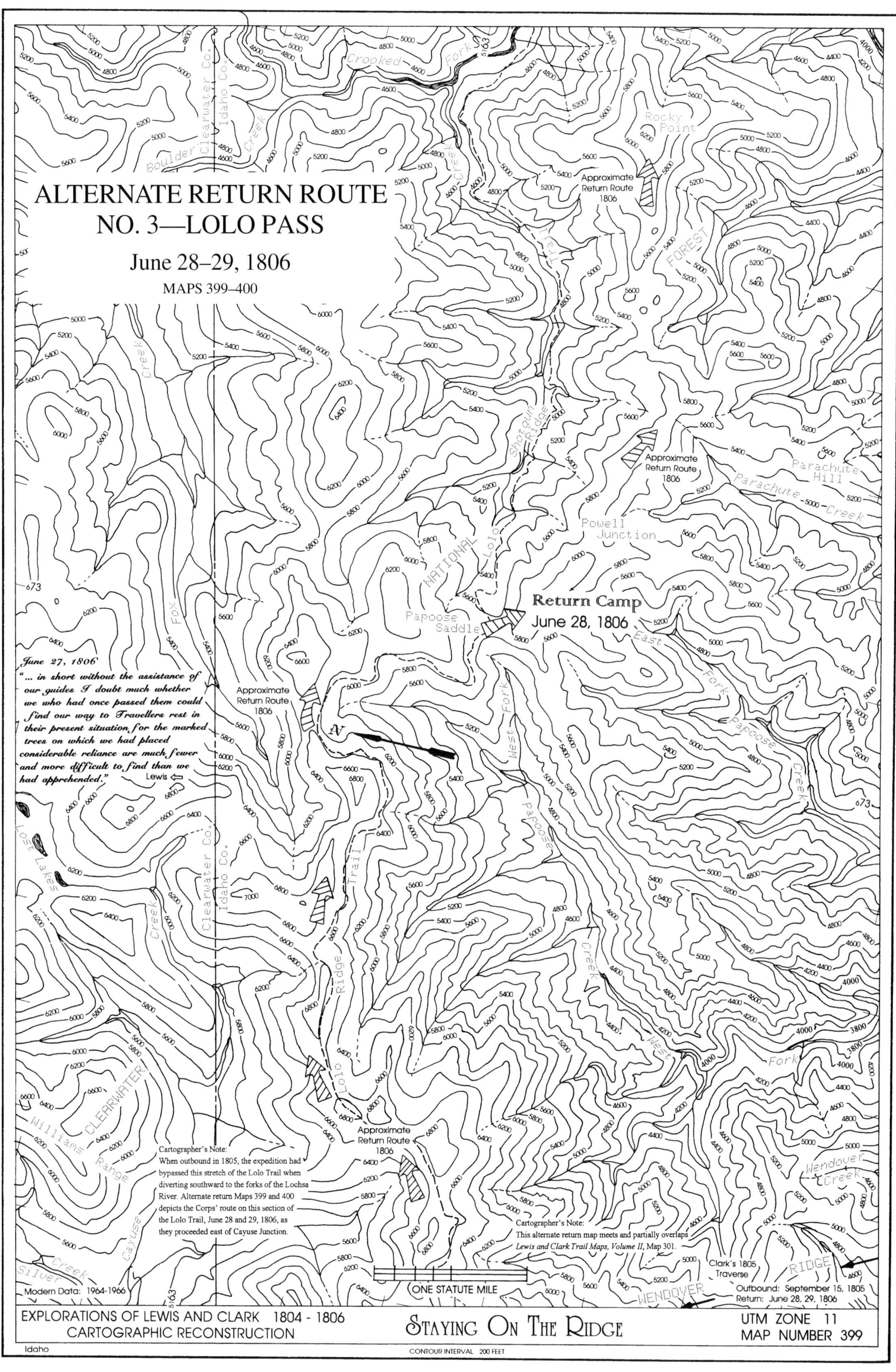

ALTERNATE RETURN ROUTE
NO. 3—LOLO PASS
June 28–29, 1806
MAPS 399–400
June 27, 1806
"... in short without the assistance of our guides I doubt much whether we who had once passed them could find our way to Travellers rest in their present situation, for the marked trees on which we had placed considerable reliance are much fewer and more difficult to find than we had apprehended."
Lewis
Return Camp
June 28, 1806
Approximate Return Route 1806
Crooked Fork
Boulder Creek
Clearwater Co.
Idaho Co.
Rocky Point
Lolo Trail
Shotgun Ridge
FOREST
NATIONAL
Parachute Hill
Parachute Creek
Powell Junction
Papoose Saddle
East Fork Papoose Creek
West Fork Papoose Creek
Fox Creek
Lost Lakes
Lolo Ridge Trail
West Fork
Williams Range
CLEARWATER
Cayuse Creek
Silver Creek
Wendover Creek
WENDOVER RIDGE
Cartographer's Note:
When outbound in 1805, the expedition had bypassed this stretch of the Lolo Trail when diverting southward to the forks of the Lochsa River. Alternate return Maps 399 and 400 depicts the Corps' route on this section of the Lolo Trail, June 28 and 29, 1806, as they proceeded east of Cayuse Junction.
Cartographer's Note:
This alternate return map meets and partially overlaps Lewis and Clark Trail Maps, Volume II, Map 301.
Clark's 1805 Traverse
Outbound: September 15, 1805
Return: June 28, 29, 1806
ONE STATUTE MILE
Modern Data: 1964-1966
EXPLORATIONS OF LEWIS AND CLARK 1804 - 1806
CARTOGRAPHIC RECONSTRUCTION
STAYING ON THE RIDGE
UTM ZONE 11
MAP NUMBER 399
Idaho
CONTOUR INTERVAL 200 FEET

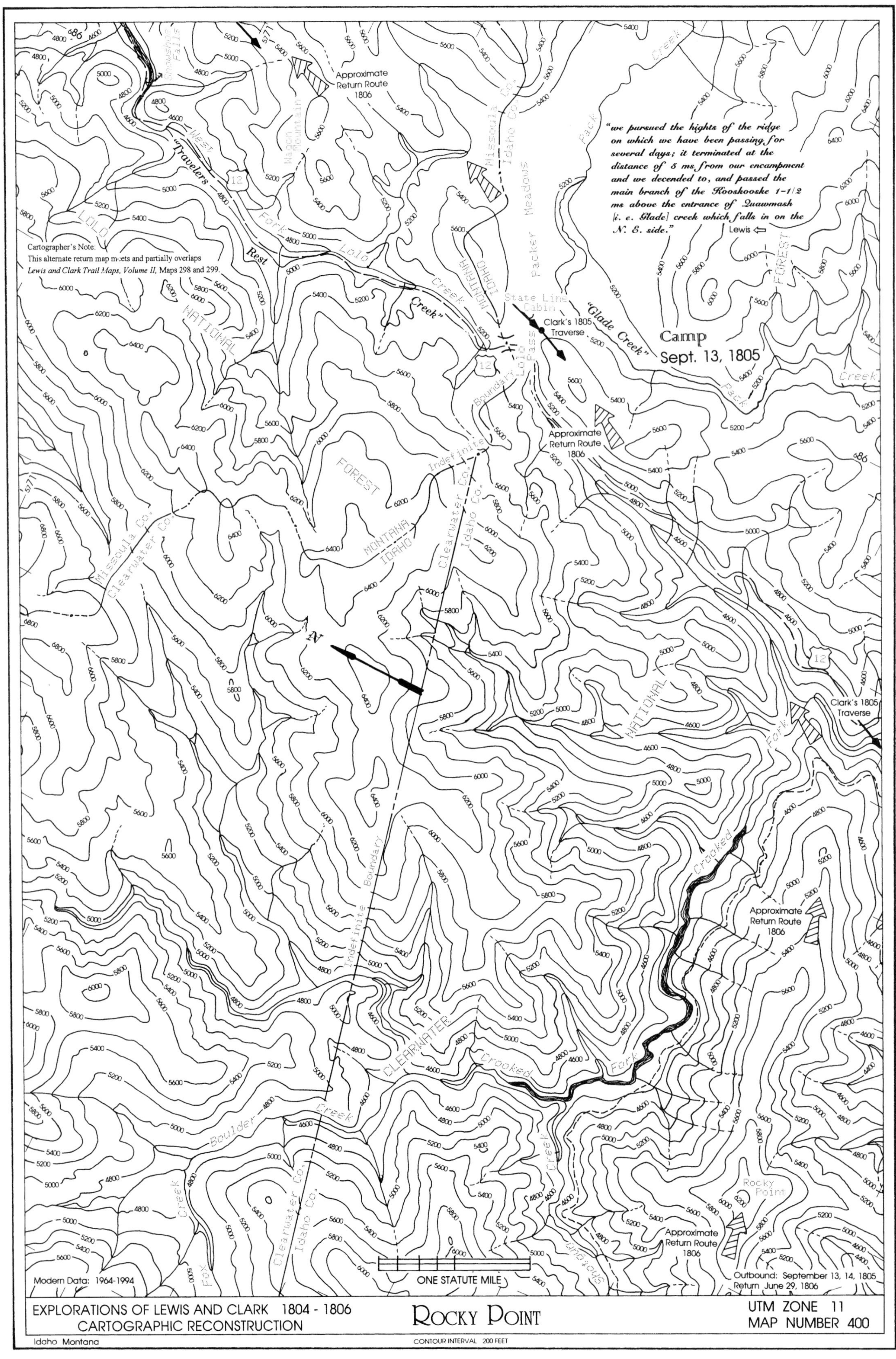
"we pursued the hights of the ridge on which we have been passing for several days; it terminated at the distance of 5 ms from our encampment and we decended to, and passed the main branch of the Kooskooske 1-1/2 ms above the entrance of Quawmash [i. e. Glade] creek which falls in on the N. E. side."
Lewis
Cartographer's Note:
This alternate return map meets and partially overlaps Lewis and Clark Trail Maps, Volume II, Maps 298 and 299.
Camp
Sept. 13, 1805
Clark's 1805 Traverse
Approximate Return Route 1806
State Line Cabin
"Glade Creek"
Packer Meadows
Lolo Pass
Travelers Rest
West Fork Lolo Creek
Rocky Point
Crooked Fork
Boulder Creek
Modern Data: 1964-1994
ONE STATUTE MILE
Outbound: September 13, 14, 1805
Return June 29, 1806
EXPLORATIONS OF LEWIS AND CLARK 1804 - 1806
CARTOGRAPHIC RECONSTRUCTION
ROCKY POINT
UTM ZONE 11
MAP NUMBER 400
Idaho Montana
CONTOUR INTERVAL 200 FEET

Alternate Return Route No. 4

MAPS 401 - 411

CLARK PARTY, BIG HOLE, MONTANA

Capt. William Clark (journalist)
Toussaint Charbonneau (interpreter)
Sacagawea
Jean Baptiste Charbonneau
York
Sgt. John Ordway (journalist)
Sgt. Nathaniel Pryor
Pvt. William Bratton
Pvt. John Collins
Pvt. John Colter
Pvt. Pierre Cruzatte
Pvt. George Gibson,
Pvt. Hugh Hall
Pvt. Thomas Howard
Pvt. François Labiche
Pvt. Baptiste Lepage
Pvt. John Potts
Pvt. George Shannon
Pvt. John Shields
Pvt. Joseph Whitehouse
Pvt. Alexander Willard
Pvt. Richard Windsor
Pvt. Peter Wiser
49 horses and a colt

At Travelers Rest on July 3, 1806, as Lewis's 10-man contingent headed north to today's Clark Fork, Clark's larger party of 23 people, plus 49 horses and a colt, set out south toward the Beaverhead country, where canoes and baggage had been cached in August 1805.

With the division of the expedition at Travelers Rest, the four journalists whose diaries are known to exist for this period were evenly split between the two parties. Sergeant Patrick Gass would go with Meriwether Lewis as far as the Great Falls, whereas Sergeant Nathaniel Ordway went with William Clark to the Beaverhead country and the Three Forks. (Private Joseph Whitehouse's *known* journal entries had abruptly ended on April 2, 1806, at the Willamette River.)

After separating on July 3, the two contingents would not reunite until 40 days later on the Missouri River, August 12, 1806.

Clark initially retraced the Corps's 1805 route in the Bitterroot Valley, before taking a shortcut that had been described to him by the mountain tribes. This new route proceeded to Gibbons Pass, continued through the Big Hole valley, crossed Big Hole Pass to Grasshopper Creek, and entered "Shoshone Cove" (today's Horse Prairie) near the "Camp Fortunate" caches. This new alternate route bypassed the longer and more difficult trail followed in the Lemhi country during the late summer, 1805.

The Big Hole, a great natural opening in the Rocky Mountains, is crescent shaped—approximately 45 to 50 miles in length, and about 4 to 25 miles wide. Many small streams draining from the high, snow-bound mountains surrounding the Big Hole form the Big Hole River, named the "Wisdom River" by Lewis and Clark.

The continental divide follows the crest of the Beaverhead Mountains on the west side of the valley. The Pioneer Mountains border the east side, leaving a break for the exit of the Big Hole River eastward to the Beaverhead. Numerous mountain passes lead to and from the valley, including two located approximately 25 miles apart, both called Big Hole Pass.

During the 1877 Nez Perce War, a hard-fought battle occurred at the north end of the valley between U.S. troops and Nez Perces (Big Hole National Battlefield).

Clark summed up the July 5–8, 1806, journey through the Big Hole country in his usual straightforward style:

"The road which we have traveled from travellers rest Creek to *this place* (*this place is the head of Jeffer river where we left our canoes*) [is] an excellent road...[This] road and with only a few trees being cut out of the way would be an excellent waggon road one Mountain of about 4 miles over excepted which would require a little digging The distance is 164 Miles."

The captain obviously was pleased with this better route. It is clear, too, that Clark well understood the geography of the many interrelated watersheds and mountain ridges in this region.

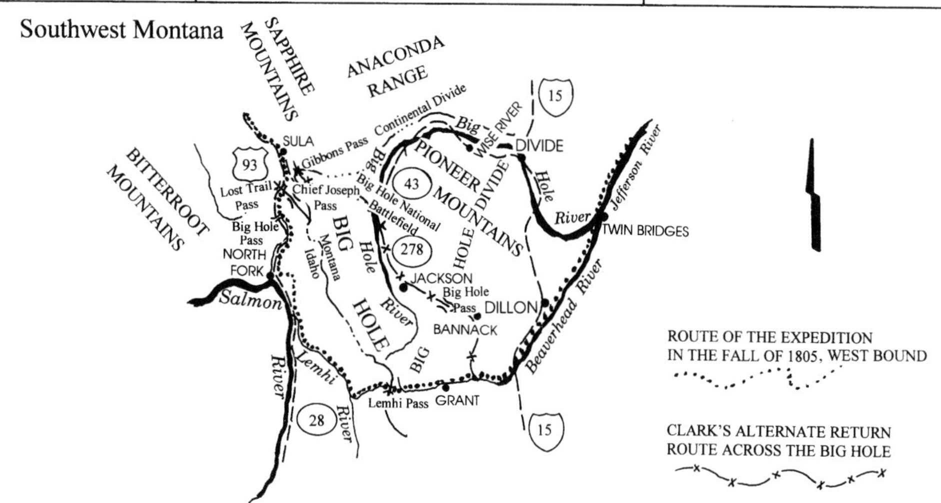

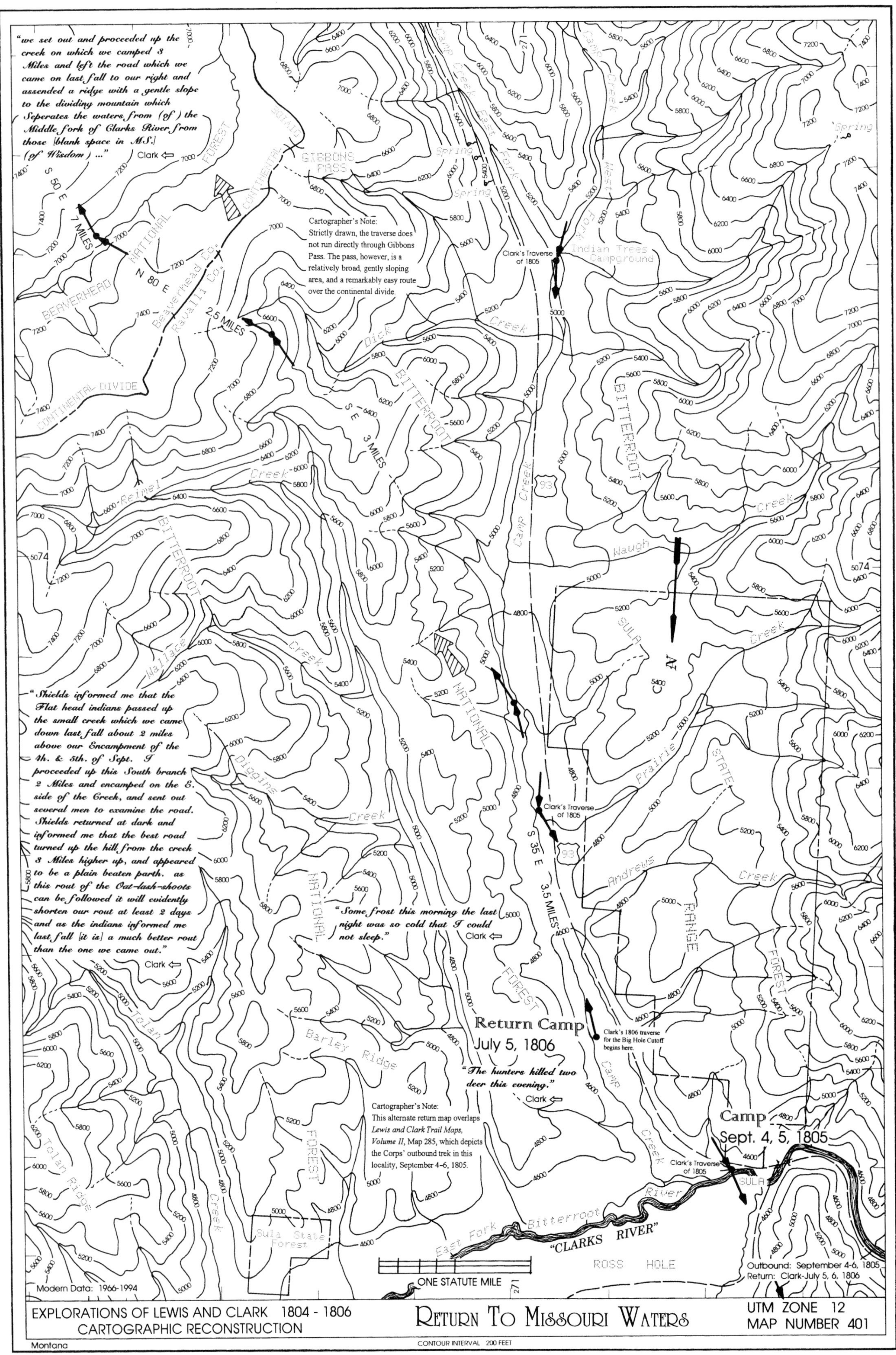
"we set out and proceeded up the creek on which we camped 3 Miles and left the road which we came on last fall to our right and assended a ridge with a gentle slope to the dividing mountain which Seperates the waters from (of) the Middle fork of Clarks River from those [blank space in MS.] (of Wisdom) ..." Clark
Cartographer's Note: Strictly drawn, the traverse does not run directly through Gibbons Pass. The pass, however, is a relatively broad, gently sloping area, and a remarkably easy route over the continental divide.
GIBBONS PASS
Clark's Traverse of 1805
Indian Trees Campground
"Shields informed me that the Flat head indians passed up the small creek which we came down last fall about 2 miles above our Encampment of the 4th. & 5th. of Sept. I proceeded up this South branch 2 Miles and encamped on the E. side of the Creek, and sent out several men to examine the road. Shields returned at dark and informed me that the best road turned up the hill from the creek 3 Miles higher up, and appeared to be a plain beaten parth. as this rout of the Oat-lash-shoots can be followed it will evidently shorten our rout at least 2 days and as the indians informed me last fall [it is] a much better rout than the one we came out." Clark
"Some frost this morning the last night was so cold that I could not sleep." Clark
Return Camp
July 5, 1806
"The hunters killed two deer this evening." Clark
Clark's 1806 traverse for the Big Hole Cutoff begins here.
Cartographer's Note: This alternate return map overlaps Lewis and Clark Trail Maps, Volume II, Map 285, which depicts the Corps' outbound trek in this locality, September 4-6, 1805.
Camp
Sept. 4, 5, 1805
"CLARKS RIVER"
ROSS HOLE
Outbound: September 4-6, 1805
Return: Clark-July 5, 6, 1806
ONE STATUTE MILE
Modern Data: 1966-1994
EXPLORATIONS OF LEWIS AND CLARK 1804 - 1806
CARTOGRAPHIC RECONSTRUCTION
RETURN TO MISSOURI WATERS
UTM ZONE 12
MAP NUMBER 401
Montana
CONTOUR INTERVAL 200 FEET

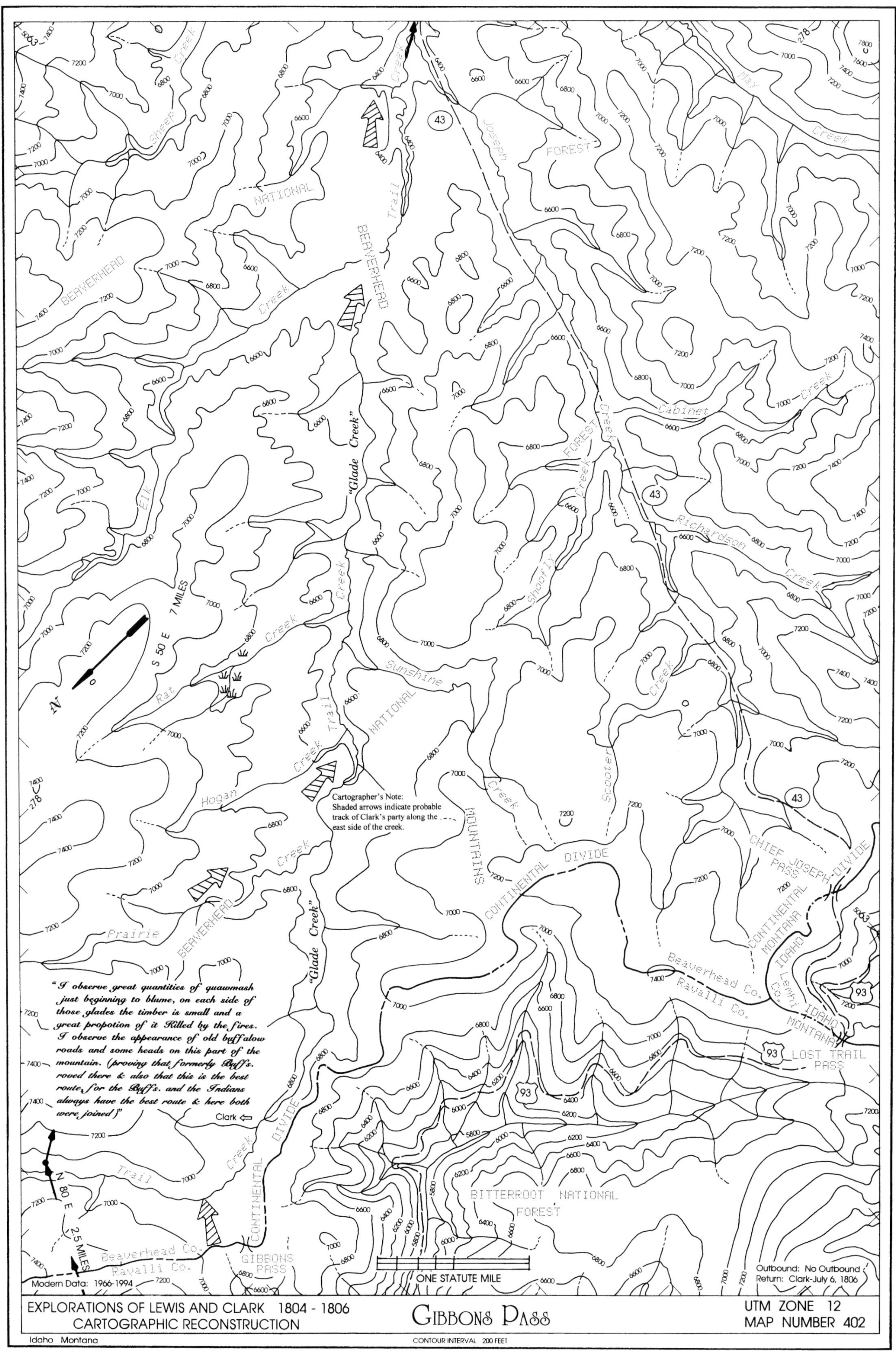
EXPLORATIONS OF LEWIS AND CLARK 1804 - 1806
CARTOGRAPHIC RECONSTRUCTION
GIBBONS PASS
UTM ZONE 12
MAP NUMBER 402
Idaho Montana
CONTOUR INTERVAL 200 FEET
Modern Data: 1966-1994
ONE STATUTE MILE
Outbound: No Outbound
Return: Clark-July 6, 1806
Cartographer's Note:
Shaded arrows indicate probable track of Clark's party along the east side of the creek.
"I observe great quantities of quawmash just beginning to blume, on each side of those glades the timber is small and a great propotion of it Killed by the fires. I observe the appearance of old buffalow roads and some heads on this part of the mountain. (proving that formerly Buff's. roved there & also that this is the best route, for the Buff's. and the Indians always have the best route & here both were joined)"
Clark ⇦
S 50 E 7 MILES
N 80 E 2.5 MILES
N
"Glade Creek"
"Glade Creek"
Trail Creek
BEAVERHEAD NATIONAL FOREST
Sheep Creek
Joseph Creek
May Creek
Cabinet Creek
Richardson Creek
Forest Creek
Shoofly Creek
Sunshine Creek
Scooter Creek
Elk Creek
Rat Creek
Hogan Creek
Prairie Creek
BEAVERHEAD MOUNTAINS
CONTINENTAL DIVIDE
CHIEF JOSEPH PASS
CONTINENTAL DIVIDE MONTANA IDAHO
Beaverhead Co.
Ravalli Co.
Lemhi Co.
IDAHO
MONTANA
LOST TRAIL PASS
BITTERROOT NATIONAL FOREST
GIBBONS PASS
43
93
278

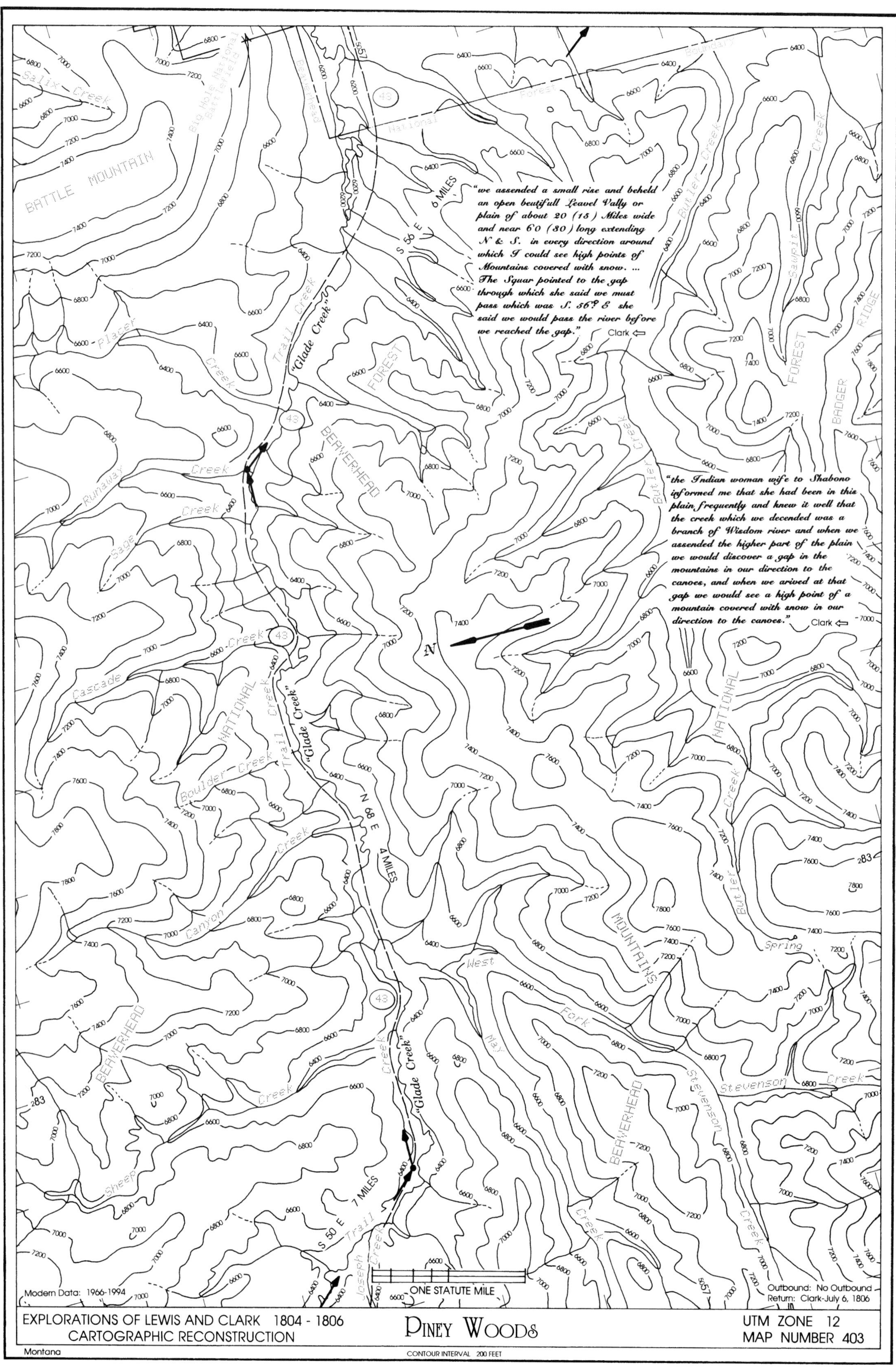
"we assended a small rise and beheld an open beutifull Leavel Vally or plain of about 20 (15) Miles wide and near 60 (30) long extending N & S. in every direction around which I could see high points of Mountains covered with snow. ... The Squar pointed to the gap through which she said we must pass which was S. 56° E she said we would pass the river before we reached the gap."
Clark
"the Indian woman wife to Shabono informed me that she had been in this plain frequently and knew it well that the creek which we decended was a branch of Wisdom river and when we assended the higher part of the plain we would discover a gap in the mountains in our direction to the canoes, and when we arived at that gap we would see a high point of a mountain covered with snow in our direction to the canoes."
Clark
BATTLE MOUNTAIN
Big Hole National Battlefield
National Forest Boundary
Salix Creek
Placer Creek
Trail Creek
"Glade Creek"
Runaway Creek
Sage Creek
Cascade Creek
Boulder Creek
Canyon Creek
Sheep Creek
Joseph Creek
West Fork
May Creek
Stevenson Creek
Butler Creek
Sawpit Creek
Spring
BADGER RIDGE
FOREST
BEAVERHEAD NATIONAL FOREST
BEAVERHEAD MOUNTAINS
S 50 E 6 MILES
N 68 E 4 MILES
S 50 E 7 MILES
N
ONE STATUTE MILE
Modern Data: 1966-1994
Outbound: No Outbound
Return: Clark-July 6, 1806
EXPLORATIONS OF LEWIS AND CLARK 1804 - 1806
CARTOGRAPHIC RECONSTRUCTION
PINEY WOODS
UTM ZONE 12
MAP NUMBER 403
Montana
CONTOUR INTERVAL 200 FEET

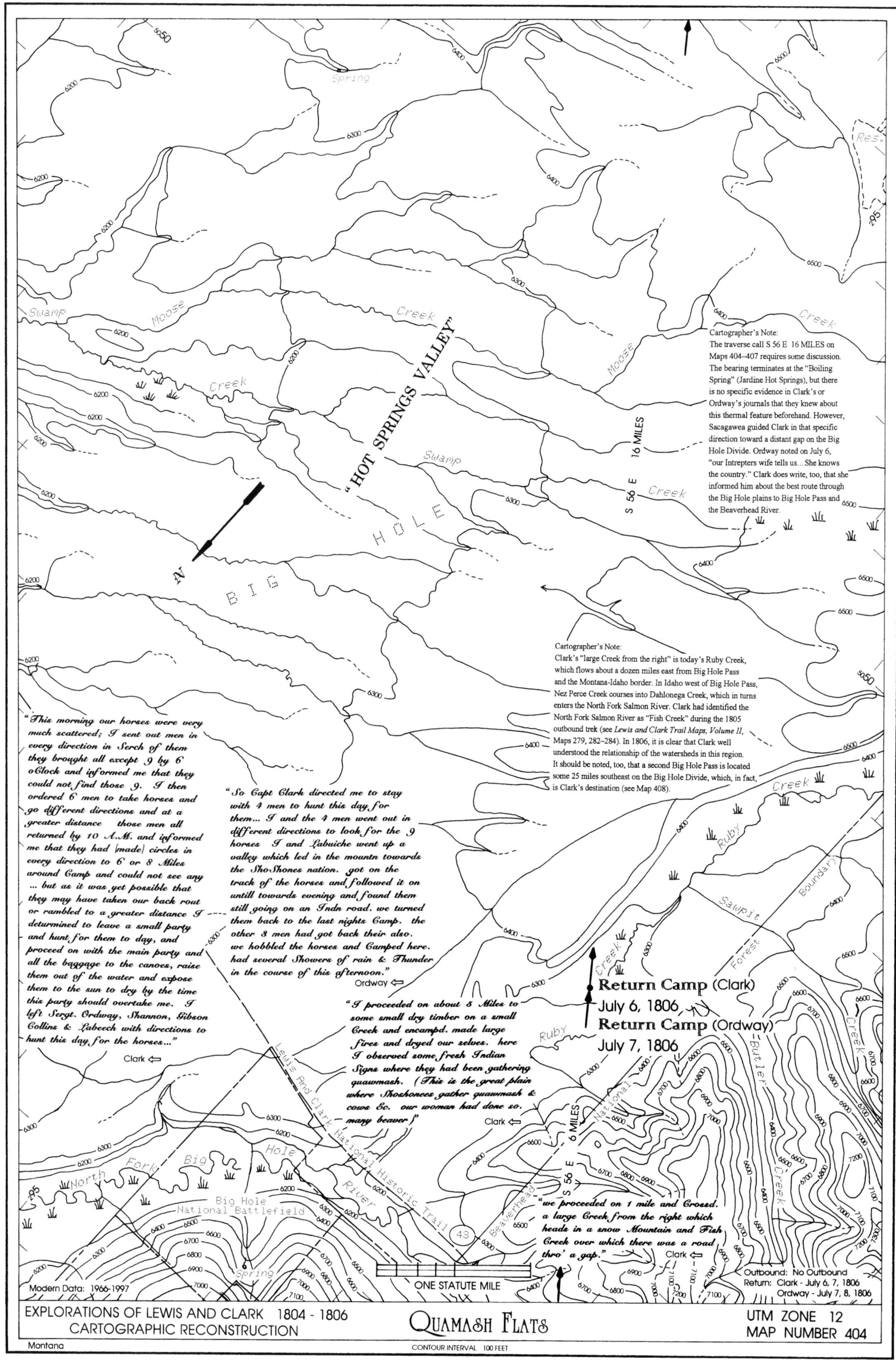
"HOT SPRINGS VALLEY"
BIG HOLE
Swamp Moose Creek
Moose Creek
Swamp Creek
Creek
Spring
Res.
S 56 E 16 MILES
Cartographer's Note:
The traverse call S 56 E 16 MILES on Maps 404–407 requires some discussion. The bearing terminates at the "Boiling Spring" (Jardine Hot Springs), but there is no specific evidence in Clark's or Ordway's journals that they knew about this thermal feature beforehand. However, Sacagawea guided Clark in that specific direction toward a distant gap on the Big Hole Divide. Ordway noted on July 6, "our Intrepters wife tells us... She knows the country." Clark does write, too, that she informed him about the best route through the Big Hole plains to Big Hole Pass and the Beaverhead River.
Cartographer's Note:
Clark's "large Creek from the right" is today's Ruby Creek, which flows about a dozen miles east from Big Hole Pass and the Montana-Idaho border. In Idaho west of Big Hole Pass, Nez Perce Creek courses into Dahlonega Creek, which in turns enters the North Fork Salmon River. Clark had identified the North Fork Salmon River as "Fish Creek" during the 1805 outbound trek (see Lewis and Clark Trail Maps, Volume II, Maps 279, 282–284). In 1806, it is clear that Clark well understood the relationship of the watersheds in this region. It should be noted, too, that a second Big Hole Pass is located some 25 miles southeast on the Big Hole Divide, which, in fact, is Clark's destination (see Map 408).
"This morning our horses were very much scattered; I sent out men in every direction in Serch of them they brought all except 9 by 6 oClock and informed me that they could not find those 9. I then ordered 6 men to take horses and go different directions and at a greater distance those men all returned by 10 A.M. and informed me that they had [made] circles in every direction to 6 or 8 Miles around Camp and could not see any ... but as it was yet possible that they may have taken our back rout or rambled to a greater distance I determined to leave a small party and hunt for them to day, and proceed on with the main party and all the baggage to the canoes, raise them out of the water and expose them to the sun to dry by the time this party should overtake me. I left Sergt. Ordway, Shannon, Gibson Collins & Labeech with directions to hunt this day for the horses..."
Clark ⇦
"So Capt Clark directed me to stay with 4 men to hunt this day for them... I and the 4 men went out in different directions to look for the 9 horses I and Labuiche went up a valley which led in the mountn towards the ShoShones nation. got on the track of the horses and followed it on untill towards evening and found them still going on an Indn road. we turned them back to the last nights Camp. the other 3 men had got back their also. we hobbled the horses and Camped here. had several Showers of rain & Thunder in the course of this afternoon."
Ordway ⇦
"I proceeded on about 5 Miles to some small dry timber on a small Creek and encampd. made large fires and dryed our selves. here I observed some fresh Indian Signs where they had been gathering quawmash. (This is the great plain where Shoshonees gather quawmash & cows &c. our woman had done so. many beaver)"
Clark ⇦
"we proceeded on 1 mile and Crossd. a large Creek from the right which heads in a snow Mountain and Fish Creek over which there was a road thro' a gap."
Clark ⇦
Return Camp (Clark)
July 6, 1806
Return Camp (Ordway)
July 7, 1806
S 56 E 6 MILES
Ruby Creek
Sawpit Creek
Forest Boundary
National
Butler Creek
Creek
North Fork Big Hole River
Big Hole National Battlefield
Lewis And Clark National Historic Trail
Beaverhead
Spring
43
ONE STATUTE MILE
Modern Data: 1966-1997
Outbound: No Outbound
Return: Clark - July 6, 7, 1806
Ordway - July 7, 8, 1806
EXPLORATIONS OF LEWIS AND CLARK 1804 - 1806
CARTOGRAPHIC RECONSTRUCTION
QUAMASH FLATS
UTM ZONE 12
MAP NUMBER 404
Montana
CONTOUR INTERVAL 100 FEET

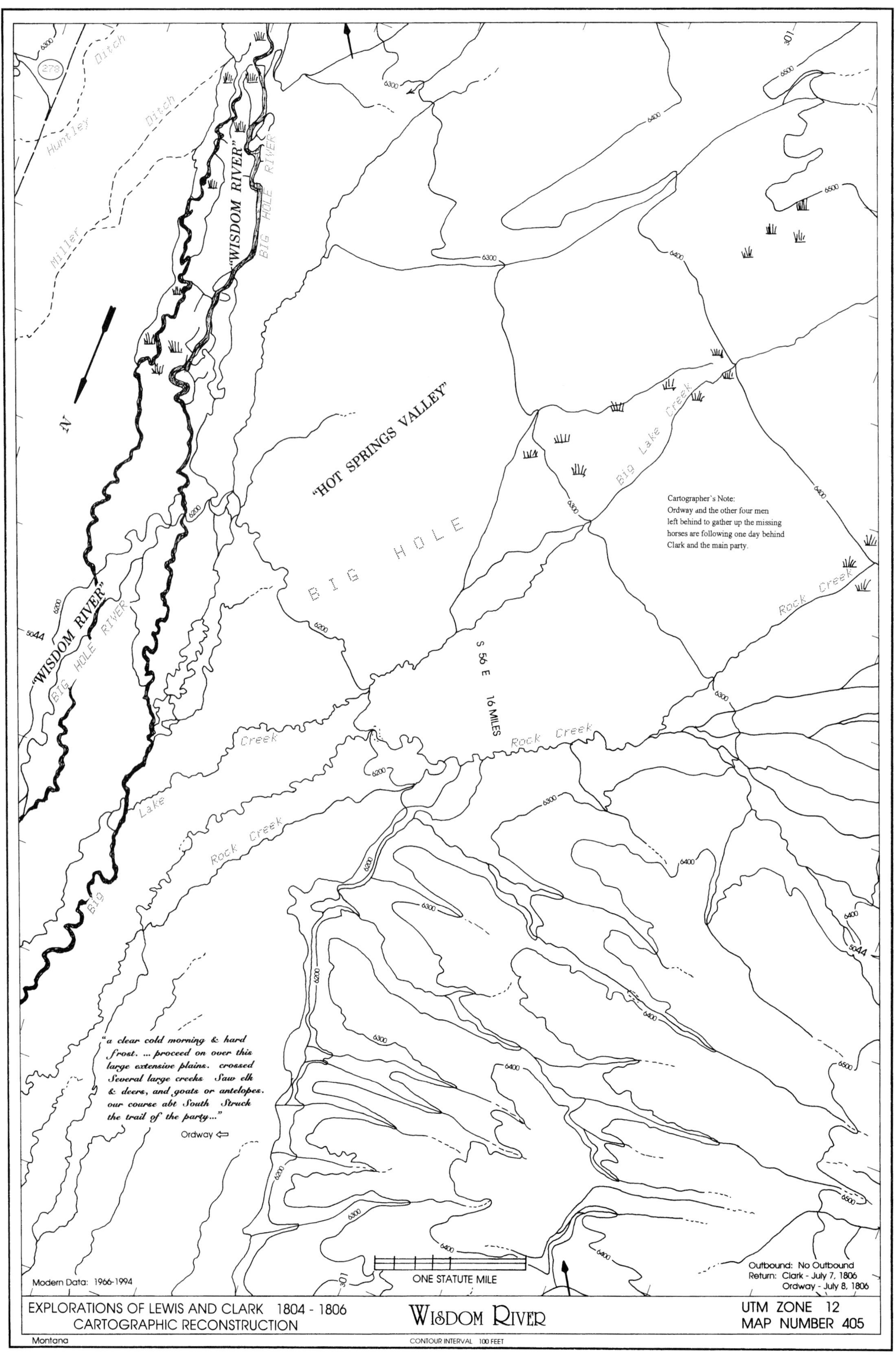
"WISDOM RIVER"
BIG HOLE RIVER
"HOT SPRINGS VALLEY"
BIG HOLE
Big Lake Creek
Rock Creek
Huntley Ditch
Miller Ditch
S 56 E 16 MILES
Cartographer's Note:
Ordway and the other four men left behind to gather up the missing horses are following one day behind Clark and the main party.
"a clear cold morning & hard frost. ... proceed on over this large extensive plains. crossed Several large creeks Saw elk & deers, and goats or antelopes. our course abt South Struck the trail of the party..."
Ordway ⇦
Modern Data: 1966-1994
ONE STATUTE MILE
Outbound: No Outbound
Return: Clark - July 7, 1806
Ordway - July 8, 1806
EXPLORATIONS OF LEWIS AND CLARK 1804 - 1806
CARTOGRAPHIC RECONSTRUCTION
WISDOM RIVER
UTM ZONE 12
MAP NUMBER 405
Montana
CONTOUR INTERVAL 100 FEET

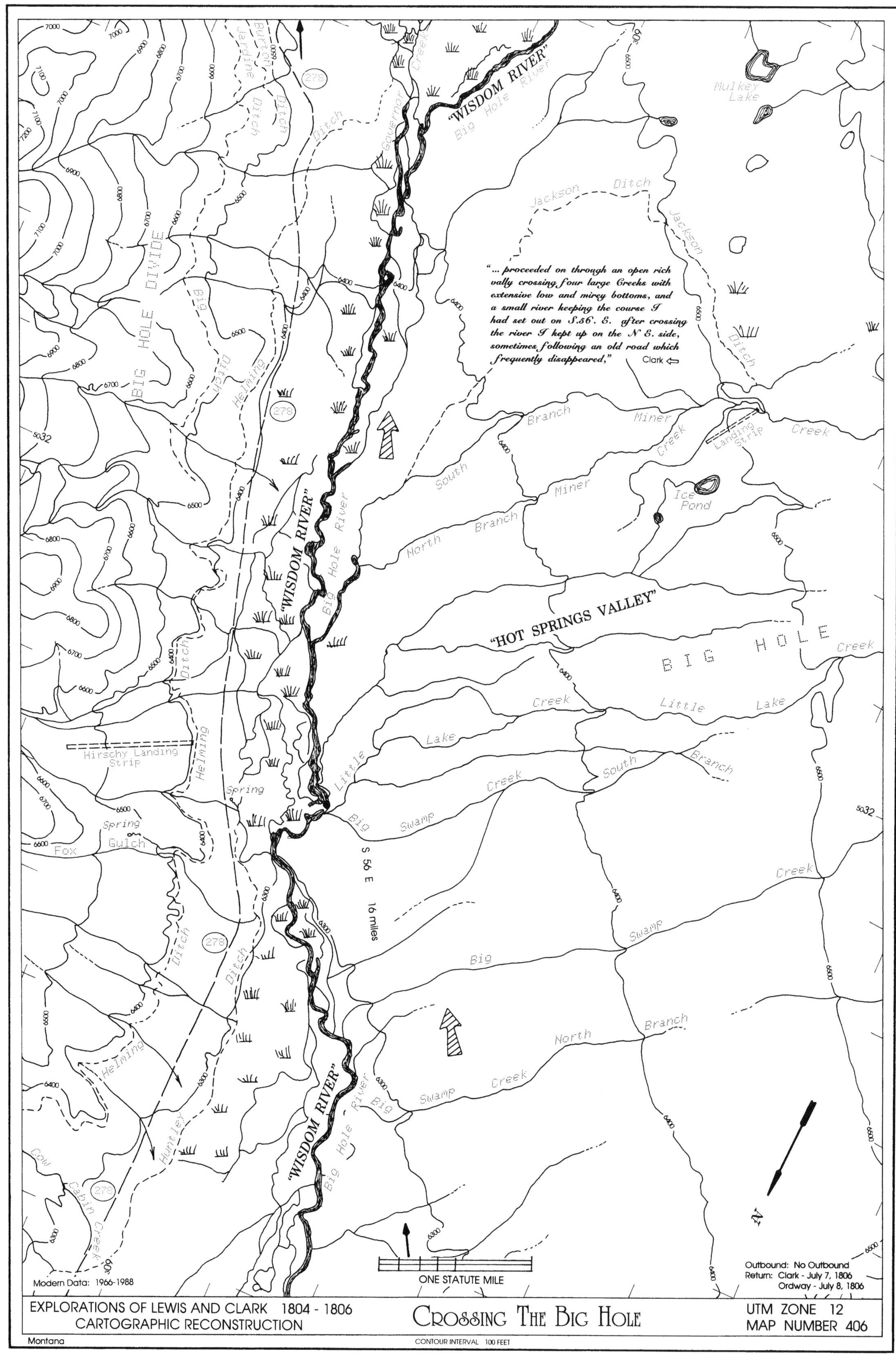
"WISDOM RIVER"
Big Hole River
Mulkey Lake
Jackson Ditch
Jackson Ditch
Burton Ditch
Jardine Ditch
Governor Creek
Ditch
BIG HOLE DIVIDE
Big Helming Ditch
"... proceeded on through an open rich vally crossing four large Creeks with extensive low and mirey bottoms, and a small river keeping the course I had set out on S.56°. E. after crossing the river I kept up on the N. E. side, sometimes following an old road which frequently disappeared,"
Clark
Branch Miner Creek
South
Miner Creek
Landing Strip
North Branch
Ice Pond
"WISDOM RIVER"
Big Hole River
"HOT SPRINGS VALLEY"
BIG HOLE
Creek
Creek
Little Lake
Lake
Little Lake Creek
Hirschy Landing Strip
Helming
Spring
South Branch
Spring Gulch
Fox
Big Swamp Creek
S 56 E 16 miles
Creek
Swamp
Big
Ditch
Ditch
Helming
North Branch
Big Swamp Creek
"WISDOM RIVER"
Big Hole River
Huntley
Cow Cabin Creek
278
Modern Data: 1966-1988
ONE STATUTE MILE
N
Outbound: No Outbound
Return: Clark - July 7, 1806
Ordway - July 8, 1806
EXPLORATIONS OF LEWIS AND CLARK 1804 - 1806
CARTOGRAPHIC RECONSTRUCTION
CROSSING THE BIG HOLE
UTM ZONE 12
MAP NUMBER 406
Montana
CONTOUR INTERVAL 100 FEET

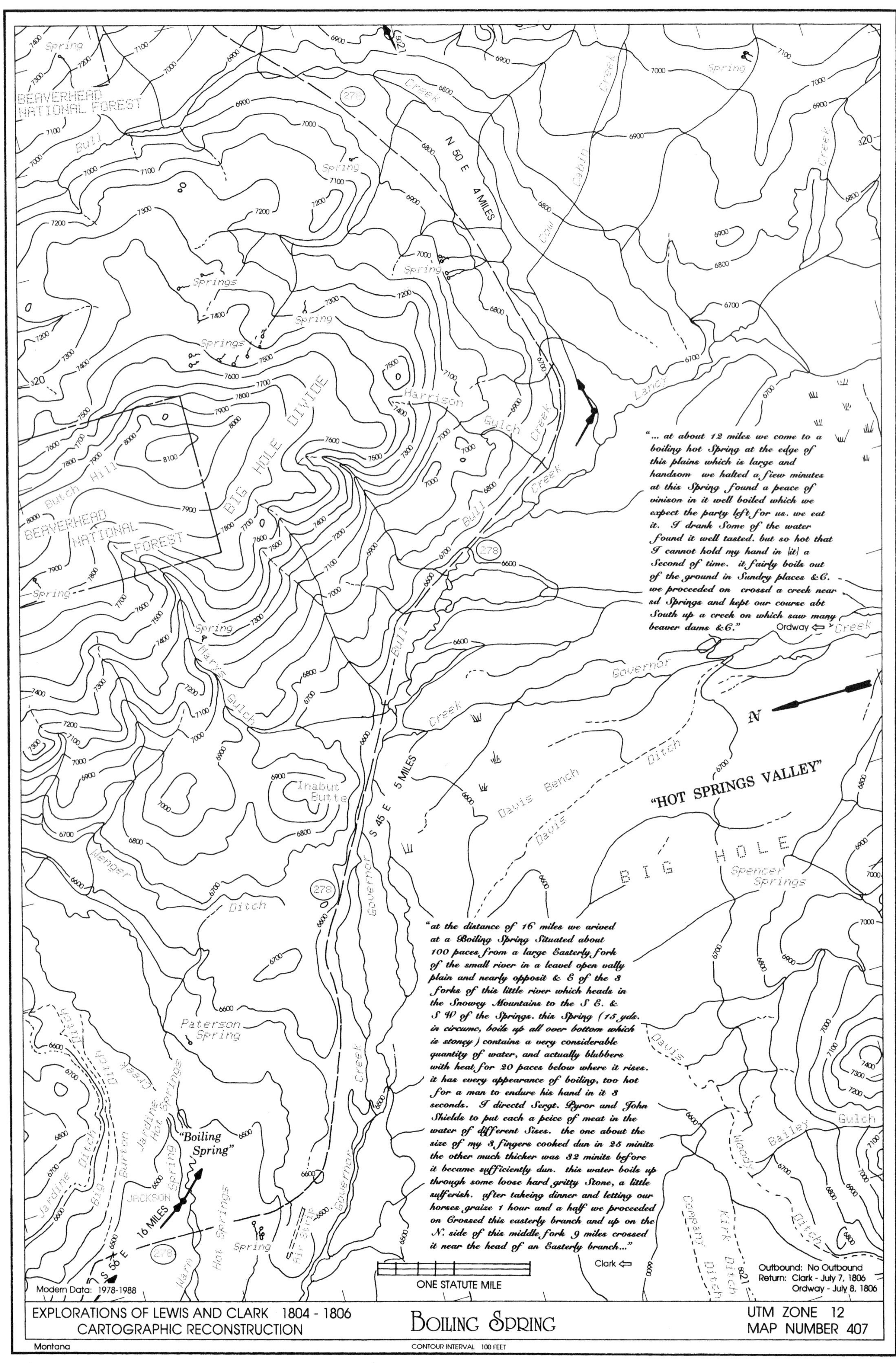
"... at about 12 miles we come to a boiling hot Spring at the edge of this plains which is large and handsom we halted a fiew minutes at this Spring found a peace of vinison in it well boiled which we expect the party left for us. we eat it. I drank Some of the water found it well tasted. but so hot that I cannot hold my hand in [it] a Second of time. it fairly boils out of the ground in Sundry places &C. we proceeded on crossd a creek near sd Springs and kept our course abt South up a creek on which saw many beaver dams &C."
Ordway ⇦
"at the distance of 16 miles we arived at a Boiling Spring Situated about 100 paces from a large Easterly fork of the small river in a leavel open vally plain and nearly opposit & E of the 3 forks of this little river which heads in the Snowey Mountains to the S E. & S W of the Springs. this Spring (15 yds. in circumc, boils up all over bottom which is stoney) contains a very considerable quantity of water, and actually blubbers with heat for 20 paces below where it rises. it has every appearance of boiling, too hot for a man to endure his hand in it 3 seconds. I directd Sergt. Pyror and John Shields to put each a peice of meat in the water of different Sises. the one about the size of my 3 fingers cooked dun in 25 minits the other much thicker was 32 minits before it became sufficiently dun. this water boils up through some loose hard gritty Stone, a little sulferish. after takeing dinner and letting our horses graize 1 hour and a half we proceeded on Crossed this easterly branch and up on the N. side of this middle fork 9 miles crossed it near the head of an Easterly branch..."
Clark ⇦
"HOT SPRINGS VALLEY"
BIG HOLE
BIG HOLE DIVIDE
BEAVERHEAD NATIONAL FOREST
"Boiling Spring"
N 50 E 4 MILES
S 45 E 5 MILES
S 56 E 16 MILES
N
ONE STATUTE MILE
Modern Data: 1978-1988
Outbound: No Outbound
Return: Clark - July 7, 1806
Ordway - July 8, 1806
EXPLORATIONS OF LEWIS AND CLARK 1804 - 1806
CARTOGRAPHIC RECONSTRUCTION
BOILING SPRING
UTM ZONE 12
MAP NUMBER 407
Montana
CONTOUR INTERVAL 100 FEET

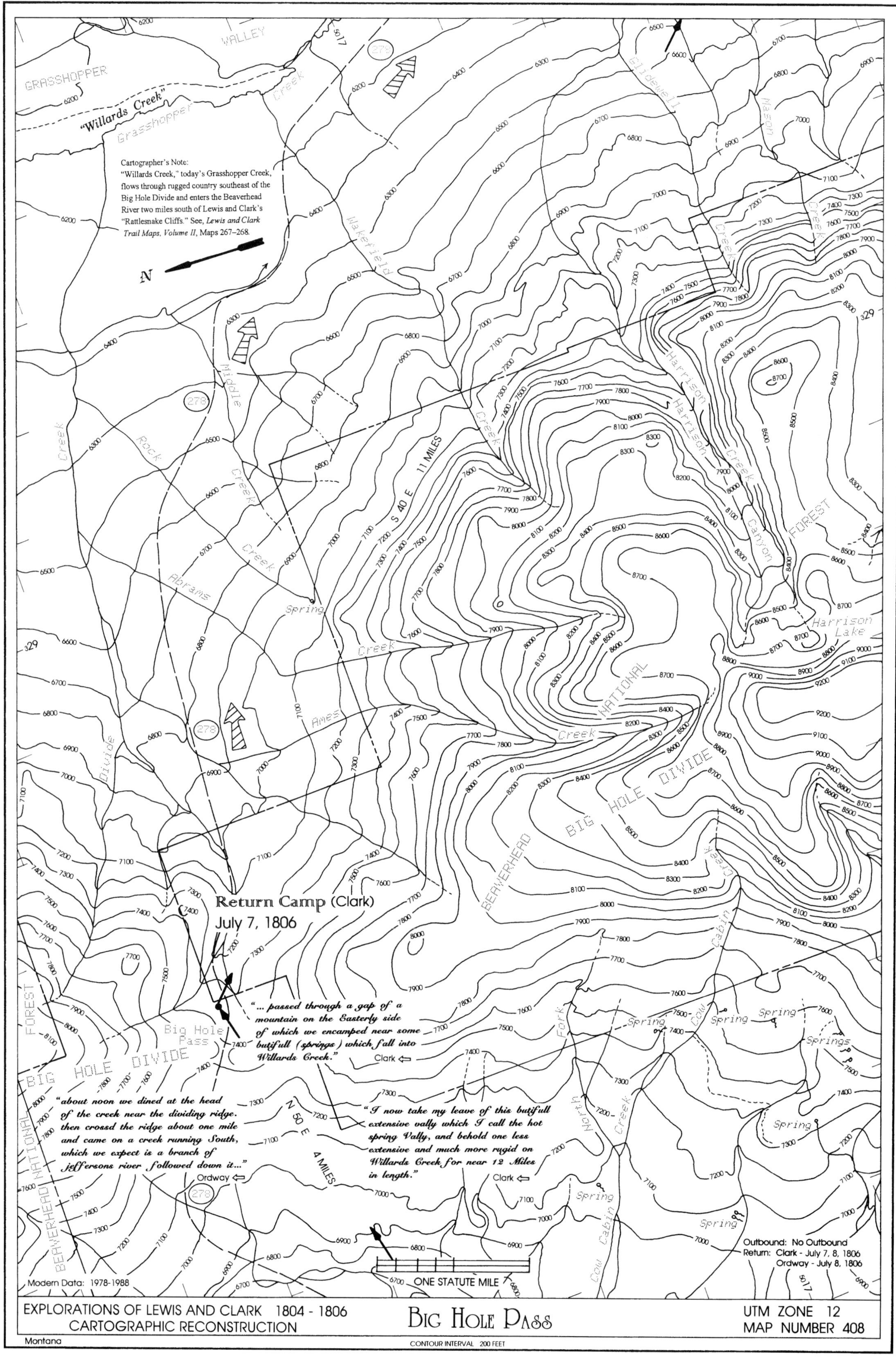
Cartographer's Note:
"Willards Creek," today's Grasshopper Creek, flows through rugged country southeast of the Big Hole Divide and enters the Beaverhead River two miles south of Lewis and Clark's "Rattlesnake Cliffs." See, *Lewis and Clark Trail Maps, Volume II*, Maps 267–268.
"Willards Creek"
Return Camp (Clark)
July 7, 1806
"... passed through a gap of a mountain on the Easterly side of which we encamped near some butifull (springs) which fall into Willards Creek."
Clark
"about noon we dined at the head of the creek near the dividing ridge. then crossd the ridge about one mile and came on a creek running South, which we expect is a branch of jeffersons river followed down it..."
Ordway
"I now take my leave of this butifull extensive vally which I call the hot spring Vally, and behold one less extensive and much more rugid on Willards Creek for near 12 Miles in length."
Clark
S 40 E 11 MILES
N 50 E 4 MILES
Big Hole Pass
BIG HOLE DIVIDE
BEAVERHEAD NATIONAL FOREST
Harrison Lake
Outbound: No Outbound
Return: Clark - July 7, 8, 1806
Ordway - July 8, 1806
ONE STATUTE MILE
Modern Data: 1978-1988
EXPLORATIONS OF LEWIS AND CLARK 1804 - 1806
CARTOGRAPHIC RECONSTRUCTION
BIG HOLE PASS
UTM ZONE 12
MAP NUMBER 408
Montana
CONTOUR INTERVAL 200 FEET

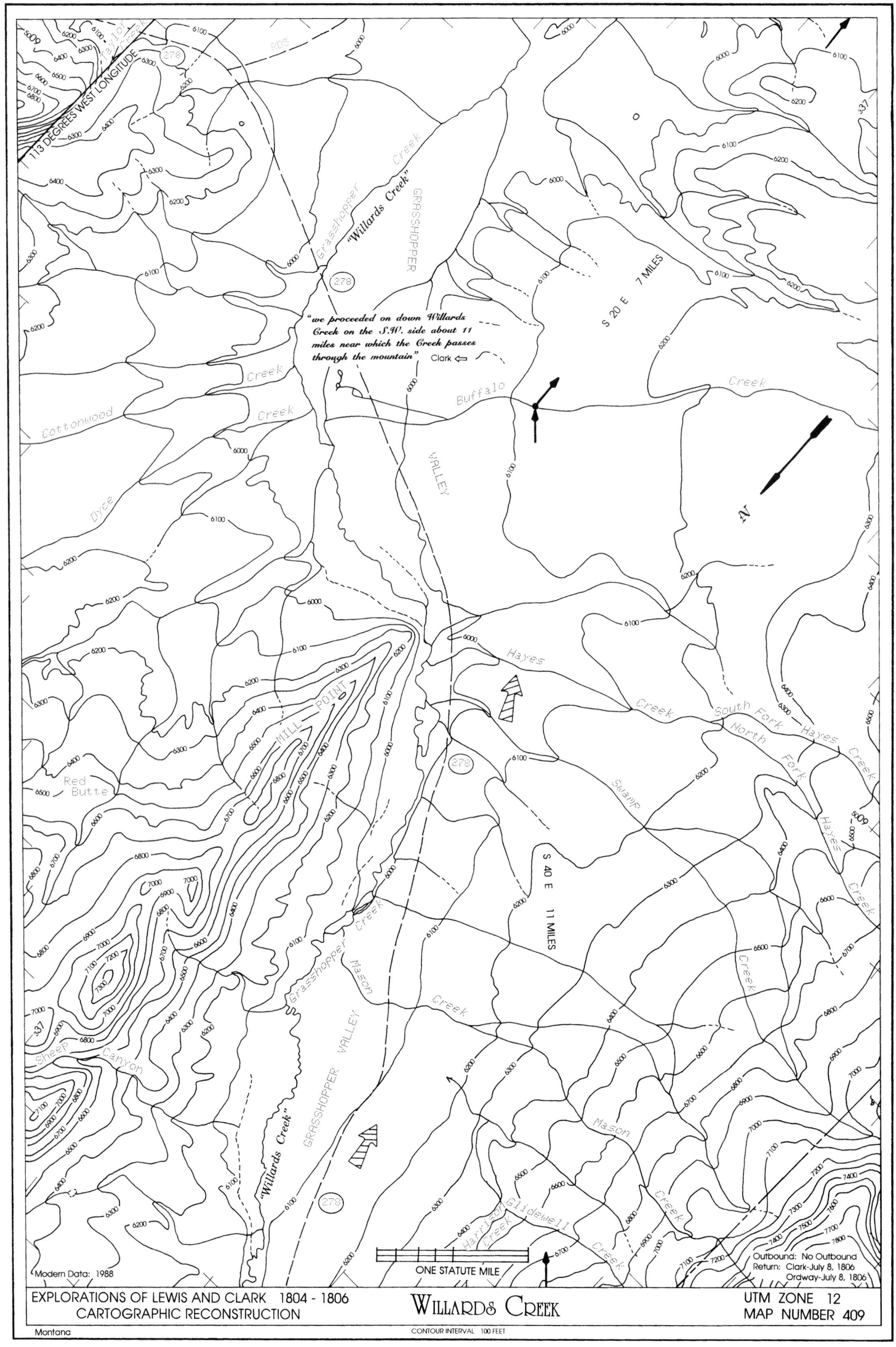

"we proceeded on down Willards Creek on the S.W. side about 11 miles near which the Creek passes through the mountain" Clark ⇦
"Willards Creek"
Grasshopper Creek
GRASSHOPPER VALLEY
Buffalo Creek
Cottonwood Creek
Dyce Creek
Hayes Creek
South Fork Hayes Creek
North Fork Hayes Creek
Swamp Creek
Mason Creek
Glidewell Creek
Harrison Creek
Sheep Canyon
Red Butte
MILL POINT
113 DEGREES WEST LONGITUDE
S 20 E 7 MILES
S 40 E 11 MILES
N
ONE STATUTE MILE
Modern Data: 1988
Outbound: No Outbound
Return: Clark-July 8, 1806
Ordway-July 8, 1806
EXPLORATIONS OF LEWIS AND CLARK 1804 - 1806
CARTOGRAPHIC RECONSTRUCTION
WILLARDS CREEK
UTM ZONE 12
MAP NUMBER 409
Montana
CONTOUR INTERVAL 100 FEET

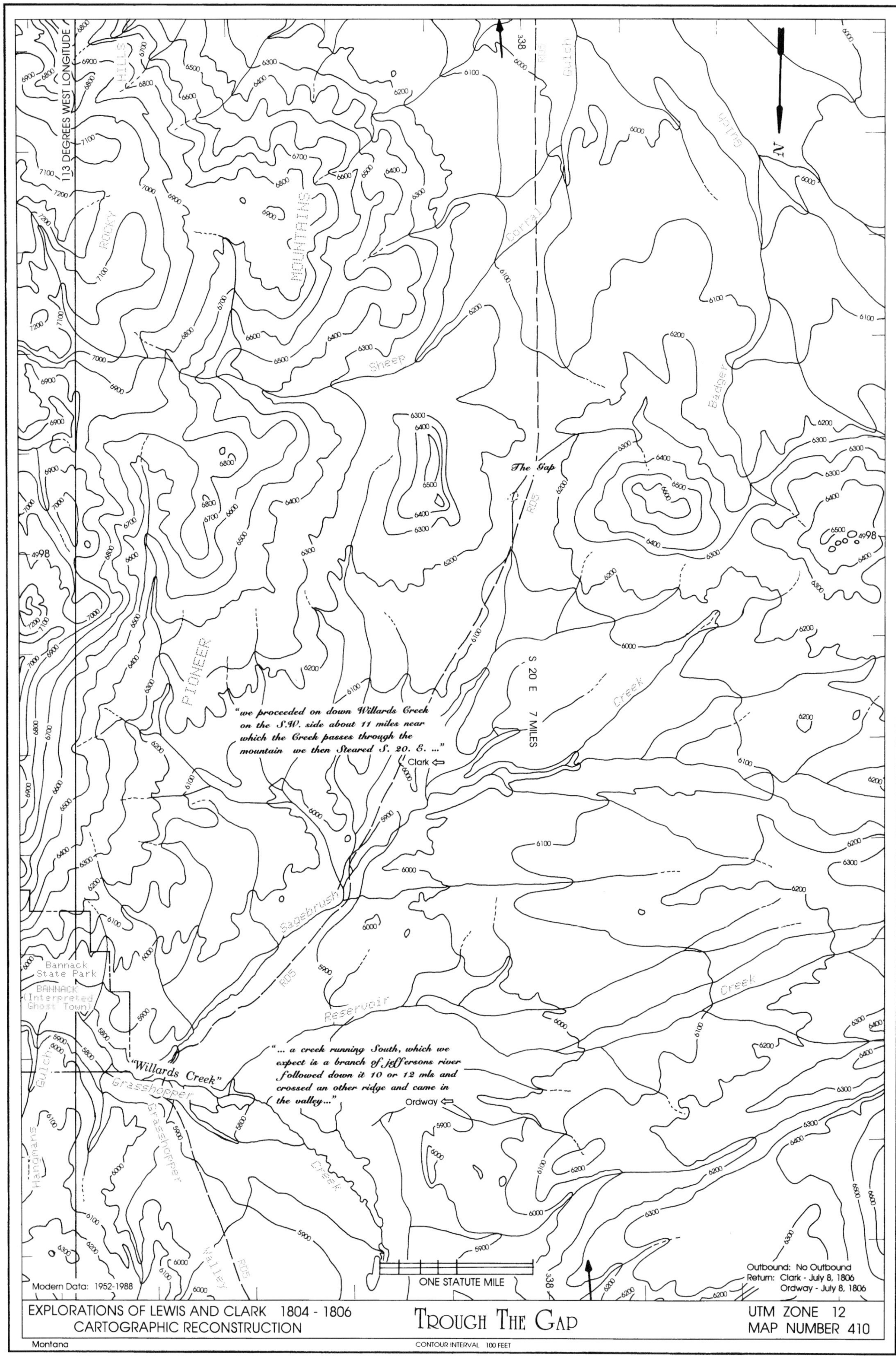
"we proceeded on down Willards Creek on the S.W. side about 11 miles near which the Creek passes through the mountain we then Steared S. 20. E. ..."
Clark
"... a creek running South, which we expect is a branch of jeffersons river followed down it 10 or 12 mls and crossed an other ridge and came in the valley..."
Ordway
The Gap
S 20 E 7 MILES
"Willards Creek"
Bannack State Park
BANNACK (Interpreted Ghost Town)
113 DEGREES WEST LONGITUDE
ROCKY
MOUNTAINS
PIONEER
HILLS
Corral
Gulch
Sheep
Badger
Creek
Sagebrush
Reservoir
Grasshopper
Hangmans
Valley
RD5
338
N
ONE STATUTE MILE
Modern Data: 1952-1988
Outbound: No Outbound
Return: Clark - July 8, 1806
Ordway - July 8, 1806
EXPLORATIONS OF LEWIS AND CLARK 1804 - 1806
CARTOGRAPHIC RECONSTRUCTION
TROUGH THE GAP
UTM ZONE 12
MAP NUMBER 410
Montana
CONTOUR INTERVAL 100 FEET

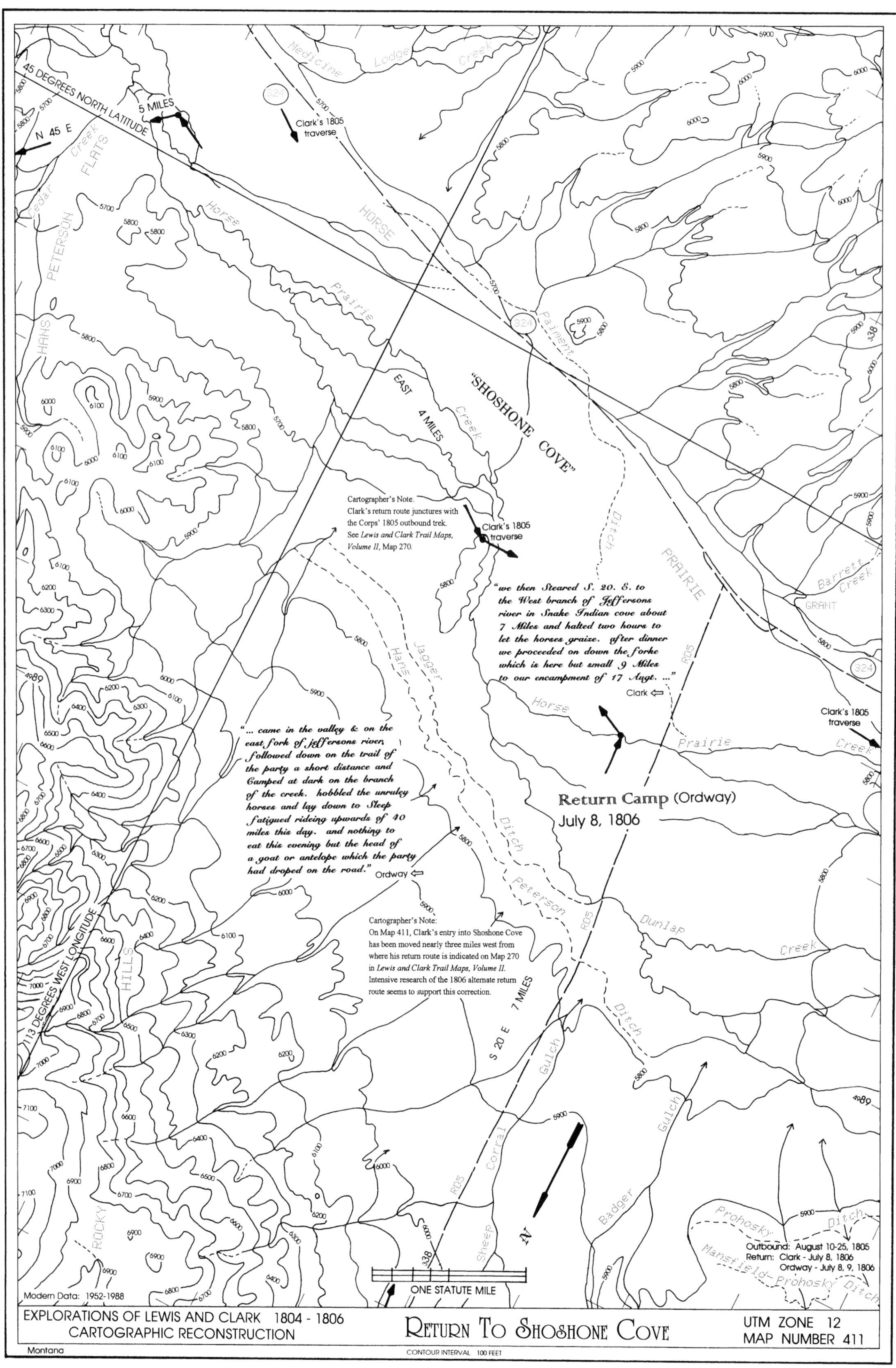
45 DEGREES NORTH LATITUDE
5 MILES
N 45 E
Medicine Lodge Creek
324
Clark's 1805 traverse
Cedar Creek FLATS
PETERSON
Horse
HORSE
Prairie
HANS
Paiment
"SHOSHONE COVE"
EAST 4 MILES
Creek
Cartographer's Note.
Clark's return route junctures with the Corps' 1805 outbound trek. See *Lewis and Clark Trail Maps, Volume II*, Map 270.
Clark's 1805 traverse
Ditch
PRAIRIE
Barrett Creek
GRANT
338
"we then Steared S. 20. E. to the West branch of Jeffersons river in Snake Indian cove about 7 Miles and halted two hours to let the horses graize. after dinner we proceeded on down the forke which is here but small 9 Miles to our encampment of 17 Augt. ..."
Clark ⇦
Hans
Jagger
RDS
Horse
Prairie
Creek
Clark's 1805 traverse
"... came in the valley & on the east fork of jeffersons river, followed down on the trail of the party a short distance and Camped at dark on the branch of the creek. hobbled the unruley horses and lay down to Sleep fatigued rideing upwards of 40 miles this day. and nothing to eat this evening but the head of a goat or antelope which the party had droped on the road."
Ordway ⇦
Return Camp (Ordway)
July 8, 1806
Ditch
Peterson
Dunlap
Creek
Cartographer's Note:
On Map 411, Clark's entry into Shoshone Cove has been moved nearly three miles west from where his return route is indicated on Map 270 in *Lewis and Clark Trail Maps, Volume II*. Intensive research of the 1806 alternate return route seems to support this correction.
113 DEGREES WEST LONGITUDE
HILLS
S 20 E 7 MILES
Gulch
Ditch
Gulch
Corral
Badger
ROCKY
Sheep
N
Prohosky Ditch
Mansfield-Prohosky Ditch
Outbound: August 10-25, 1805
Return: Clark - July 8, 1806
Ordway - July 8, 9, 1806
Modern Data: 1952-1988
ONE STATUTE MILE
EXPLORATIONS OF LEWIS AND CLARK 1804 - 1806
CARTOGRAPHIC RECONSTRUCTION
RETURN TO SHOSHONE COVE
UTM ZONE 12
MAP NUMBER 411
Montana
CONTOUR INTERVAL 100 FEET

Alternate Return Route No. 5

MAPS 412 - 434

LEWIS PARTY,
LEWIS AND CLARK PASS, MONTANA

Capt. Meriwether Lewis (journalist)
George Drouillard (interpreter)
Sgt. Patrick Gass (journalist)
Pvt. Joseph Field
Pvt. Reuben Field
Pvt. Robert Frazer
Pvt. Silas Goodrich
Pvt. Hugh McNeal
Pvt. John Thompson
Pvt. William Werner
5 Plateau Indians (short term)
17 expedition horses

BEST ROAD TO THE GREAT FALLS

From Travelers Rest on July 3, 1806, Clark's party of 23 people and 50 horses had set out south toward the Beaverhead country. At the same time, Lewis's 10-man detachment with accompanying Plateau Indian guides proceeded north down the Bitterroot River.

The enlisted men realized the northerly Rockies and Plains explorations that would be led by Lewis potentially were the most hazardous, being in the haunts of powerful Blackfeet bands. Lewis had called for volunteers and many stepped forward. Lewis had chosen George Drouillard, Sergeant Patrick Gass, and privates Joseph Field, Reuben Field, Robert Frazer, and William Werner for an upper Marias investigation. This would not occur, however, until after the Lewis party had completed their exploratory travel along the "best" route to the Great Falls.

Also, Lewis had selected Silas Goodrich, Hugh McNeal, and John Thompson to open the Great Falls caches and to portage canoes and baggage below the rapids, which would be a laborious, fairly lengthy task. (They eventually would be joined by a detached squad from Clark's party, coming down the Missouri River in canoes from the Three Forks.)

Consequently, when setting out from Travelers Rest on July 3, 1806, Lewis had a combined party of 10 men, 5 Indian guides, and 17 of the Corps' horses.

They continued a few miles north to the Bitterroot-Clark Fork confluence and forded the river, before turning a short distance east up the Clark Fork in the direction of the Blackfoot River, or River of the Road to Buffalo. This night's campsite was near modern-day Missoula, Montana.

On the morning of July 4, a sixth Plateau native joined the Nez Perces. Observing the Hellgate Canyon (situated just east of Missoula, Montana), the Plateau guides indicated that the defile and the Blackfoot River country beyond it were dangerous because of frequent enemy war parties. After noting their concern for the Lewis party's safety, the guides left to go visit friendly tribesmen elsewhere along the Clark Fork; afterward, they planned to return west over the Lolo Trail to their people in Nez Perce country.

Continuing up the extensive Blackfoot River watershed and following behind a series of recently abandoned Indian campsites, Lewis's detachment eventually crossed the Rockies and the continental divide at today's Lewis and Clark Pass. They then traveled out onto the Great Plains, coming to the "Medicine" or Sun River, which they followed down to the Great Falls and the site of their 1805 upper portage camp.

Eight days, July 3–11, 1806, were expended on the journey from Travelers Rest to the Missouri River.

Lewis's investigation of this "best" route between Travelers Rest and the Great Falls was of exceptional importance to the captains. When Clark was completing the map work at Fort Clatsop, the captains were well aware that an all-water route (or one with a short portage) between Missouri and Columbia waters had not been found. Discovering a practicable water route, with a short Rocky Mountain land portage, had been one of Jefferson's and the Corps' most desired goals. The captains were well aware that this had not been accomplished.

Making the best of the situation, they now intended to identify the most acceptable land route through the Rockies. Largely based on Indian advice, the natives' well-traveled road along the Blackfoot River seemed to best fit expectations.

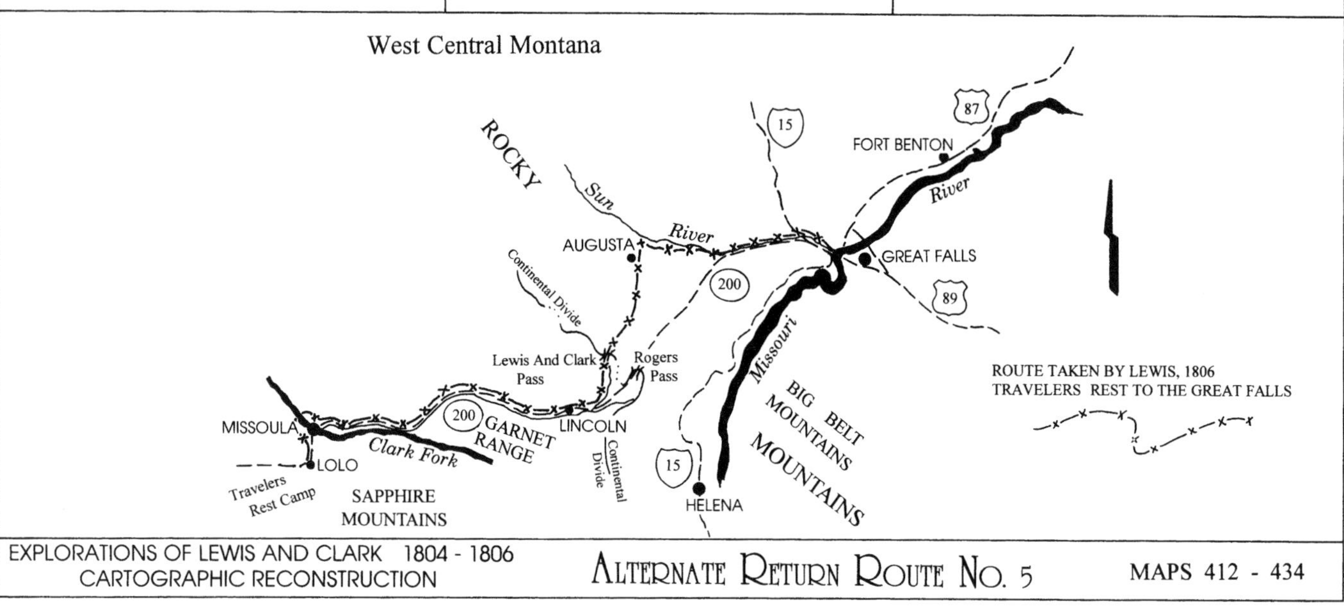

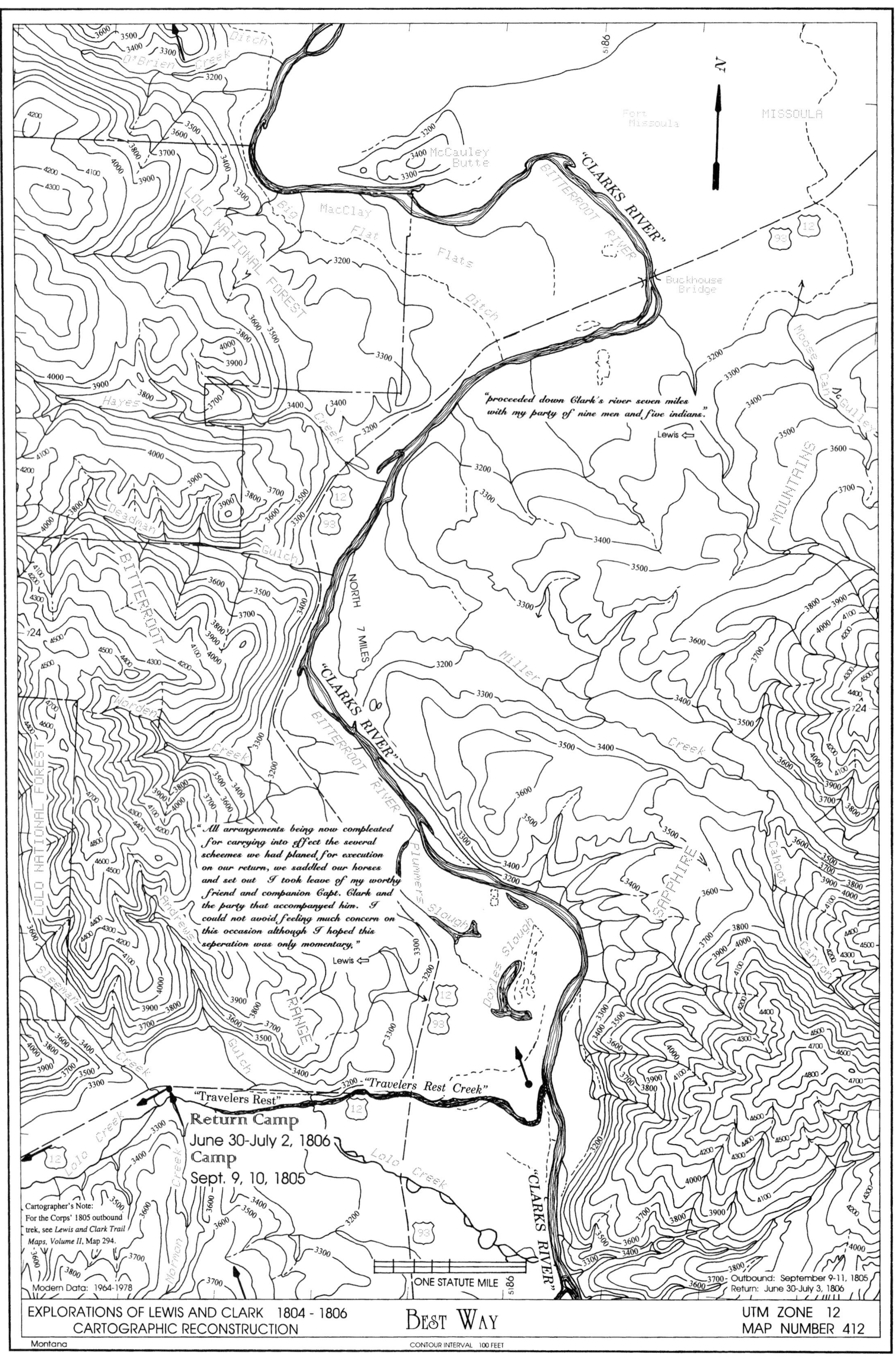

"CLARKS RIVER"
BITTERROOT RIVER
MISSOULA
Fort Missoula
McCauley Butte
Buckhouse Bridge
LOLO NATIONAL FOREST
Big MacClay Flat
Flats
Ditch
O'Brien Creek
Hayes Creek
Deadman Gulch
BITTERROOT RANGE
Worden Creek
Andrews Creek
Sleeman Creek
Gulch
MOUNTAINS
SAPPHIRE
Miller Creek
Moose Can Gulley
Cahoot Canyon
Plummers Slough
Doyles Slough
"proceeded down Clark's river seven miles with my party of nine men and five indians."
Lewis
"All arrangements being now compleated for carrying into effect the several scheemes we had planed for execution on our return, we saddled our horses and set out I took leave of my worthy friend and companion Capt. Clark and the party that accompanyed him. I could not avoid feeling much concern on this occasion although I hoped this seperation was only momentary."
Lewis
NORTH 7 MILES
"Travelers Rest"
"Travelers Rest Creek"
Return Camp June 30-July 2, 1806
Camp Sept. 9, 10, 1805
Lolo Creek
Norman Creek
Cartographer's Note: For the Corps' 1805 outbound trek, see Lewis and Clark Trail Maps, Volume II, Map 294.
Modern Data: 1964-1978
ONE STATUTE MILE
Outbound: September 9-11, 1805
Return: June 30-July 3, 1806
EXPLORATIONS OF LEWIS AND CLARK 1804 - 1806
CARTOGRAPHIC RECONSTRUCTION
BEST WAY
UTM ZONE 12
MAP NUMBER 412
Montana
CONTOUR INTERVAL 100 FEET

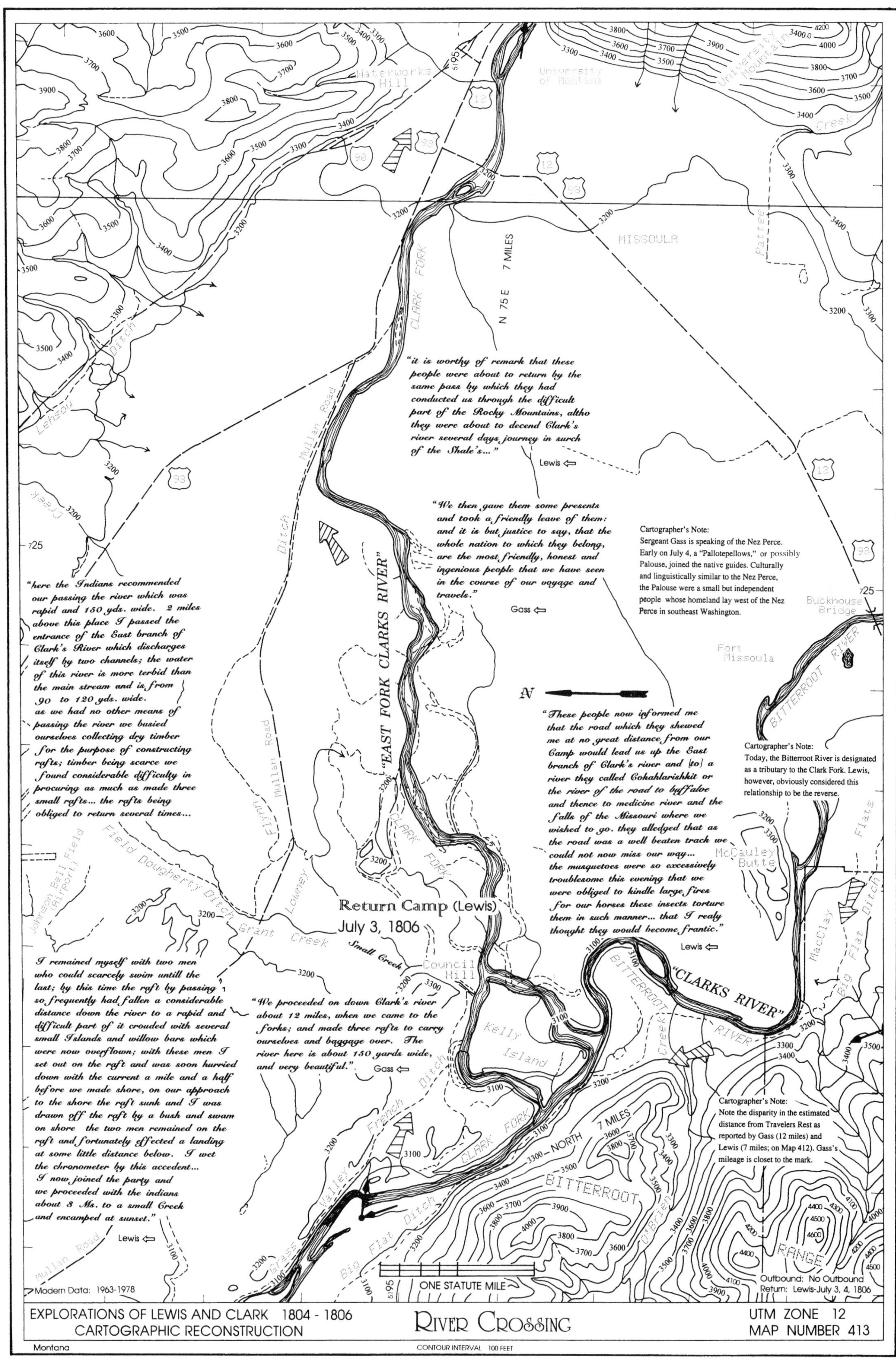

Waterworks Hill
University of Montana
University Mountain
MISSOULA
N 75 E 7 MILES
CLARK FORK
Mullan Road
"it is worthy of remark that these people were about to return by the same pass by which they had conducted us through the difficult part of the Rocky Mountains, altho they were about to decend Clark's river several days journey in surch of the Shale's..."
Lewis ⇦
"We then gave them some presents and took a friendly leave of them: and it is but justice to say, that the whole nation to which they belong, are the most friendly, honest and ingenious people that we have seen in the course of our voyage and travels."
Gass ⇦
Cartographer's Note:
Sergeant Gass is speaking of the Nez Perce. Early on July 4, a "Pallotepellows," or possibly Palouse, joined the native guides. Culturally and linguistically similar to the Nez Perce, the Palouse were a small but independent people whose homeland lay west of the Nez Perce in southeast Washington.
Buckhouse Bridge
Fort Missoula
BITTERROOT RIVER
"here the Indians recommended our passing the river which was rapid and 150 yds. wide. 2 miles above this place I passed the entrance of the East branch of Clark's River which discharges itself by two channels; the water of this river is more terbid than the main stream and is from 90 to 120 yds. wide. as we had no other means of passing the river we busied ourselves collecting dry timber for the purpose of constructing rafts; timber being scarce we found considerable difficulty in procuring as much as made three small rafts... the rafts being obliged to return several times...
"EAST FORK CLARKS RIVER"
N
"These people now informed me that the road which they shewed me at no great distance from our Camp would lead us up the East branch of Clark's river and [to] a river they called Cohahlarishkit or the river of the road to buffaloe and thence to medicine river and the falls of the Missouri where we wished to go. they alledged that as the road was a well beaten track we could not now miss our way... the musquetoes were so excessively troublesome this evening that we were obliged to kindle large fires for our horses these insects torture them in such manner... that I realy thought they would become frantic."
Lewis ⇦
Cartographer's Note:
Today, the Bitterroot River is designated as a tributary to the Clark Fork. Lewis, however, obviously considered this relationship to be the reverse.
Ditch
Flynn Mullan Road
Lowney
Johnson Bell Field (Airport)
Field Dougherty Ditch
Grant Creek
Return Camp (Lewis)
July 3, 1806
Small Creek
Council Hill
McCauley Butte
"CLARKS RIVER"
Flats
MacClay
Big Flat Ditch
"I remained myself with two men who could scarcely swim untill the last; by this time the raft by passing so frequently had fallen a considerable distance down the river to a rapid and difficult part of it crouded with several small Islands and willow bars which were now overflown; with these men I set out on the raft and was soon hurried down with the current a mile and a half before we made shore, on our approach to the shore the raft sunk and I was drawn off the raft by a bush and swam on shore the two men remained on the raft and fortunately effected a landing at some little distance below. I wet the chronometer by this accedent... I now joined the party and we proceeded with the indians about 3 Ms. to a small Creek and encamped at sunset."
Lewis ⇦
"We proceeded on down Clark's river about 12 miles, when we came to the forks; and made three rafts to carry ourselves and baggage over. The river here is about 150 yards wide, and very beautiful."
Gass ⇦
Kelly Island
BITTERROOT RIVER
Creek
French Ditch
Grass Valley
Big Flat Ditch
NORTH 7 MILES
BITTERROOT RANGE
O'Brien
Cartographer's Note:
Note the disparity in the estimated distance from Travelers Rest as reported by Gass (12 miles) and Lewis (7 miles; on Map 412). Gass's mileage is closet to the mark.
Mullan Road
ONE STATUTE MILE
Modern Data: 1963-1978
Outbound: No Outbound
Return: Lewis-July 3, 4, 1806
EXPLORATIONS OF LEWIS AND CLARK 1804 - 1806
CARTOGRAPHIC RECONSTRUCTION
RIVER CROSSING
UTM ZONE 12
MAP NUMBER 413
Montana
CONTOUR INTERVAL 100 FEET

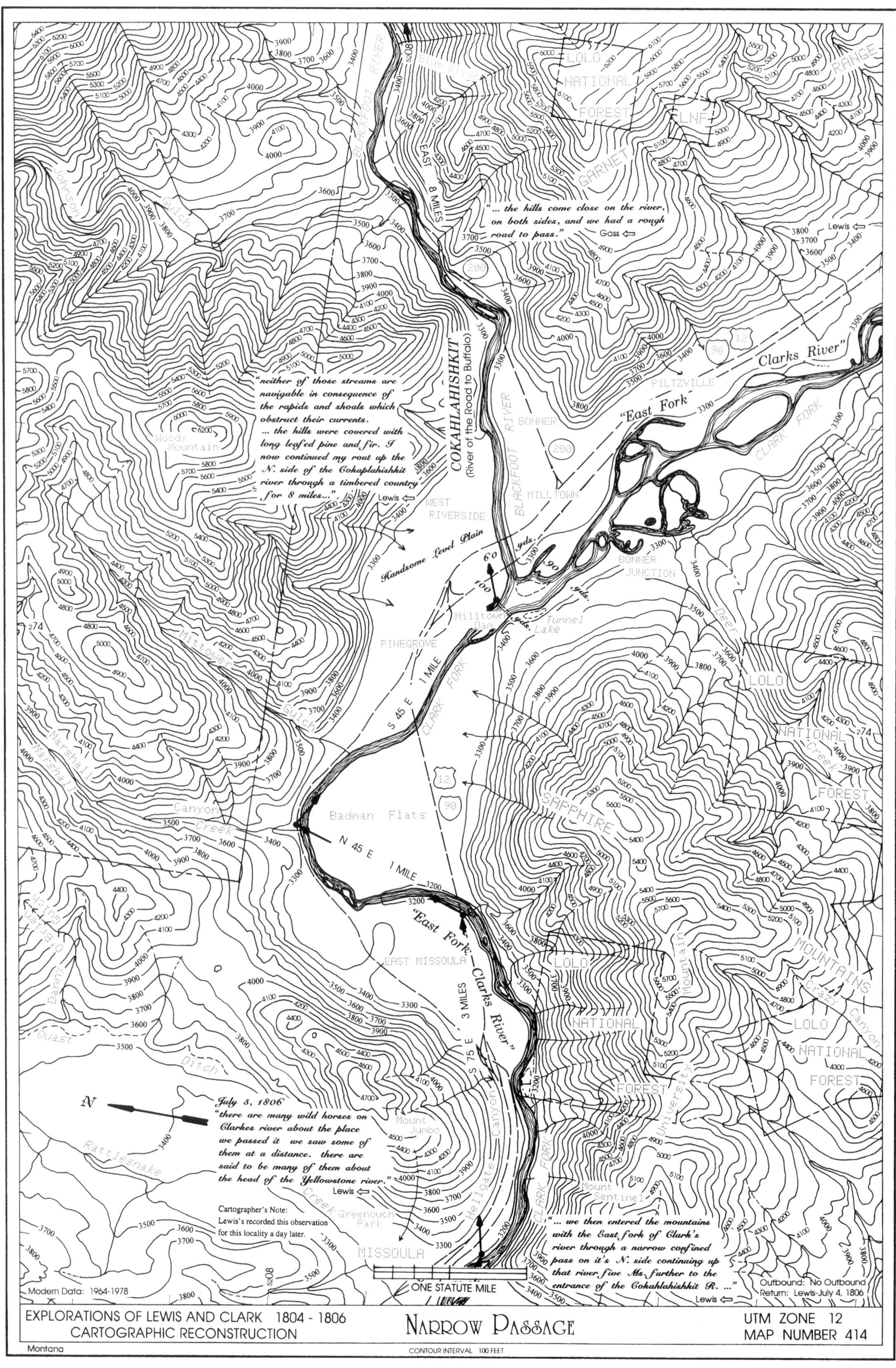
"... the hills come close on the river, on both sides, and we had a rough road to pass." Gass
"neither of those streams are navigable in consequence of the rapids and shoals which obstruct their currents. ... the hills were covered with long leafed pine and fir. I now continued my rout up the N. side of the Cokahlahishkit river through a timbered country for 8 miles..." Lewis
COKAHLAHISHKIT
(River of the Road to Buffalo)
"Clarks River"
"East Fork"
Handsome Level Plain
"East Fork Clarks River"
July 5, 1806
"there are many wild horses on Clarkes river about the place we passed it we saw some of them at a distance. there are said to be many of them about the head of the Yellowstone river." Lewis
Cartographer's Note:
Lewis's recorded this observation for this locality a day later.
"... we then entered the mountains with the East fork of Clark's river through a narrow confined pass on it's N. side continuing up that river five Ms. further to the entrance of the Cokahlahishkit R. ..." Lewis
Outbound: No Outbound
Return: Lewis-July 4, 1806
Modern Data: 1964-1978
ONE STATUTE MILE
8 MILES
S 45 E 1 MILE
N 45 E 1 MILE
S 75 E 3 MILES
LOLO NATIONAL FOREST
GARNET RANGE
SAPPHIRE MOUNTAINS
BLACKFOOT RIVER
CLARK FORK
MILLTOWN
BONNER
BONNER JUNCTION
PILTZVILLE
WEST RIVERSIDE
PINEGROVE
EAST MISSOULA
MISSOULA
Badman Flats
Woody Mountain
Mount Jumbo
Mount Sentinel
Hellgate Canyon
EXPLORATIONS OF LEWIS AND CLARK 1804 - 1806
CARTOGRAPHIC RECONSTRUCTION
NARROW PASSAGE
UTM ZONE 12
MAP NUMBER 414
Montana
CONTOUR INTERVAL 100 FEET

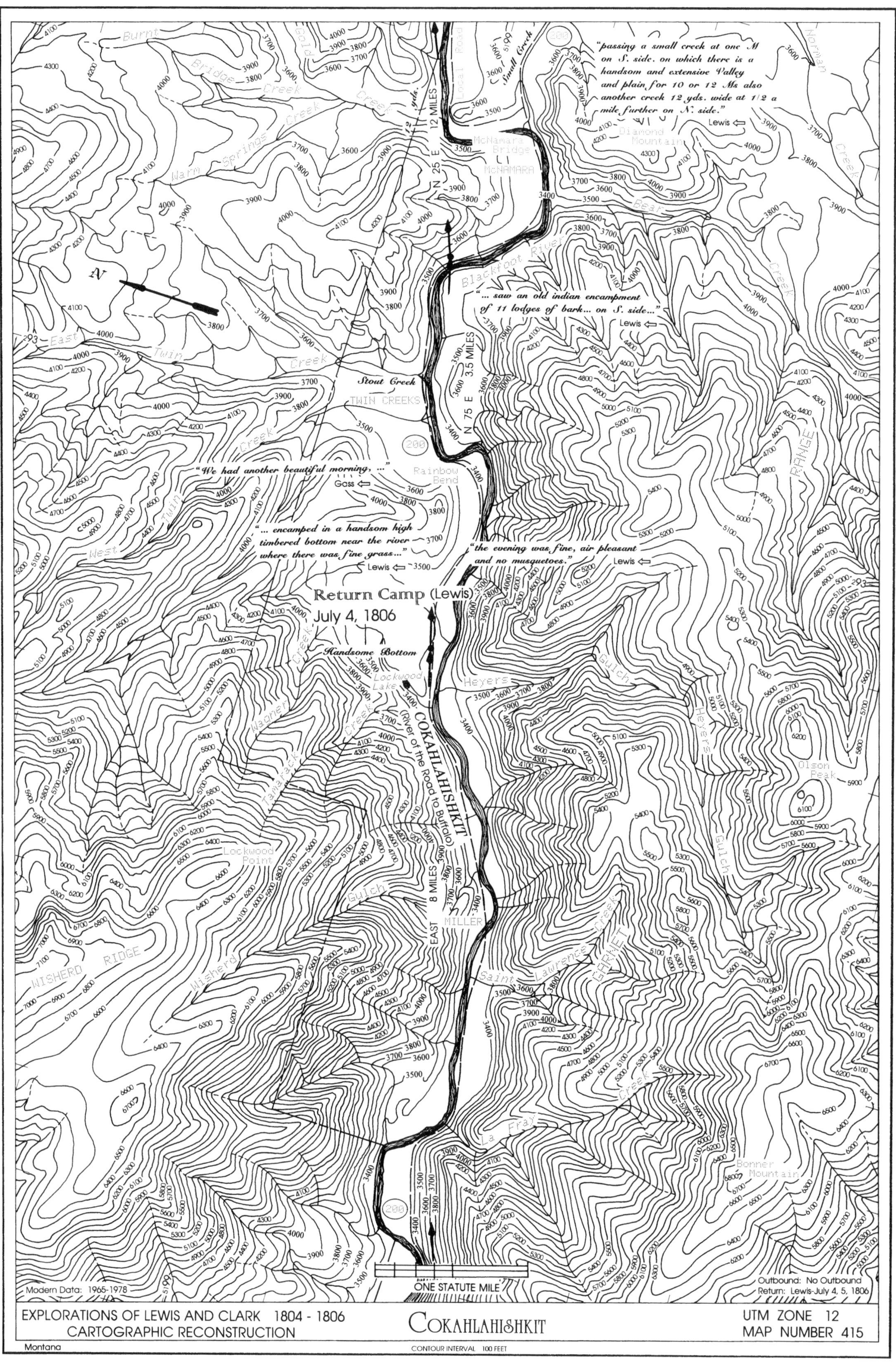
"passing a small creek at one M on S. side. on which there is a handsom and extensive Valley and plain for 10 or 12 Ms also another creek 12 yds. wide at 1/2 a mile further on N. side."
Lewis
"... saw an old indian encampment of 11 lodges of bark... on S. side..."
Lewis
"We had another beautiful morning, ..."
Gass
"... encamped in a handsom high timbered bottom near the river where there was fine grass..."
Lewis
"the evening was fine, air pleasant and no musquetoes."
Lewis
Return Camp (Lewis)
July 4, 1806
Handsome Bottom
Stout Creek
Small Creek
12 yds.
N 25 E 12 MILES
N 75 E 3.5 MILES
EAST 8 MILES
COKAHLAHISHKIT
(River of the Road to Buffalo)
Blackfoot River
McNamara Bridge
McNAMARA
Diamond Mountain
TWIN CREEKS
Rainbow Bend
Lockwood Lake
Heyers
MILLER
Lockwood Point
Olson Peak
Bonner Mountain
WISHERD RIDGE
GARNET
RANGE
Burnt Bridge Creek
Gold Creek
Warm Springs Creek
East Twin Creek
West Twin Creek
Wagner Creek
Tamarack Creek
Wisherd Gulch
Bear Creek
Heyers Gulch
Saint Lawrence Creek
La Fray Creek
Norman Creek
Local Road
N
Modern Data: 1965-1978
ONE STATUTE MILE
Outbound: No Outbound
Return: Lewis-July 4, 5, 1806
EXPLORATIONS OF LEWIS AND CLARK 1804 - 1806
CARTOGRAPHIC RECONSTRUCTION
COKAHLAHISHKIT
UTM ZONE 12
MAP NUMBER 415
Montana
CONTOUR INTERVAL 100 FEET

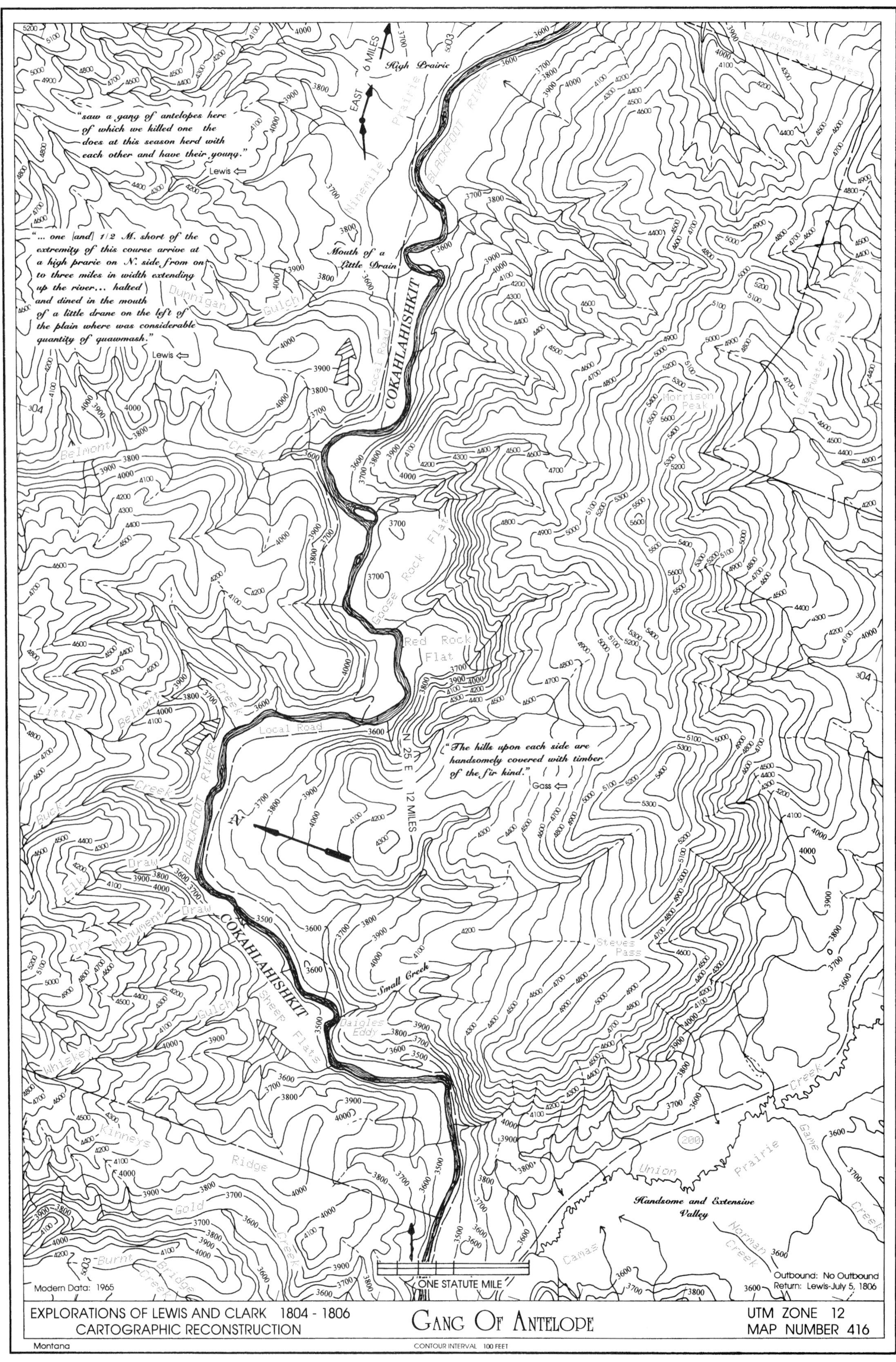
"saw a gang of antelopes here of which we killed one the does at this season herd with each other and have their young."
Lewis
"... one [and] 1/2 M. short of the extremity of this course arrive at a high prarie on N. side from one to three miles in width extending up the river... halted and dined in the mouth of a little drane on the left of the plain where was considerable quantity of quawmash."
Lewis
High Prairie
EAST 6 MILES
Mouth of a Little Drain
Ninemile Prairie
BLACKFOOT RIVER
COKAHLAHISHKIT
Dunnigan Gulch
Local Road
Belmont Creek
Goose Rock Flat
Red Rock Flat
Little Belmont Creek
Buck Creek
Elk Draw
Dry Monument Draw
Whiskey Gulch
Sheep Flats
Daigles Eddy
Small Creek
N 25 E 12 MILES
"The hills upon each side are handsomely covered with timber of the fir kind."
Gass
Morrison Peak
Lubrecht State Experimental Forest
Clearwater State Forest
Steves Pass
Kinneys Ridge
Gold Creek
Burnt Bridge Creek
Union Prairie
Game Creek
Norman Creek
Camas
200
Handsome and Extensive Valley
ONE STATUTE MILE
Modern Data: 1965
Outbound: No Outbound
Return: Lewis-July 5, 1806
EXPLORATIONS OF LEWIS AND CLARK 1804 - 1806
CARTOGRAPHIC RECONSTRUCTION
GANG OF ANTELOPE
UTM ZONE 12
MAP NUMBER 416
Montana
CONTOUR INTERVAL 100 FEET

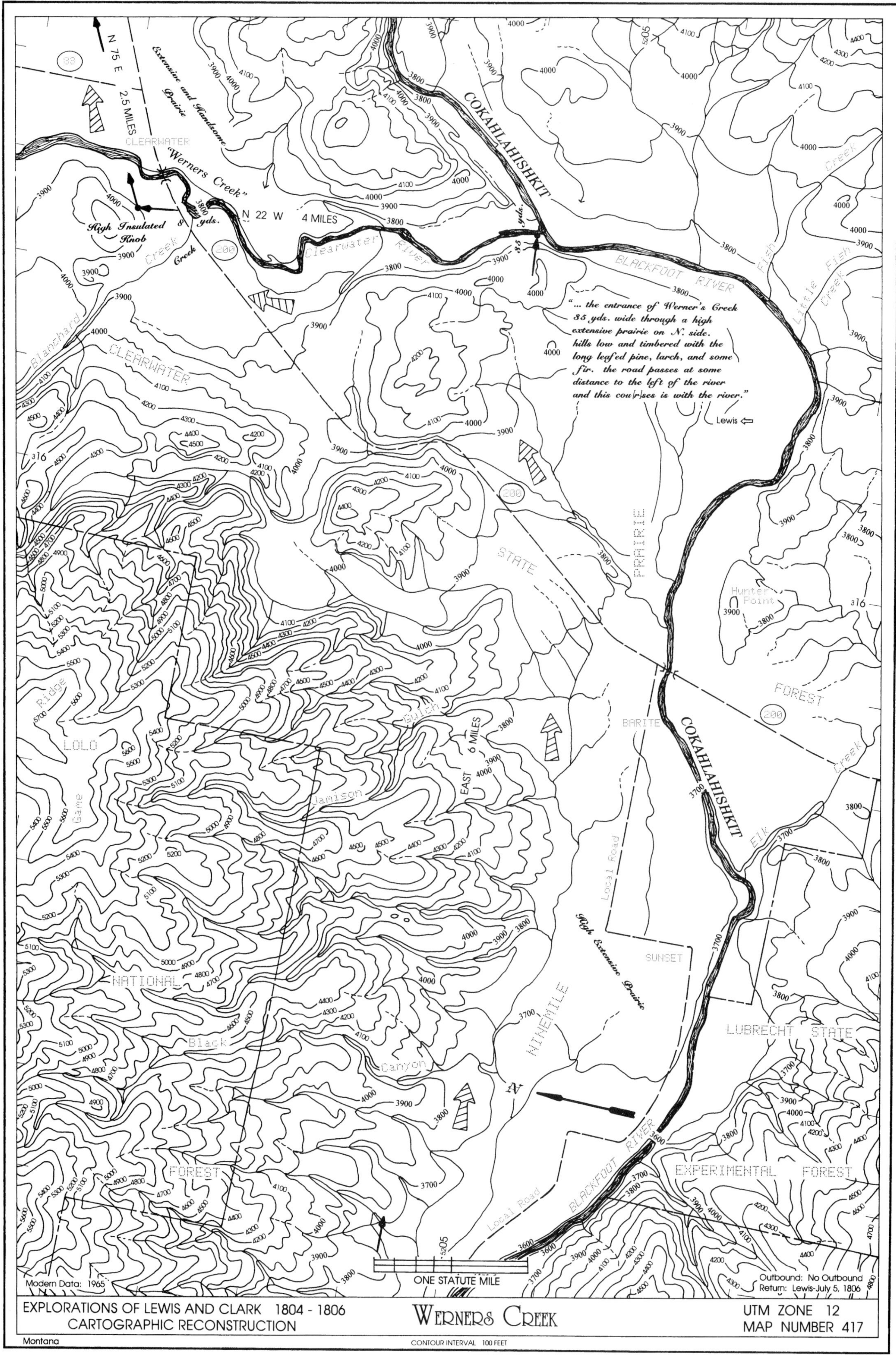

"... the entrance of Werner's Creek 85 yds. wide through a high extensive prairie on N. side. hills low and timbered with the long leafed pine, larch, and some fir. the road passes at some distance to the left of the river and this cou[r]ses is with the river."
Lewis
Extensive and Handsome Prairie
"Werners Creek"
High Insulated Knob
High Extensive Prairie
N 75 E 2.5 MILES
N 22 W 4 MILES
EAST 6 MILES
COKAHLAHISHKIT
BLACKFOOT RIVER
Clearwater River
ONE STATUTE MILE
Modern Data: 1965
Outbound: No Outbound
Return: Lewis-July 5, 1806
EXPLORATIONS OF LEWIS AND CLARK 1804 - 1806
CARTOGRAPHIC RECONSTRUCTION
WERNERS CREEK
UTM ZONE 12
MAP NUMBER 417
Montana
CONTOUR INTERVAL 100 FEET

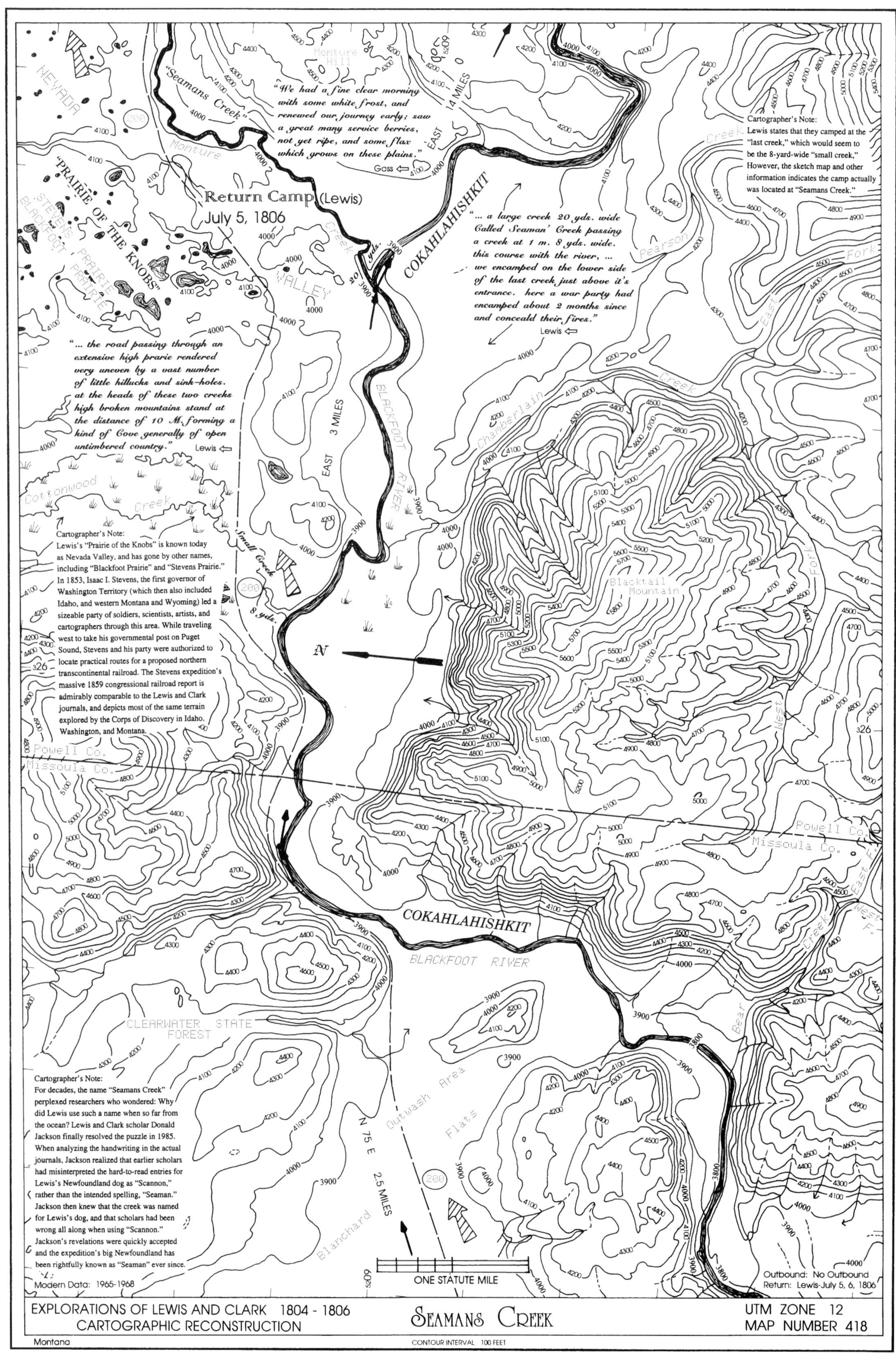
"We had a fine clear morning with some white frost, and renewed our journey early; saw a great many service berries, not yet ripe, and some flax which grows on these plains."
Gass ⇦
"Seamans Creek"
Monture Hill
Monture
NEVADA
14 MILES
EAST
Return Camp (Lewis)
July 5, 1806
"PRAIRIE OF THE KNOBS"
STEVENS PRAIRIE
BLACKFOOT PRAIRIE
VALLEY
Creek
20 yds.
COKAHLAHISHKIT
"... a large creek 20 yds. wide Called Seaman' Creek passing a creek at 1 m. 8 yds. wide. this course with the river, ... we encamped on the lower side of the last creek just above it's entrance. here a war party had encamped about 2 months since and conceald their fires."
Lewis ⇦
Cartographer's Note:
Lewis states that they camped at the "last creek," which would seem to be the 8-yard-wide "small creek." However, the sketch map and other information indicates the camp actually was located at "Seamans Creek."
Pearson
East Fork
Chamberlain Creek
"... the road passing through an extensive high prarie rendered very uneven by a vast number of little hillucks and sink-holes. at the heads of these two creeks high broken mountains stand at the distance of 10 M. forming a kind of Cove generally of open untimbered country."
Lewis ⇦
BLACKFOOT RIVER
3 MILES
EAST
Cottonwood Creek
Small Creek
8 yds.
Cartographer's Note:
Lewis's "Prairie of the Knobs" is known today as Nevada Valley, and has gone by other names, including "Blackfoot Prairie" and "Stevens Prairie." In 1853, Isaac I. Stevens, the first governor of Washington Territory (which then also included Idaho, and western Montana and Wyoming) led a sizeable party of soldiers, scientists, artists, and cartographers through this area. While traveling west to take his governmental post on Puget Sound, Stevens and his party were authorized to locate practical routes for a proposed northern transcontinental railroad. The Stevens expedition's massive 1859 congressional railroad report is admirably comparable to the Lewis and Clark journals, and depicts most of the same terrain explored by the Corps of Discovery in Idaho, Washington, and Montana.
N
Blacktail Mountain
West Fork
Powell Co.
Missoula Co.
COKAHLAHISHKIT
BLACKFOOT RIVER
Bear Creek
CLEARWATER STATE FOREST
Outwash Area
Flats
N 75 E
2.5 MILES
Blanchard
Cartographer's Note:
For decades, the name "Seamans Creek" perplexed researchers who wondered: Why did Lewis use such a name when so far from the ocean? Lewis and Clark scholar Donald Jackson finally resolved the puzzle in 1985. When analyzing the handwriting in the actual journals, Jackson realized that earlier scholars had misinterpreted the hard-to-read entries for Lewis's Newfoundland dog as "Scannon," rather than the intended spelling, "Seaman." Jackson then knew that the creek was named for Lewis's dog, and that scholars had been wrong all along when using "Scannon." Jackson's revelations were quickly accepted and the expedition's big Newfoundland has been rightfully known as "Seaman" ever since.
Modern Data: 1965-1968
ONE STATUTE MILE
Outbound: No Outbound
Return: Lewis-July 5, 6, 1806
EXPLORATIONS OF LEWIS AND CLARK 1804 - 1806
CARTOGRAPHIC RECONSTRUCTION
SEAMANS CREEK
UTM ZONE 12
MAP NUMBER 418
Montana
CONTOUR INTERVAL 100 FEET

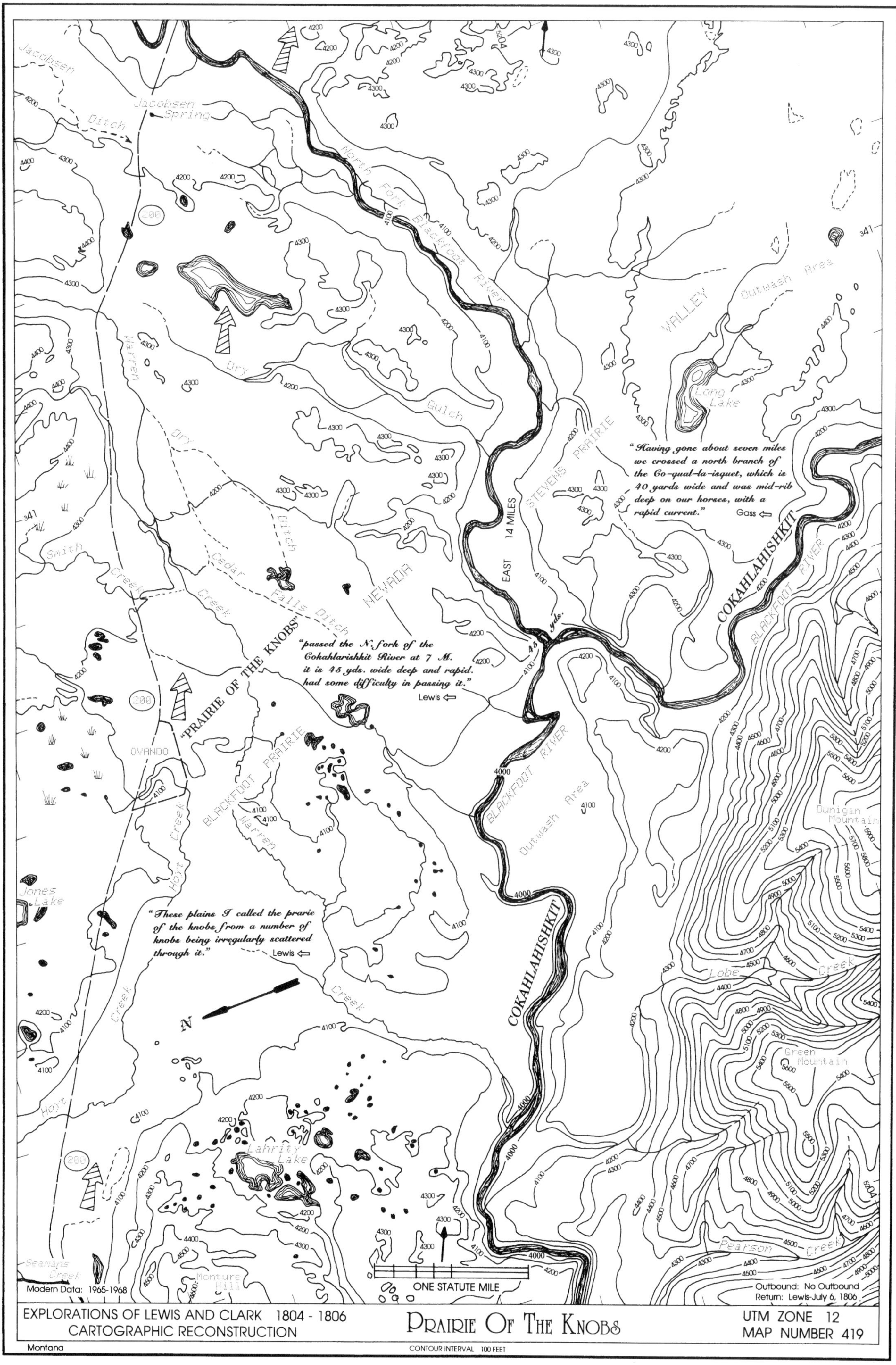
"Having gone about seven miles we crossed a north branch of the Co-qual-la-isquet, which is 40 yards wide and was mid-rib deep on our horses, with a rapid current."
Gass
"passed the N. fork of the Cohahlarishkit River at 7 M. it is 45 yds. wide deep and rapid. had some difficulty in passing it."
Lewis
"These plains I called the prarie of the knobs from a number of knobs being irregularly scattered through it."
Lewis
"PRAIRIE OF THE KNOBS"
COKAHLAHISHKIT
BLACKFOOT RIVER
North Fork Blackfoot River
STEVENS PRAIRIE
BLACKFOOT PRAIRIE
EAST 14 MILES
45 yds.
NEVADA
VALLEY
Outwash Area
Long Lake
Lahrity Lake
Jones Lake
Jacobsen Spring
Jacobsen Ditch
Dry Gulch
Dry Ditch
Cedar Falls Ditch
Cedar Creek
Warren Creek
Smith Creek
Hoyt Creek
Lobe Creek
Pearson Creek
Seamans Creek
Dunigan Mountain
Green Mountain
Monture Hill
OVANDO
200
141
N
ONE STATUTE MILE
Modern Data: 1965-1968
Outbound: No Outbound
Return: Lewis-July 6, 1806
EXPLORATIONS OF LEWIS AND CLARK 1804 - 1806
CARTOGRAPHIC RECONSTRUCTION
PRAIRIE OF THE KNOBS
UTM ZONE 12
MAP NUMBER 419
Montana
CONTOUR INTERVAL 100 FEET

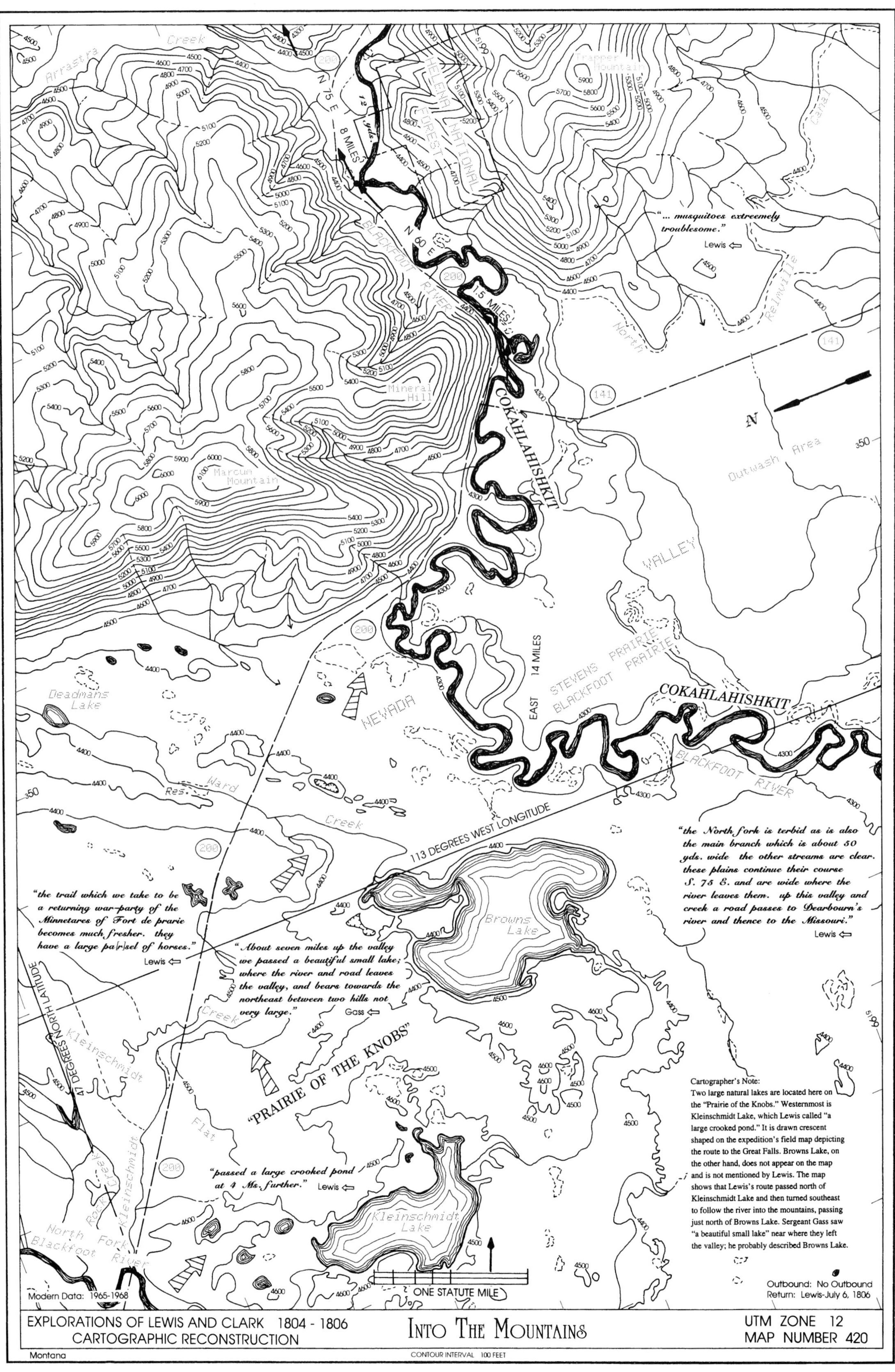
"... musquitoes extreemely troublesome."
Lewis
"the North fork is terbid as is also the main branch which is about 50 yds. wide the other streams are clear. these plains continue their course S. 75 E. and are wide where the river leaves them. up this valley and creek a road passes to Dearbourn's river and thence to the Missouri."
Lewis
"the trail which we take to be a returning war-party of the Minnetares of Fort de prarie becomes much fresher. they have a large pa[r]sel of horses."
Lewis
"About seven miles up the valley we passed a beautiful small lake; where the river and road leaves the valley, and bears towards the northeast between two hills not very large."
Gass
"passed a large crooked pond at 4 Ms. further."
Lewis
Cartographer's Note:
Two large natural lakes are located here on the "Prairie of the Knobs." Westernmost is Kleinschmidt Lake, which Lewis called "a large crooked pond." It is drawn crescent shaped on the expedition's field map depicting the route to the Great Falls. Browns Lake, on the other hand, does not appear on the map and is not mentioned by Lewis. The map shows that Lewis's route passed north of Kleinschmidt Lake and then turned southeast to follow the river into the mountains, passing just north of Browns Lake. Sergeant Gass saw "a beautiful small lake" near where they left the valley; he probably described Browns Lake.
Outbound: No Outbound
Return: Lewis-July 6, 1806
Modern Data: 1965-1968
ONE STATUTE MILE
EXPLORATIONS OF LEWIS AND CLARK 1804 - 1806
CARTOGRAPHIC RECONSTRUCTION
INTO THE MOUNTAINS
UTM ZONE 12
MAP NUMBER 420
Montana
CONTOUR INTERVAL 100 FEET
COKAHLAHISHKIT
BLACKFOOT RIVER
HELENA NATIONAL FOREST
Trapper Mountain
Mineral Hill
Marcum Mountain
Arrastra Creek
Outwash Area
VALLEY
STEVENS PRAIRIE
BLACKFOOT PRAIRIE
NEVADA
Deadmans Lake
Ward Creek
Res.
Browns Lake
Kleinschmidt Lake
Kleinschmidt Flat
Kleinschmidt Creek
Rock Creek
North Fork Blackfoot River
North Fork
Reinmuller
Canal
"PRAIRIE OF THE KNOBS"
113 DEGREES WEST LONGITUDE
47 DEGREES NORTH LATITUDE
N 75 E
8 MILES
N 60 E
1.5 MILES
EAST 14 MILES
N

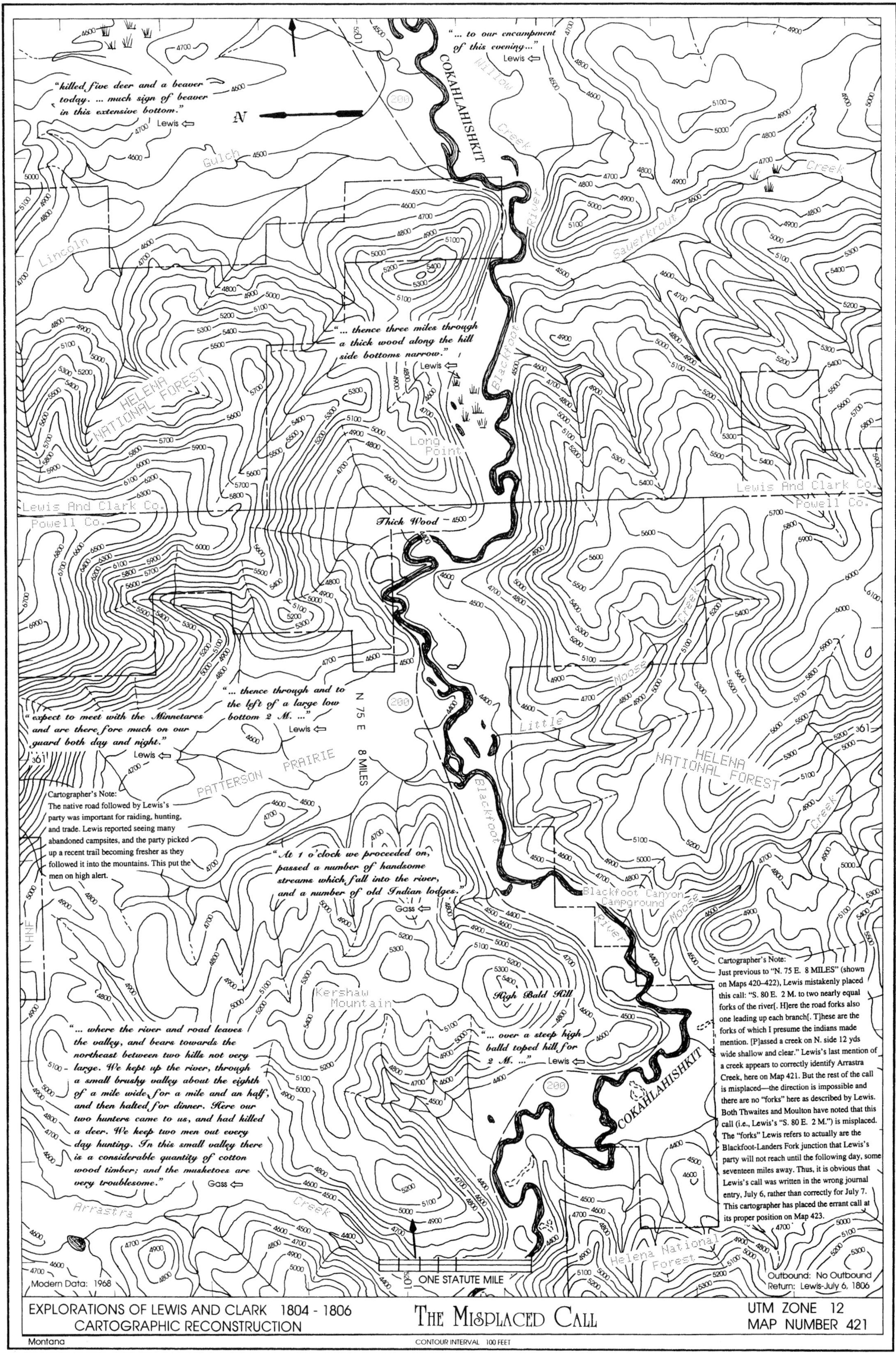
"... to our encampment of this evening..."
Lewis ⇦
"killed five deer and a beaver today. ... much sign of beaver in this extensive bottom."
Lewis ⇦
N
COKAHLAHISHKIT
200
Willow Creek
Gulch
Creek
Sauerkrout
River
Lincoln
"... thence three miles through a thick wood along the hill side bottoms narrow."
Lewis ⇦
HELENA NATIONAL FOREST
Blackfoot
Long Point
Lewis And Clark Co.
Powell Co.
Lewis And Clark Co.
Powell Co.
Thick Wood
Moose Creek
Little
"... thence through and to the left of a large low bottom 2 M. ..."
Lewis ⇦
N 75 E 8 MILES
"expect to meet with the Minnetares and are there fore much on our guard both day and night."
Lewis ⇦
361
PATTERSON PRAIRIE
HELENA NATIONAL FOREST
361
Cartographer's Note:
The native road followed by Lewis's party was important for raiding, hunting, and trade. Lewis reported seeing many abandoned campsites, and the party picked up a recent trail becoming fresher as they followed it into the mountains. This put the men on high alert.
"At 1 o'clock we proceeded on, passed a number of handsome streams which fall into the river, and a number of old Indian lodges."
Gass ⇦
Blackfoot
Blackfoot Canyon Campground
River
Moose
HNF
Kershaw Mountain
High Bald Hill
Cartographer's Note:
Just previous to "N. 75 E. 8 MILES" (shown on Maps 420–422), Lewis mistakenly placed this call: "S. 80 E. 2 M. to two nearly equal forks of the river[. H]ere the road forks also one leading up each branch[. T]hese are the forks of which I presume the indians made mention. [P]assed a creek on N. side 12 yds wide shallow and clear." Lewis's last mention of a creek appears to correctly identify Arrastra Creek, here on Map 421. But the rest of the call is misplaced—the direction is impossible and there are no "forks" here as described by Lewis. Both Thwaites and Moulton have noted that this call (i.e., Lewis's "S. 80 E. 2 M.") is misplaced. The "forks" Lewis refers to actually are the Blackfoot-Landers Fork junction that Lewis's party will not reach until the following day, some seventeen miles away. Thus, it is obvious that Lewis's call was written in the wrong journal entry, July 6, rather than correctly for July 7. This cartographer has placed the errant call at its proper position on Map 423.
"... where the river and road leaves the valley, and bears towards the northeast between two hills not very large. We kept up the river, through a small brushy valley about the eighth of a mile wide, for a mile and an half, and then halted for dinner. Here our two hunters came to us, and had killed a deer. We keep two men out every day hunting. In this small valley there is a considerable quantity of cotton wood timber; and the musketoes are very troublesome."
Gass ⇦
"... over a steep high balld toped hill for 2 M. ..."
Lewis ⇦
COKAHLAHISHKIT
200
Arrastra Creek
Helena National Forest
Modern Data: 1968
ONE STATUTE MILE
Outbound: No Outbound
Return: Lewis-July 6, 1806
EXPLORATIONS OF LEWIS AND CLARK 1804 - 1806
CARTOGRAPHIC RECONSTRUCTION
THE MISPLACED CALL
UTM ZONE 12
MAP NUMBER 421
Montana
CONTOUR INTERVAL 100 FEET

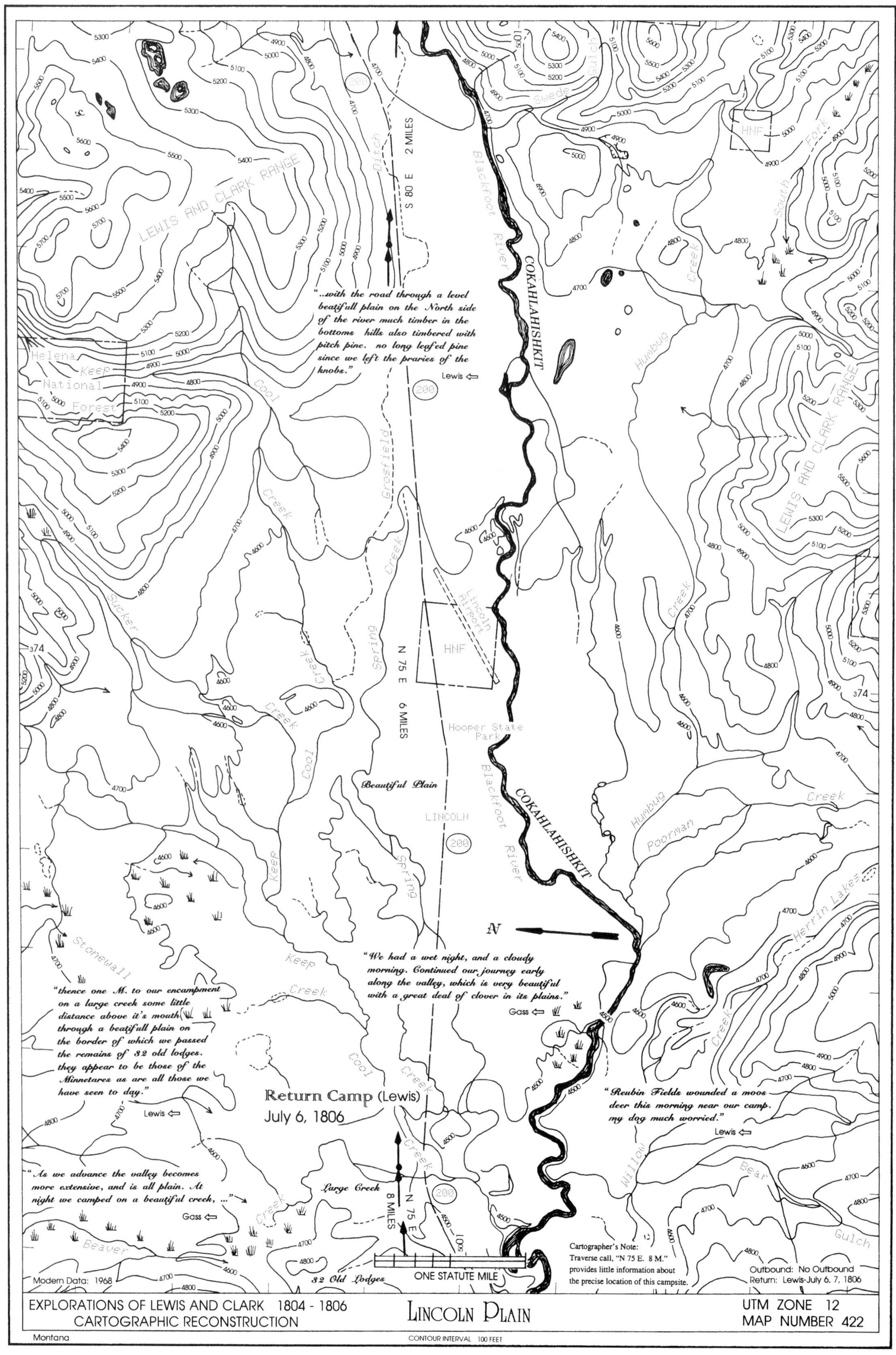
"...with the road through a level beatifull plain on the North side of the river much timber in the bottoms hills also timbered with pitch pine. no long leafed pine since we left the praries of the knobs."
Lewis
"We had a wet night, and a cloudy morning. Continued our journey early along the valley, which is very beautiful with a great deal of clover in its plains."
Gass
"thence one M. to our encampment on a large creek some little distance above it's mouth through a beatifull plain on the border of which we passed the remains of 32 old lodges. they appear to be those of the Minnetares as are all those we have seen to day."
Lewis
"Reubin Fields wounded a moos deer this morning near our camp. my dog much worried."
Lewis
"As we advance the valley becomes more extensive, and is all plain. At night we camped on a beautiful creek, ..."
Gass
Return Camp (Lewis)
July 6, 1806
Beautiful Plain
Large Creek
32 Old Lodges
LEWIS AND CLARK RANGE
COKAHLAHISHKIT
Blackfoot River
Helena National Forest
HNF
LINCOLN
Lincoln Airport
Hooper State Park
S 80 E 2 MILES
N 75 E 6 MILES
N 75 E 8 MILES
N
ONE STATUTE MILE
Cartographer's Note:
Traverse call, "N 75 E. 8 M."
provides little information about the precise location of this campsite.
Outbound: No Outbound
Return: Lewis-July 6, 7, 1806
Modern Data: 1968
EXPLORATIONS OF LEWIS AND CLARK 1804 - 1806
CARTOGRAPHIC RECONSTRUCTION
LINCOLN PLAIN
UTM ZONE 12
MAP NUMBER 422
Montana
CONTOUR INTERVAL 100 FEET

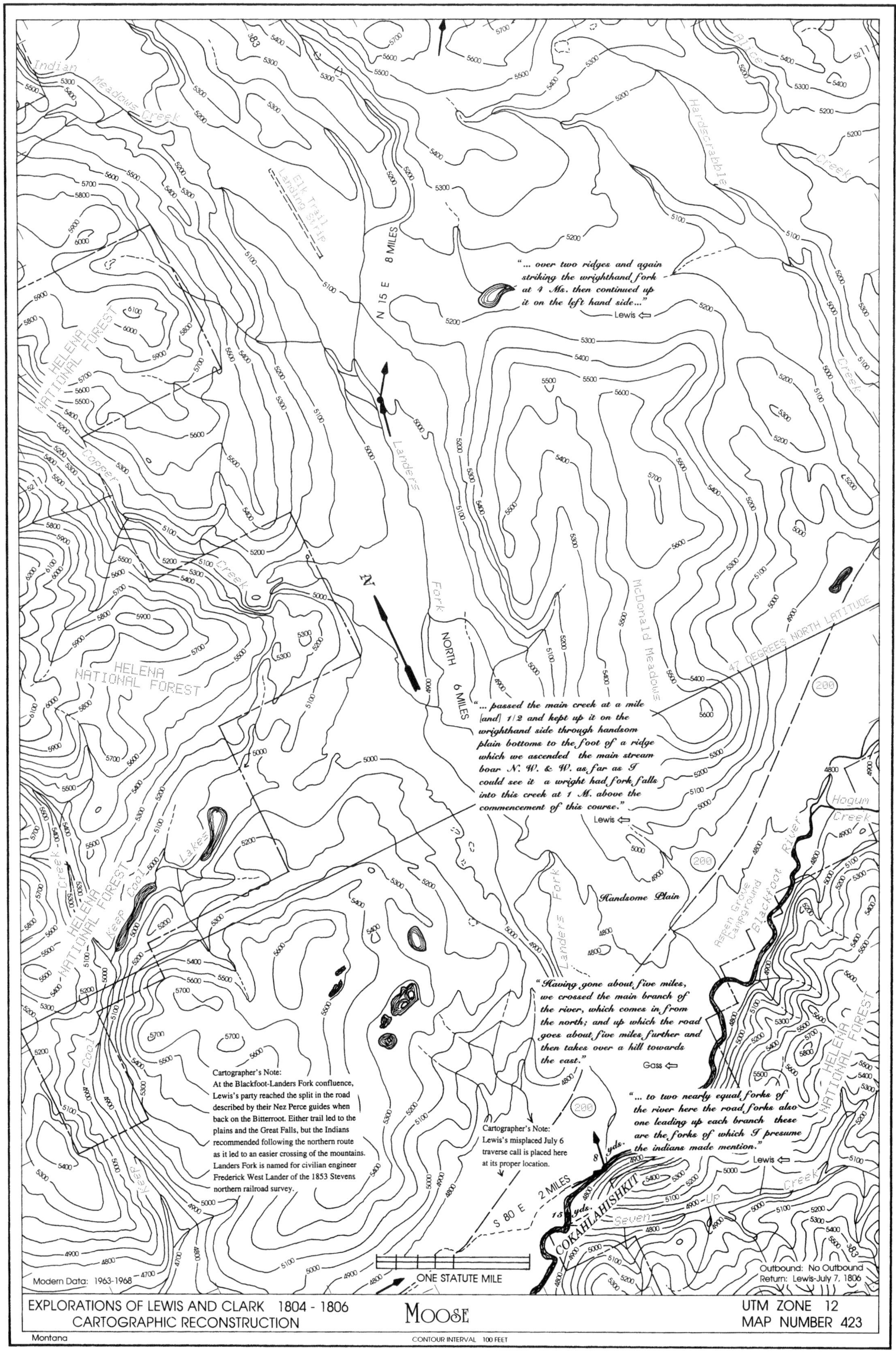
"... over two ridges and again striking the wrighthand fork at 4 Ms. then continued up it on the left hand side..."
Lewis ⇦
"... passed the main creek at a mile [and] 1/2 and kept up it on the wrighthand side through handsom plain bottoms to the foot of a ridge which we ascended the main stream boar N. W. & W. as far as I could see it a wright had fork falls into this creek at 1 M. above the commencement of this course."
Lewis ⇦
"Having gone about five miles, we crossed the main branch of the river, which comes in from the north; and up which the road goes about five miles further and then takes over a hill towards the east."
Gass ⇦
"... to two nearly equal forks of the river here the road forks also one leading up each branch these are the forks of which I presume the indians made mention."
Lewis ⇦
Cartographer's Note:
At the Blackfoot-Landers Fork confluence, Lewis's party reached the split in the road described by their Nez Perce guides when back on the Bitterroot. Either trail led to the plains and the Great Falls, but the Indians recommended following the northern route as it led to an easier crossing of the mountains. Landers Fork is named for civilian engineer Frederick West Lander of the 1853 Stevens northern railroad survey.
Cartographer's Note:
Lewis's misplaced July 6 traverse call is placed here at its proper location.
N 15 E 8 MILES
NORTH 6 MILES
S 80 E 2 MILES
Handsome Plain
COKAHLAHISHKIT
47 DEGREES NORTH LATITUDE
HELENA NATIONAL FOREST
Elk Trail Landing Strip
Landers Fork
McDonald Meadows
Copper Creek
Indian Meadows Creek
Alice Creek
Hardscrabble Creek
Hogum Creek
Blackfoot River
Aspen Grove Campground
Seven Up Creek
Keep Cool Lakes
Keep Cool Creek
Modern Data: 1963-1968
ONE STATUTE MILE
Outbound: No Outbound
Return: Lewis-July 7, 1806
EXPLORATIONS OF LEWIS AND CLARK 1804 - 1806
CARTOGRAPHIC RECONSTRUCTION
MOOSE
UTM ZONE 12
MAP NUMBER 423
Montana
CONTOUR INTERVAL 100 FEET

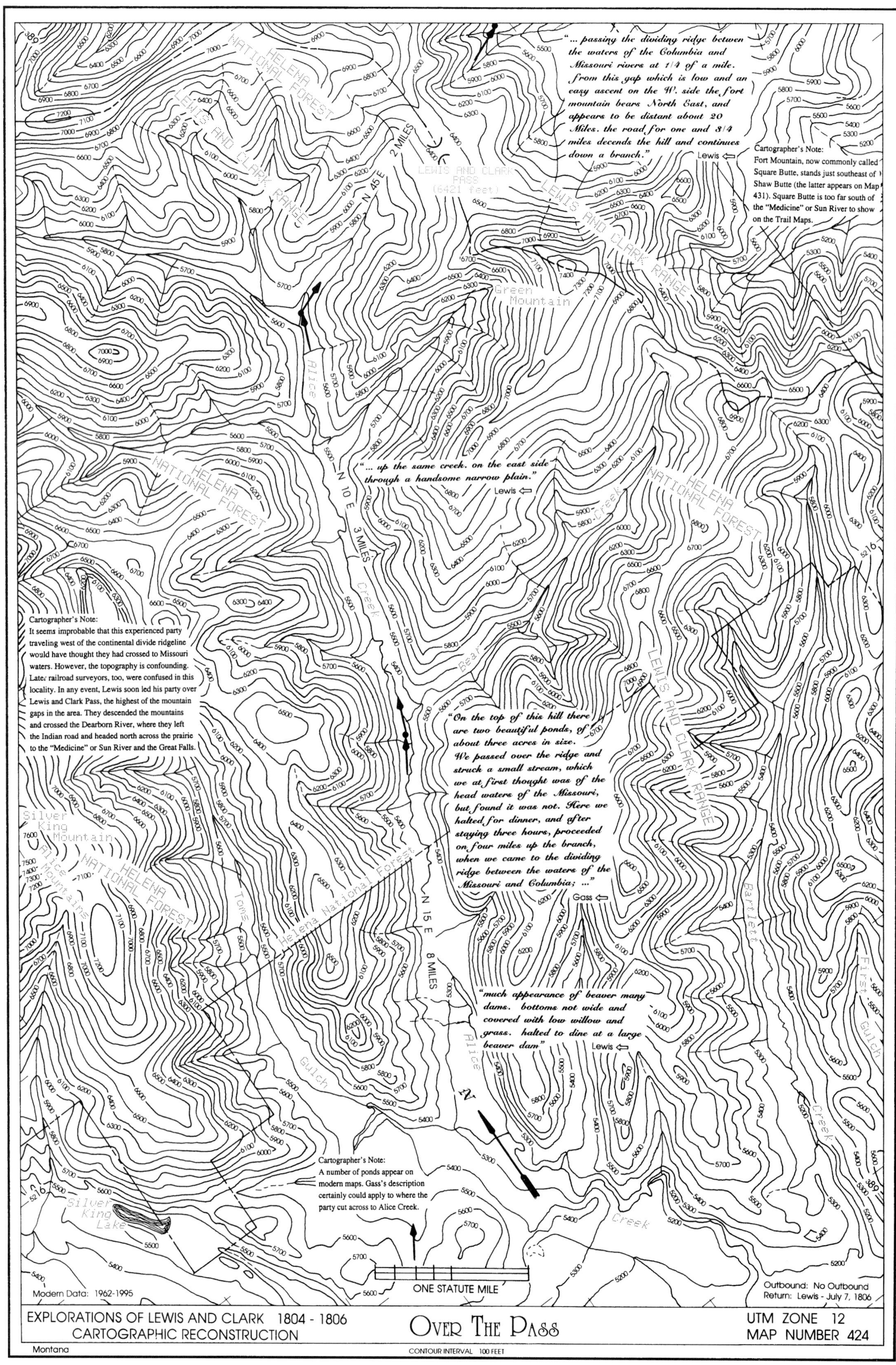
"... passing the dividing ridge betwen the waters of the Columbia and Missouri rivers at 1/4 of a mile. from this gap which is low and an easy ascent on the W. side the fort mountain bears North East, and appears to be distant about 20 Miles. the road for one and 3/4 miles decends the hill and continues down a branch."
Lewis
Cartographer's Note:
Fort Mountain, now commonly called Square Butte, stands just southeast of Shaw Butte (the latter appears on Map 431). Square Butte is too far south of the "Medicine" or Sun River to show on the Trail Maps.
HELENA NATIONAL FOREST
LEWIS AND CLARK RANGE
LEWIS AND CLARK PASS (6421 feet)
2 MILES
N 45 E
Green Mountain
Alice
"... up the same creek. on the east side through a handsome narrow plain."
Lewis
N 10 E
3 MILES
Creek
Bear
Cartographer's Note:
It seems improbable that this experienced party traveling west of the continental divide ridgeline would have thought they had crossed to Missouri waters. However, the topography is confounding. Later railroad surveyors, too, were confused in this locality. In any event, Lewis soon led his party over Lewis and Clark Pass, the highest of the mountain gaps in the area. They descended the mountains and crossed the Dearborn River, where they left the Indian road and headed north across the prairie to the "Medicine" or Sun River and the Great Falls.
"On the top of this hill there are two beautiful ponds, of about three acres in size. We passed over the ridge and struck a small stream, which we at first thought was of the head waters of the Missouri, but found it was not. Here we halted for dinner, and after staying three hours, proceeded on four miles up the branch, when we came to the dividing ridge between the waters of the Missouri and Columbia; ..."
Gass
Silver King Mountain
Alice Mountains
Helena National Forest
Toms
N 15 E
8 MILES
Bartlett
First
Gulch
"much appearance of beaver many dams. bottoms not wide and covered with low willow and grass. halted to dine at a large beaver dam"
Lewis
Gulch
Creek
Cartographer's Note:
A number of ponds appear on modern maps. Gass's description certainly could apply to where the party cut across to Alice Creek.
Silver King Lake
ONE STATUTE MILE
Modern Data: 1962-1995
Outbound: No Outbound
Return: Lewis - July 7, 1806
EXPLORATIONS OF LEWIS AND CLARK 1804 - 1806
CARTOGRAPHIC RECONSTRUCTION
OVER THE PASS
UTM ZONE 12
MAP NUMBER 424
Montana
CONTOUR INTERVAL 100 FEET

"The morning was pleasant with some white frost. We started early and proceeded on nearly north; ..."
Gass
"Set out at 6 A. M."
Lewis
Return Camp (Lewis)
July 7, 1806
"Drewyer killed two beaver and shot a third which bit his knee very badly and escaped"
Lewis
3.5 MILES
N 25 W
N 20 W 7 MILES
"saw some sighn of buffaloe early this morning in the valley where we encamped last evening from which it appears that the buffaloe do sometimes penetrate these mountains a few miles. we saw no buffaloe this evening."
Lewis
"... passed over the ridge and came to a fine spring the waters of which run into the Missouri. We then kept down this stream or branch about a mile; then turned a north course along the side of the dividing ridge for eight miles, passing a number of small streams or branches, and at 9 o'clock at night encamped after coming 32 miles."
Gass
"... over several hills and hollows along the foot of the mountain hights passing five small rivulets running to the wright."
Lewis
2 MILES
N 45 E
ONE STATUTE MILE
Wrangle Creek
Big Skunk Creek
Little Skunk Creek
Sugarloaf Mountain
Cuniff Basin
Falls Creek Ridge
East Fork
Hardgrove Creek
Bedrock Creek
Sunset Hill
Sunrise Hill
North Fork
Green Creek
Burned Point
Red Mountain
LEWIS AND CLARK NATIONAL FOREST
Modern Data: 1962-1995
Outbound: No Outbound
Return: Lewis-July 7, 8, 1806
EXPLORATIONS OF LEWIS AND CLARK 1804 - 1806
CARTOGRAPHIC RECONSTRUCTION
EAST OF THE DIVIDE
UTM ZONE 12
MAP NUMBER 425
Montana
CONTOUR INTERVAL 100 FEET

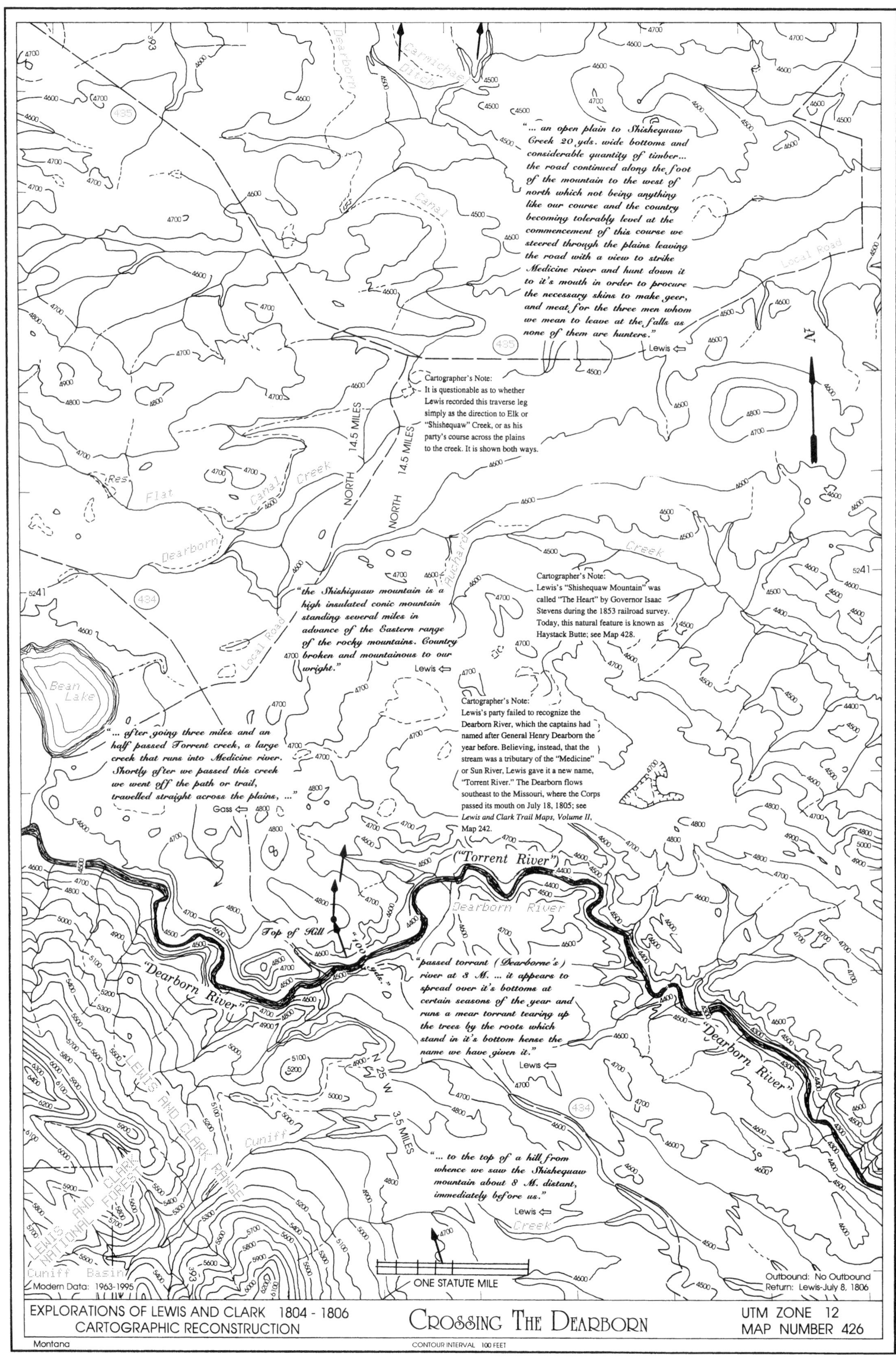

"... an open plain to Shishequaw Creek 20 yds. wide bottoms and considerable quantity of timber... the road continued along the foot of the mountain to the west of north which not being anything like our course and the country becoming tolerably level at the commencement of this course we steered through the plains leaving the road with a view to strike Medicine river and hunt down it to it's mouth in order to procure the necessary skins to make geer, and meat for the three men whom we mean to leave at the falls as none of them are hunters."
Lewis
Cartographer's Note:
It is questionable as to whether Lewis recorded this traverse leg simply as the direction to Elk or "Shishequaw" Creek, or as his party's course across the plains to the creek. It is shown both ways.
NORTH 14.5 MILES
NORTH 14.5 MILES
"the Shishiquaw mountain is a high insulated conic mountain standing several miles in advance of the Eastern range of the rocky mountains. Country broken and mountainous to our wright."
Lewis
Cartographer's Note:
Lewis's "Shishequaw Mountain" was called "The Heart" by Governor Isaac Stevens during the 1853 railroad survey. Today, this natural feature is known as Haystack Butte; see Map 428.
Cartographer's Note:
Lewis's party failed to recognize the Dearborn River, which the captains had named after General Henry Dearborn the year before. Believing, instead, that the stream was a tributary of the "Medicine" or Sun River, Lewis gave it a new name, "Torrent River." The Dearborn flows southeast to the Missouri, where the Corps passed its mouth on July 18, 1805; see Lewis and Clark Trail Maps, Volume II, Map 242.
"... after going three miles and an half passed Torrent creek, a large creek that runs into Medicine river. Shortly after we passed this creek we went off the path or trail, travelled straight across the plains, ..."
Gass
("Torrent River")
Top of Hill
"Dearborn River"
"passed torrant (Dearborne's) river at 8 M. ... it appears to spread over it's bottoms at certain seasons of the year and runs a mear torrant tearing up the trees by the roots which stand in it's bottom hense the name we have given it."
Lewis
"Dearborn River"
N 25 W 3.5 MILES
"... to the top of a hill from whence we saw the Shishequaw mountain about 8 M. distant, immediately before us."
Lewis
Dearborn
Carmichael Ditch
Canal
Local Road
Res
Flat
Canal Creek
Dearborn
Auchard
Creek
Bean Lake
Local Road
Dearborn River
Cuniff
Creek
LEWIS AND CLARK RANGE
LEWIS AND CLARK NATIONAL FOREST
Cuniff Basin
Modern Data: 1963-1995
ONE STATUTE MILE
Outbound: No Outbound
Return: Lewis-July 8, 1806
EXPLORATIONS OF LEWIS AND CLARK 1804 - 1806
CARTOGRAPHIC RECONSTRUCTION
CROSSING THE DEARBORN
UTM ZONE 12
MAP NUMBER 426
Montana
CONTOUR INTERVAL 100 FEET

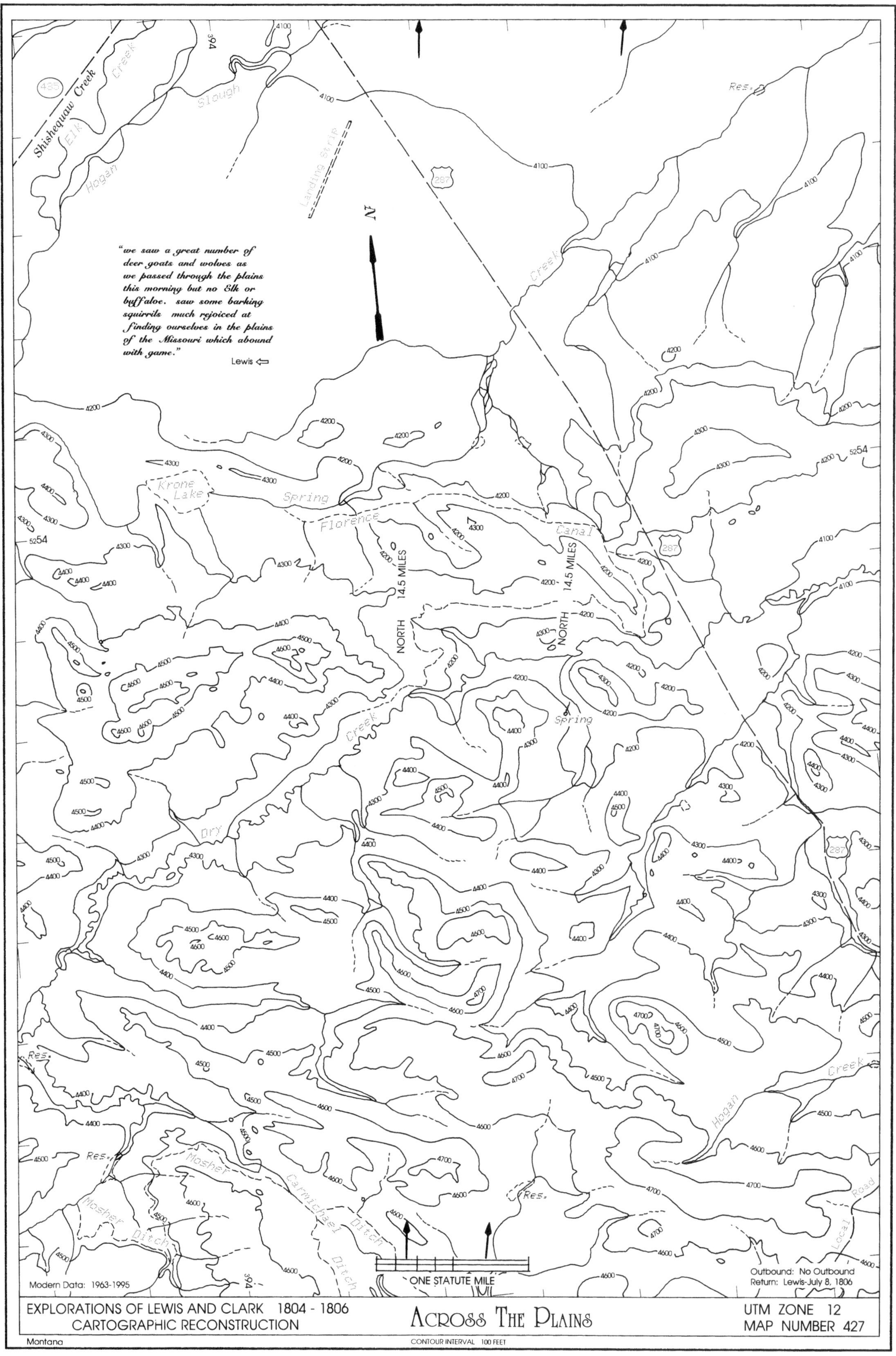
"we saw a great number of deer goats and wolves as we passed through the plains this morning but no Elk or buffaloe. saw some barking squirrils much rejoiced at finding ourselves in the plains of the Missouri which abound with game."
Lewis
Shishequaw Creek
Elk Creek
Hogan
Slough
Landing strip
Krone Lake
Spring
Florence Canal
Creek
Dry
Spring
Hogan Creek
Res.
Mosher
Mosher Ditch
Carmichael Ditch
Local Road
NORTH 14.5 MILES
NORTH 14.5 MILES
ONE STATUTE MILE
Modern Data: 1963-1995
Outbound: No Outbound
Return: Lewis-July 8, 1806
EXPLORATIONS OF LEWIS AND CLARK 1804 - 1806
CARTOGRAPHIC RECONSTRUCTION
Across The Plains
UTM ZONE 12
MAP NUMBER 427
Montana
CONTOUR INTERVAL 100 FEET

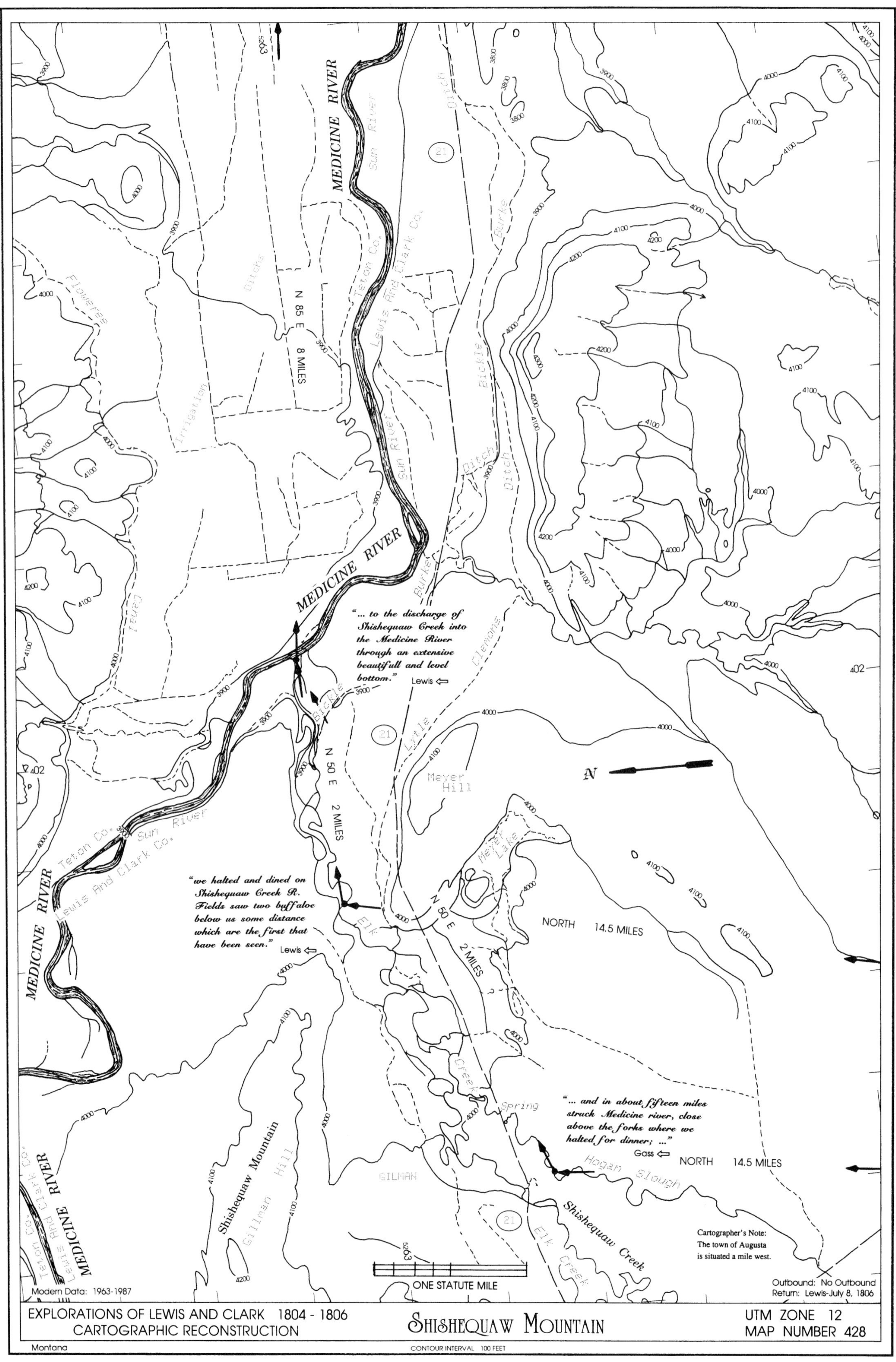

MEDICINE RIVER
"... to the discharge of Shishequaw Creek into the Medicine River through an extensive beautifull and level bottom."
Lewis
"we halted and dined on Shishequaw Creek R. Fields saw two buffaloe below us some distance which are the first that have been seen."
Lewis
"... and in about fifteen miles struck Medicine river, close above the forks where we halted for dinner; ..."
Gass
N 85 E 8 MILES
N 50 E 2 MILES
N 50 E 2 MILES
NORTH 14.5 MILES
NORTH 14.5 MILES
N
Shishequaw Mountain
Shishequaw Creek
Meyer Hill
Meyer Lake
Gillman Hill
GILMAN
Hogan Slough
Spring
Elk Creek
Sun River
Teton Co.
Lewis And Clark Co.
Cartographer's Note:
The town of Augusta is situated a mile west.
ONE STATUTE MILE
Modern Data: 1963-1987
Outbound: No Outbound
Return: Lewis-July 8, 1806
EXPLORATIONS OF LEWIS AND CLARK 1804 - 1806
CARTOGRAPHIC RECONSTRUCTION
SHISHEQUAW MOUNTAIN
UTM ZONE 12
MAP NUMBER 428
Montana
CONTOUR INTERVAL 100 FEET

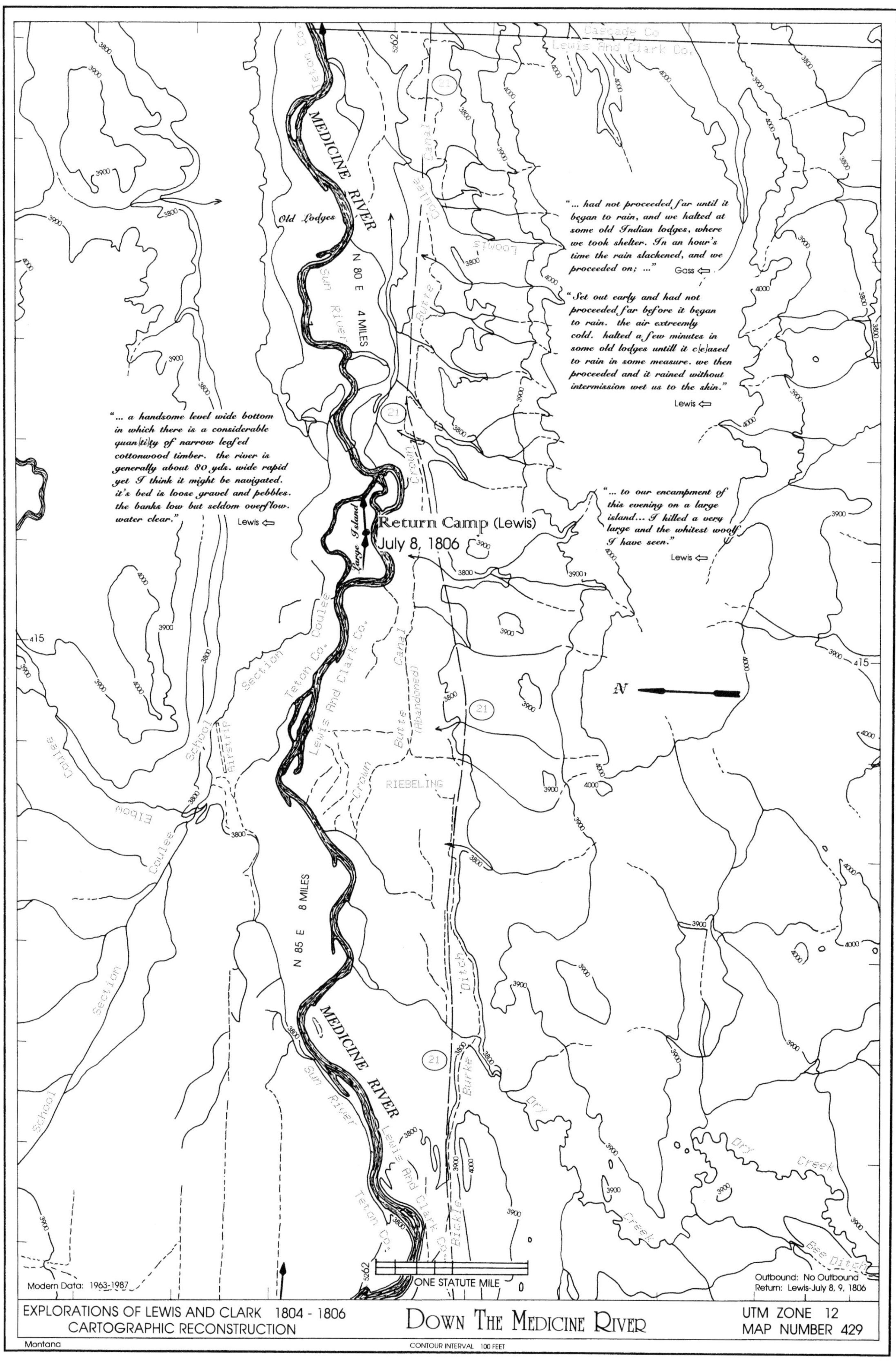
MEDICINE RIVER
Old Lodges
N 80 E 4 MILES
"... had not proceeded far until it began to rain, and we halted at some old Indian lodges, where we took shelter. In an hour's time the rain slackened, and we proceeded on; ..." Gass
"Set out early and had not proceeded far before it began to rain. the air extreemly cold. halted a few minutes in some old lodges untill it c[e]ased to rain in some measure. we then proceeded and it rained without intermission wet us to the skin." Lewis
"... a handsome level wide bottom in which there is a considerable quan[ti]ty of narrow leafed cottonwood timber. the river is generally about 80 yds. wide rapid yet I think it might be navigated. it's bed is loose gravel and pebbles. the banks low but seldom overflow. water clear." Lewis
Large Island
Return Camp (Lewis)
July 8, 1806
"... to our encampment of this evening on a large island... I killed a very large and the whitest woolf I have seen." Lewis
N 85 E 8 MILES
MEDICINE RIVER
N
ONE STATUTE MILE
Modern Data: 1963-1987
Outbound: No Outbound
Return: Lewis-July 8, 9, 1806
EXPLORATIONS OF LEWIS AND CLARK 1804 - 1806
CARTOGRAPHIC RECONSTRUCTION
DOWN THE MEDICINE RIVER
UTM ZONE 12
MAP NUMBER 429
Montana
CONTOUR INTERVAL 100 FEET

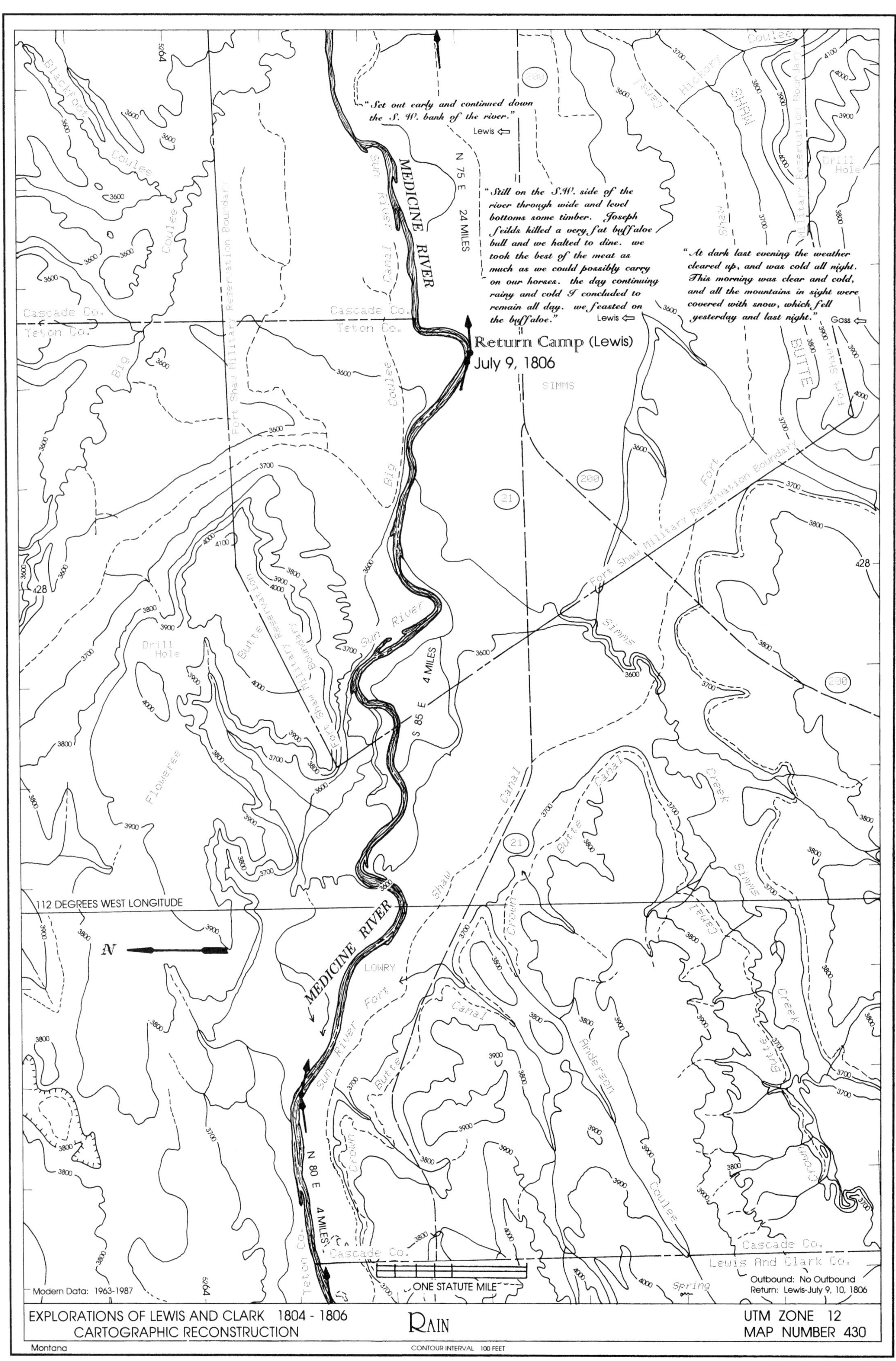
"Set out early and continued down the S. W. bank of the river."
Lewis
"Still on the S.W. side of the river through wide and level bottoms some timber. Joseph feilds killed a very fat buffaloe bull and we halted to dine. we took the best of the meat as much as we could possibly carry on our horses. the day continuing rainy and cold I concluded to remain all day. we feasted on the buffaloe."
Lewis
"At dark last evening the weather cleared up, and was cold all night. This morning was clear and cold, and all the mountains in sight were covered with snow, which fell yesterday and last night."
Gass
Return Camp (Lewis)
July 9, 1806
MEDICINE RIVER
N 75 E 24 MILES
S 85 E 4 MILES
N 80 E 4 MILES
112 DEGREES WEST LONGITUDE
N
ONE STATUTE MILE
Modern Data: 1963-1987
Outbound: No Outbound
Return: Lewis-July 9, 10, 1806
EXPLORATIONS OF LEWIS AND CLARK 1804 - 1806
CARTOGRAPHIC RECONSTRUCTION
RAIN
UTM ZONE 12
MAP NUMBER 430
Montana
CONTOUR INTERVAL 100 FEET

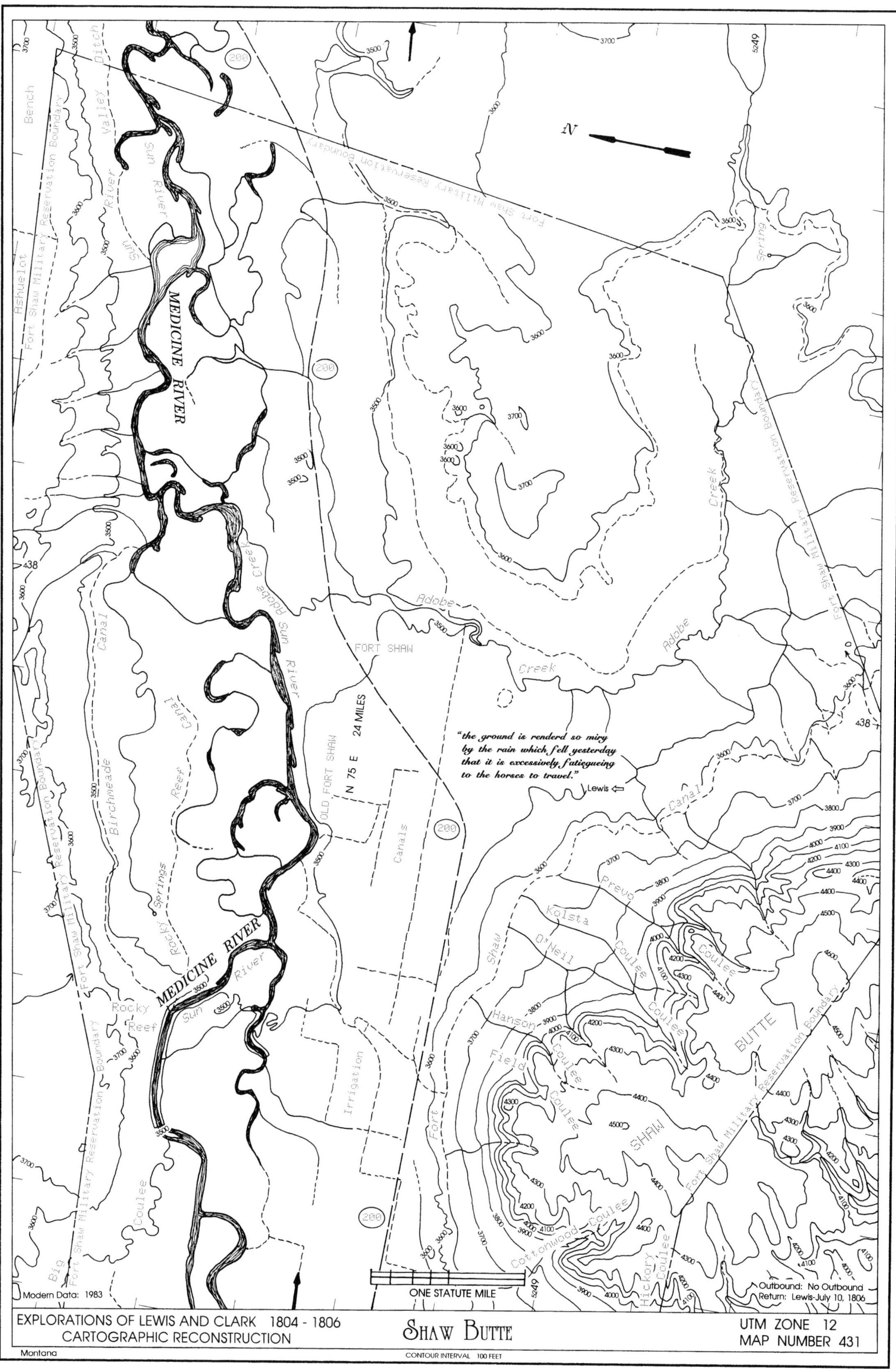

EXPLORATIONS OF LEWIS AND CLARK 1804 - 1806
CARTOGRAPHIC RECONSTRUCTION
SHAW BUTTE
UTM ZONE 12
MAP NUMBER 431
Montana
CONTOUR INTERVAL 100 FEET
ONE STATUTE MILE
Modern Data: 1983
Outbound: No Outbound
Return: Lewis-July 10, 1806
"the ground is renderd so miry by the rain which fell yesterday that it is excessively fatiegueing to the horses to travel."
Lewis
MEDICINE RIVER
Sun River
Adobe Creek
FORT SHAW
OLD FORT SHAW
N 75 E 24 MILES
Canals
Irrigation
Fort Shaw Military Reservation Boundary
Sun River Valley Ditch
Ashuelot Bench
Birchneade Canal
Rocky Reef Canal
Rocky Springs
Rocky Reef
Big Coulee
Shaw Coulee
Kolsta Coulee
O'Neil Coulee
Prevo Coulee
Hanson Coulee
Field Coulee
Cottonwood Coulee
Hickory Coulee
Spring
SHAW BUTTE
Fort
N

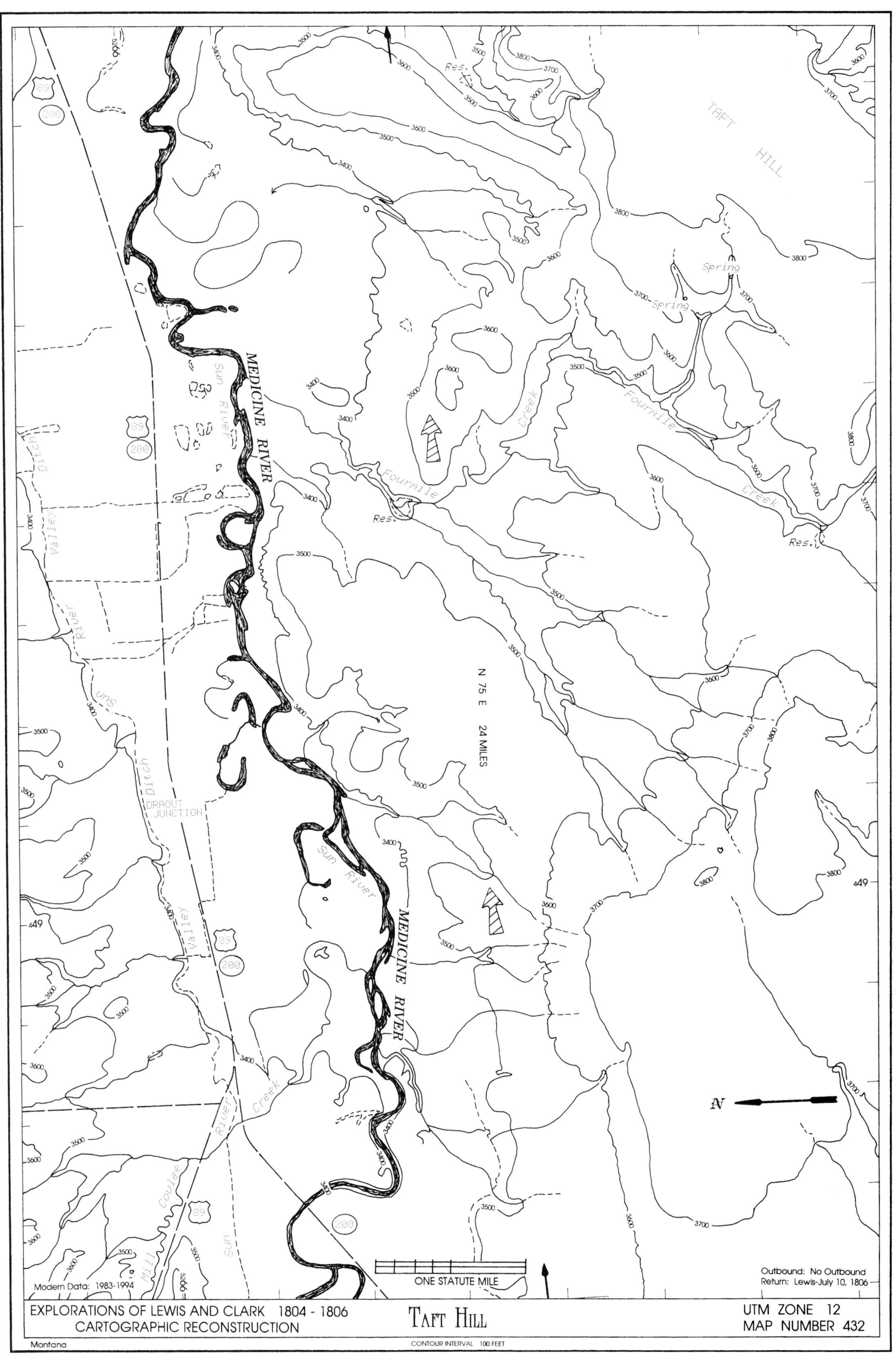
MEDICINE RIVER
Sun River
Fourmile Creek
TAFT HILL
Spring
Res.
DRAOUT JUNCTION
N 75 E 24 MILES
ONE STATUTE MILE
Modern Data: 1983-1994
Outbound: No Outbound
Return: Lewis-July 10, 1806
EXPLORATIONS OF LEWIS AND CLARK 1804 - 1806
CARTOGRAPHIC RECONSTRUCTION
TAFT HILL
UTM ZONE 12
MAP NUMBER 432
Montana
CONTOUR INTERVAL 100 FEET

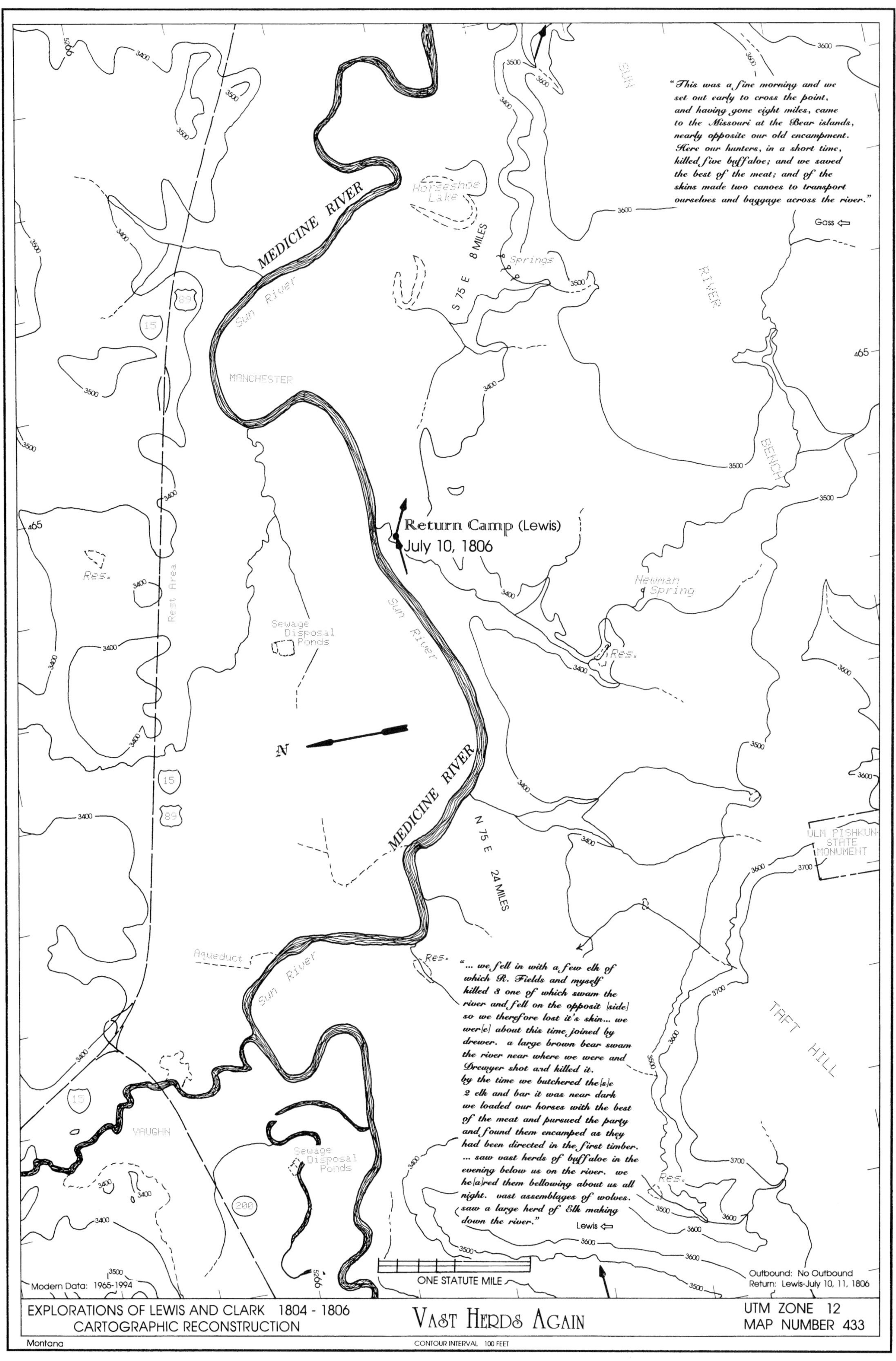
"This was a fine morning and we
set out early to cross the point,
and having gone eight miles, came
to the Missouri at the Bear islands,
nearly opposite our old encampment.
Here our hunters, in a short time,
killed five buffaloe; and we saved
the best of the meat; and of the
skins made two canoes to transport
ourselves and baggage across the river."
Gass ⇦
MEDICINE RIVER
Sun River
Horseshoe Lake
Springs
S 75 E 8 MILES
SUN RIVER BENCH
MANCHESTER
Return Camp (Lewis)
July 10, 1806
Res.
Rest Area
Sewage Disposal Ponds
Newman Spring
N
MEDICINE RIVER
N 75 E 24 MILES
ULM PISHKUN STATE MONUMENT
TAFT HILL
Aqueduct
VAUGHN
"... we fell in with a few elk of
which R. Fields and myself
killed 3 one of which swam the
river and fell on the opposit [side]
so we therefore lost it's skin... we
wer[e] about this time joined by
drewer. a large brown bear swam
the river near where we were and
Drewyer shot and killed it.
by the time we butchered the[s]e
2 elk and bar it was near dark
we loaded our horses with the best
of the meat and pursued the party
and found them encamped as they
had been directed in the first timber.
... saw vast herds of buffaloe in the
evening below us on the river. we
he[a]red them bellowing about us all
night. vast assemblages of wolves.
saw a large herd of Elk making
down the river."
Lewis ⇦
ONE STATUTE MILE
Modern Data: 1965-1994
Outbound: No Outbound
Return: Lewis-July 10, 11, 1806
EXPLORATIONS OF LEWIS AND CLARK 1804 - 1806
CARTOGRAPHIC RECONSTRUCTION
VAST HERDS AGAIN
UTM ZONE 12
MAP NUMBER 433
Montana
CONTOUR INTERVAL 100 FEET

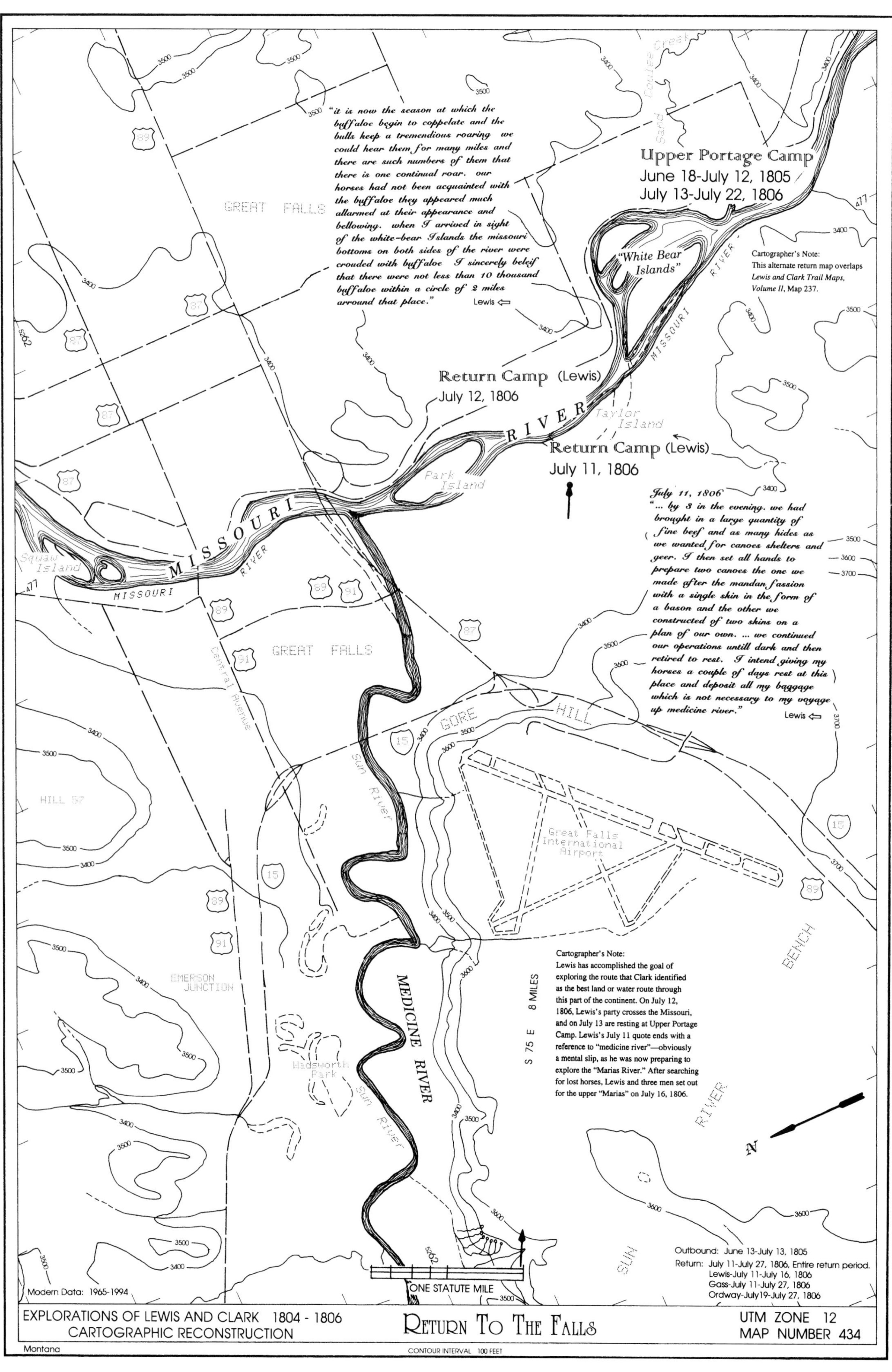
"it is now the season at which the buffaloe begin to coppelate and the bulls keep a tremendious roaring we could hear them for many miles and there are such numbers of them that there is one continual roar. our horses had not been acquainted with the buffaloe they appeared much allarmed at their appearance and bellowing. when I arrived in sight of the white-bear Islands the missouri bottoms on both sides of the river were crouded with buffaloe I sincerely belief that there were not less than 10 thousand buffaloe within a circle of 2 miles arround that place."
Lewis
Upper Portage Camp
June 18-July 12, 1805
July 13-July 22, 1806
"White Bear Islands"
Cartographer's Note:
This alternate return map overlaps *Lewis and Clark Trail Maps, Volume II*, Map 237.
Return Camp (Lewis)
July 12, 1806
Return Camp (Lewis)
July 11, 1806
July 11, 1806
"... by 3 in the evening. we had brought in a large quantity of fine beef and as many hides as we wanted for canoes shelters and geer. I then set all hands to prepare two canoes the one we made after the mandan fassion with a single skin in the form of a bason and the other we constructed of two skins on a plan of our own. ... we continued our operations untill dark and then retired to rest. I intend giving my horses a couple of days rest at this place and deposit all my baggage which is not necessary to my voyage up medicine river."
Lewis
MISSOURI RIVER
Taylor Island
Park Island
Squaw Island
GREAT FALLS
GORE HILL
Central Avenue
Sun River
HILL 57
Great Falls International Airport
EMERSON JUNCTION
Wadsworth Park
MEDICINE RIVER
BENCH
RIVER
SUN
Sand Coulee Creek
Cartographer's Note:
Lewis has accomplished the goal of exploring the route that Clark identified as the best land or water route through this part of the continent. On July 12, 1806, Lewis's party crosses the Missouri, and on July 13 are resting at Upper Portage Camp. Lewis's July 11 quote ends with a reference to "medicine river"—obviously a mental slip, as he was now preparing to explore the "Marias River." After searching for lost horses, Lewis and three men set out for the upper "Marias" on July 16, 1806.
S 75 E 8 MILES
N
Outbound: June 13-July 13, 1805
Return: July 11-July 27, 1806, Entire return period.
Lewis-July 11-July 16, 1806
Gass-July 11-July 27, 1806
Ordway-July 19-July 27, 1806
Modern Data: 1965-1994
ONE STATUTE MILE
EXPLORATIONS OF LEWIS AND CLARK 1804 - 1806
CARTOGRAPHIC RECONSTRUCTION
RETURN TO THE FALLS
UTM ZONE 12
MAP NUMBER 434
Montana
CONTOUR INTERVAL 100 FEET

Alternate Return Route No. 6

MAPS 435 - 494

CLARK PARTY,
YELLOWSTONE RIVER, MONTANA

Yellowstone contingent—
- Capt. William Clark (journalist)
- Toussaint Charbonneau (interpreter)
- Sacagawea
- Jean Baptiste Charbonneau
- York
- Sgt. Nathaniel Pryor
- Pvt. William Bratton
- Pvt. George Gibson
- Pvt. Hugh Hall
- Pvt. François Labiche
- Pvt. George Shannon
- Pvt. John Shields
- Pvt. Richard Windsor
- 49 horses and a colt

Missouri contingent—
- Sgt. John Ordway (journalist)
- Pvt. John Collins
- Pvt. John Colter
- Pvt. Pierre Cruzatte
- Pvt. Thomas Howard
- Pvt. Baptiste Lepage
- Pvt. John Potts
- Pvt. Joseph Whitehouse
- Pvt. Alexander Willard
- Pvt. Peter Wiser
- 6 canoes

A main objective for the Corps of Discovery during the return from the Pacific Ocean in 1806 was the exploration of the Yellowstone River. At the Mandan villages during the winter of 1804-5, the Indians had told Lewis and Clark about this great river, drawing a map on a hide showing its significant tributaries. Later, when Clark descended the Yellowstone in 1806, the rather crude map would cause some confusion—sometimes he found it difficult to link the Indian names to the correct tributaries.

After Clark's contingent crossed the Big Hole cutoff to open the "Camp Fortunate" caches and retrieve the canoes, they had set out on July 10, 1806, moving quickly down the Beaverhead River toward the Jefferson and the Three Forks. Sergeant Pryor and 6 men guided the horse herd, while Clark and the remaining 15 people proceeded in 6 canoes. Finding the river contingent outdistancing the horses, Clark put more of the horse baggage into the canoes.

Reaching the Three Forks on July 13, the party stopped to dine and make final preparations to divide into a canoe party under Sergeant Ordway and an overland party led by Clark. Checking the division of baggage one last time, they said their farewells.

Ordway's group of 10 men in the 6 canoes set out down the Missouri to join Lewis's men at the Great Falls, to assist in uncovering caches and portage canoes and equipment around the rapids.

Clark's party, totaling 13 people plus 49 horses and a colt, turned eastward, following Indian and buffalo paths in the stream-braided plains of the Gallatin watershed, an area Sacagawea knew well. After crossing the mountains and Bozeman Pass, the party reached the Yellowstone River at modern-day Livingston, Montana.

From here, Clark's party would proceed down the Yellowstone to its mouth, exploring the greater part of the river's overall length. Eventually, their travel by horseback was supplanted by watercraft (the Clark party's entire horse herd, too, soon was stolen by Plains warriors).

Clark's Yellowstone traverse notes are not as precisely done as on the outbound trip. An added complication is the fact that Clark began recording "water" distances, in addition to the usual land distances. Clark only occasionally had done this before. These land and water estimates can vary considerably, and thus complicate any cartographic reconstruction.

On the maps for Alternate Return Route No. 6, the Yellowstone's modern course is depicted, but modified in places where justified by Clark's observations.

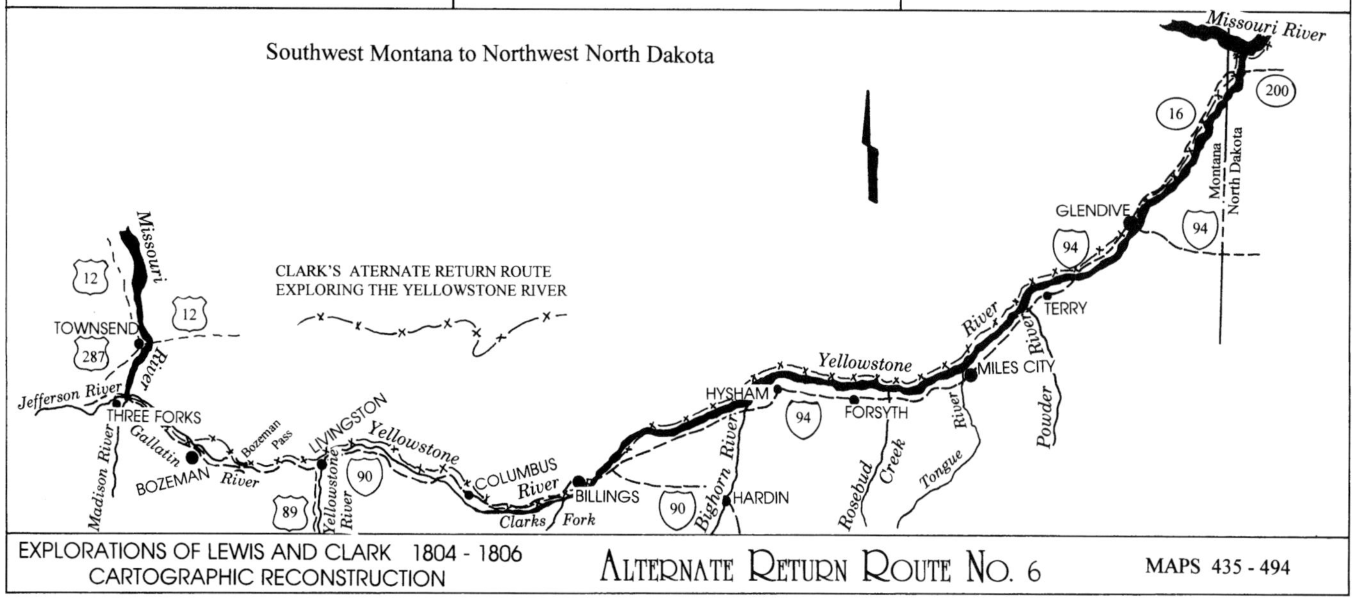

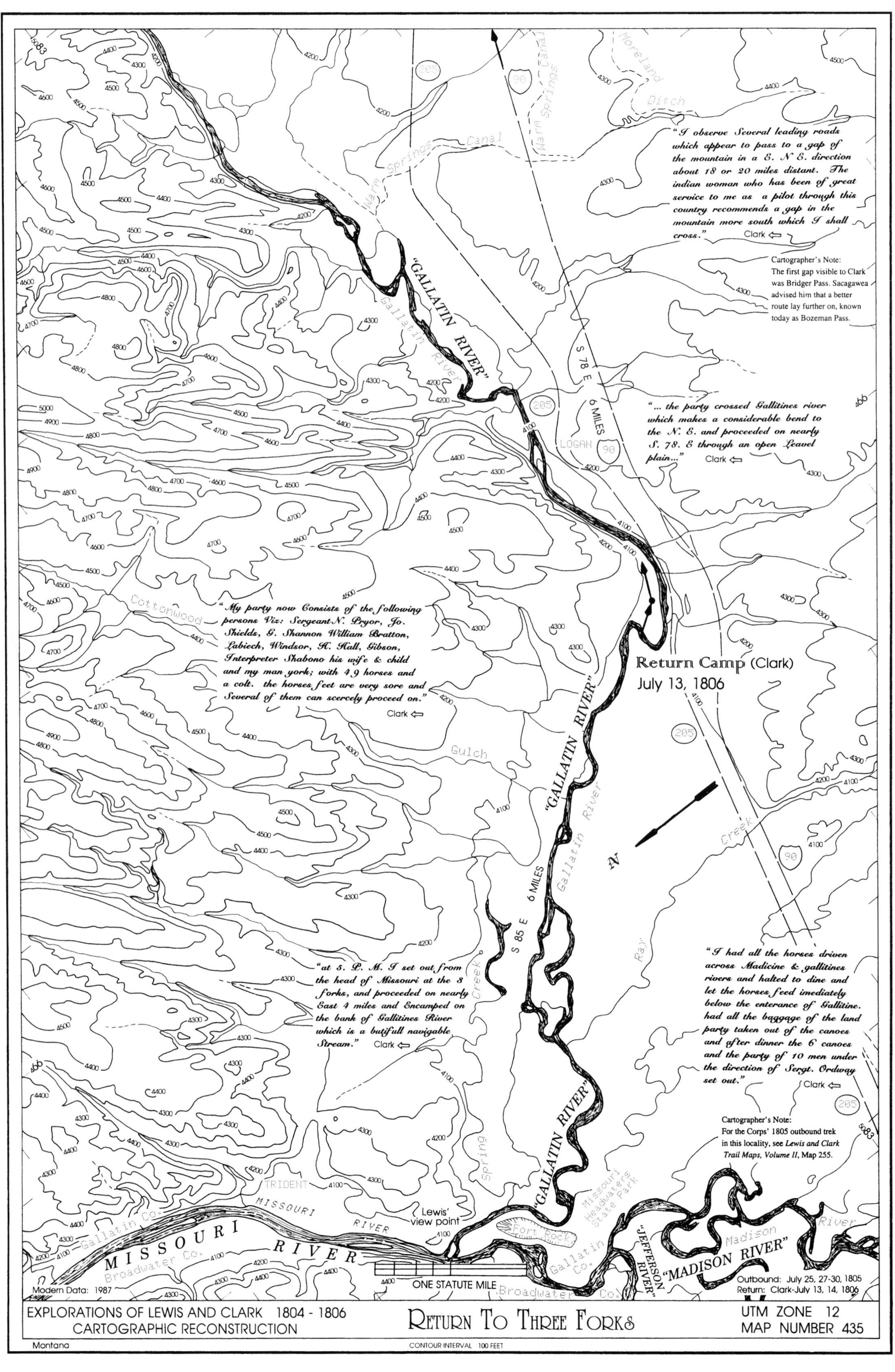
"I observe Several leading roads which appear to pass to a gap of the mountain in a E. N E. direction about 18 or 20 miles distant. The indian woman who has been of great service to me as a pilot through this country recommends a gap in the mountain more south which I shall cross." Clark
Cartographer's Note: The first gap visible to Clark was Bridger Pass. Sacagawea advised him that a better route lay further on, known today as Bozeman Pass.
"... the party crossed Gallitines river which makes a considerable bend to the N. E. and proceeded on nearly S. 78. E through an open Leavel plain..." Clark
"My party now Consists of the following persons Viz: Sergeant N. Pryor, Jo. Shields, G. Shannon William Bratton, Labiech, Windsor, H. Hall, Gibson, Interpreter Shabono his wife & child and my man york; with 49 horses and a colt. the horses feet are very sore and Several of them can scercely proceed on." Clark
Return Camp (Clark)
July 13, 1806
"at 5. P. M. I set out from the head of Missouri at the 3 forks, and proceeded on nearly East 4 miles and Encamped on the bank of Gallitines River which is a butifull navigable Stream." Clark
"I had all the horses driven across Madicine & gallitines rivers and halted to dine and let the horses feed imediately below the enterance of Gallitine. had all the baggage of the land party taken out of the canoes and after dinner the 6 canoes and the party of 10 men under the direction of Sergt. Ordway set out." Clark
Cartographer's Note: For the Corps' 1805 outbound trek in this locality, see Lewis and Clark Trail Maps, Volume II, Map 255.
"GALLATIN RIVER"
Gallatin River
Warm Springs Canal
Moreland Ditch
S 78 E
6 MILES
LOGAN
S 85 E
6 MILES
Cottonwood
Gulch
Creek
Ray Creek
Spring Creek
TRIDENT
MISSOURI RIVER
Lewis' view point
Fort Rock
Missouri Headwaters State Park
"JEFFERSON RIVER"
"MADISON RIVER"
Madison River
Gallatin Co.
Broadwater Co.
ONE STATUTE MILE
Outbound: July 25, 27-30, 1805
Return: Clark-July 13, 14, 1806
Modern Data: 1987
EXPLORATIONS OF LEWIS AND CLARK 1804 - 1806
CARTOGRAPHIC RECONSTRUCTION
RETURN TO THREE FORKS
UTM ZONE 12
MAP NUMBER 435
Montana
CONTOUR INTERVAL 100 FEET

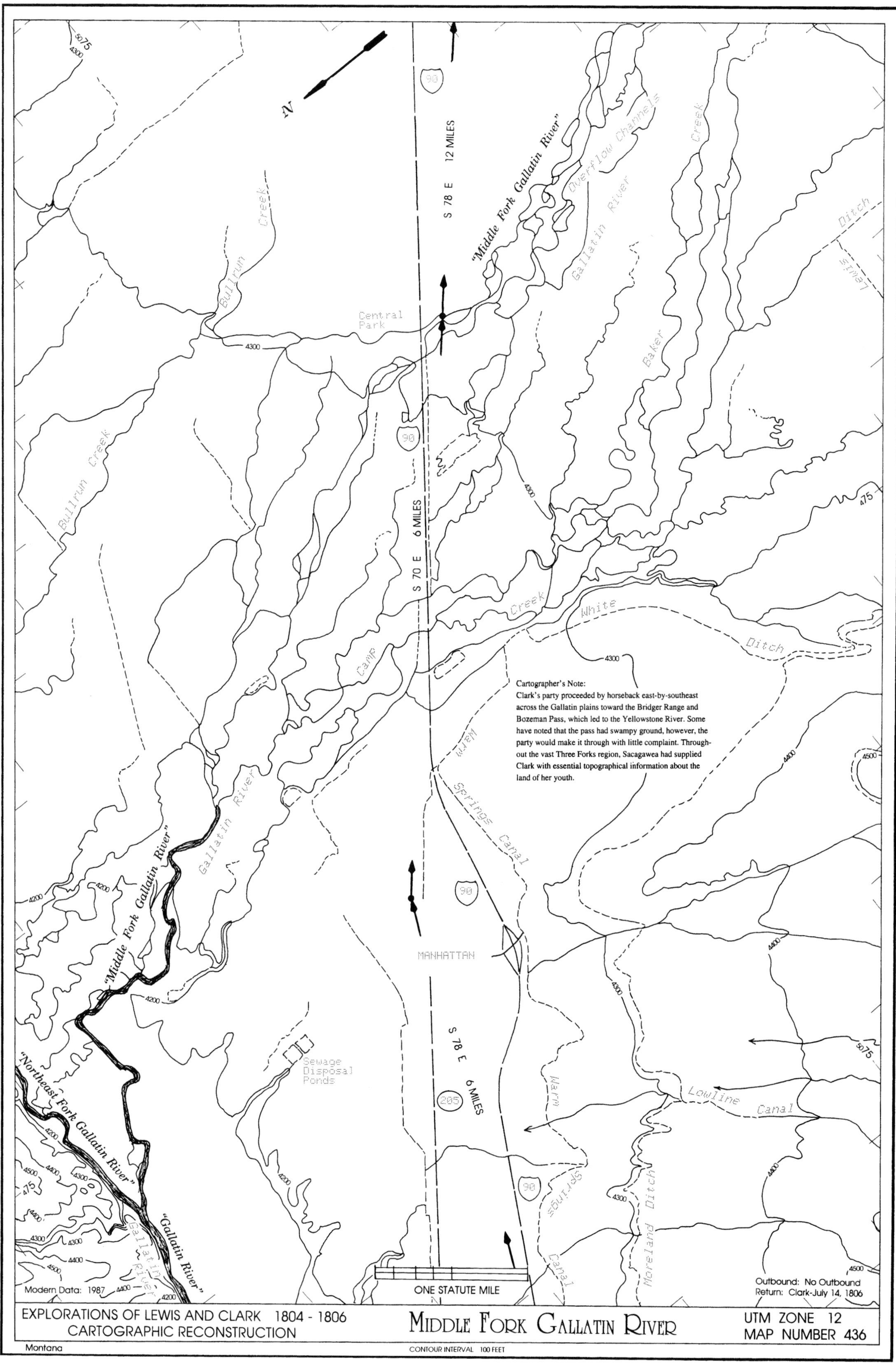
Cartographer's Note:
Clark's party proceeded by horseback east-by-southeast across the Gallatin plains toward the Bridger Range and Bozeman Pass, which led to the Yellowstone River. Some have noted that the pass had swampy ground, however, the party would make it through with little complaint. Throughout the vast Three Forks region, Sacagawea had supplied Clark with essential topographical information about the land of her youth.
N
S 78 E 12 MILES
S 70 E 6 MILES
S 78 E 6 MILES
"Middle Fork Gallatin River"
"Northeast Fork Gallatin River"
"Gallatin River"
Overflow Channels
Gallatin River
Bullrun Creek
Central Park
Baker Creek
Camp Creek
White Ditch
Lewis Ditch
Warm Springs Canal
Lowline Canal
Moreland Ditch
MANHATTAN
Sewage Disposal Ponds
90
205
ONE STATUTE MILE
Modern Data: 1987
Outbound: No Outbound
Return: Clark-July 14, 1806
EXPLORATIONS OF LEWIS AND CLARK 1804 - 1806
CARTOGRAPHIC RECONSTRUCTION
MIDDLE FORK GALLATIN RIVER
UTM ZONE 12
MAP NUMBER 436
Montana
CONTOUR INTERVAL 100 FEET

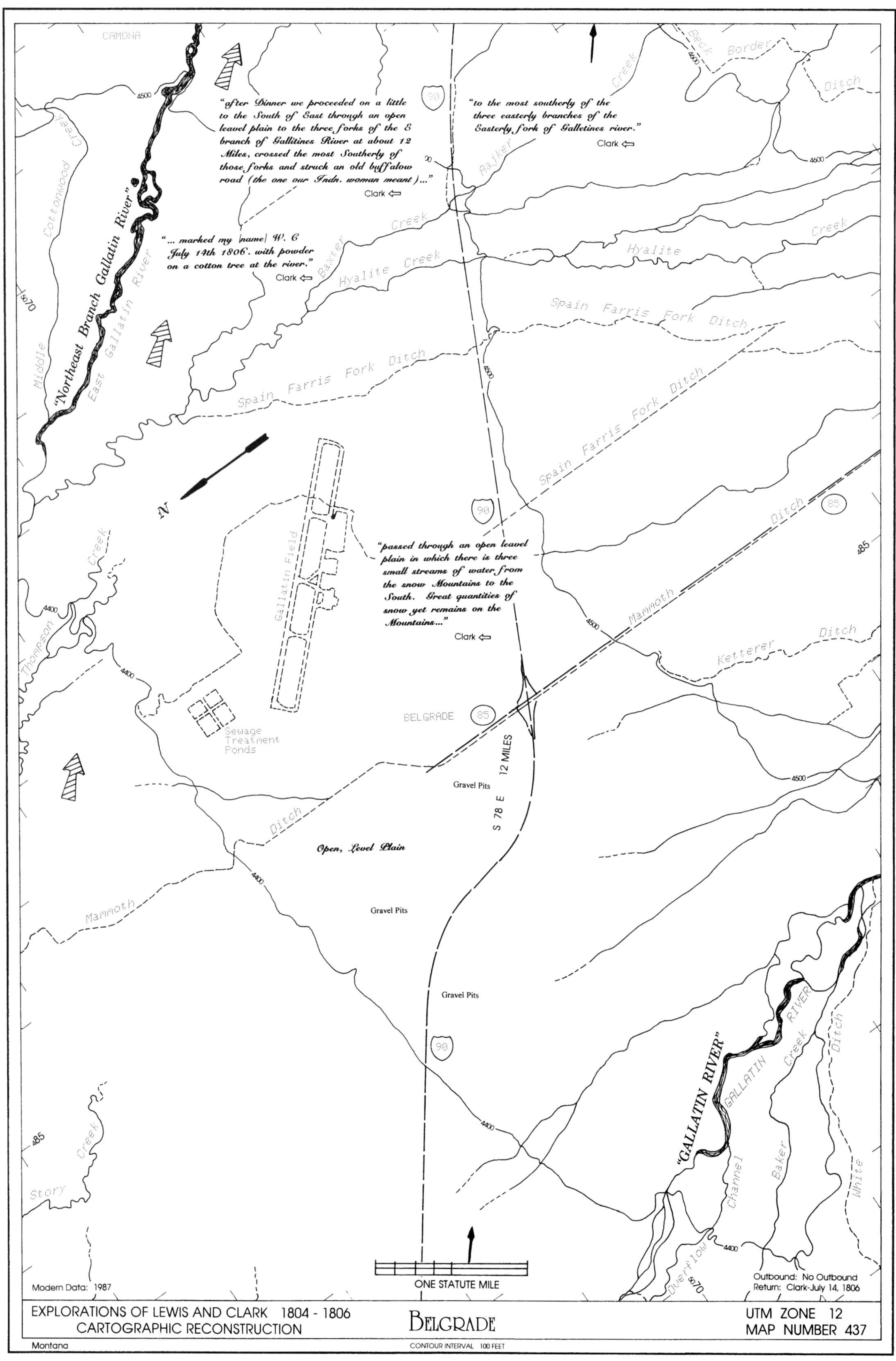

"after Dinner we proceeded on a little to the South of East through an open leavel plain to the three forks of the E branch of Gallitines River at about 12 Miles, crossed the most Southerly of those forks and struck an old buffalow road (the one our Indn. woman meant)..."
Clark
"to the most southerly of the three easterly branches of the Easterly fork of Galletines river."
Clark
"... marked my [name] W. C July 14th 1806. with powder on a cotton tree at the river."
Clark
"passed through an open leavel plain in which there is three small streams of water from the snow Mountains to the South. Great quantities of snow yet remains on the Mountains..."
Clark
"Northeast Branch Gallatin River"
"GALLATIN RIVER"
Open, Level Plain
Gravel Pits
S 78 E 12 MILES
BELGRADE
Sewage Treatment Ponds
Gallatin Field
Spain Farris Fork Ditch
Hyalite Creek
Mammoth Ditch
Ketterer Ditch
Modern Data: 1987
ONE STATUTE MILE
Outbound: No Outbound
Return: Clark-July 14, 1806
EXPLORATIONS OF LEWIS AND CLARK 1804 - 1806
CARTOGRAPHIC RECONSTRUCTION
BELGRADE
UTM ZONE 12
MAP NUMBER 437
Montana
CONTOUR INTERVAL 100 FEET

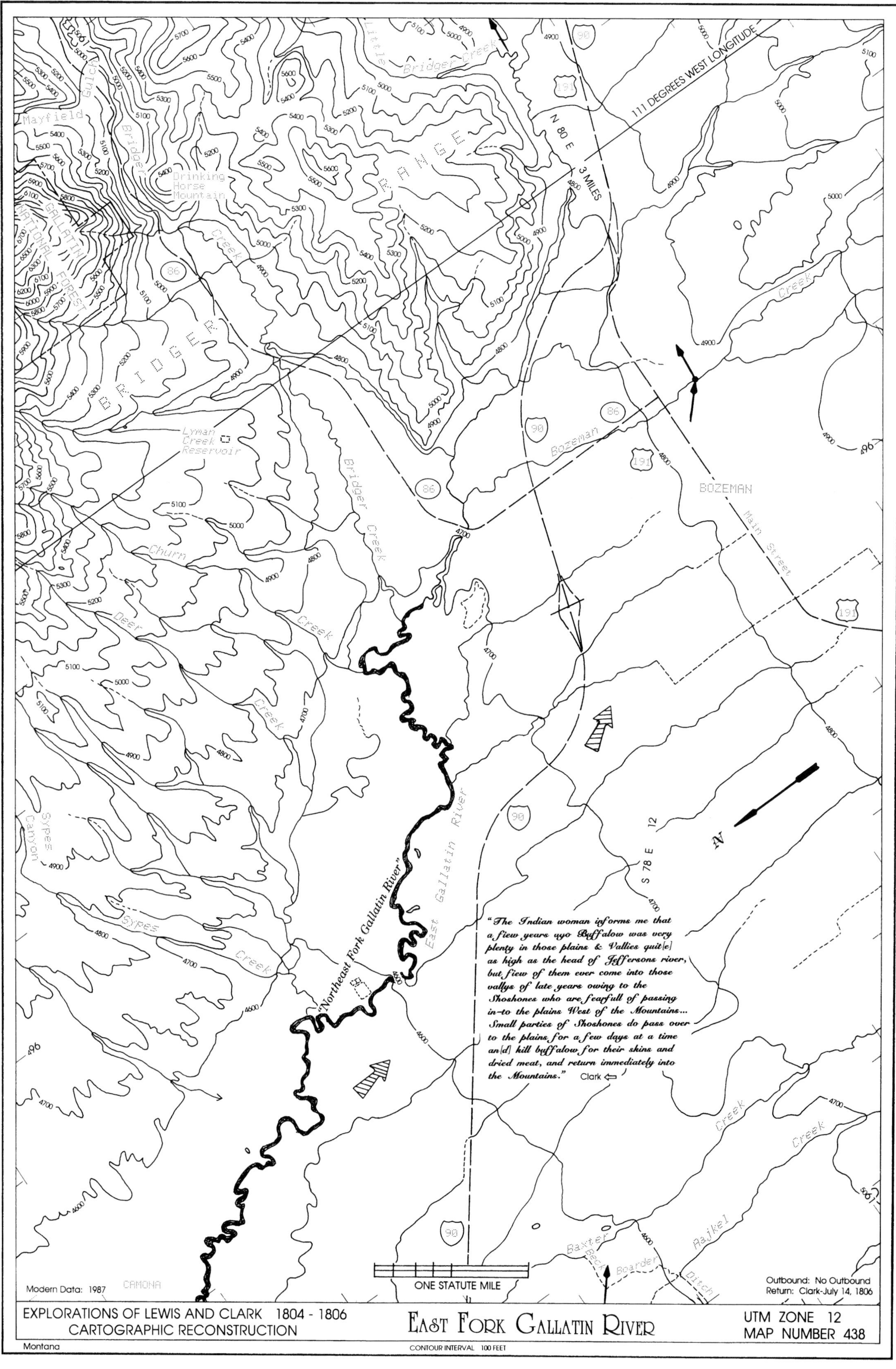

111 DEGREES WEST LONGITUDE
N 80 E
3 MILES
Little Bridger Creek
Mayfield
Gulch
Bridger
Drinking Horse Mountain
GALLATIN NATIONAL FOREST
BRIDGER RANGE
Lyman Creek Reservoir
Bridger Creek
Bozeman
BOZEMAN
Main Street
Churn Creek
Deer Creek
Sypes Canyon
Sypes Creek
"Northeast Fork Gallatin River"
East Gallatin River
S 78 E
N
"The Indian woman informs me that a fiew years ago Buffalow was very plenty in those plains & Vallies quit[e] as high as the head of Jeffersons river, but fiew of them ever come into those vallys of late years owing to the Shoshones who are fearfull of passing in-to the plains West of the Mountains... Small parties of Shoshones do pass over to the plains for a few days at a time an[d] kill buffalow for their skins and dried meat, and return immediately into the Mountains." Clark
Creek
Creek
Rajkel
Baxter
Beck
Boarder
Ditch
Modern Data: 1987
CAMONA
ONE STATUTE MILE
Outbound: No Outbound
Return: Clark-July 14, 1806
EXPLORATIONS OF LEWIS AND CLARK 1804 - 1806
CARTOGRAPHIC RECONSTRUCTION
EAST FORK GALLATIN RIVER
UTM ZONE 12
MAP NUMBER 438
Montana
CONTOUR INTERVAL 100 FEET

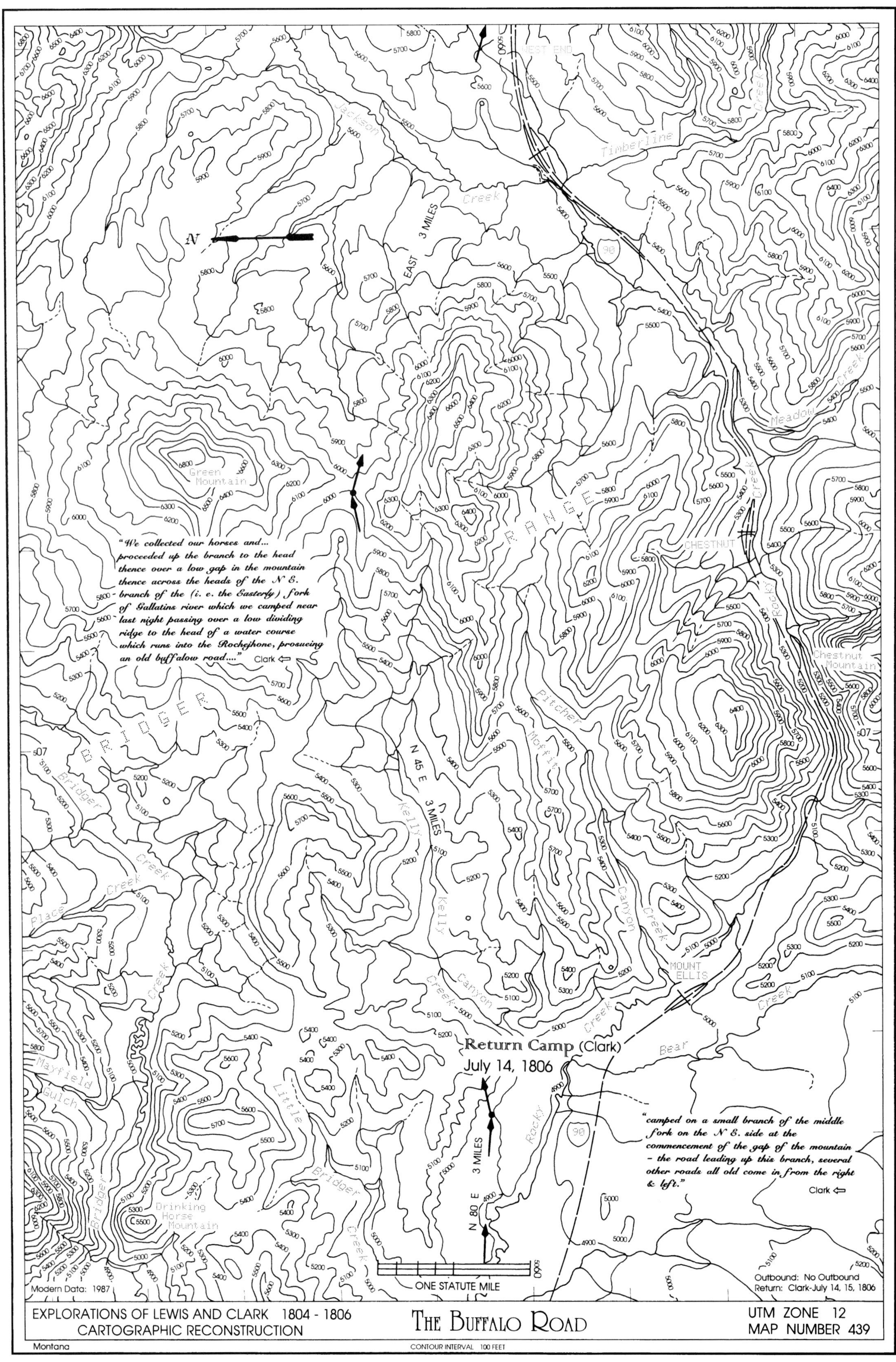
"We collected our horses and... proceeded up the branch to the head thence over a low gap in the mountain thence across the heads of the N. E. branch of the (i. e. the Easterly) fork of Gallatins river which we camped near last night passing over a low dividing ridge to the head of a water course which runs into the Rochejhone, prosueing an old buffalow road...." Clark
"camped on a small branch of the middle fork on the N. E. side at the commencement of the gap of the mountain - the road leading up this branch, several other roads all old come in from the right & left." Clark
Return Camp (Clark)
July 14, 1806
N
EAST 3 MILES
N 45 E 3 MILES
N 80 E 3 MILES
Green Mountain
Chestnut Mountain
Drinking Horse Mountain
BRIDGER
RANGE
WEST END
Timberline
Jackson
Creek
Meadow Creek
CHESTNUT
Rocky
Pitcher
Moffit
Kelly
Canyon Creek
MOUNT ELLIS
Bear
Bridger Creek
Place Creek
Mayfield Gulch
Little Bridger Creek
90
ONE STATUTE MILE
Modern Data: 1987
Outbound: No Outbound
Return: Clark-July 14, 15, 1806
EXPLORATIONS OF LEWIS AND CLARK 1804 - 1806
CARTOGRAPHIC RECONSTRUCTION
THE BUFFALO ROAD
UTM ZONE 12
MAP NUMBER 439
Montana
CONTOUR INTERVAL 100 FEET

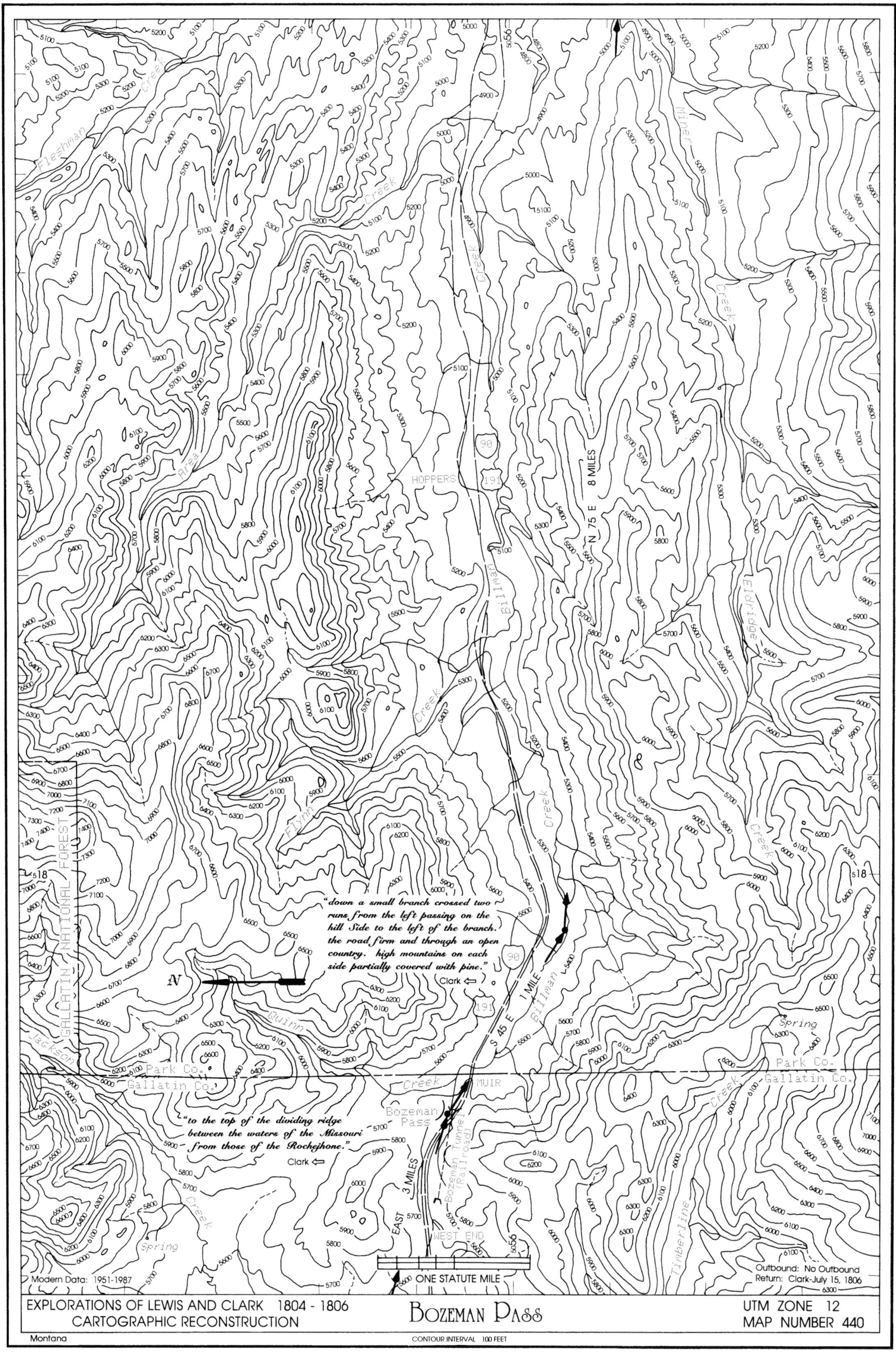
"down a small branch crossed two runs from the left passing on the hill Side to the left of the branch. the road firm and through an open country. high mountains on each side partially covered with pine."
Clark ⇦
"to the top of the dividing ridge between the waters of the Missouri from those of the Rochejhone."
Clark ⇦
N
N 75 E 8 MILES
S 45 E 1 MILE
EAST 3 MILES
HOPPERS
MUIR
Bozeman Pass
Bozeman Tunnel (Railroad)
WEST END
Park Co.
Gallatin Co.
GALLATIN NATIONAL FOREST
ONE STATUTE MILE
Modern Data: 1951-1987
Outbound: No Outbound
Return: Clark-July 15, 1806
EXPLORATIONS OF LEWIS AND CLARK 1804 - 1806
CARTOGRAPHIC RECONSTRUCTION
BOZEMAN PASS
UTM ZONE 12
MAP NUMBER 440
Montana
CONTOUR INTERVAL 100 FEET

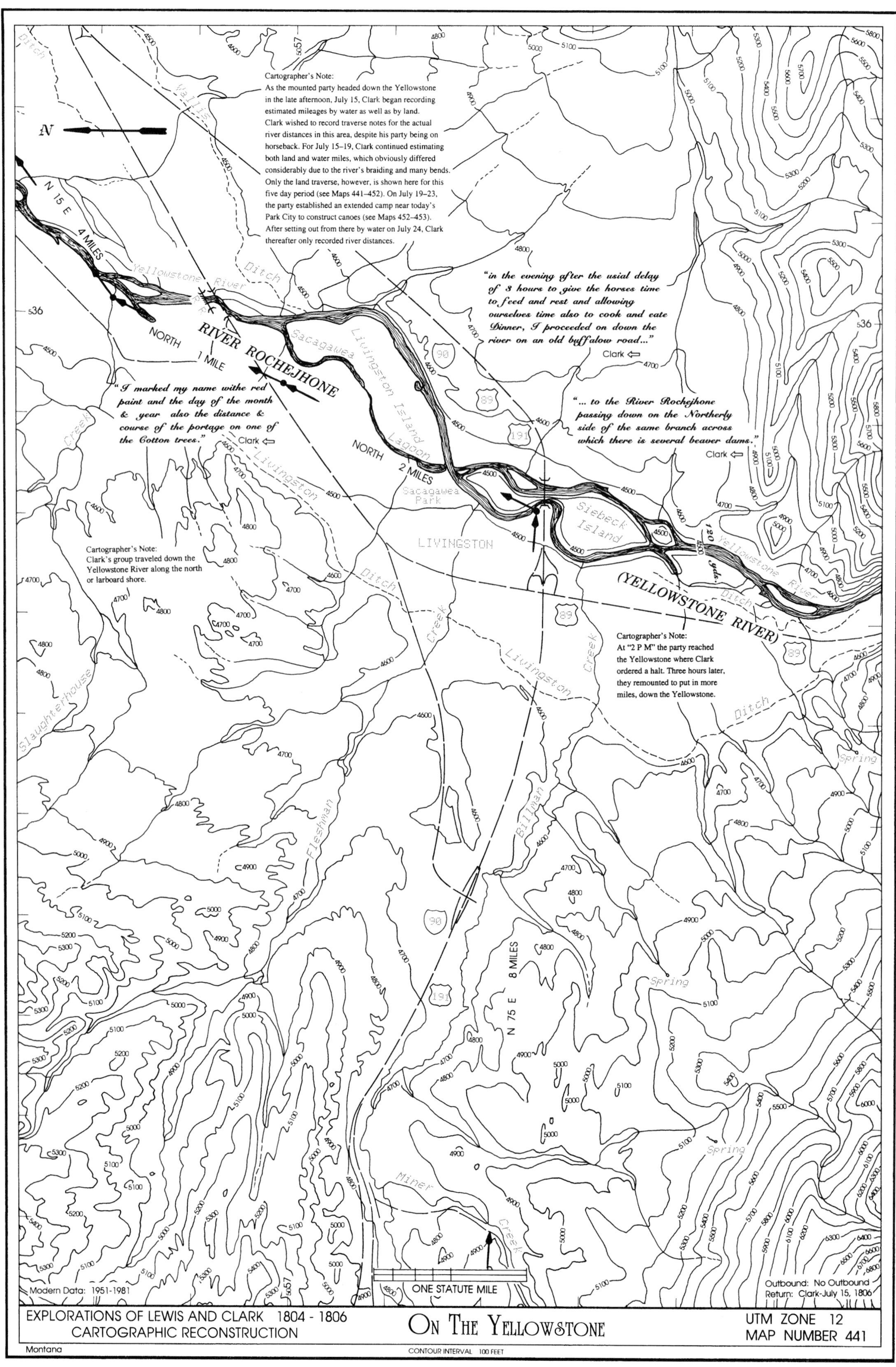
Cartographer's Note:
As the mounted party headed down the Yellowstone in the late afternoon, July 15, Clark began recording estimated mileages by water as well as by land. Clark wished to record traverse notes for the actual river distances in this area, despite his party being on horseback. For July 15–19, Clark continued estimating both land and water miles, which obviously differed considerably due to the river's braiding and many bends. Only the land traverse, however, is shown here for this five day period (see Maps 441–452). On July 19–23, the party established an extended camp near today's Park City to construct canoes (see Maps 452–453). After setting out from there by water on July 24, Clark thereafter only recorded river distances.
N
N 15 E
4 MILES
NORTH
1 MILE
RIVER ROCHEJHONE
"in the evening after the usial delay of 3 hours to give the horses time to feed and rest and allowing ourselves time also to cook and eate Dinner, I proceeded on down the river on an old buffalow road..."
Clark ⇦
"I marked my name withe red paint and the day of the month & year also the distance & course of the portage on one of the Cotton trees."
Clark ⇦
"... to the River Rochejhone passing down on the Northerly side of the same branch across which there is several beaver dams."
Clark ⇦
NORTH
2 MILES
Yellowstone River
Sacagawea
Livingston Island
Lagoon
Sacagawea Park
Siebeck Island
LIVINGSTON
120 yds.
(YELLOWSTONE RIVER)
Cartographer's Note:
Clark's group traveled down the Yellowstone River along the north or larboard shore.
Cartographer's Note:
At "2 P M" the party reached the Yellowstone where Clark ordered a halt. Three hours later, they remounted to put in more miles, down the Yellowstone.
Wallis
Ditch
Livingston
Creek
Slaughterhouse
Fleshman
Billman
Miner
Spring
N 75 E 8 MILES
90
89
191
ONE STATUTE MILE
Modern Data: 1951-1981
Outbound: No Outbound
Return: Clark-July 15, 1806
EXPLORATIONS OF LEWIS AND CLARK 1804 - 1806
CARTOGRAPHIC RECONSTRUCTION
ON THE YELLOWSTONE
UTM ZONE 12
MAP NUMBER 441
Montana
CONTOUR INTERVAL 100 FEET

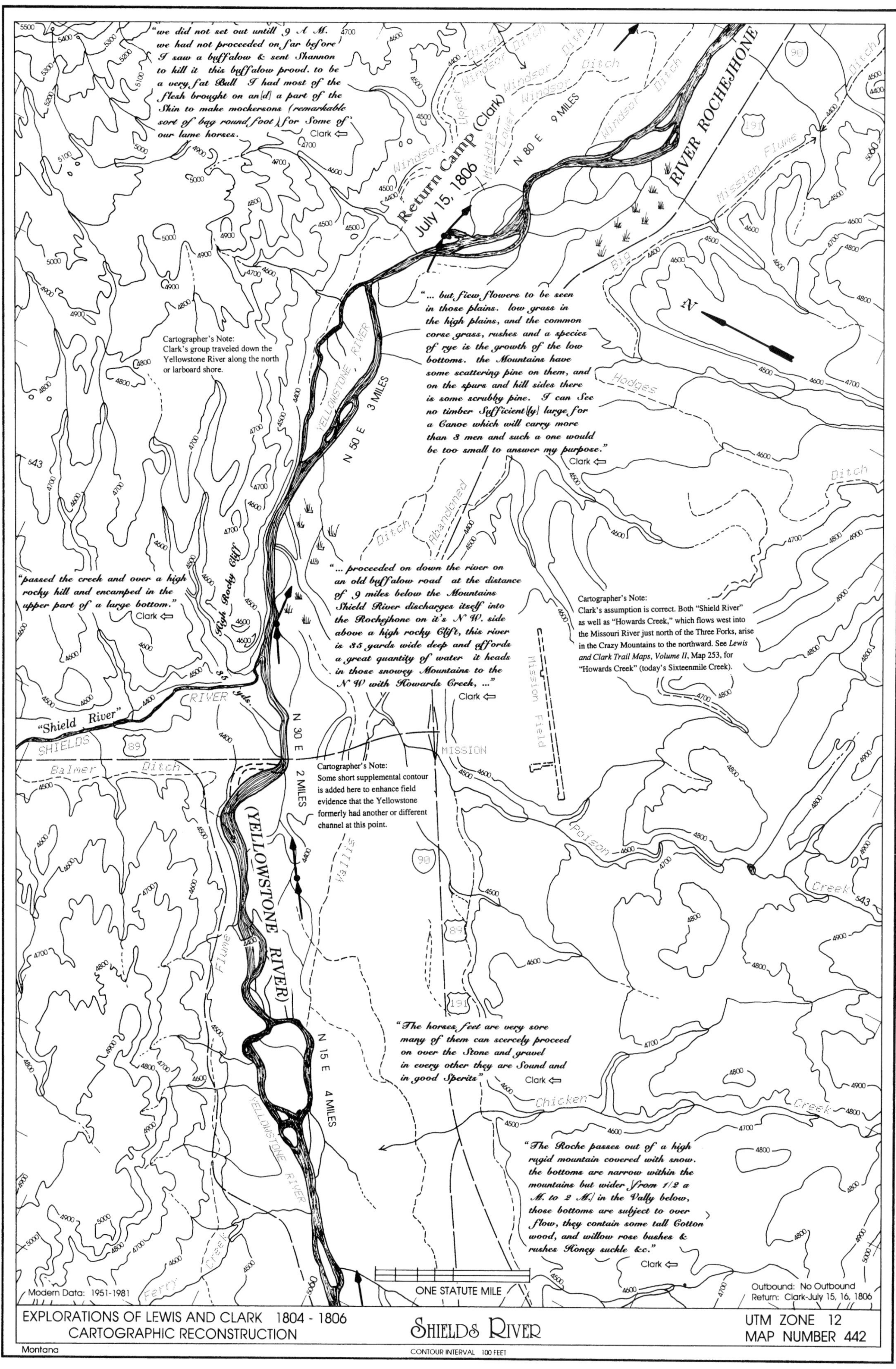
"we did not set out untill 9 A M. we had not proceeded on far before I saw a buffalow & sent Shannon to kill it this buffalow prood. to be a very fat Bull I had most of the flesh brought on an[d] a part of the Skin to make mockersons (remarkable sort of bag round foot) for Some of our lame horses.
Clark ⇦
Return Camp (Clark) July 15, 1806
N 80 E 9 MILES
RIVER ROCHEJHONE
"... but fiew flowers to be seen in those plains. low grass in the high plains, and the common corse grass, rushes and a species of rye is the growth of the low bottoms. the Mountains have some scattering pine on them, and on the spurs and hill sides there is some scrubby pine. I can See no timber Sufficient[ly] large for a Canoe which will carry more than 3 men and such a one would be too small to answer my purpose."
Clark ⇦
Cartographer's Note:
Clark's group traveled down the Yellowstone River along the north or larboard shore.
YELLOWSTONE RIVER
N 50 E 3 MILES
High Rocky Cliff
"passed the creek and over a high rocky hill and encamped in the upper part of a large bottom."
Clark ⇦
"... proceeded on down the river on an old buffalow road at the distance of 9 miles below the Mountains Shield River discharges itself into the Rochejhone on it's N W. side above a high rocky Clift, this river is 35 yards wide deep and affords a great quantity of water it heads in those snowey Mountains to the N W with Howards Creek, ..."
Clark ⇦
Cartographer's Note:
Clark's assumption is correct. Both "Shield River" as well as "Howards Creek," which flows west into the Missouri River just north of the Three Forks, arise in the Crazy Mountains to the northward. See Lewis and Clark Trail Maps, Volume II, Map 253, for "Howards Creek" (today's Sixteenmile Creek).
"Shield River"
35 yds.
SHIELDS RIVER
MISSION
Mission Field
N 30 E 2 MILES
Cartographer's Note:
Some short supplemental contour is added here to enhance field evidence that the Yellowstone formerly had another or different channel at this point.
(YELLOWSTONE RIVER)
"The horses feet are very sore many of them can scercely proceed on over the Stone and gravel in every other they are Sound and in good Sperits"
Clark ⇦
N 15 E 4 MILES
"The Roche passes out of a high rugid mountain covered with snow. the bottoms are narrow within the mountains but wider [from 1/2 a M. to 2 M.] in the Vally below, those bottoms are subject to over flow, they contain some tall Cotton wood, and willow rose bushes & rushes Honey suckle &c."
Clark ⇦
ONE STATUTE MILE
Modern Data: 1951-1981
Outbound: No Outbound
Return: Clark-July 15, 16, 1806
EXPLORATIONS OF LEWIS AND CLARK 1804 - 1806
CARTOGRAPHIC RECONSTRUCTION
SHIELDS RIVER
UTM ZONE 12
MAP NUMBER 442
Montana
CONTOUR INTERVAL 100 FEET

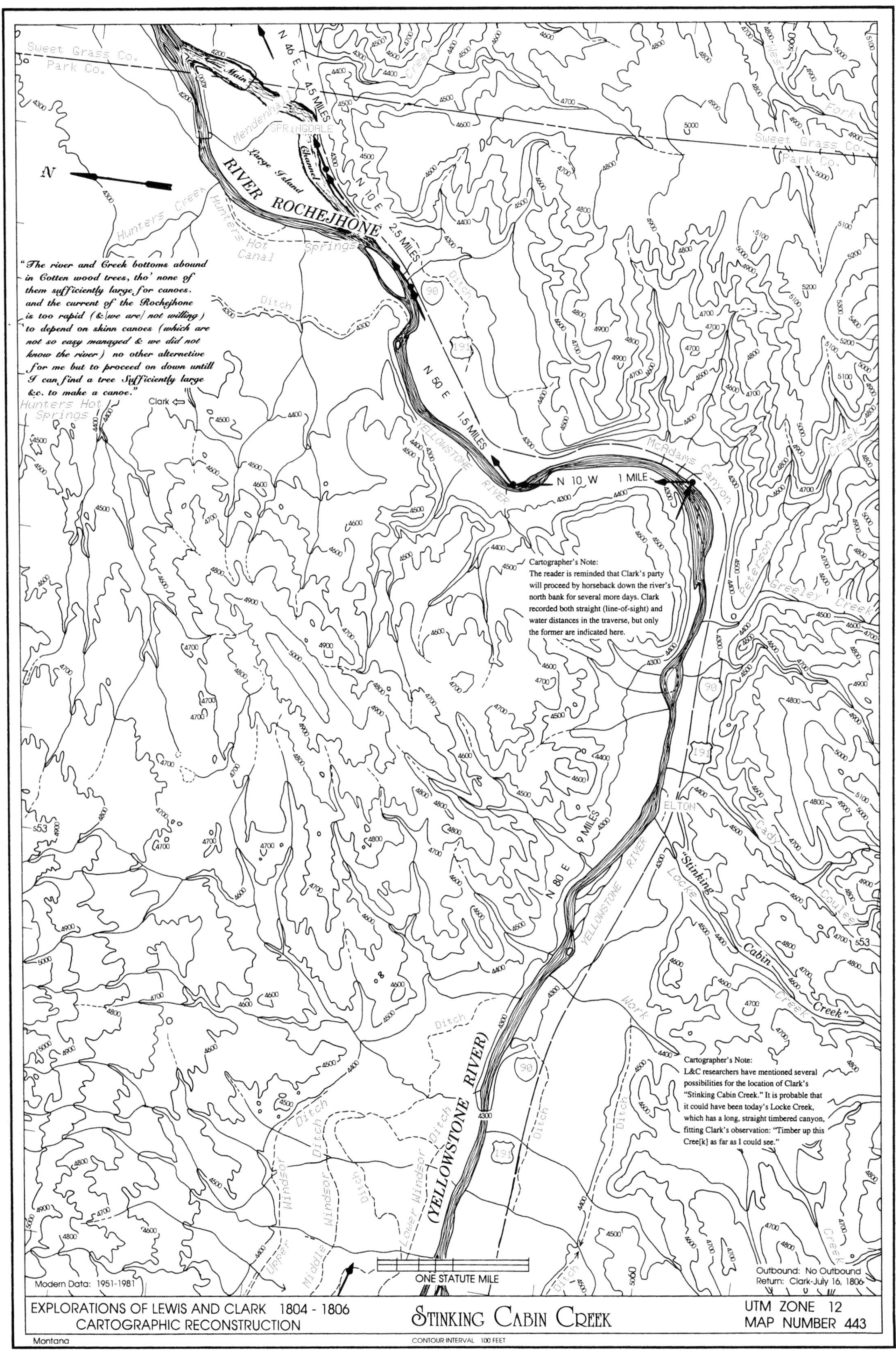

"The river and Creek bottoms abound in Cotten wood trees, tho' none of them sufficiently large for canoes. and the current of the Rochejhone is too rapid (& [we are] not willing) to depend on skinn canoes (which are not so easy managed & we did not know the river) no other alternetive for me but to proceed on down untill I can find a tree Sufficiently large &c. to make a canoe."
Clark
Cartographer's Note:
The reader is reminded that Clark's party will proceed by horseback down the river's north bank for several more days. Clark recorded both straight (line-of-sight) and water distances in the traverse, but only the former are indicated here.
Cartographer's Note:
L&C researchers have mentioned several possibilities for the location of Clark's "Stinking Cabin Creek." It is probable that it could have been today's Locke Creek, which has a long, straight timbered canyon, fitting Clark's observation: "Timber up this Cree[k] as far as I could see."
RIVER ROCHEJHONE
(YELLOWSTONE RIVER)
"Stinking Cabin Creek"
Large Island
Main Channel
SPRINGDALE
ELTON
N 46 E 4.5 MILES
N 10 E
2.5 MILES
N 50 E 1.5 MILES
N 10 W 1 MILE
N 80 E 9 MILES
Sweet Grass Co.
Park Co.
Hunters Creek
Hunters Hot Springs
Hunters Hot Springs Canal
Mendenhall Creek
McAdams Canyon
Peterson
Greeley Creek
Cady Coulee
Locke Creek
West Fork
Work Creek
Upper Windsor Ditch
Middle Windsor Ditch
Lower Windsor Ditch
Modern Data: 1951-1981
ONE STATUTE MILE
Outbound: No Outbound
Return: Clark-July 16, 1806
EXPLORATIONS OF LEWIS AND CLARK 1804 - 1806
CARTOGRAPHIC RECONSTRUCTION
STINKING CABIN CREEK
UTM ZONE 12
MAP NUMBER 443
Montana
CONTOUR INTERVAL 100 FEET

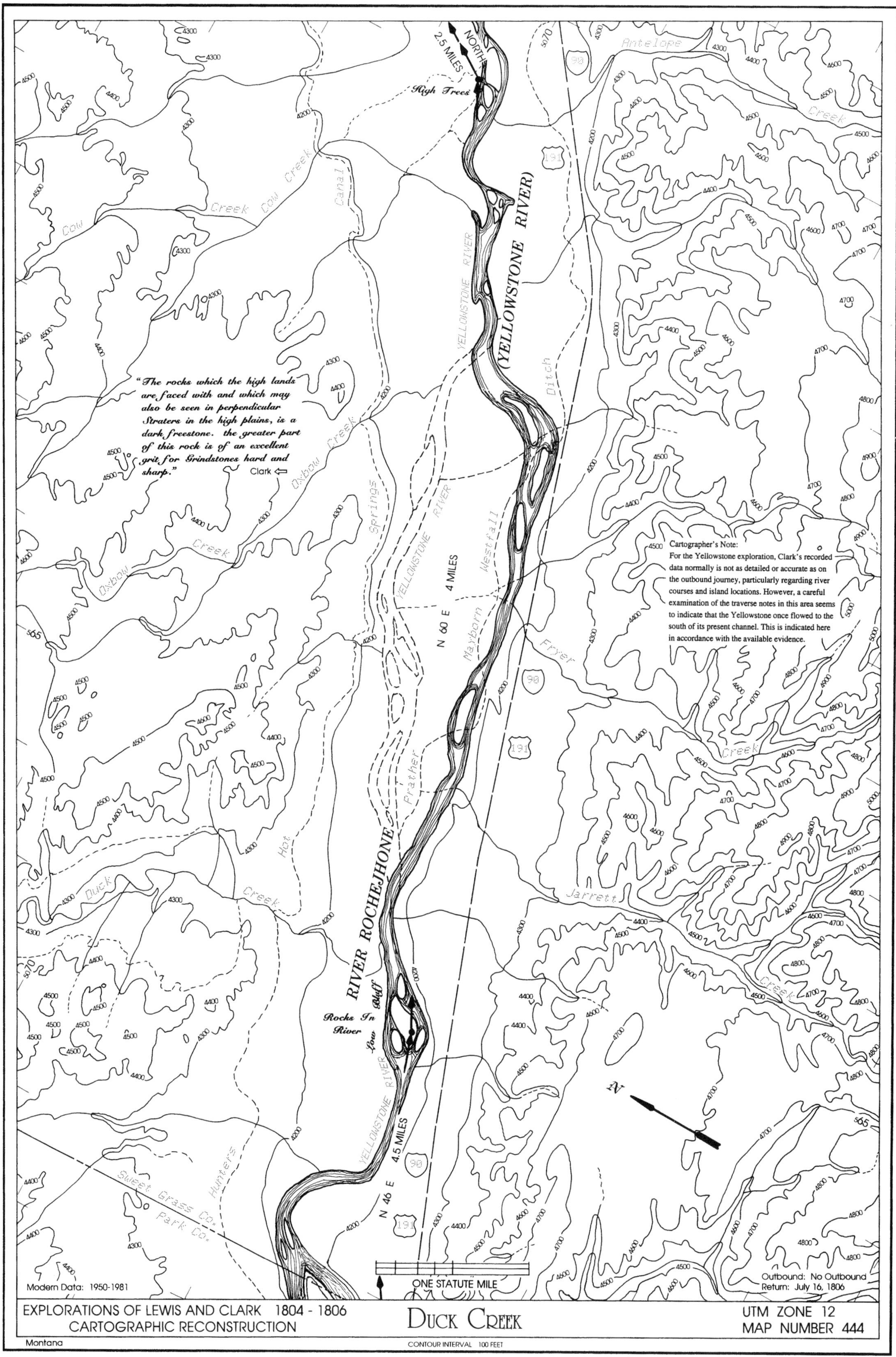
NORTH
2.5 MILES
High Trees
Antelope
Creek
Cow Creek
Cow Creek
Canal
(YELLOWSTONE RIVER)
YELLOWSTONE RIVER
Ditch
"The rocks which the high lands are faced with and which may also be seen in perpendicular Straters in the high plains, is a dark freestone. the greater part of this rock is of an excellent grit for Grindstones hard and sharp."
Clark
Oxbow Creek
Oxbow Creek
Springs
YELLOWSTONE RIVER
Westfall
N 60 E 4 MILES
Cartographer's Note:
For the Yellowstone exploration, Clark's recorded data normally is not as detailed or accurate as on the outbound journey, particularly regarding river courses and island locations. However, a careful examination of the traverse notes in this area seems to indicate that the Yellowstone once flowed to the south of its present channel. This is indicated here in accordance with the available evidence.
Mayborn
Fryer
Creek
Prather
Hot
RIVER ROCHEJHONE
Duck Creek
Jarrett
Creek
Bluff
Rocks In River
Low
YELLOWSTONE RIVER
N 46 E 4.5 MILES
N
Hunters
Sweet Grass Co.
Park Co.
ONE STATUTE MILE
Modern Data: 1950-1981
Outbound: No Outbound
Return: July 16, 1806
EXPLORATIONS OF LEWIS AND CLARK 1804 - 1806
CARTOGRAPHIC RECONSTRUCTION
DUCK CREEK
UTM ZONE 12
MAP NUMBER 444
Montana
CONTOUR INTERVAL 100 FEET

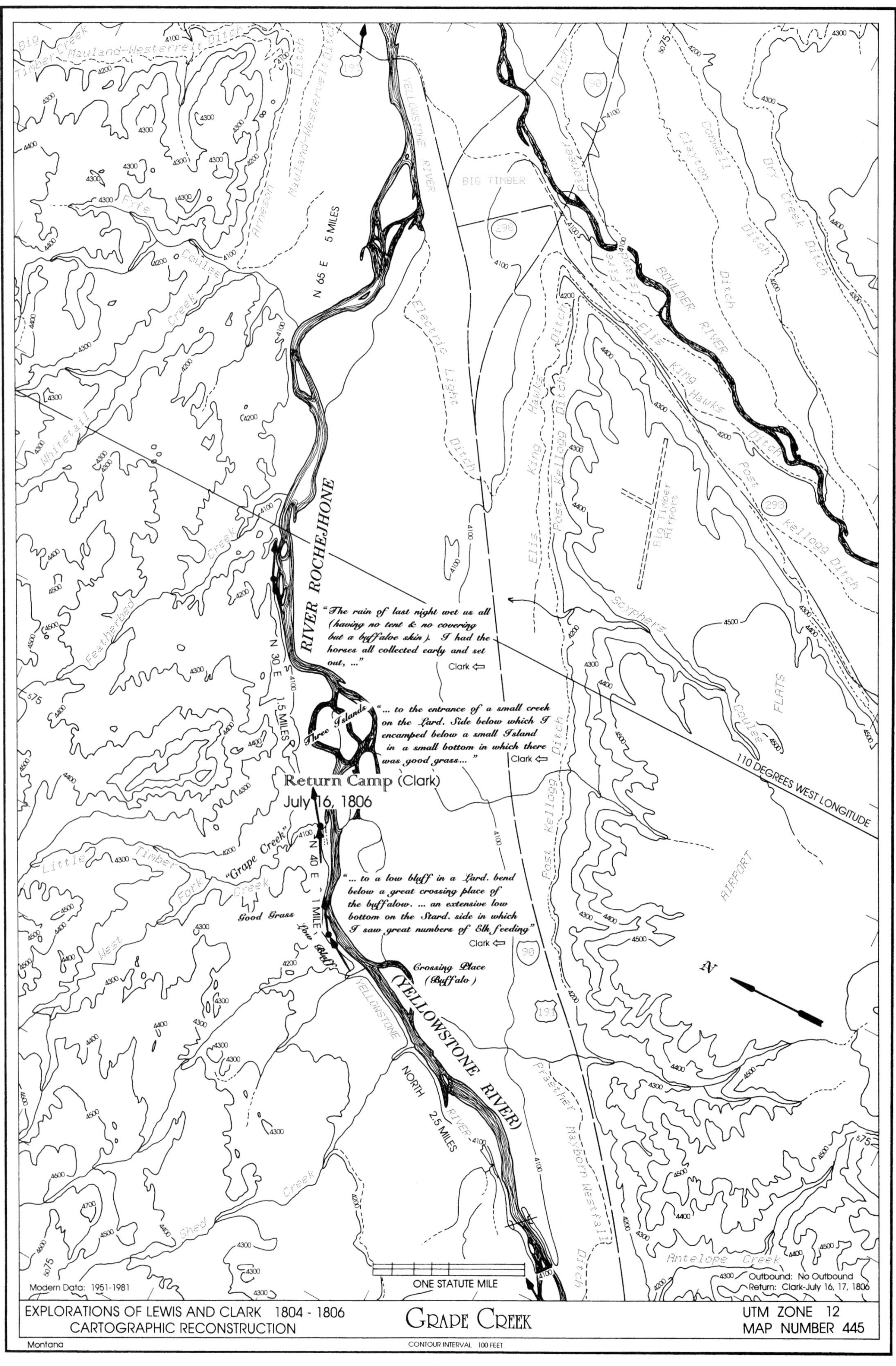

RIVER ROCHEJHONE
N 65 E 5 MILES
N 30 E 1.5 MILES
N 40 E 1 MILE
NORTH 2.5 MILES
"The rain of last night wet us all (having no tent & no covering but a buffaloe skin). I had the horses all collected early and set out, ..."
Clark ⇦
"... to the entrance of a small creek on the Lard. Side below which I encamped below a small Island in a small bottom in which there was good grass... "
Clark ⇦
Three Islands
Return Camp (Clark)
July 16, 1806
"... to a low bluff in a Lard. bend below a great crossing place of the buffalow. ... an extensive low bottom on the Stard. side in which I saw great numbers of Elk feeding"
Clark ⇦
"Grape Creek"
Good Grass
Low Bluff
Crossing Place (Buffalo)
(YELLOWSTONE RIVER)
YELLOWSTONE RIVER
BIG TIMBER
BOULDER RIVER
Big Timber Airport
110 DEGREES WEST LONGITUDE
FLATS
AIRPORT
N
ONE STATUTE MILE
Modern Data: 1951-1981
Outbound: No Outbound
Return: Clark-July 16, 17, 1806
EXPLORATIONS OF LEWIS AND CLARK 1804 - 1806
CARTOGRAPHIC RECONSTRUCTION
GRAPE CREEK
UTM ZONE 12
MAP NUMBER 445
Montana
CONTOUR INTERVAL 100 FEET

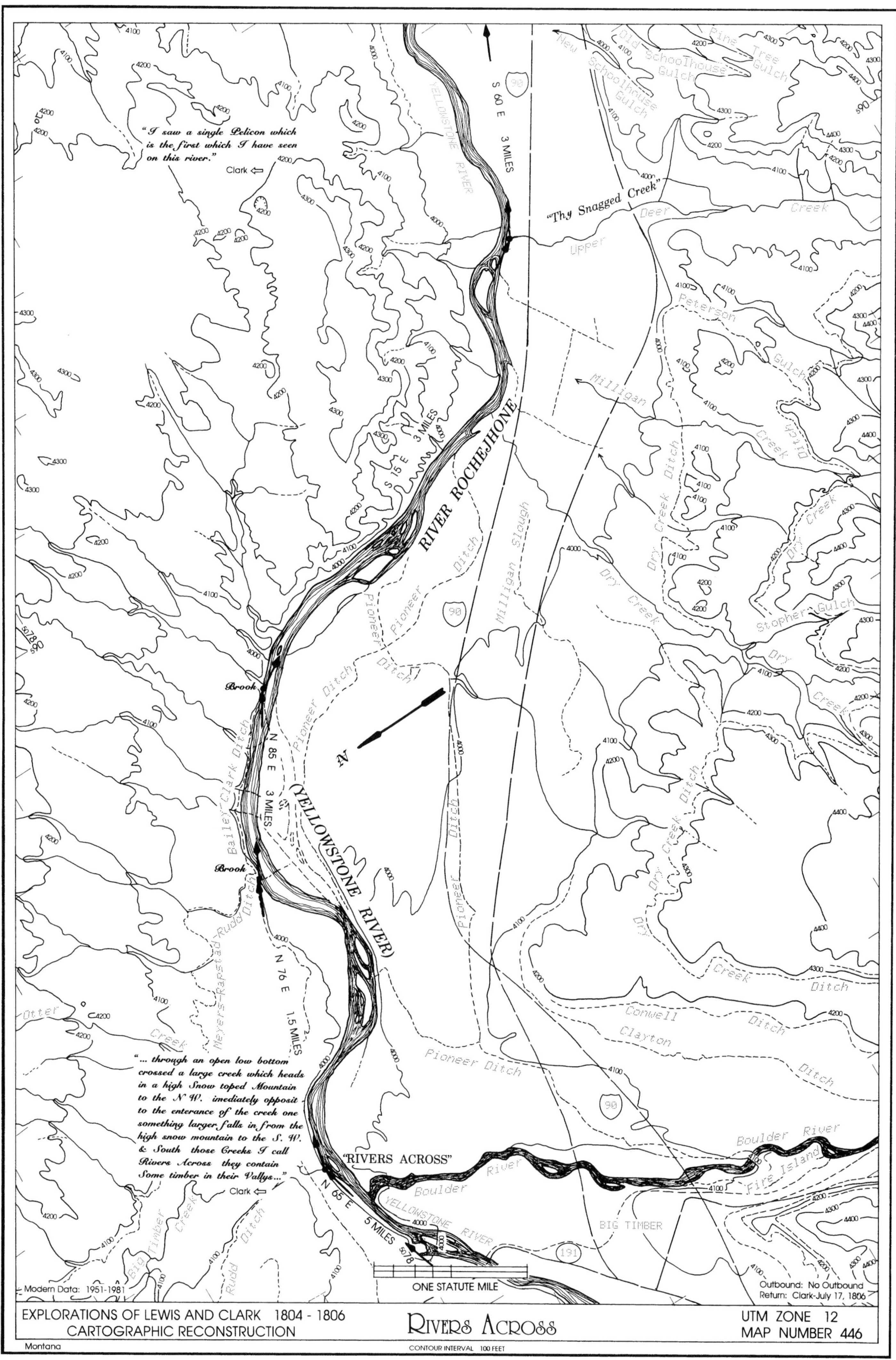
"I saw a single Pelicon which is the first which I have seen on this river."
Clark ⇦
"Thy Snagged Creek"
S 60 E 3 MILES
S 15 E 3 MILES
N 85 E 3 MILES
N 76 E 1.5 MILES
N 65 E 5 MILES
RIVER ROCHEJHONE
(YELLOWSTONE RIVER)
YELLOWSTONE RIVER
Brook
Brook
"RIVERS ACROSS"
"... through an open low bottom crossed a large creek which heads in a high Snow toped Mountain to the N W. imediately opposit to the enterance of the creek one something larger falls in from the high snow mountain to the S. W. & South those Creeks I call Rivers Across they contain Some timber in their Vallys..."
Clark ⇦
Boulder River
BIG TIMBER
ONE STATUTE MILE
Modern Data: 1951-1981
Outbound: No Outbound
Return: Clark-July 17, 1806
EXPLORATIONS OF LEWIS AND CLARK 1804 - 1806
CARTOGRAPHIC RECONSTRUCTION
RIVERS ACROSS
UTM ZONE 12
MAP NUMBER 446
Montana
CONTOUR INTERVAL 100 FEET

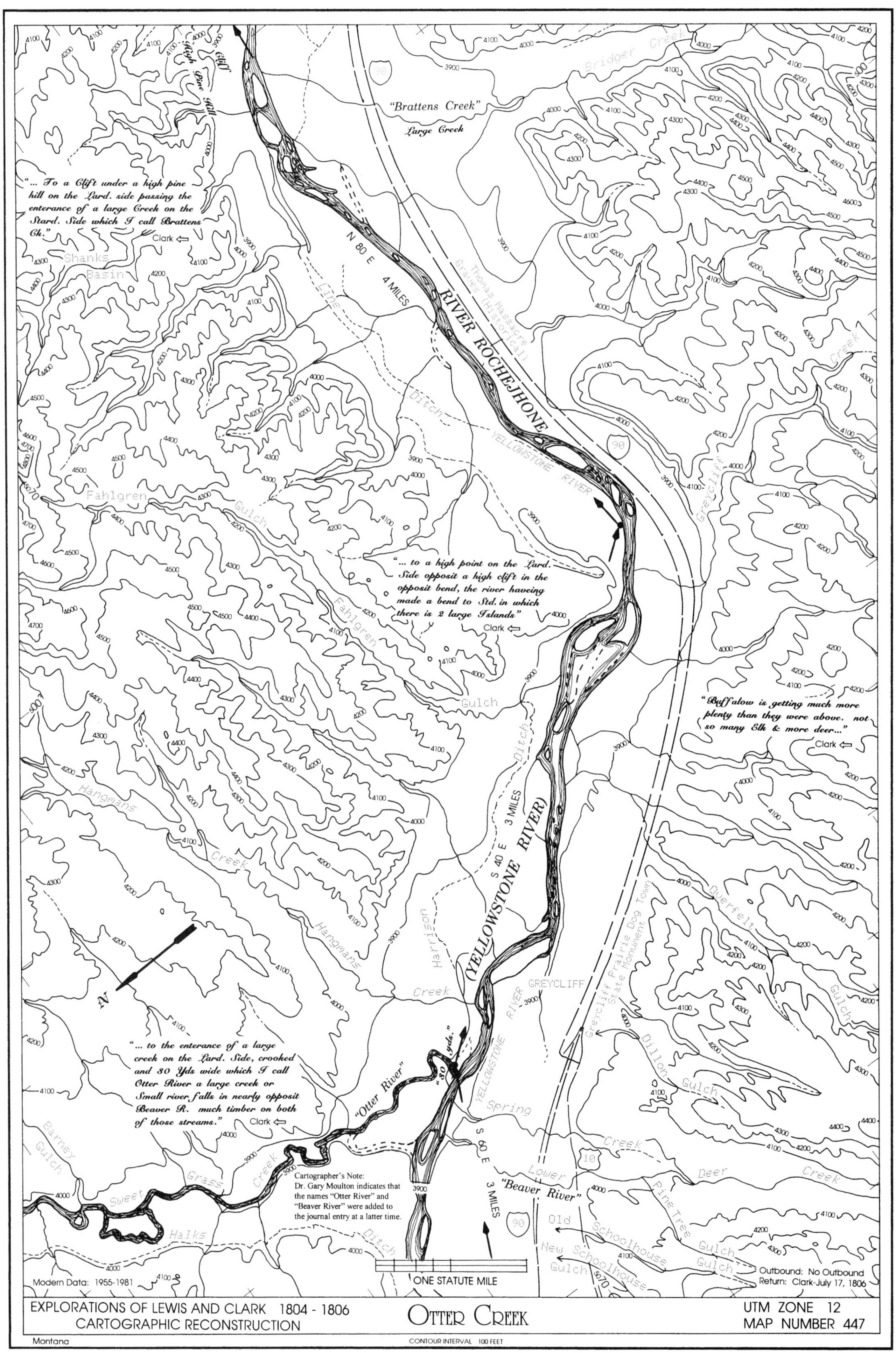
"... To a Clift under a high pine hill on the Lard. side passing the enterance of a large Creek on the Stard. Side which I call Brattens Ck." Clark
"Brattens Creek"
Large Creek
High Pine Hill
Cliff
RIVER ROCHEJHONE
N 80 E 4 MILES
"... to a high point on the Lard. Side opposit a high clift in the opposit bend, the river haveing made a bend to Std. in which there is 2 large Islands" Clark
"Buffalow is getting much more plenty than they were above. not so many Elk & more deer..." Clark
S 40 E 3 MILES
(YELLOWSTONE RIVER)
"... to the enterance of a large creek on the Lard. Side, crooked and 80 Yds wide which I call Otter River a large creek or Small river falls in nearly opposit Beaver R. much timber on both of those streams." Clark
"Otter River"
"80 yds."
"Beaver River"
S 60 E 3 MILES
Cartographer's Note: Dr. Gary Moulton indicates that the names "Otter River" and "Beaver River" were added to the journal entry at a latter time.
Shanks Basin
Fahlgren
Fahlgren Gulch
Hangmans Creek
Hangmans
Harrison
Sweet Grass Creek
Barney Gulch
Halks Ditch
Thomas Massacre Graves (Historical)
Bridger Creek
Greycliff
GREYCLIFF
Greycliff Prairie Dog Town State Monument
Duerfelt Gulch
Dillon Gulch
Spring Creek
Lower Deer Creek
Pine Tree Gulch
Old Schoolhouse Gulch
New Schoolhouse Gulch
YELLOWSTONE RIVER
ONE STATUTE MILE
Modern Data: 1955-1981
Outbound: No Outbound
Return: Clark-July 17, 1806
EXPLORATIONS OF LEWIS AND CLARK 1804 - 1806
CARTOGRAPHIC RECONSTRUCTION
OTTER CREEK
UTM ZONE 12
MAP NUMBER 447
Montana
CONTOUR INTERVAL 100 FEET

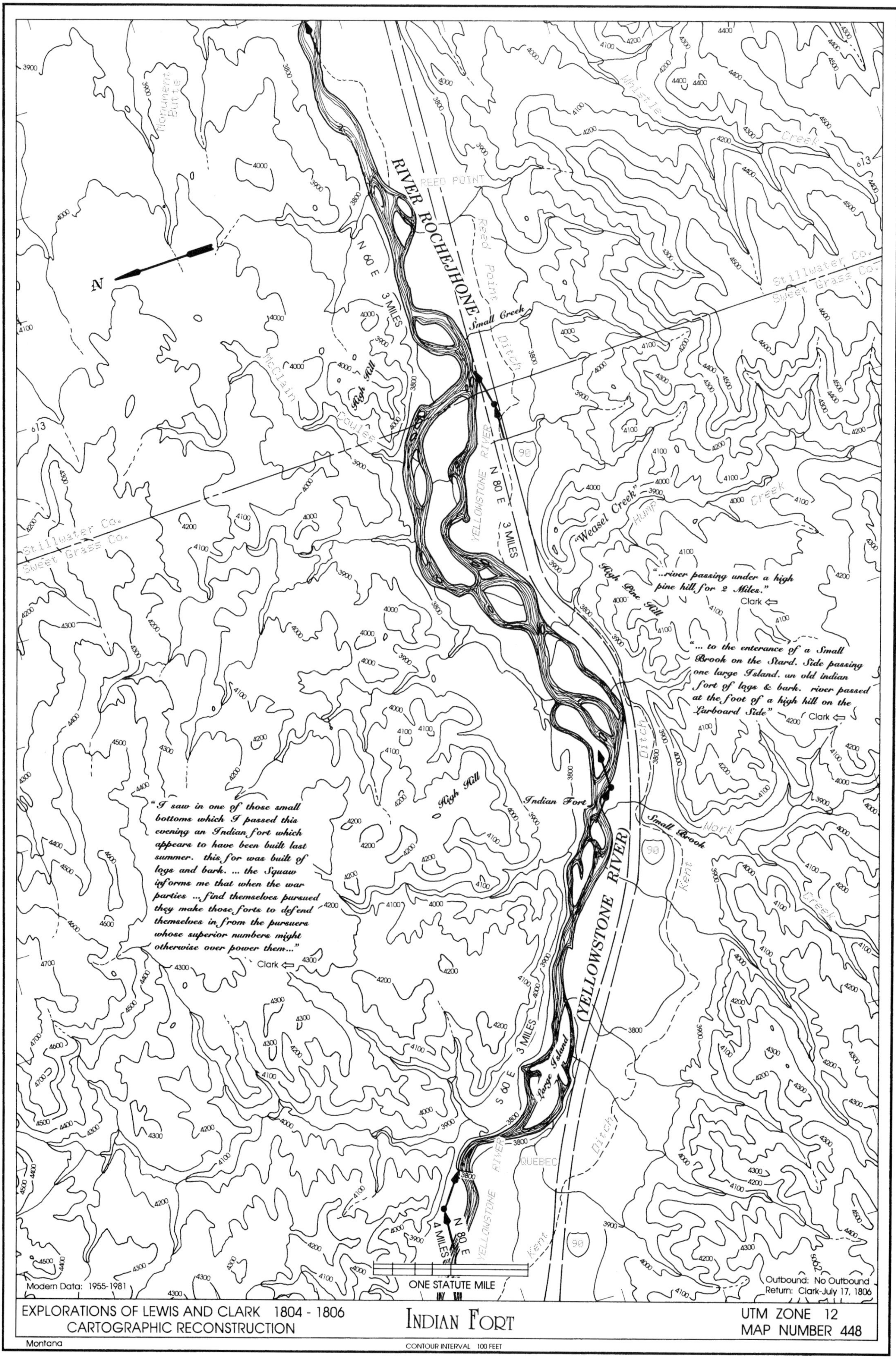
RIVER ROCHEJHONE
(YELLOWSTONE RIVER)
N 60 E 3 MILES
N 80 E 3 MILES
S 60 E 3 MILES
N 80 E 4 MILES
REED POINT
Reed Point Ditch
Small Creek
High Hill
McClain Coulee
Monument Butte
Whistle Creek
Stillwater Co.
Sweet Grass Co.
"Weasel Creek"
Hump Creek
High Pine Hill
Indian Fort
Small Brook
Work Creek
Kent
Large Island
QUEBEC
"...river passing under a high pine hill for 2 Miles." Clark
"... to the enterance of a Small Brook on the Stard. Side passing one large Island. an old indian fort of logs & bark. river passed at the foot of a high hill on the Larboard Side" Clark
"I saw in one of those small bottoms which I passed this evening an Indian fort which appears to have been built last summer. this for was built of logs and bark. ... the Squaw informs me that when the war parties ...find themselves pursued they make those forts to defend themselves in from the pursuers whose superior numbers might otherwise over power them..." Clark
ONE STATUTE MILE
Modern Data: 1955-1981
Outbound: No Outbound
Return: Clark-July 17, 1806
EXPLORATIONS OF LEWIS AND CLARK 1804 - 1806
CARTOGRAPHIC RECONSTRUCTION
INDIAN FORT
UTM ZONE 12
MAP NUMBER 448
Montana
CONTOUR INTERVAL 100 FEET

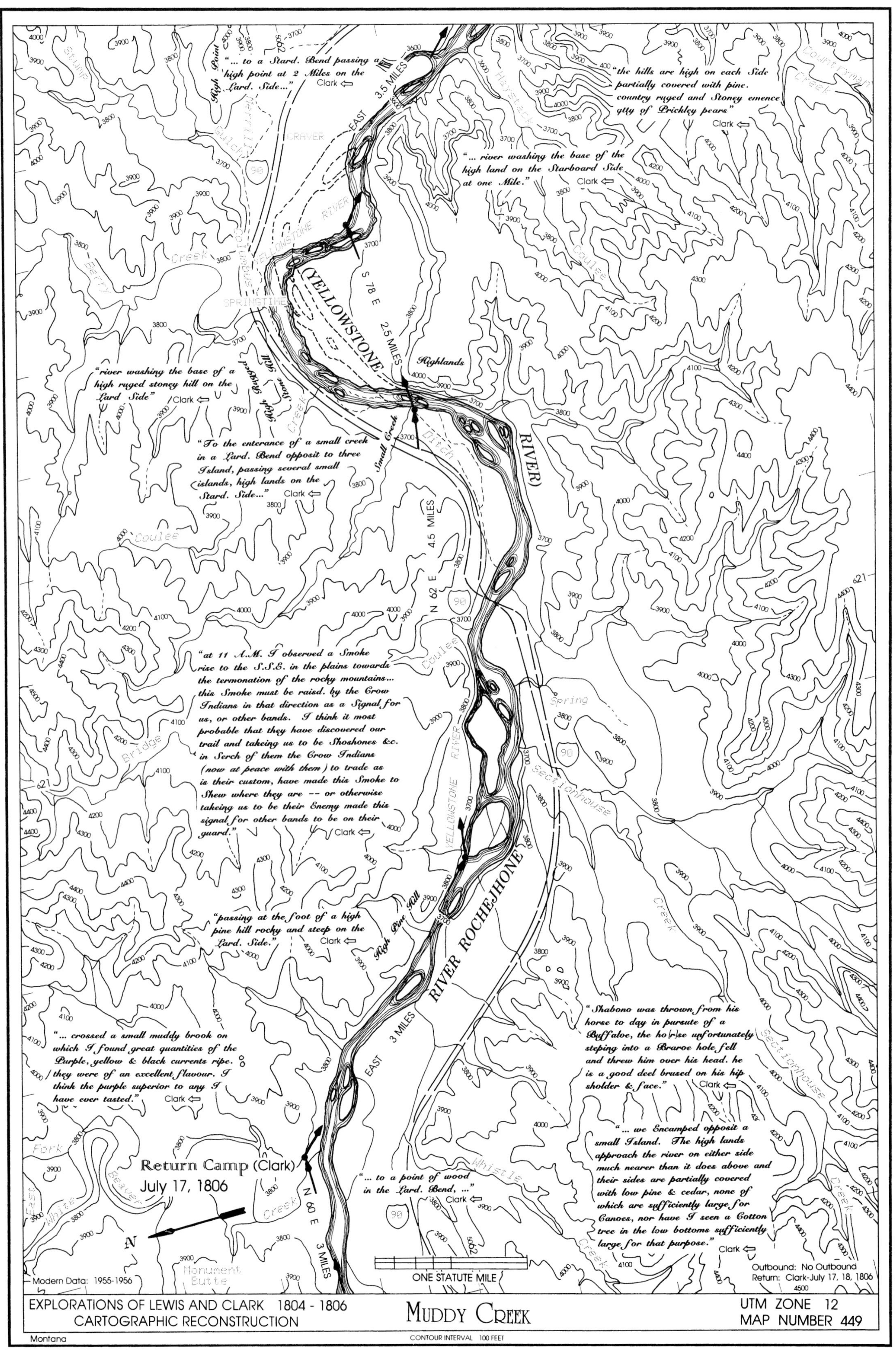
"... to a Stard. Bend passing a high point at 2 Miles on the Lard. Side..." Clark
"the hills are high on each Side partially covered with pine. country ruged and Stoney emence qty of Prickley pears" Clark
"... river washing the base of the high land on the Starboard Side at one Mile." Clark
"river washing the base of a high ruged stoney hill on the Lard Side" Clark
"To the enterance of a small creek in a Lard. Bend opposit to three Island, passing several small islands, high lands on the Stard. Side..." Clark
"at 11 A.M. I observed a Smoke rise to the S.S.E. in the plains towards the termonation of the rocky mountains... this Smoke must be raisd. by the Crow Indians in that direction as a Signal for us, or other bands. I think it most probable that they have discovered our trail and takeing us to be Shoshones &c. in Serch of them the Crow Indians (now at peace with them) to trade as is their custom, have made this Smoke to Shew where they are -- or otherwise takeing us to be their Enemy made this signal for other bands to be on their guard." Clark
"passing at the foot of a high pine hill rocky and steep on the Lard. Side." Clark
"Shabono was thrown from his horse to day in pursute of a Buffaloe, the ho[r]se unfortunately steping into a Braroe hole fell and threw him over his head. he is a good deel brused on his hip sholder & face." Clark
"... crossed a small muddy brook on which I found great quantities of the Purple, yellow & black currents ripe. they were of an excellent flavour. I think the purple superior to any I have ever tasted." Clark
"... we Encamped opposit a small Island. The high lands approach the river on either side much nearer than it does above and their sides are partially covered with low pine & cedar, none of which are sufficiently large for Canoes, nor have I seen a Cotton tree in the low bottoms sufficiently large for that purpose." Clark
"... to a point of wood in the Lard. Bend, ..." Clark
Return Camp (Clark)
July 17, 1806
(YELLOWSTONE RIVER)
RIVER ROCHEJHONE
YELLOWSTONE RIVER
3.5 MILES
EAST
S 78 E
2.5 MILES
4.5 MILES
N 62 E
3 MILES
EAST
N 60 E
3 MILES
High Point
Highlands
High Ragged Hill
High Stone Hill
Small Creek
High Pine Hill
Stump
Merrill Gulch
Columbus
Berry Creek
Springtime
Craver
Haystack
Countryman Creek
Coulee
Ditch
Spring
Bridge
Sectionhouse
Sectionhouse Creek
Whistle Creek
Fork
East White Beaver Creek
Monument Butte
N
ONE STATUTE MILE
Modern Data: 1955-1956
Outbound: No Outbound
Return: Clark-July 17, 18, 1806
EXPLORATIONS OF LEWIS AND CLARK 1804 - 1806
CARTOGRAPHIC RECONSTRUCTION
MUDDY CREEK
UTM ZONE 12
MAP NUMBER 449
Montana
CONTOUR INTERVAL 100 FEET

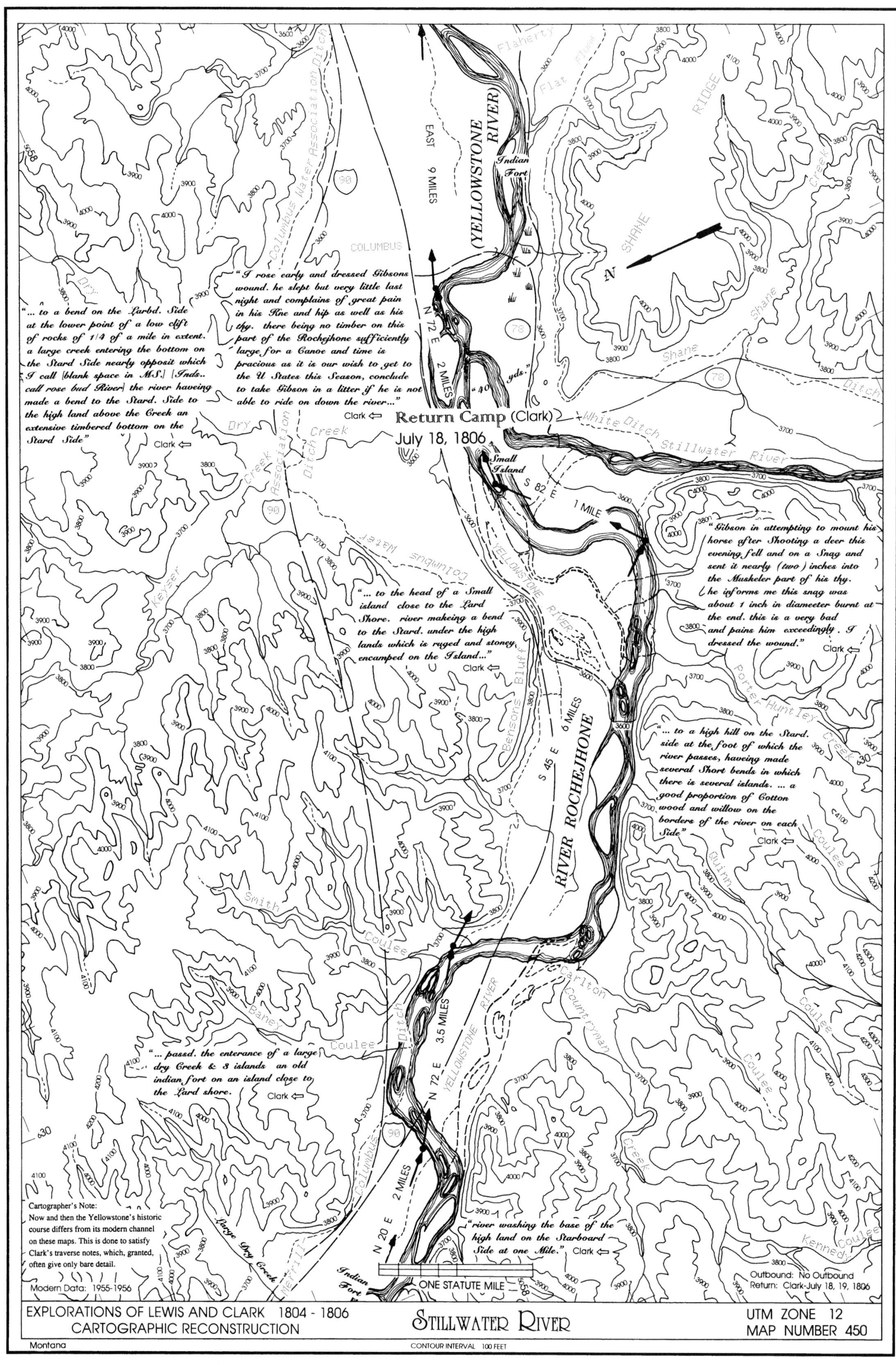

(YELLOWSTONE RIVER)
EAST 9 MILES
Indian Fort
COLUMBUS
Columbus Water Association Ditch
Flaherty
Flat
SHANE
RIDGE
N
"I rose early and dressed Gibsons wound. he slept but very little last night and complains of great pain in his Kne and hip as well as his thy. there being no timber on this part of the Rochejhone sufficiently large for a Canoe and time is pracious as it is our wish to get to the U States this Season, conclude to take Gibson in a litter if he is not able to ride on down the river..."
Clark
"... to a bend on the Larbd. Side at the lower point of a low clift of rocks of 1/4 of a mile in extent. a large creek entering the bottom on the Stard Side nearly opposit which I call [blank space in MS.] [Inds.. call rose bud River] the river haveing made a bend to the Stard. Side to the high land above the Creek an extensive timbered bottom on the Stard Side"
Clark
N 72 E
2 MILES
"40 yds."
Return Camp (Clark)
July 18, 1806
White Ditch
Stillwater River
Small Island
S 82 E
1 MILE
Dry Creek
"Gibson in attempting to mount his horse after Shooting a deer this evening fell and on a Snag and sent it nearly (two) inches into the Muskeler part of his thy. he informs me this snag was about 1 inch in diameter burnt at the end. this is a very bad and pains him exceedingly. I dressed the wound."
Clark
"... to the head of a Small island close to the Lard Shore. river makeing a bend to the Stard. under the high lands which is ruged and stoney encamped on the Island..."
Clark
YELLOWSTONE RIVER
Bensons Bluff
S 45 E
6 MILES
RIVER ROCHEJHONE
Keyser
Porter Huntley Creek
"... to a high hill on the Stard. side at the foot of which the river passes, haveing made several Short bends in which there is several islands. ... a good proportion of Cotton wood and willow on the borders of the river on each Side"
Clark
Quinn
Smith Coulee
Carlton
Countryman
Baney Coulee
3.5 MILES
N 72 E
YELLOWSTONE RIVER
"... passd. the enterance of a large dry Creek & 3 islands an old indian fort on an island close to the Lard shore."
Clark
Columbus
Merrill
Large Dry Creek
Indian Fort
N 20 E
2 MILES
"river washing the base of the high land on the Starboard Side at one Mile."
Clark
Kennedy Coulee
Creek
Cartographer's Note:
Now and then the Yellowstone's historic course differs from its modern channel on these maps. This is done to satisfy Clark's traverse notes, which, granted, often give only bare detail.
Modern Data: 1955-1956
ONE STATUTE MILE
Outbound: No Outbound
Return: Clark-July 18, 19, 1806
EXPLORATIONS OF LEWIS AND CLARK 1804 - 1806
CARTOGRAPHIC RECONSTRUCTION
STILLWATER RIVER
UTM ZONE 12
MAP NUMBER 450
Montana
CONTOUR INTERVAL 100 FEET

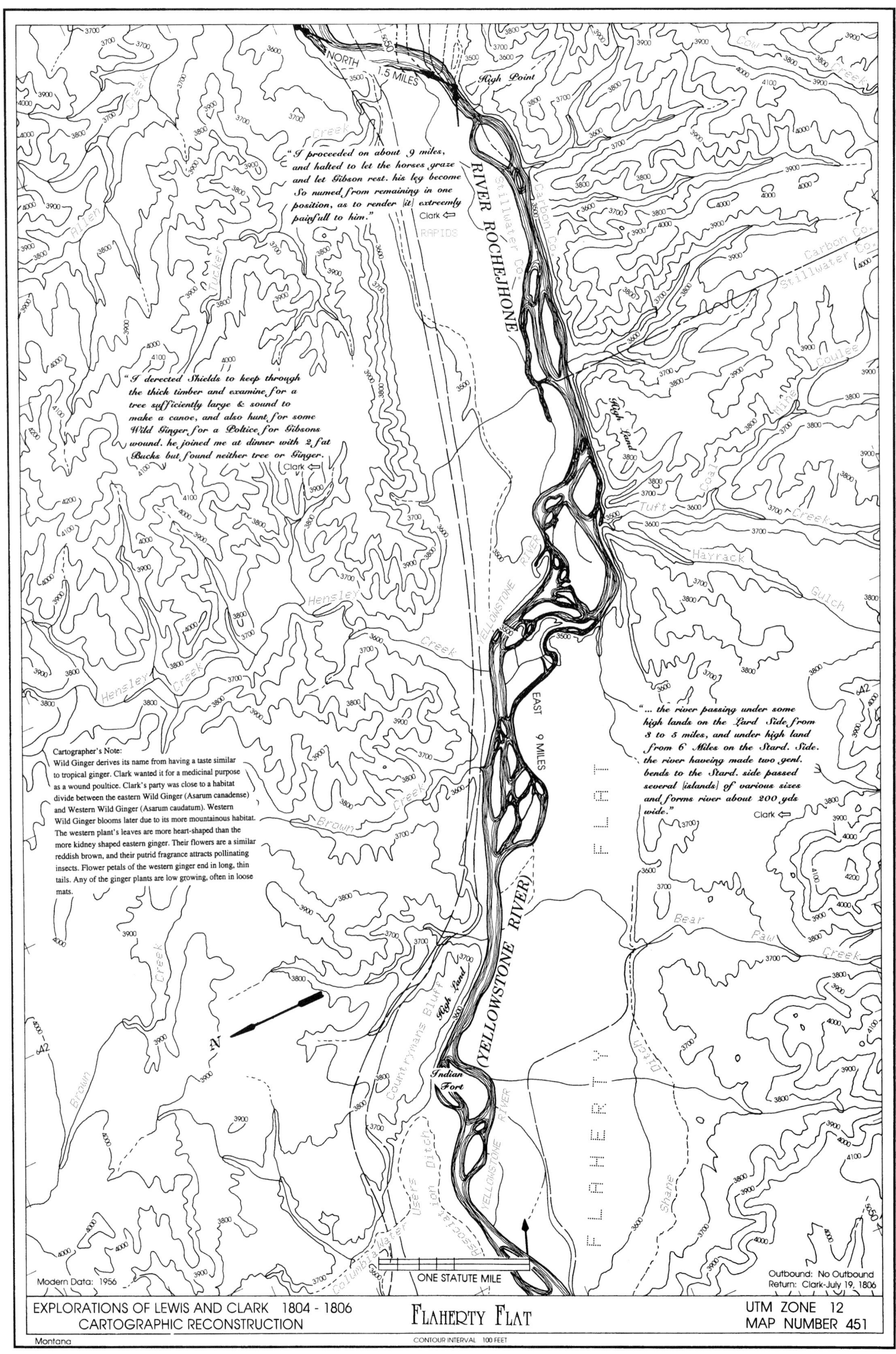
"I proceeded on about 9 miles, and halted to let the horses graze and let Gibson rest. his leg become So numed from remaining in one position, as to render [it] extreemly painfull to him."
Clark
"I derected Shields to keep through the thick timber and examine for a tree sufficiently large & sound to make a canoe, and also hunt for some Wild Ginger for a Poltice for Gibsons wound. he joined me at dinner with 2 fat Bucks but found neither tree or Ginger."
Clark
"... the river passing under some high lands on the Lard Side from 3 to 5 miles, and under high land from 6 Miles on the Stard. Side. the river haveing made two genl. bends to the Stard. side passed several [islands] of various sizes and forms river about 200 yds wide."
Clark
Cartographer's Note:
Wild Ginger derives its name from having a taste similar to tropical ginger. Clark wanted it for a medicinal purpose as a wound poultice. Clark's party was close to a habitat divide between the eastern Wild Ginger (Asarum canadense) and Western Wild Ginger (Asarum caudatum). Western Wild Ginger blooms later due to its more mountainous habitat. The western plant's leaves are more heart-shaped than the more kidney shaped eastern ginger. Their flowers are a similar reddish brown, and their putrid fragrance attracts pollinating insects. Flower petals of the western ginger end in long, thin tails. Any of the ginger plants are low growing, often in loose mats.
RIVER ROCHEJHONE
(YELLOWSTONE RIVER)
YELLOWSTONE RIVER
High Point
High Land
Indian Fort
Countrymans Bluff
FLAHERTY FLAT
NORTH 1.5 MILES
EAST 9 MILES
RAPIDS
Stillwater Co.
Carbon Co.
Allen Creek
Tucker Creek
Hensley Creek
Brown Creek
Cow Creek
Coulee
Mine
Coal
Tuft Creek
Hayrack Gulch
Bear Paw Creek
Ditch
Shane
Columbia Water Users Ditch
N
ONE STATUTE MILE
Modern Data: 1956
Outbound: No Outbound
Return: Clark-July 19, 1806
EXPLORATIONS OF LEWIS AND CLARK 1804 - 1806
CARTOGRAPHIC RECONSTRUCTION
FLAHERTY FLAT
UTM ZONE 12
MAP NUMBER 451
Montana
CONTOUR INTERVAL 100 FEET

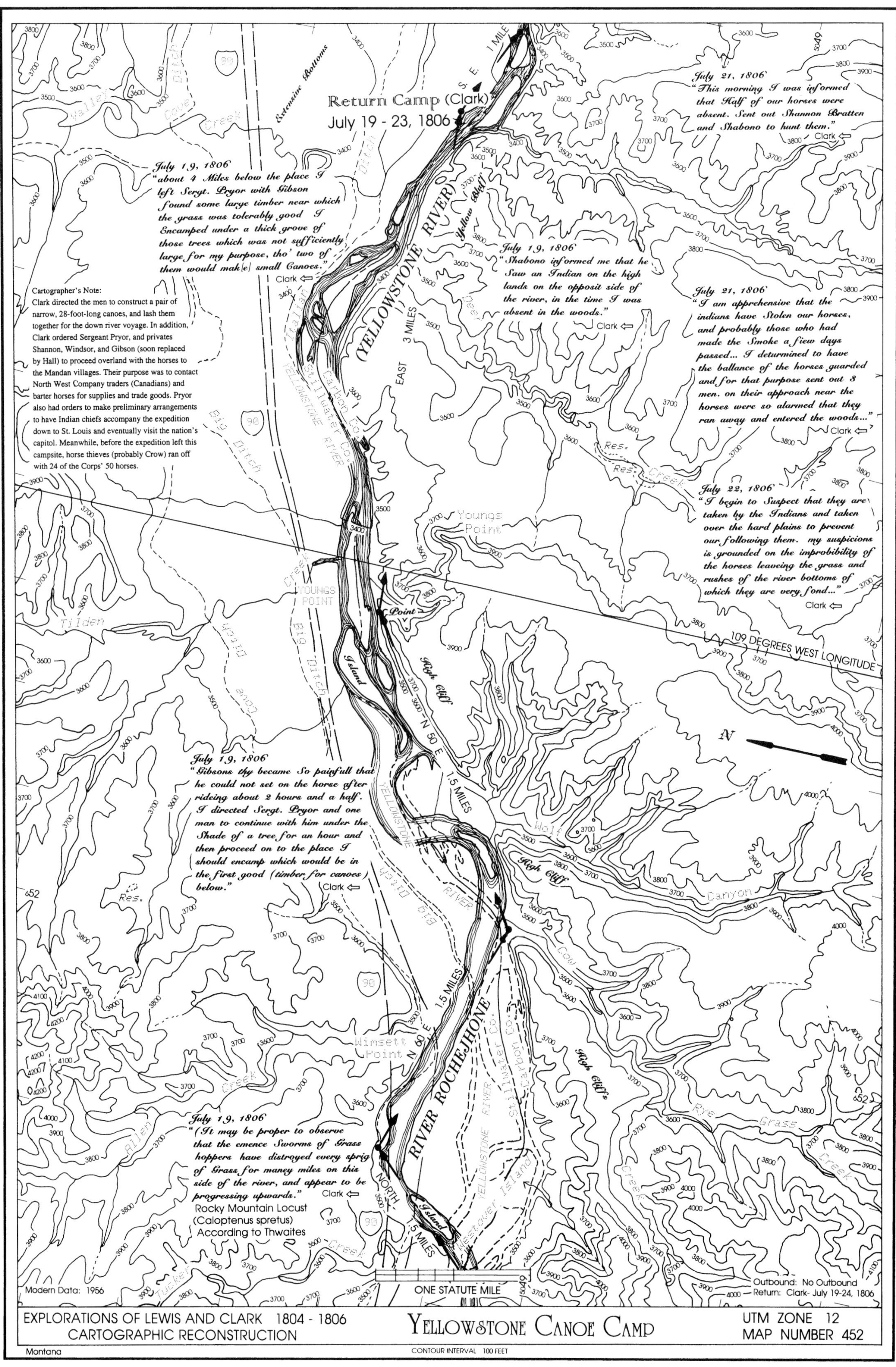

Return Camp (Clark)
July 19 - 23, 1806
July 19, 1806
"about 4 Miles below the place I left Sergt. Pryor with Gibson found some large timber near which the grass was tolerably good I Encamped under a thick grove of those trees which was not sufficiently large for my purpose, tho' two of them would mak[e] small Canoes."
Clark ⇦
Cartographer's Note:
Clark directed the men to construct a pair of narrow, 28-foot-long canoes, and lash them together for the down river voyage. In addition, Clark ordered Sergeant Pryor, and privates Shannon, Windsor, and Gibson (soon replaced by Hall) to proceed overland with the horses to the Mandan villages. Their purpose was to contact North West Company traders (Canadians) and barter horses for supplies and trade goods. Pryor also had orders to make preliminary arrangements to have Indian chiefs accompany the expedition down to St. Louis and eventually visit the nation's capitol. Meanwhile, before the expedition left this campsite, horse thieves (probably Crow) ran off with 24 of the Corps' 50 horses.
July 21, 1806
"This morning I was informed that Half of our horses were absent. Sent out Shannon Bratten and Shabono to hunt them."
Clark ⇦
July 19, 1806
"Shabono informed me that he Saw an Indian on the high lands on the opposit side of the river, in the time I was absent in the woods."
Clark ⇦
July 21, 1806
"I am apprehensive that the indians have Stolen our horses, and probably those who had made the Smoke a fiew days passed... I determined to have the ballance of the horses guarded and for that purpose sent out 3 men. on their approach near the horses were so alarmed that they ran away and entered the woods..."
Clark ⇦
July 22, 1806
"I begin to Suspect that they are taken by the Indians and taken over the hard plains to prevent our following them. my suspicions is grounded on the improbibility of the horses leaveing the grass and rushes of the river bottoms of which they are very fond..."
Clark ⇦
July 19, 1806
"Gibsons thy became So painfull that he could not set on the horse after rideing about 2 hours and a half. I directed Sergt. Pryor and one man to continue with him under the Shade of a tree for an hour and then proceed on to the place I should encamp which would be in the first good (timber for canoes) below."
Clark ⇦
July 19, 1806
"(It may be proper to observe that the emence Sworms of Grass hoppers have distroyed every sprig of Grass for maney miles on this side of the river, and appear to be progressing upwards."
Clark ⇦
Rocky Mountain Locust
(Caloptenus spretus)
According to Thwaites
(YELLOWSTONE RIVER)
RIVER ROCHEJHONE
Yellow Bluff
High Cliff
High Cliffs
Extensive Bottoms
Island
Point
Westover Island
S. E. 1 MILE
EAST 3 MILES
N 50 E 1.5 MILES
N 60 E 1.5 MILES
NORTH 1.5 MILES
109 DEGREES WEST LONGITUDE
N
Youngs Point
Wimsett Point
Valley Creek
Cove Ditch
Big Ditch
Italian Ditch
Deer Creek
Res. Creek
Wolf Canyon
Cow Creek
Rye Grass Creek
Allen Creek
Tucker Creek
Tilden
Stillwater Co.
Carbon Co.
Modern Data: 1956
ONE STATUTE MILE
Outbound: No Outbound
Return: Clark- July 19-24, 1806
EXPLORATIONS OF LEWIS AND CLARK 1804 - 1806
CARTOGRAPHIC RECONSTRUCTION
YELLOWSTONE CANOE CAMP
UTM ZONE 12
MAP NUMBER 452
Montana
CONTOUR INTERVAL 100 FEET

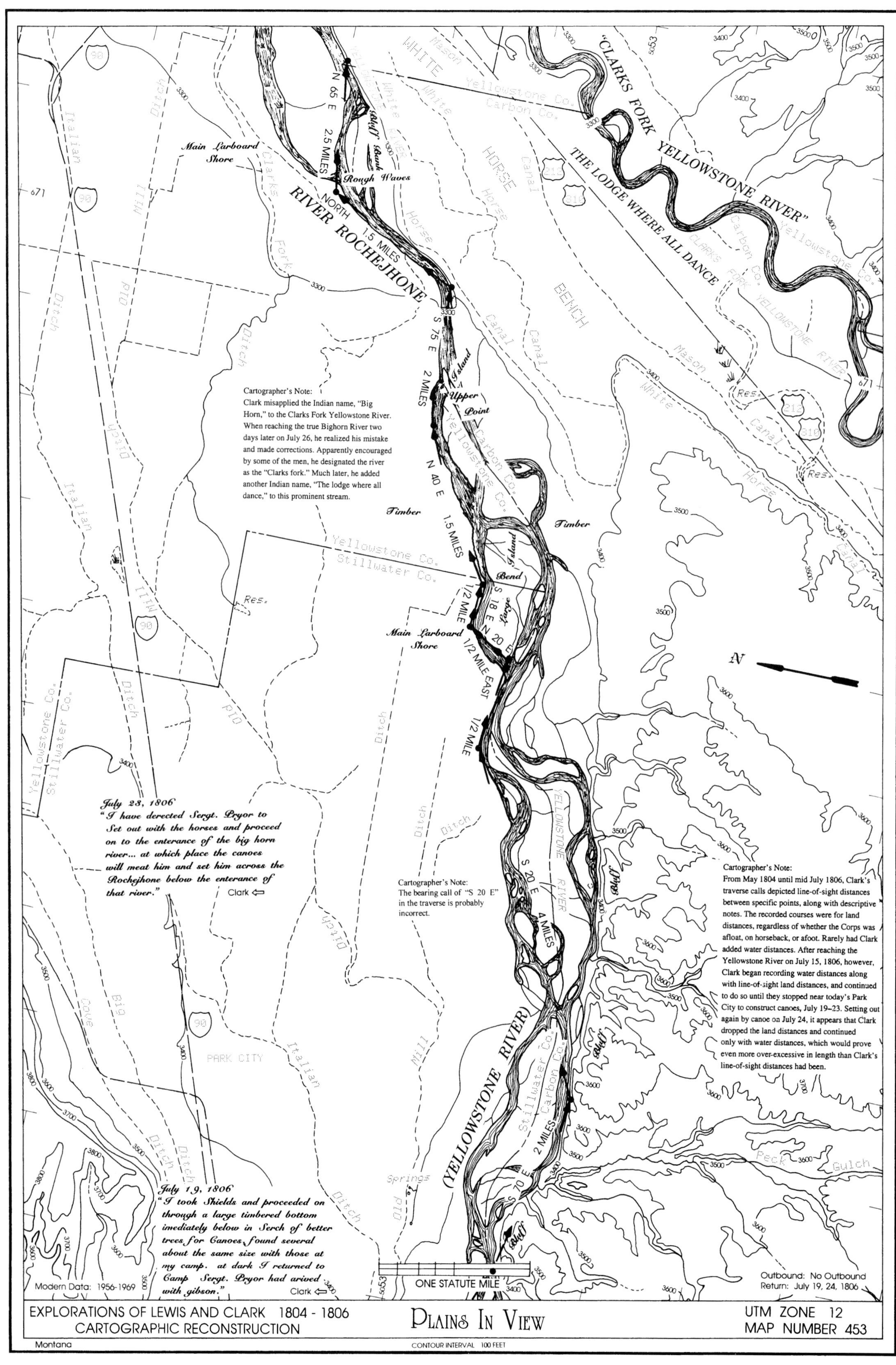

Cartographer's Note:
Clark misapplied the Indian name, "Big Horn," to the Clarks Fork Yellowstone River. When reaching the true Bighorn River two days later on July 26, he realized his mistake and made corrections. Apparently encouraged by some of the men, he designated the river as the "Clarks fork." Much later, he added another Indian name, "The lodge where all dance," to this prominent stream.
Cartographer's Note:
The bearing call of "S 20 E" in the traverse is probably incorrect.
Cartographer's Note:
From May 1804 until mid July 1806, Clark's traverse calls depicted line-of-sight distances between specific points, along with descriptive notes. The recorded courses were for land distances, regardless of whether the Corps was afloat, on horseback, or afoot. Rarely had Clark added water distances. After reaching the Yellowstone River on July 15, 1806, however, Clark began recording water distances along with line-of-sight land distances, and continued to do so until they stopped near today's Park City to construct canoes, July 19–23. Setting out again by canoe on July 24, it appears that Clark dropped the land distances and continued only with water distances, which would prove even more over-excessive in length than Clark's line-of-sight distances had been.
July 23, 1806
"I have derected Sergt. Pryor to Set out with the horses and proceed on to the enterance of the big horn river... at which place the canoes will meat him and set him across the Rochejhone below the enterance of that river."
Clark ⇦
July 19, 1806
"I took Shields and proceeded on through a large timbered bottom imediately below in Serch of better trees for Canoes, found several about the same size with those at my camp. at dark I returned to Camp Sergt. Pryor had arived with gibson."
Clark ⇦
RIVER ROCHEJHONE
"CLARKS FORK YELLOWSTONE RIVER"
THE LODGE WHERE ALL DANCE
(YELLOWSTONE RIVER)
Main Larboard Shore
Rough Waves
Bluff
Bank
Island
Upper Point
Timber
Bend
Large
N 65 E
2.5 MILES
NORTH
1.5 MILES
S 75 E
2 MILES
N 40 E
1.5 MILES
1/2 MILE
S 18 E
N 20 E
1/2 MILE EAST
1/2 MILE
S 20 E
4 MILES
2 MILES
S 70 E
PARK CITY
Yellowstone Co.
Stillwater Co.
Carbon Co.
Springs
Peck Gulch
Italian Ditch
Mill Ditch
Big Ditch
White Horse Canal
Mason Canal
BENCH
Res.
Modern Data: 1956-1969
ONE STATUTE MILE
Outbound: No Outbound
Return: July 19, 24, 1806
EXPLORATIONS OF LEWIS AND CLARK 1804 - 1806
CARTOGRAPHIC RECONSTRUCTION
PLAINS IN VIEW
UTM ZONE 12
MAP NUMBER 453
Montana
CONTOUR INTERVAL 100 FEET

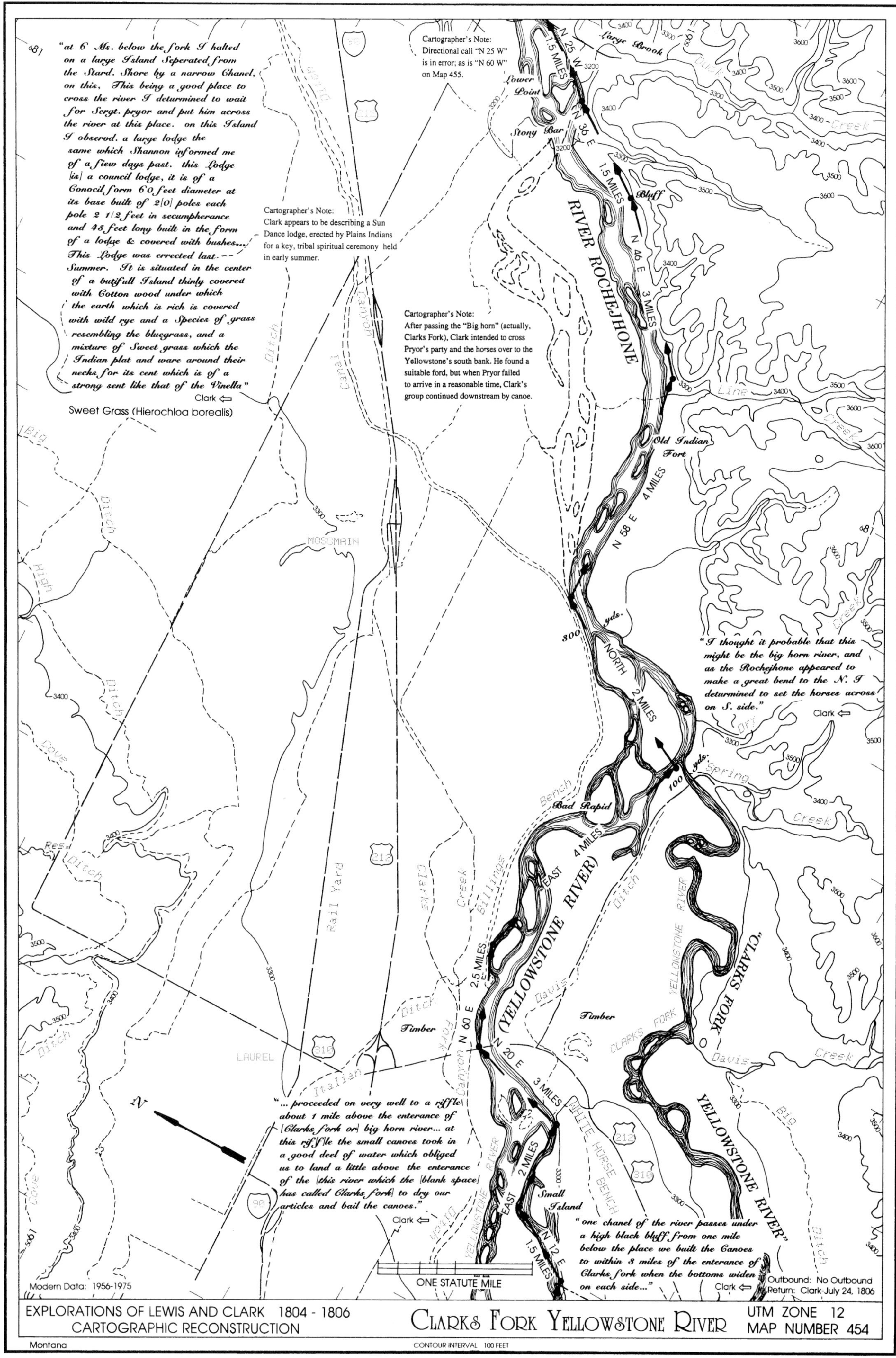
"at 6 Ms. below the fork I halted on a large Island Seperated from the Stard. Shore by a narrow Chanel, on this, This being a good place to cross the river I determined to wait for Sergt. pryor and put him across the river at this place. on this Island I obsevrd. a large lodge the same which Shannon informed me of a fiew days past. this Lodge [is] a council lodge, it is of a Conocil form 60 feet diameter at its base built of 2[0] poles each pole 2 1/2 feet in secumpherance and 45 feet long built in the form of a lodge & covered with bushes... This Lodge was errected last Summer. It is situated in the center of a butifull Island thinly covered with Cotton wood under which the earth which is rich is covered with wild rye and a Species of grass resembling the bluegrass, and a mixture of Sweet grass which the Indian plat and ware around their necks for its cent which is of a strong sent like that of the Vinella"
Clark ⇦
Sweet Grass (Hierochloa borealis)
Cartographer's Note:
Directional call "N 25 W" is in error; as is "N 60 W" on Map 455.
Cartographer's Note:
Clark appears to be describing a Sun Dance lodge, erected by Plains Indians for a key, tribal spiritual ceremony held in early summer.
Cartographer's Note:
After passing the "Big horn" (actually, Clarks Fork), Clark intended to cross Pryor's party and the horses over to the Yellowstone's south bank. He found a suitable ford, but when Pryor failed to arrive in a reasonable time, Clark's group continued downstream by canoe.
Large Brook
Lower Point
Stony Bar
Bluff
RIVER ROCHEJHONE
N 25 W
1.5 MILES
N 36 E
N 46 E
3 MILES
Old Indian Fort
4 MILES
N 58 E
300 yds.
NORTH
2 MILES
100 yds.
Bad Rapid
4 MILES
EAST
(YELLOWSTONE RIVER)
2.5 MILES
N 60 E
N 20 E
3 MILES
2 MILES
EAST
Small Island
N 12 E
1.5 MILES
Timber
"CLARKS FORK"
"YELLOWSTONE RIVER"
MOSSMAIN
LAUREL
Rail Yard
Canal
Ditch
Creek
Line Creek
Dry
Spring Creek
Duck Creek
Big Ditch
High Ditch
Cove Ditch
Res Ditch
Billings Bench
White Horse Bench
Davis Creek
Italian
Canyon Creek
Clarks Fork
Yellowstone River
"I thought it probable that this might be the big horn river, and as the Rochejhone appeared to make a great bend to the N. I determined to set the horses across on S. side."
Clark ⇦
"... proceeded on very well to a riffle about 1 mile above the enterance of [Clarks fork or] big horn river... at this rif[f]le the small canoes took in a good deel of water which obliged us to land a little above the enterance of the [this river which the [blank space] has called Clarks fork] to dry our articles and bail the canoes."
Clark ⇦
"one chanel of the river passes under a high black bluff from one mile below the place we built the Canoes to within 3 miles of the enterance of Clarks fork when the bottoms widen on each side..."
Clark ⇦
N
Modern Data: 1956-1975
ONE STATUTE MILE
Outbound: No Outbound
Return: Clark-July 24, 1806
EXPLORATIONS OF LEWIS AND CLARK 1804 - 1806
CARTOGRAPHIC RECONSTRUCTION
CLARKS FORK YELLOWSTONE RIVER
UTM ZONE 12
MAP NUMBER 454
Montana
CONTOUR INTERVAL 100 FEET

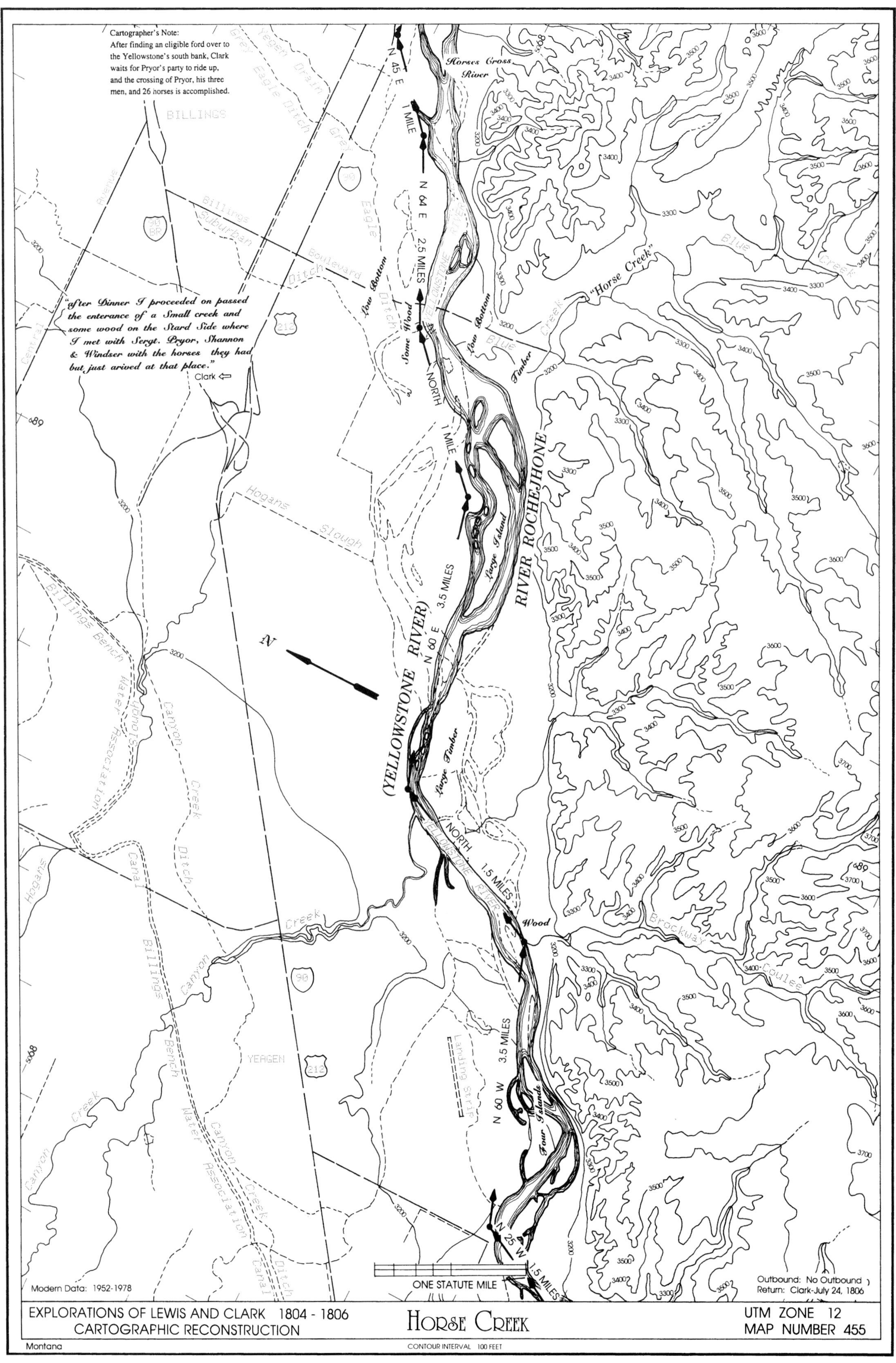

Cartographer's Note:
After finding an eligible ford over to the Yellowstone's south bank, Clark waits for Pryor's party to ride up, and the crossing of Pryor, his three men, and 26 horses is accomplished.
"after Dinner I proceeded on passed the enterance of a Small creek and some wood on the Stard Side where I met with Sergt. Pryor, Shannon & Windser with the horses they had but just arived at that place."
Clark
Horses Cross River
"Horse Creek"
Blue Creek
RIVER ROCHEJHONE
(YELLOWSTONE RIVER)
Large Island
Large Timber
Four Islands
Low Bottom
Some Wood
Timber
Wood
N 45 E
1 MILE
N 64 E
2.5 MILES
NORTH
3.5 MILES
N 60 E
1.5 MILES
N 60 W
N 25 W
BILLINGS
YEAGEN
Brockway Coulee
Hogans Slough
Canyon Creek
Billings Bench
Landing Strip
ONE STATUTE MILE
Modern Data: 1952-1978
Outbound: No Outbound
Return: Clark-July 24, 1806
EXPLORATIONS OF LEWIS AND CLARK 1804 - 1806
CARTOGRAPHIC RECONSTRUCTION
HORSE CREEK
UTM ZONE 12
MAP NUMBER 455
Montana
CONTOUR INTERVAL 100 FEET

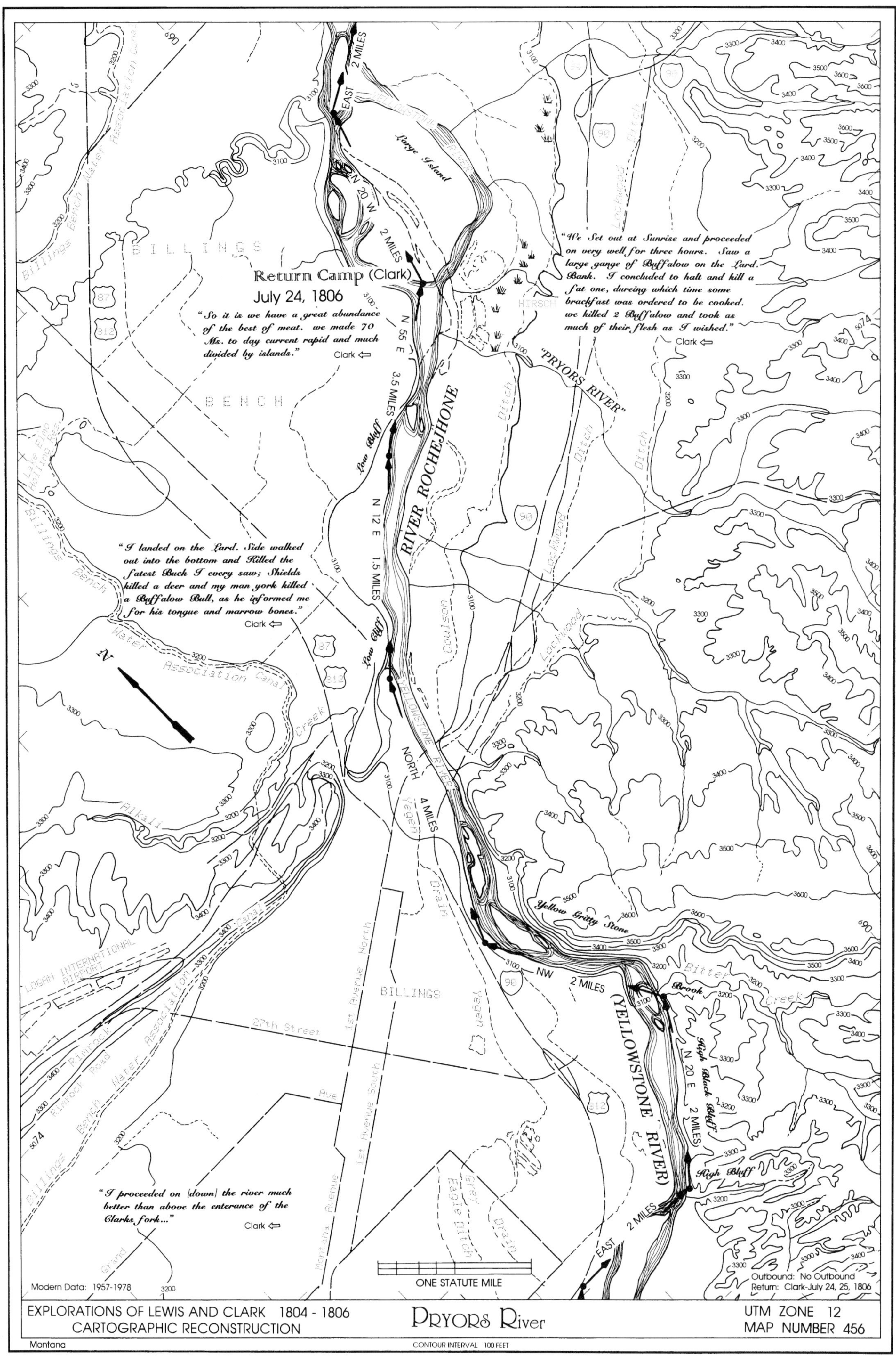
Return Camp (Clark)
July 24, 1806
"So it is we have a great abundance of the best of meat. we made 70 Ms. to day current rapid and much divided by islands."
Clark
"We Set out at Sunrise and proceeded on very well for three hours. Saw a large gange of Buffalow on the Lard. Bank. I concluded to halt and kill a fat one, dureing which time some brackfast was ordered to be cooked. we killed 2 Buffalow and took as much of their flesh as I wished."
Clark
"I landed on the Lard. Side walked out into the bottom and Killed the fatest Buck I every saw; Shields killed a deer and my man york killed a Buffalow Bull, as he informed me for his tongue and marrow bones."
Clark
"I proceeded on [down] the river much better than above the enterance of the Clarks fork..."
Clark
Large Island
"PRYORS RIVER"
RIVER ROCHEJHONE
YELLOWSTONE RIVER
BILLINGS
BENCH
Low Bluff
Yellow Gritty Stone
Bitter Creek
Brook
High Black Bluff
High Bluff
EAST 2 MILES
N 20 W 2 MILES
N 55 E 3.5 MILES
N 12 E 1.5 MILES
NORTH 4 MILES
NW 2 MILES
N 20 E 2 MILES
EAST 2 MILES
HIRSCH
Lockwood Ditch
Coulson
Alkali Creek
Billings Bench Water Association Canal
Lake Elmo
LOGAN INTERNATIONAL AIRPORT
Rimrock Road
27th Street
1st Avenue North
1st Avenue South
Montana Avenue
Yegen Drain
Grey Eagle Ditch
N
ONE STATUTE MILE
Modern Data: 1957-1978
Outbound: No Outbound
Return: Clark-July 24, 25, 1806
EXPLORATIONS OF LEWIS AND CLARK 1804 - 1806
CARTOGRAPHIC RECONSTRUCTION
PRYORS River
UTM ZONE 12
MAP NUMBER 456
Montana
CONTOUR INTERVAL 100 FEET

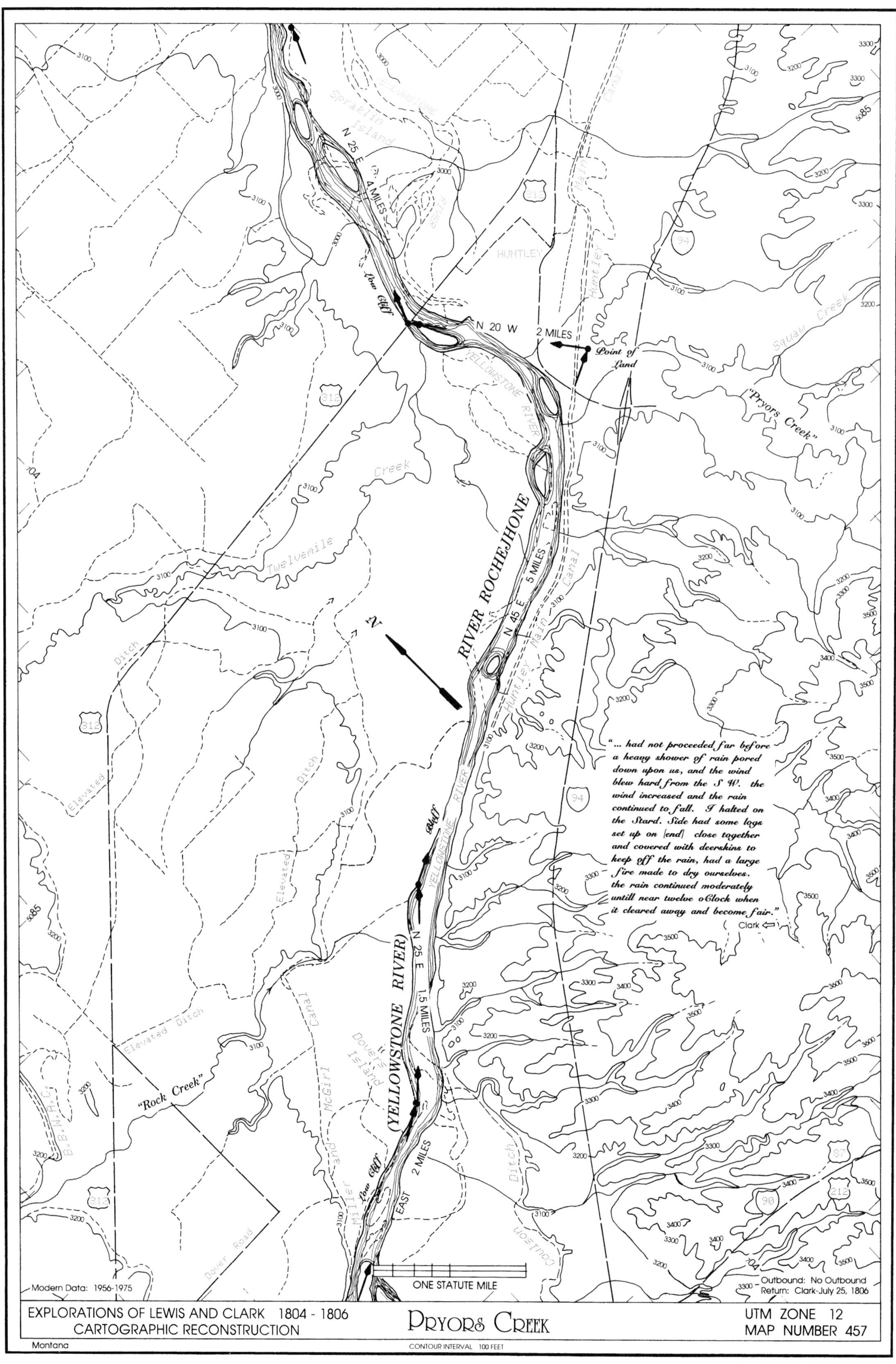
"... had not proceeded far before a heavy shower of rain pored down upon us, and the wind blew hard from the S W. the wind increased and the rain continued to fall. I halted on the Stard. Side had some logs set up on [end] close together and covered with deerskins to keep off the rain, had a large fire made to dry ourselves. the rain continued moderately untill near twelve oClock when it cleared away and become fair."
Clark
RIVER ROCHEJHONE
(YELLOWSTONE RIVER)
"Pryors Creek"
"Rock Creek"
Point of Land
Low Cliff
Cliff
N 25 E 4 MILES
N 20 W 2 MILES
N 45 E 5 MILES
N 25 E 1.5 MILES
EAST 2 MILES
HUNTLEY
Huntley Main Canal
Twelvemile Creek
Squaw Creek
Elevated Ditch
Dover Road
Miller and McGirl Canal
Spraklin Island
Dovers Island
ONE STATUTE MILE
Modern Data: 1956-1975
Outbound: No Outbound
Return: Clark-July 25, 1806
EXPLORATIONS OF LEWIS AND CLARK 1804 - 1806
CARTOGRAPHIC RECONSTRUCTION
PRYORS CREEK
UTM ZONE 12
MAP NUMBER 457
Montana
CONTOUR INTERVAL 100 FEET

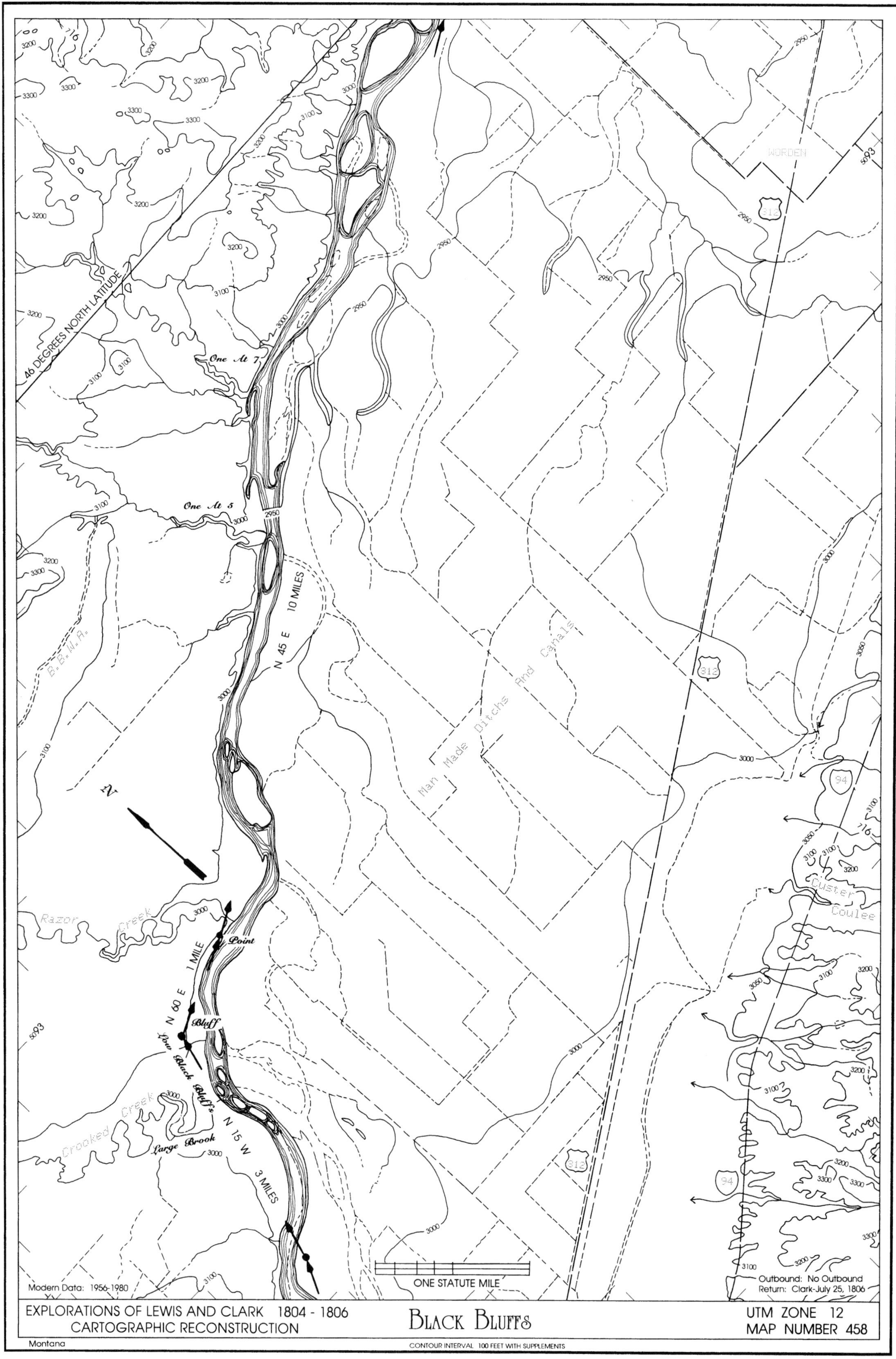

46 DEGREES NORTH LATITUDE
WORDEN
One At 7
One At 5
N 45 E 10 MILES
B.B.W.M.A.
Man Made Ditchs And Canals
Razor Creek
Point
N 60 E 1 MILE
Bluff
Low Black Bluffs
Crooked Creek
Large Brook
N 15 W 3 MILES
Custer Coulee
ONE STATUTE MILE
Modern Data: 1956-1980
Outbound: No Outbound
Return: Clark-July 25, 1806
EXPLORATIONS OF LEWIS AND CLARK 1804 - 1806
CARTOGRAPHIC RECONSTRUCTION
BLACK BLUFFS
UTM ZONE 12
MAP NUMBER 458
Montana
CONTOUR INTERVAL 100 FEET WITH SUPPLEMENTS

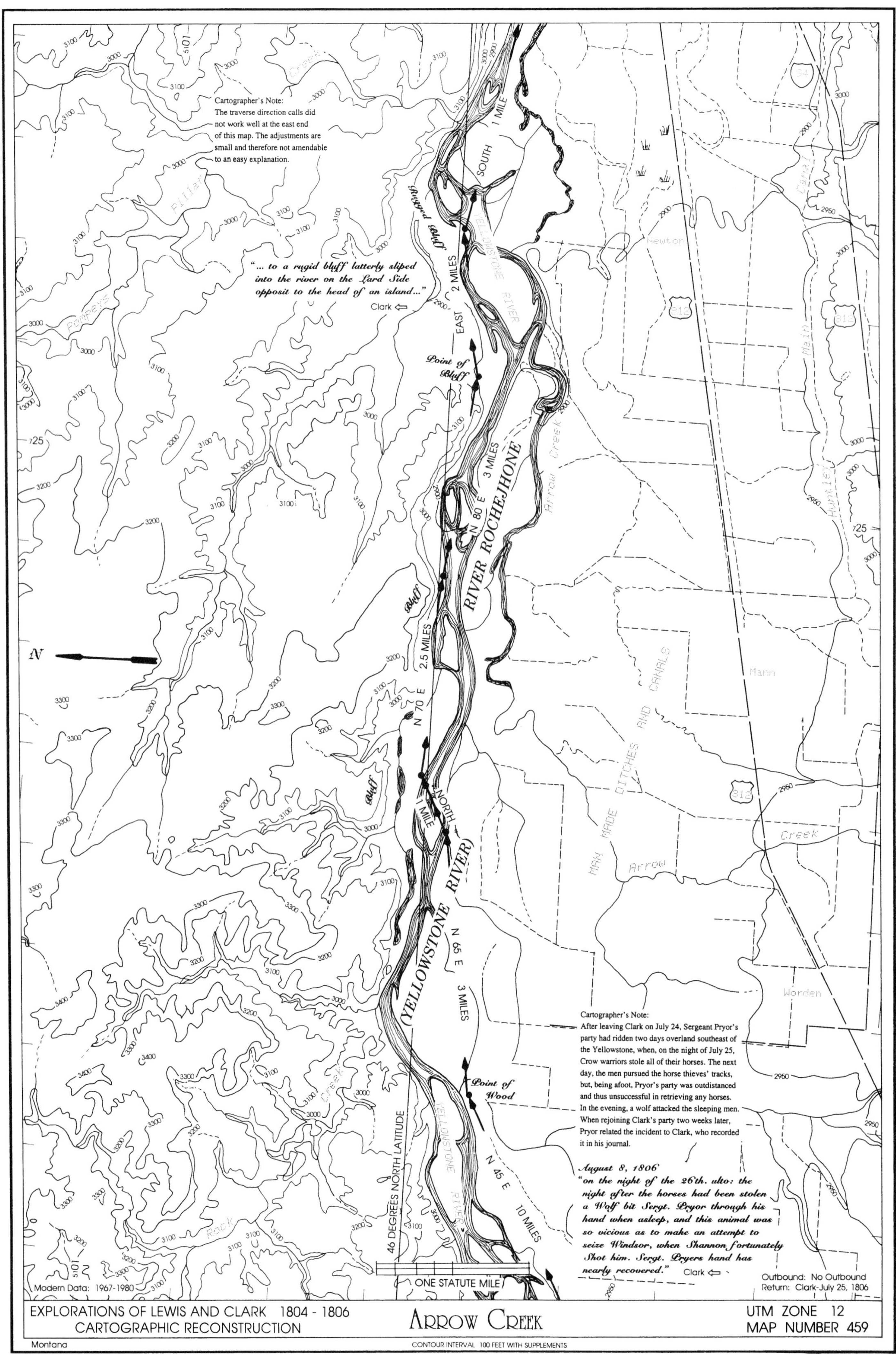
Cartographer's Note:
The traverse direction calls did not work well at the east end of this map. The adjustments are small and therefore not amendable to an easy explanation.
"... to a ragid bluff latterly sliped into the river on the Lard Side opposit to the head of an island..."
Clark
Point of Bluff
RIVER ROCHEJHONE
YELLOWSTONE RIVER
(YELLOWSTONE RIVER)
Point of Wood
Cartographer's Note:
After leaving Clark on July 24, Sergeant Pryor's party had ridden two days overland southeast of the Yellowstone, when, on the night of July 25, Crow warriors stole all of their horses. The next day, the men pursued the horse thieves' tracks, but, being afoot, Pryor's party was outdistanced and thus unsuccessful in retrieving any horses. In the evening, a wolf attacked the sleeping men. When rejoining Clark's party two weeks later, Pryor related the incident to Clark, who recorded it in his journal.
August 8, 1806
"on the night of the 26th. ulto: the night after the horses had been stolen a Wolf bit Sergt. Pryor through his hand when asleep, and this animal was so vicious as to make an attempt to seize Windsor, when Shannon fortunately Shot him. Sergt. Pryers hand has nearly recovered."
Clark
Outbound: No Outbound
Return: Clark-July 25, 1806
Modern Data: 1967-1980
ONE STATUTE MILE
46 DEGREES NORTH LATITUDE
EXPLORATIONS OF LEWIS AND CLARK 1804 - 1806
CARTOGRAPHIC RECONSTRUCTION
ARROW CREEK
UTM ZONE 12
MAP NUMBER 459
Montana
CONTOUR INTERVAL 100 FEET WITH SUPPLEMENTS

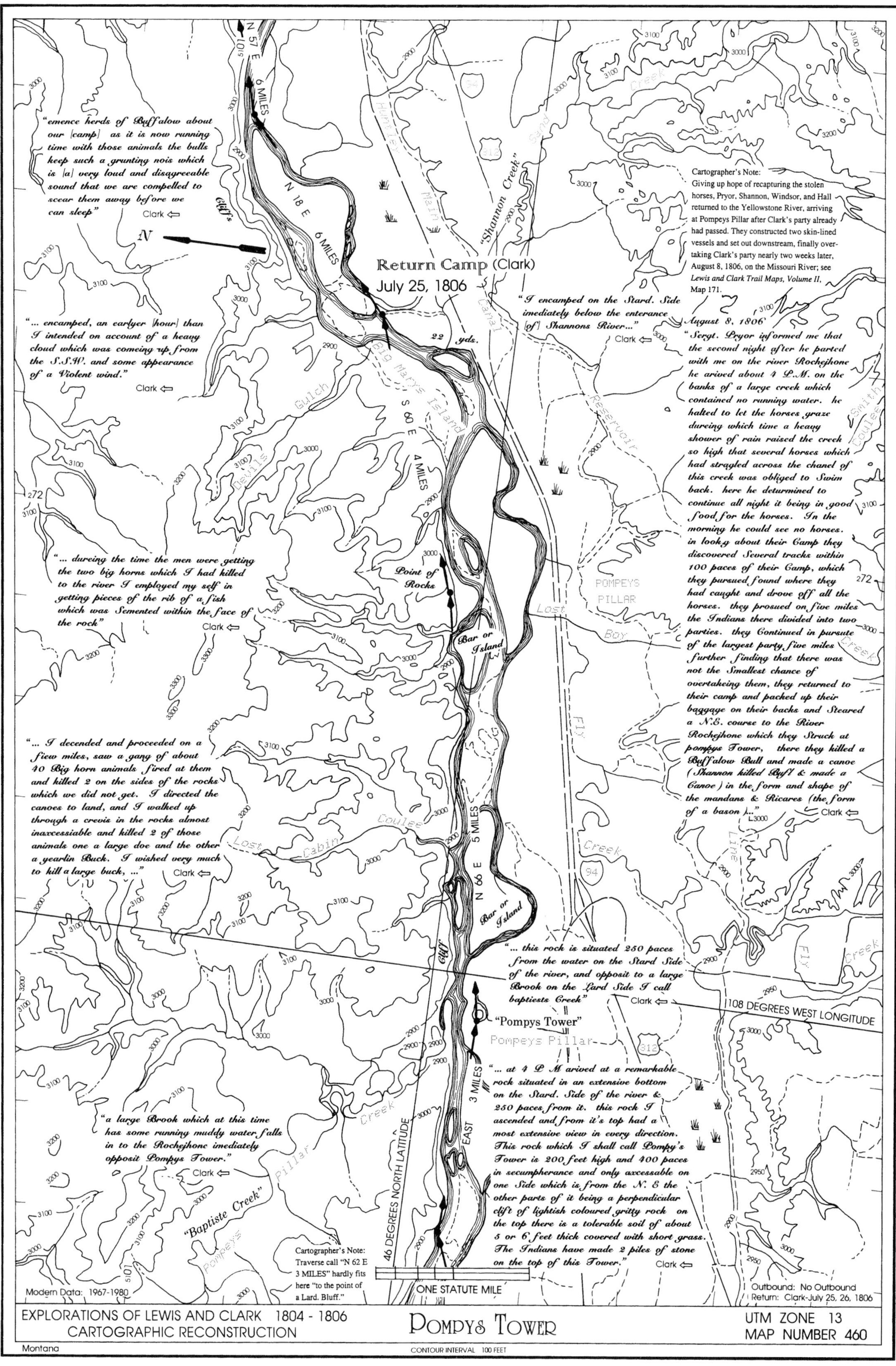

"emence herds of Buffalow about our [camp] as it is now running time with those animals the bulls keep such a grunting nois which is [a] very loud and disagreeable sound that we are compelled to scear them away before we can sleep" Clark ⇦
N
N 57 E
6 MILES
N 18 E 6 MILES
Cliffs
Return Camp (Clark)
July 25, 1806
"... encamped, an earlyer [hour] than I intended on account of a heavy cloud which was comeing up from the S.S.W. and some appearance of a Violent wind." Clark ⇦
22 yds.
"Shannon Creek"
"I encamped on the Stard. Side imediately below the enterance [of] Shannons River..." Clark ⇦
Cartographer's Note:
Giving up hope of recapturing the stolen horses, Pryor, Shannon, Windsor, and Hall returned to the Yellowstone River, arriving at Pompeys Pillar after Clark's party already had passed. They constructed two skin-lined vessels and set out downstream, finally overtaking Clark's party nearly two weeks later, August 8, 1806, on the Missouri River; see Lewis and Clark Trail Maps, Volume II, Map 171.
August 8, 1806
"Sergt. Pryor informed me that the second night after he parted with me on the river Rochejhone he arived about 4 P.M. on the banks of a large creek which contained no running water. he halted to let the horses graze dureing which time a heavy shower of rain raised the creek so high that several horses which had stragled across the chanel of this creek was obliged to Swim back. here he determined to continue all night it being in good food for the horses. In the morning he could see no horses. in lookg about their Camp they discovered Several tracks within 100 paces of their Camp, which they pursued found where they had caught and drove off all the horses. they prosued on five miles the Indians there divided into two parties. they Continued in pursute of the largest party five miles further finding that there was not the Smallest chance of overtakeing them, they returned to their camp and packed up their baggage on their backs and Steared a N.E. course to the River Rochejhone which they Struck at pompys Tower, there they killed a Buffalow Bull and made a canoe (Shannon killed Buff & made a Canoe) in the form and shape of the mandans & Ricares (the form of a bason)..." Clark ⇦
Humtley Main Canal
Sand Creek
Big Marys Island
S 60 E
4 MILES
Gulch
Devils
Reservoir
Smith Coulee
"... dureing the time the men were getting the two big horns which I had killed to the river I employed my self in getting pieces of the rib of a fish which was Semented within the face of the rock" Clark ⇦
Point of Rocks
Bar or Island
POMPEYS PILLAR
Lost Boy Creek
Fly Creek
"... I decended and proceeded on a fiew miles, saw a gang of about 40 Big horn animals fired at them and killed 2 on the sides of the rocks which we did not get. I directed the canoes to land, and I walked up through a crevis in the rocks almost inaccessiable and killed 2 of those animals one a large doe and the other a yearlin Buck. I wished very much to kill a large buck, ..." Clark ⇦
Lost Cabin Coulee
5 MILES
N 66 E
Bar or Island
Line Creek
Cliff
"... this rock is situated 250 paces from the water on the Stard Side of the river, and opposit to a large Brook on the Lard Side I call baptiests Creek" Clark ⇦
108 DEGREES WEST LONGITUDE
"Pompys Tower"
Pompeys Pillar
"... at 4 P M arived at a remarkable rock situated in an extensive bottom on the Stard. Side of the river & 250 paces from it. this rock I ascended and from it's top had a most extensive view in every direction. This rock which I shall call Pompy's Tower is 200 feet high and 400 paces in secumpherance and only axcessable on one Side which is from the N. E the other parts of it being a perpendicular clift of lightish coloured gritty rock on the top there is a tolerable soil of about 5 or 6 feet thick covered with short grass. The Indians have made 2 piles of stone on the top of this Tower." Clark ⇦
3 MILES
EAST
46 DEGREES NORTH LATITUDE
"a large Brook which at this time has some running muddy water falls in to the Rochejhone imediately opposit Pompys Tower." Clark ⇦
"Baptiste Creek"
Pompeys Pillar Creek
Cartographer's Note:
Traverse call "N 62 E 3 MILES" hardly fits here "to the point of a Lard. Bluff."
ONE STATUTE MILE
Modern Data: 1967-1980
Outbound: No Outbound
Return: Clark-July 25, 26, 1806
EXPLORATIONS OF LEWIS AND CLARK 1804 - 1806
CARTOGRAPHIC RECONSTRUCTION
POMPYS TOWER
UTM ZONE 13
MAP NUMBER 460
Montana
CONTOUR INTERVAL 100 FEET

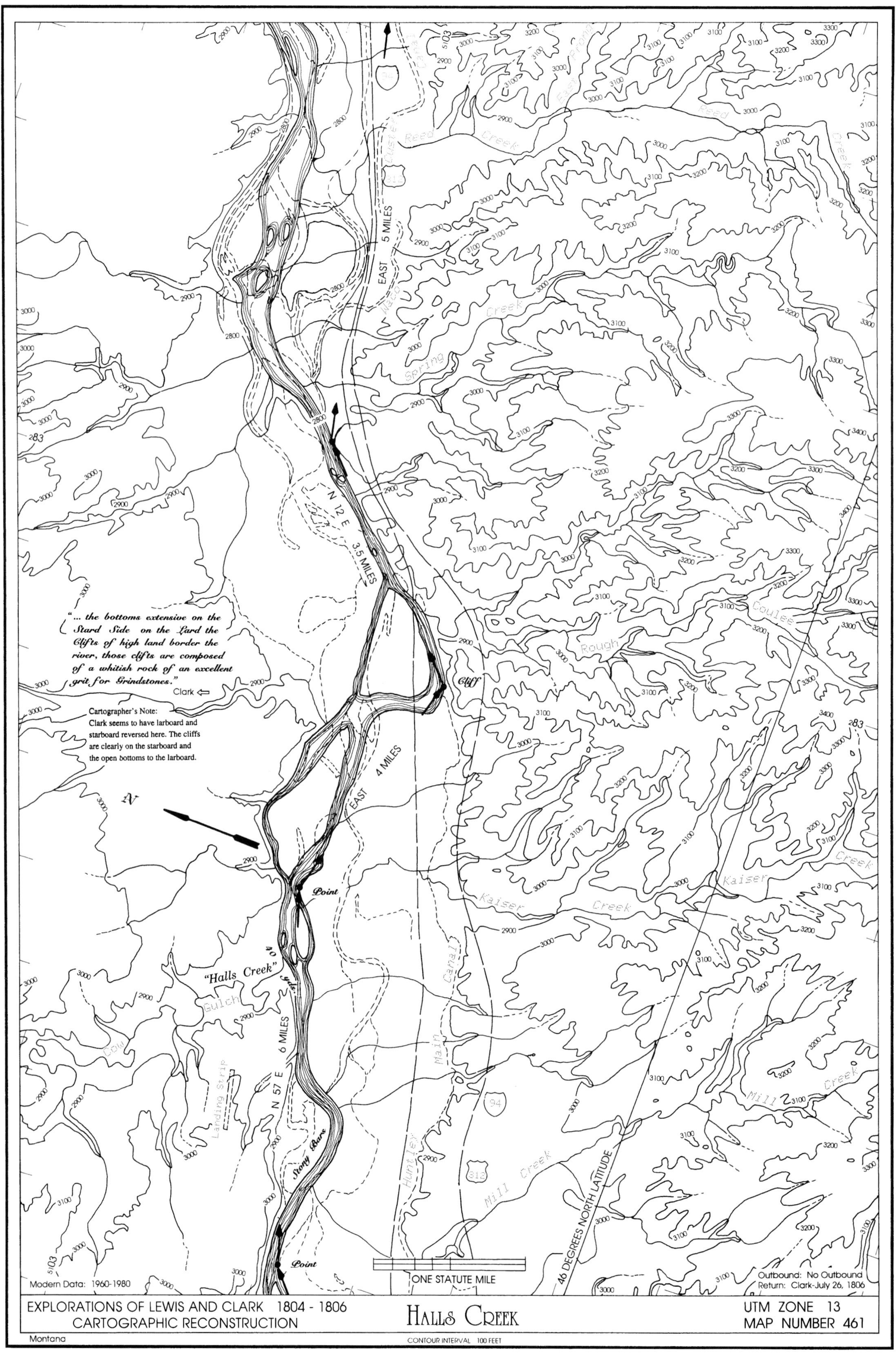
"... the bottoms extensive on the Stard Side on the Lard the Clifts of high land border the river, those clifts are composed of a whitish rock of an excellent grit for Grindstones."
Clark
Cartographer's Note:
Clark seems to have larboard and starboard reversed here. The cliffs are clearly on the starboard and the open bottoms to the larboard.
N 12 E 3.5 MILES
EAST 5 MILES
EAST 4 MILES
N 57 E 6 MILES
Cliff
Point
"Halls Creek"
Stony Bars
Point
ONE STATUTE MILE
Modern Data: 1960-1980
Outbound: No Outbound
Return: Clark-July 26, 1806
46 DEGREES NORTH LATITUDE
EXPLORATIONS OF LEWIS AND CLARK 1804 - 1806
CARTOGRAPHIC RECONSTRUCTION
HALLS CREEK
UTM ZONE 13
MAP NUMBER 461
Montana
CONTOUR INTERVAL 100 FEET

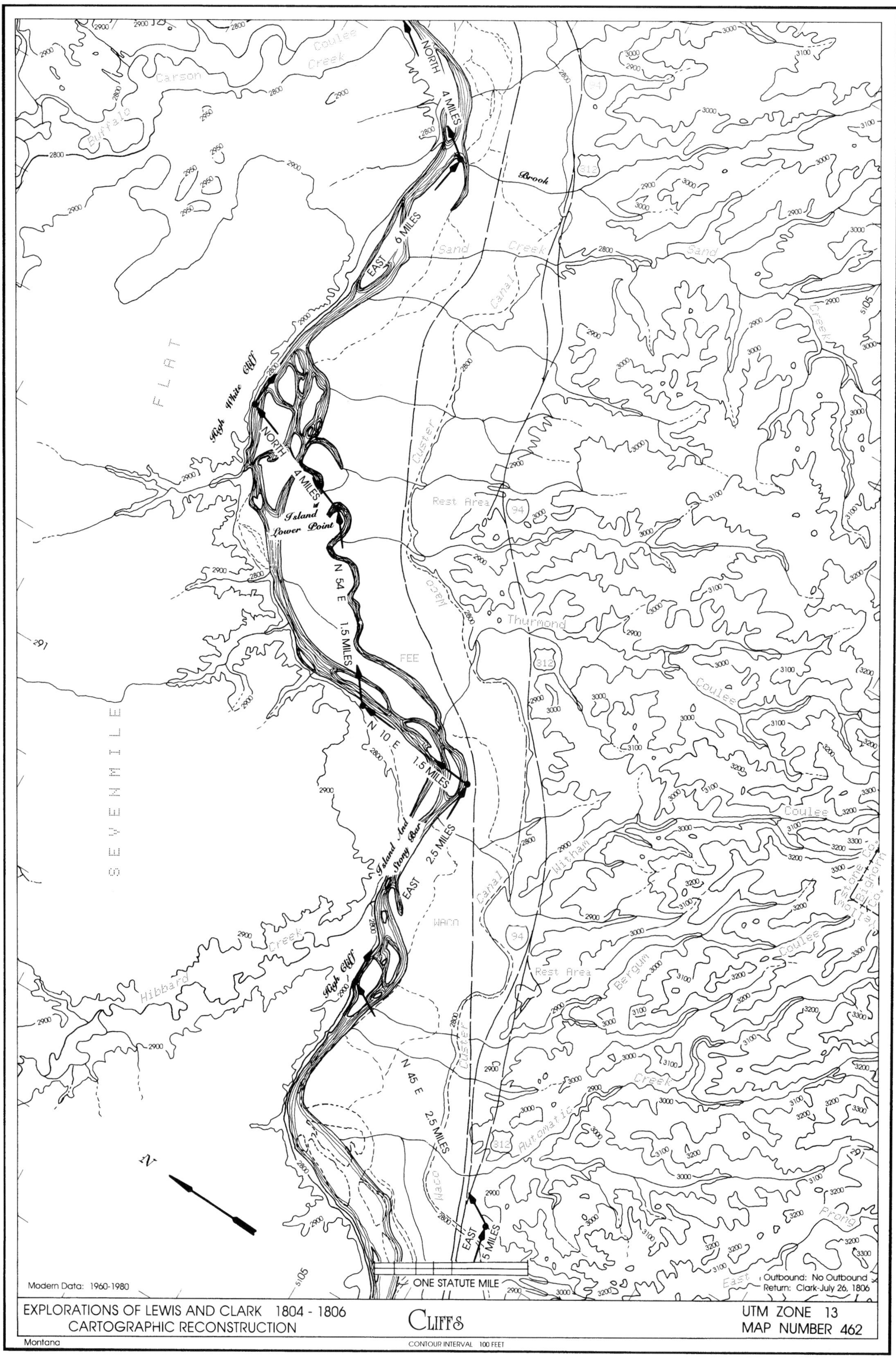

NORTH 4 MILES
Brook
EAST 6 MILES
Sand Creek
High White Cliff
NORTH 4 MILES
Island Lower Point
N 54 E 1.5 MILES
N 10 E
1.5 MILES
Island And Stony Bar
EAST 2.5 MILES
High Cliff
N 45 E 2.5 MILES
EAST 5 MILES
SEVENMILE FLAT
Hibbard Creek
Carson
Buffalo
Coulee Creek
Custer Canal
Thurmond
Rest Area
Witham
Bergum
Automatic Creek
Prong
Modern Data: 1960-1980
ONE STATUTE MILE
Outbound: No Outbound
Return: Clark-July 26, 1806
EXPLORATIONS OF LEWIS AND CLARK 1804 - 1806
CARTOGRAPHIC RECONSTRUCTION
CLIFFS
UTM ZONE 13
MAP NUMBER 462
Montana
CONTOUR INTERVAL 100 FEET

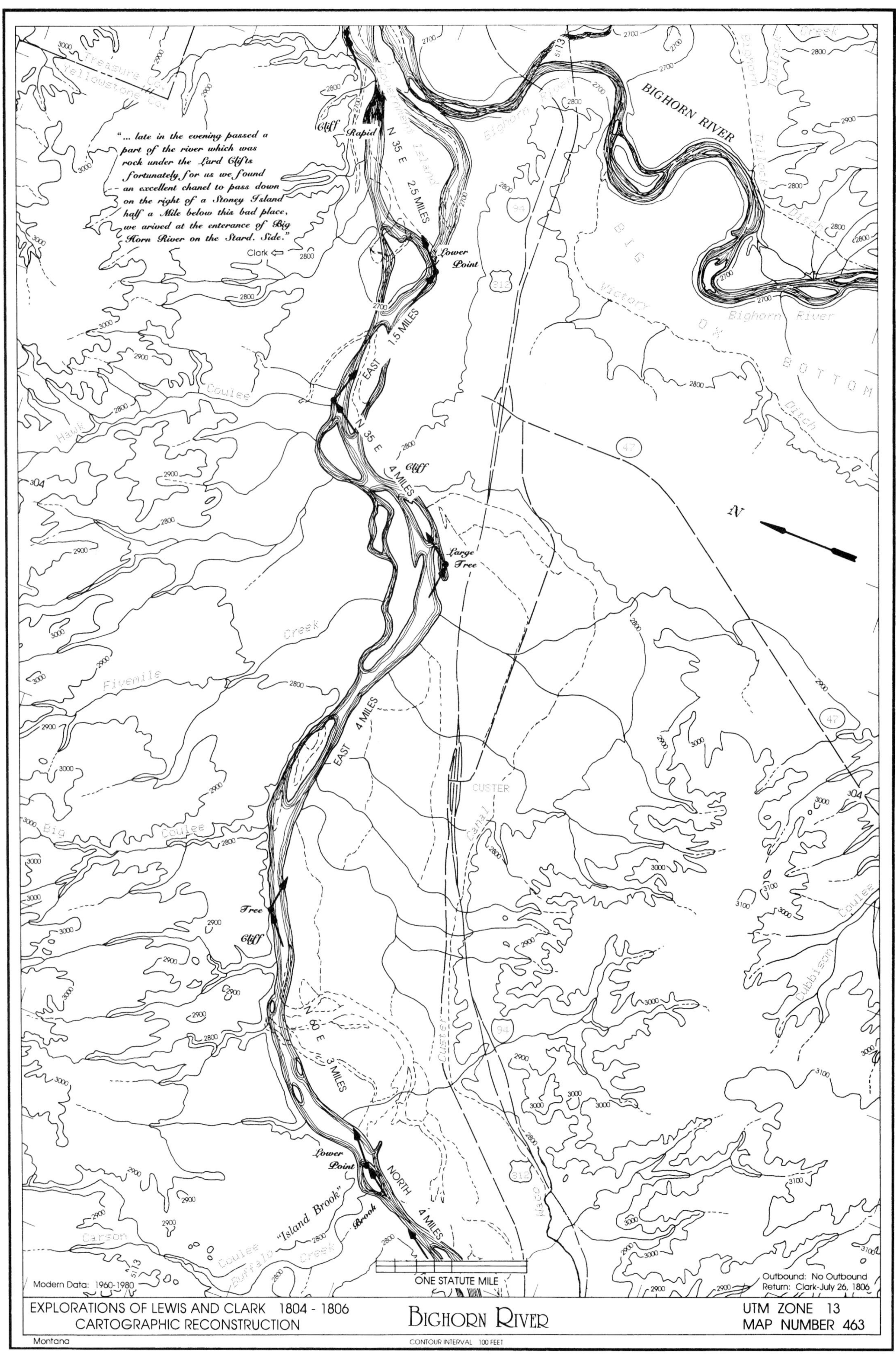
"... late in the evening passed a part of the river which was rock under the Lard Clifts fortunately for us we found an excellent chanel to pass down on the right of a Stoney Island half a Mile below this bad place, we arived at the enterance of Big Horn River on the Stard. Side."
Clark
BIGHORN RIVER
N
Cliff
Rapid
Lower Point
Large Tree
Tree
Cliff
Lower Point
Brook
"Island Brook"
N 35 E 2.5 MILES
EAST 1.5 MILES
N 35 E 4 MILES
EAST 4 MILES
N 60 E 3 MILES
NORTH 4 MILES
ONE STATUTE MILE
Modern Data: 1960-1980
Outbound: No Outbound
Return: Clark-July 26, 1806
EXPLORATIONS OF LEWIS AND CLARK 1804 - 1806
CARTOGRAPHIC RECONSTRUCTION
BIGHORN RIVER
UTM ZONE 13
MAP NUMBER 463
Montana
CONTOUR INTERVAL 100 FEET

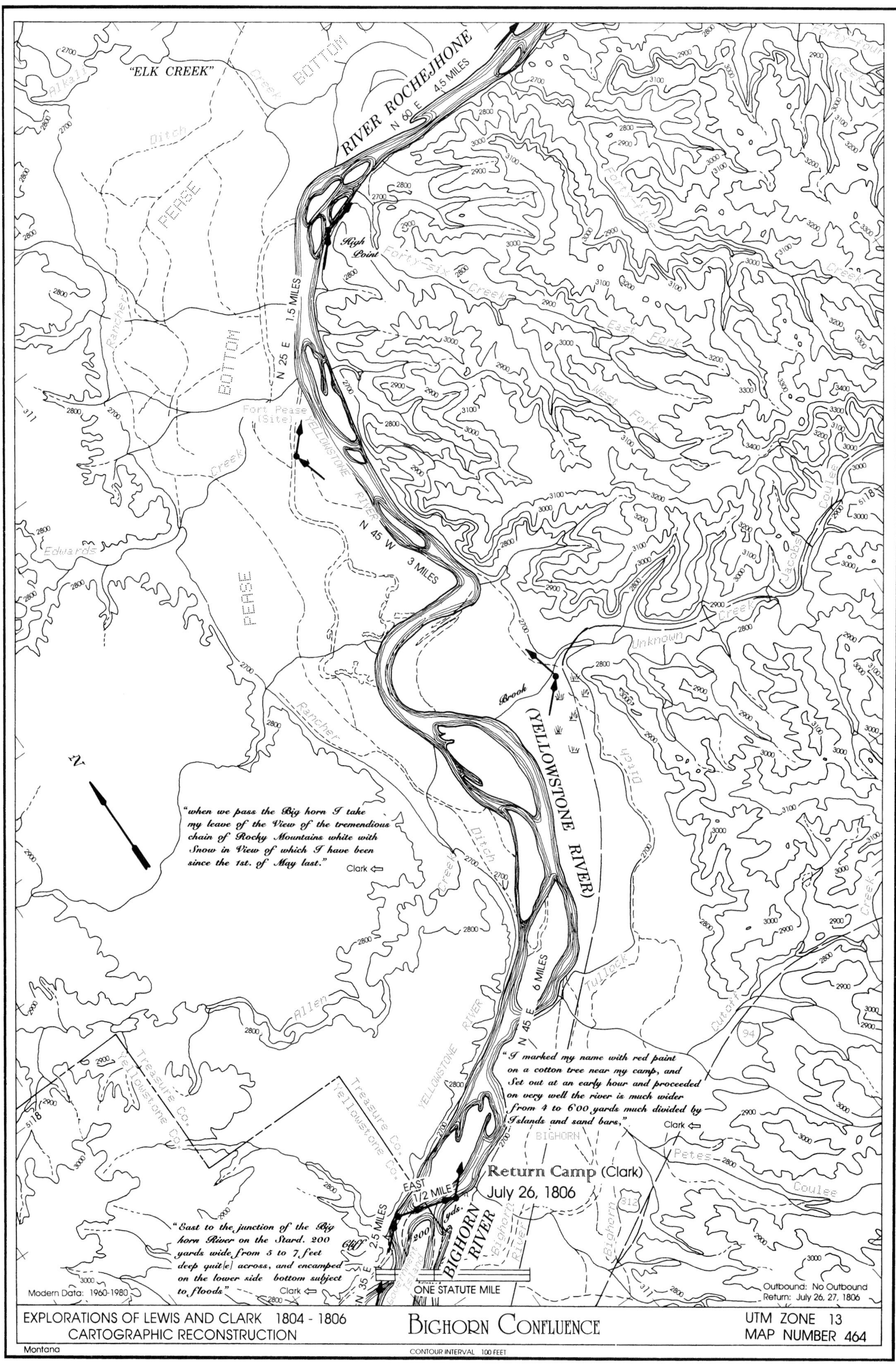
"ELK CREEK"
RIVER ROCHEJHONE
N 60 E 4.5 MILES
High Point
N 25 E 1.5 MILES
Fort Pease (Site)
YELLOWSTONE RIVER
N 45 W 3 MILES
Brook
(YELLOWSTONE RIVER)
N
"when we pass the Big horn I take my leave of the View of the tremendious chain of Rocky Mountains white with Snow in View of which I have been since the 1st. of May last."
Clark ⇦
N 45 E 6 MILES
"I marked my name with red paint on a cotton tree near my camp, and Set out at an early hour and proceeded on very well the river is much wider from 4 to 6'00 yards much divided by Islands and sand bars,"
Clark ⇦
Return Camp (Clark)
July 26, 1806
EAST 1/2 MILE
BIGHORN RIVER
"East to the junction of the Big horn River on the Stard. 200 yards wide from 5 to 7 feet deep quit[e] across, and encamped on the lower side bottom subject to floods"
Clark ⇦
Cliff
200 yds.
2.5 MILES
N 35 E
ONE STATUTE MILE
Modern Data: 1960-1980
Outbound: No Outbound
Return: July 26, 27, 1806
EXPLORATIONS OF LEWIS AND CLARK 1804 - 1806
CARTOGRAPHIC RECONSTRUCTION
BIGHORN CONFLUENCE
UTM ZONE 13
MAP NUMBER 464
Montana
CONTOUR INTERVAL 100 FEET

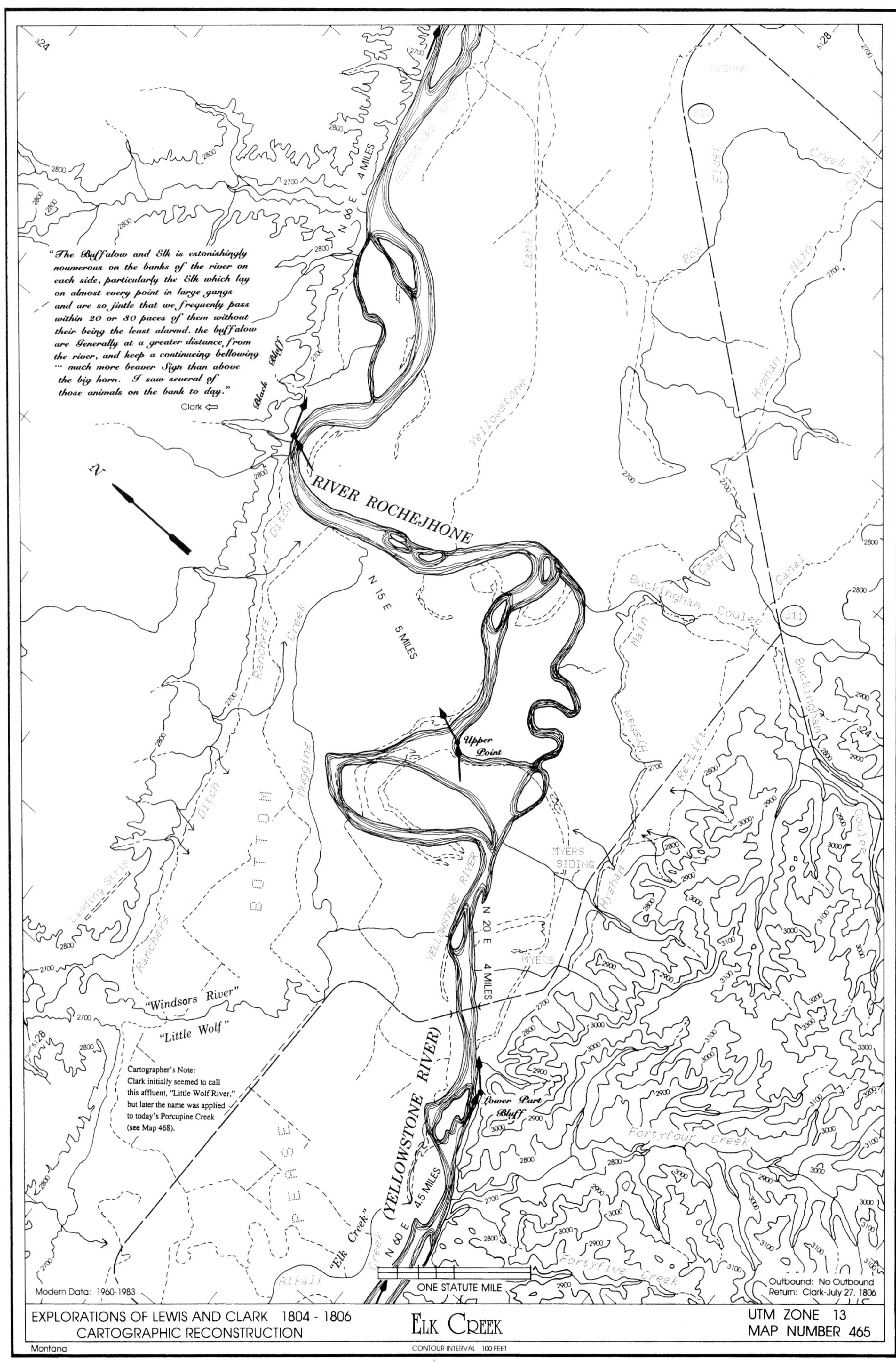
"The Buffalow and Elk is estonishingly noumerous on the banks of the river on each side, particularly the Elk which lay on almost every point in large gangs and are so jintle that we frequenly pass within 20 or 30 paces of them without their being the least alarmd. the buffalow are Generally at a greater distance from the river, and keep a continueing bellowing ··· much more beaver Sign than above the big horn. I saw several of those animals on the bank to day."
Clark ⇦
Black Bluff
RIVER ROCHEJHONE
Upper Point
Lower Part Bluff
"Windsors River"
"Little Wolf"
"Elk Creek"
(YELLOWSTONE RIVER)
Cartographer's Note:
Clark initially seemed to call this affluent, "Little Wolf River," but later the name was applied to today's Porcupine Creek (see Map 468).
N 66 E 4 MILES
N 15 E 5 MILES
N 20 E 4 MILES
N 60 E 4.5 MILES
YELLOWSTONE RIVER
Yellowstone
Canal
Box Elder Creek
Main Canal
Hysham
Buckingham Coulee
Buckingham Canal
Main Canal
Re-Lift
MYERS SIDING
MYERS
Hysham
HYSHAM
Ranchers Ditch
Muggins Creek
BOTTOM
PEASE
Landing Strip
Fortyfour Creek
Fortyfive Creek
Alkali Creek
Buckingham Coulee
ONE STATUTE MILE
Modern Data: 1960-1983
Outbound: No Outbound
Return: Clark-July 27, 1806
EXPLORATIONS OF LEWIS AND CLARK 1804 - 1806
CARTOGRAPHIC RECONSTRUCTION
ELK CREEK
UTM ZONE 13
MAP NUMBER 465
Montana
CONTOUR INTERVAL 100 FEET

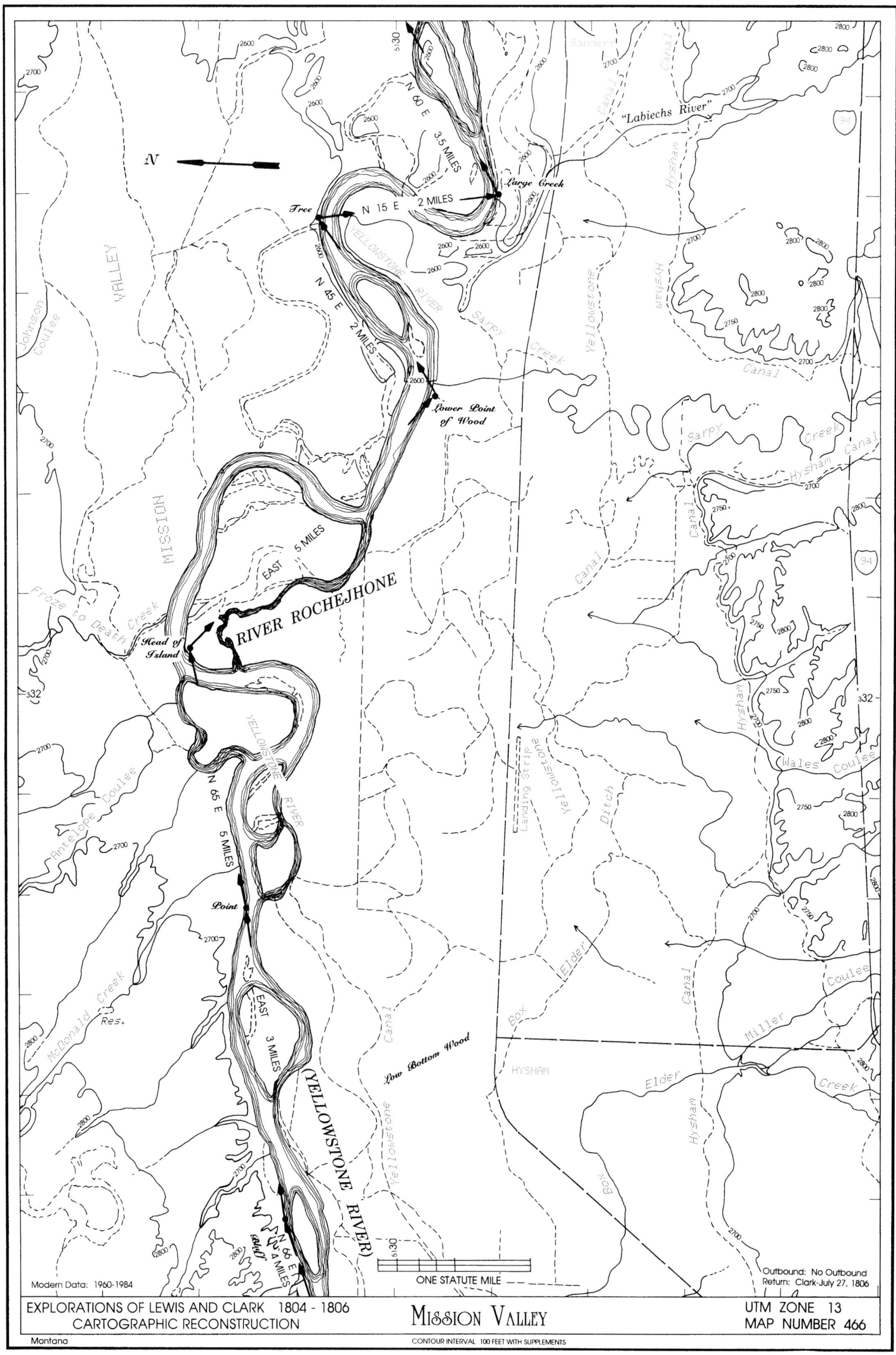
N
"Labiechs River"
Large Creek
N 60 E
3.5 MILES
N 15 E
2 MILES
Tree
N 45 E
2 MILES
YELLOWSTONE RIVER
Lower Point of Wood
EAST
5 MILES
RIVER ROCHEJHONE
Head of Island
N 65 E
5 MILES
Point
EAST
3 MILES
Low Bottom Wood
(YELLOWSTONE RIVER)
N 66 E
4 MILES
Bluff
ONE STATUTE MILE
Modern Data: 1960-1984
Outbound: No Outbound
Return: Clark-July 27, 1806
EXPLORATIONS OF LEWIS AND CLARK 1804 - 1806
CARTOGRAPHIC RECONSTRUCTION
MISSION VALLEY
UTM ZONE 13
MAP NUMBER 466
Montana
CONTOUR INTERVAL 100 FEET WITH SUPPLEMENTS

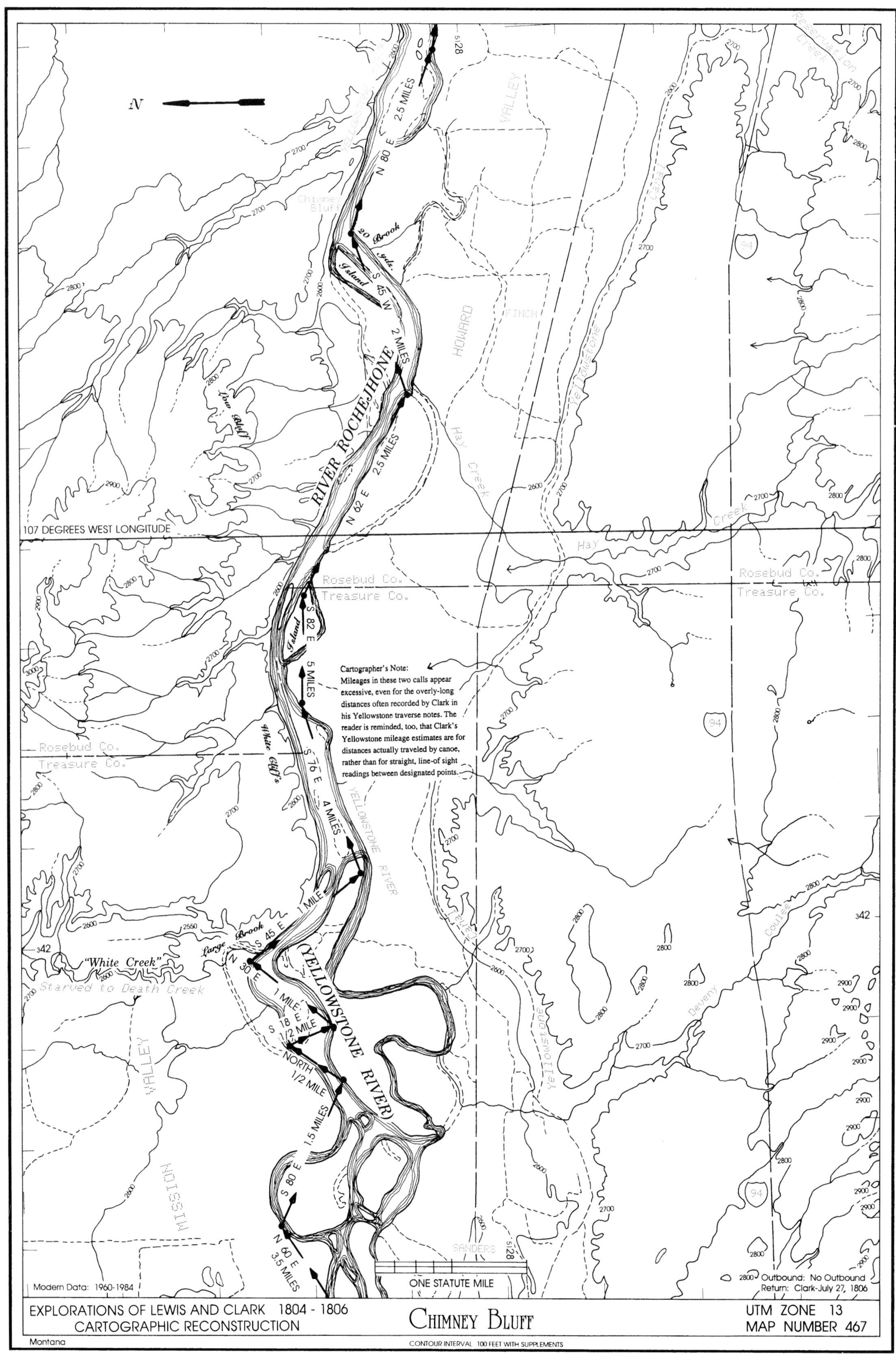
N
RIVER ROCHEJHONE
(YELLOWSTONE RIVER)
Chimney Bluff
Low Bluff
White Cliff's
"White Creek"
Large Brook
20 Brook yds.
Island
107 DEGREES WEST LONGITUDE
Rosebud Co.
Treasure Co.
Starved to Death Creek
Hay Creek
Cartographer's Note:
Mileages in these two calls appear excessive, even for the overly-long distances often recorded by Clark in his Yellowstone traverse notes. The reader is reminded, too, that Clark's Yellowstone mileage estimates are for distances actually traveled by canoe, rather than for straight, line-of sight readings between designated points.
ONE STATUTE MILE
Modern Data: 1960-1984
Outbound: No Outbound
Return: Clark-July 27, 1806
EXPLORATIONS OF LEWIS AND CLARK 1804 - 1806
CARTOGRAPHIC RECONSTRUCTION
CHIMNEY BLUFF
UTM ZONE 13
MAP NUMBER 467
Montana
CONTOUR INTERVAL 100 FEET WITH SUPPLEMENTS

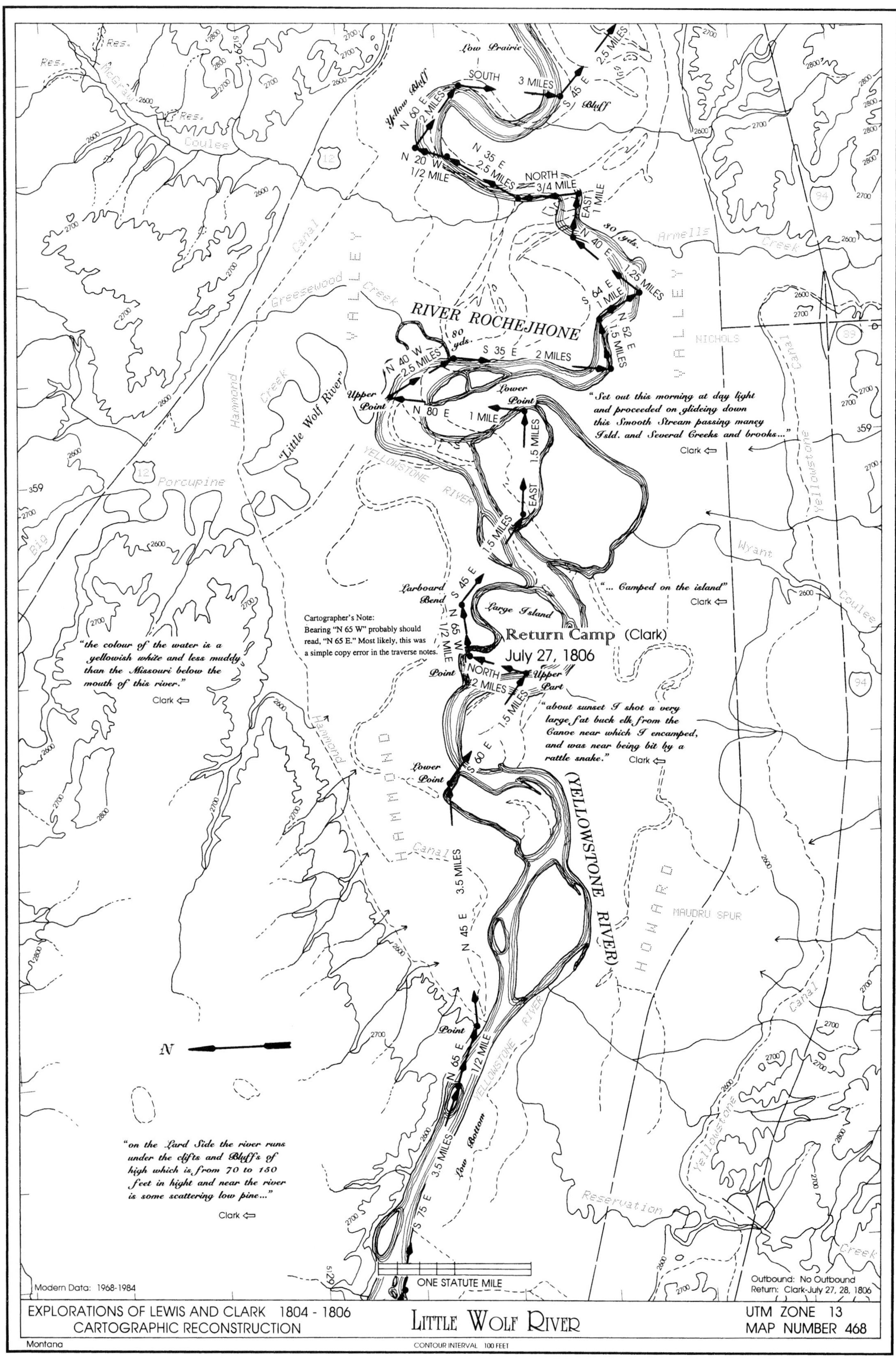
EXPLORATIONS OF LEWIS AND CLARK 1804 - 1806
CARTOGRAPHIC RECONSTRUCTION
LITTLE WOLF RIVER
UTM ZONE 13
MAP NUMBER 468
Montana
CONTOUR INTERVAL 100 FEET
Modern Data: 1968-1984
ONE STATUTE MILE
Outbound: No Outbound
Return: Clark-July 27, 28, 1806
RIVER ROCHEJHONE
(YELLOWSTONE RIVER)
Return Camp (Clark)
July 27, 1806
"Little Wolf River"
Low Prairie
Yellow Bluff
Bluff
Upper Point
Lower Point
Larboard Bend
Large Island
Point
Upper Part
Lower Point
Point
Low Bottom
80 yds.
80 yds.
"Set out this morning at day light and proceeded on glideing down this Smooth Stream passing maney Isld. and Several Creeks and brooks..."
Clark ⇦
"... Camped on the island"
Clark ⇦
"about sunset I shot a very large fat buck elk from the Canoe near which I encamped, and was near being bit by a rattle snake."
Clark ⇦
"the colour of the water is a yellowish white and less muddy than the Missouri below the mouth of this river."
Clark ⇦
"on the Lard Side the river runs under the clifts and Bluff's of high which is from 70 to 150 feet in hight and near the river is some scattering low pine..."
Clark ⇦
Cartographer's Note:
Bearing "N 65 W" probably should read, "N 65 E." Most likely, this was a simple copy error in the traverse notes.
SOUTH
3 MILES
S 45 E
2.5 MILES
N 60 E
2 MILES
N 20 W
1/2 MILE
N 35 E
2.5 MILES
NORTH
3/4 MILE
EAST
1 MILE
N 40 E
1.25 MILES
S 64 E
1 MILE
N 52 E
1.5 MILES
S 35 E
2 MILES
N 40 W
2.5 MILES
N 80 E
1 MILE
1.5 MILES
EAST
1.5 MILES
S 45 E
N 65 W
1/2 MILE
NORTH
2 MILES
1.5 MILES
S 60 E
3.5 MILES
N 45 E
N 65 E
1/2 MILE
3.5 MILES
S 75 E
N
YELLOWSTONE RIVER
Armells Creek
Greasewood Creek
Hammond Creek
Porcupine
Big
McGraw Coulee
Res.
VALLEY
NICHOLS
HAMMOND
Canal
HOWARD
MAUDRU SPUR
Wyant Coulee
Yellowstone Canal
Reservation Creek
12
94
39
359

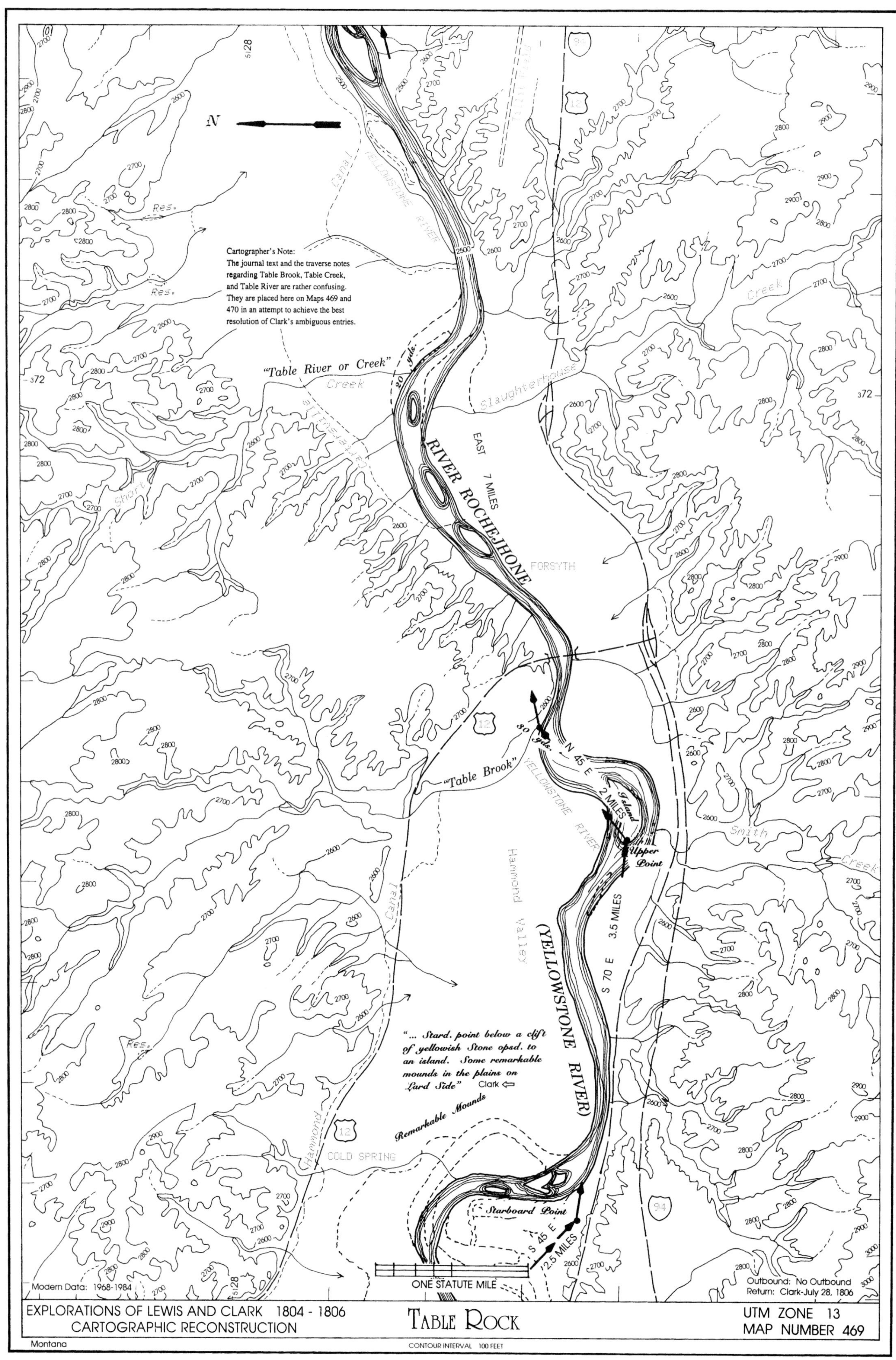
N
Cartographer's Note:
The journal text and the traverse notes regarding Table Brook, Table Creek, and Table River are rather confusing. They are placed here on Maps 469 and 470 in an attempt to achieve the best resolution of Clark's ambiguous entries.
"Table River or Creek"
20 yds.
RIVER ROCHEJHONE
EAST 7 MILES
FORSYTH
Slaughterhouse
Creek
YELLOWSTONE RIVER
"Table Brook"
30 yds.
N 45 E
2 MILES
Island
Upper Point
3.5 MILES
S 70 E
(YELLOWSTONE RIVER)
Hammond Valley
"... Stard. point below a clift of yellowish Stone opsd. to an island. Some remarkable mounds in the plains on Lard Side" Clark ⇦
Remarkable Mounds
COLD SPRING
Starboard Point
S 45 E
2.5 MILES
ONE STATUTE MILE
Modern Data: 1968-1984
Outbound: No Outbound
Return: Clark-July 28, 1806
EXPLORATIONS OF LEWIS AND CLARK 1804 - 1806
CARTOGRAPHIC RECONSTRUCTION
TABLE ROCK
UTM ZONE 13
MAP NUMBER 469
Montana
CONTOUR INTERVAL 100 FEET

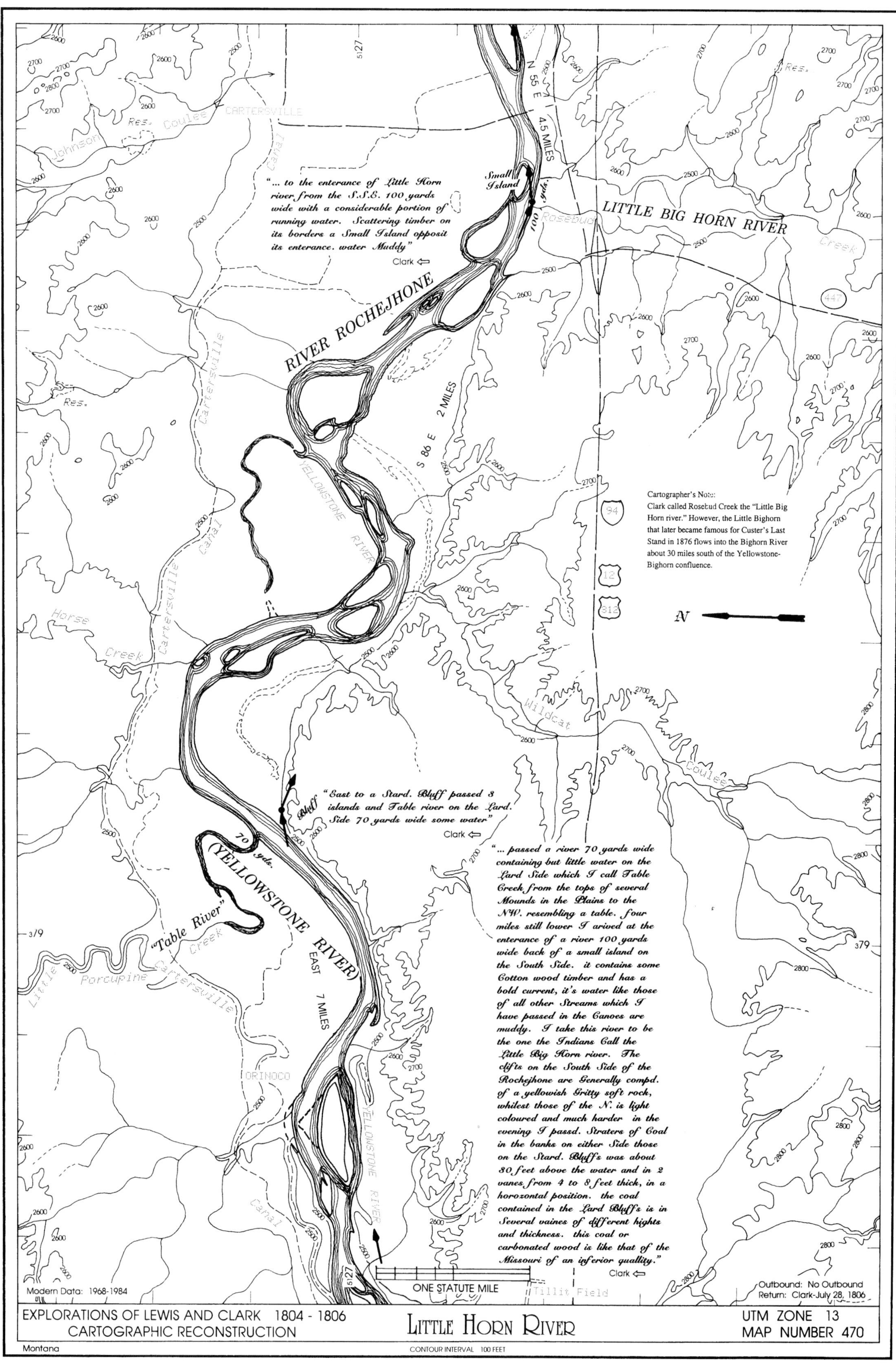
"... to the enterance of Little Horn river from the S.S.E. 100 yards wide with a considerable portion of running water. Scattering timber on its borders a Small Island opposit its enterance. water Muddy"
Clark ⇦
Small Island
100 yds.
N 55 E 4.5 MILES
LITTLE BIG HORN RIVER
Rosebud
Creek
RIVER ROCHEJHONE
S 86 E 2 MILES
YELLOWSTONE RIVER
Cartographer's Note:
Clark called Rosebud Creek the "Little Big Horn river." However, the Little Bighorn that later became famous for Custer's Last Stand in 1876 flows into the Bighorn River about 30 miles south of the Yellowstone-Bighorn confluence.
N
Johnson
Res.
Coulee
CARTERSVILLE
Canal
Cartersville
Horse
Creek
Wildcat
Coulee
"East to a Stard. Bluff passed 3 islands and Table river on the Lard. Side 70 yards wide some water"
Clark ⇦
Bluff
70 yds.
(YELLOWSTONE RIVER)
"Table River"
Creek
Little
Porcupine
Cartersville
EAST 7 MILES
"... passed a river 70 yards wide containing but little water on the Lard Side which I call Table Creek from the tops of several Mounds in the Plains to the N.W. resembling a table. four miles still lower I arived at the enterance of a river 100 yards wide back of a small island on the South Side. it contains some Cotton wood timber and has a bold current, it's water like those of all other Streams which I have passed in the Canoes are muddy. I take this river to be the one the Indians Call the Little Big Horn river. The clifts on the South Side of the Rochejhone are Generally compd. of a yellowish Gritty soft rock, whilest those of the N. is light coloured and much harder in the evening I passd. Straters of Coal in the banks on either Side those on the Stard. Bluffs was about 30 feet above the water and in 2 vanes from 4 to 8 feet thick, in a horozontal position. the coal contained in the Lard Bluffs is in Several vaines of different hights and thickness. this coal or carbonated wood is like that of the Missouri of an inferior quallity."
Clark ⇦
ORINOCO
Canal
ONE STATUTE MILE
Tillit Field
Modern Data: 1968-1984
Outbound: No Outbound
Return: Clark-July 28, 1806
EXPLORATIONS OF LEWIS AND CLARK 1804 - 1806
CARTOGRAPHIC RECONSTRUCTION
LITTLE HORN RIVER
UTM ZONE 13
MAP NUMBER 470
Montana
CONTOUR INTERVAL 100 FEET

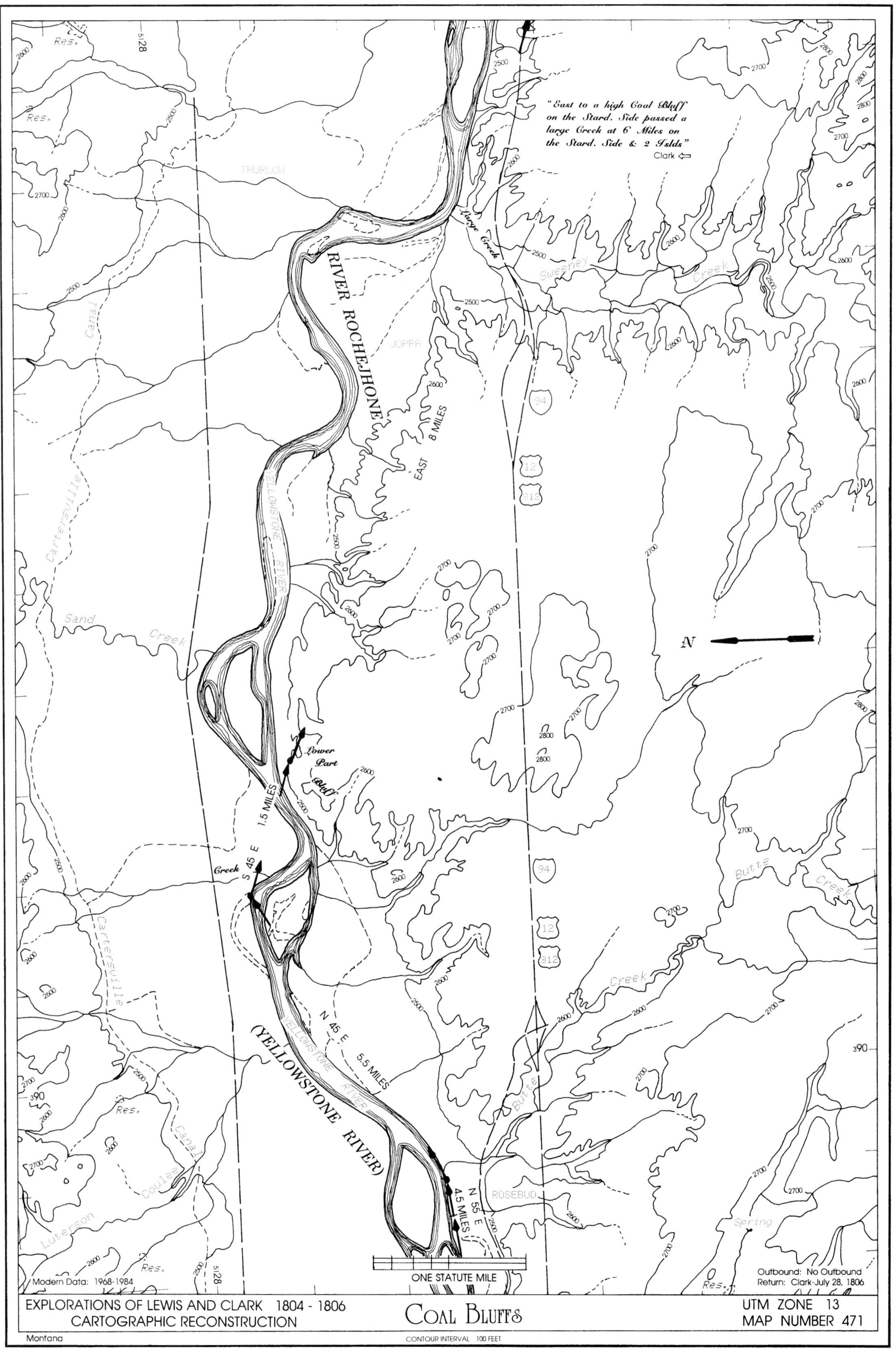
"East to a high Coal Bluff on the Stard. Side passed a large Creek at 6 Miles on the Stard. Side & 2 Islds"
Clark
RIVER ROCHEJHONE
(YELLOWSTONE RIVER)
YELLOWSTONE RIVER
EAST 8 MILES
1.5 MILES
S 45 E
N 45 E 5.5 MILES
N 55 E 4.5 MILES
Lower Part Bluff
Large Creek
Creek
Sweeney Creek
Butte Creek
Sand Creek
Cartersville
Canal
Coulee
Luterson
Res.
Spring
THURLOW
JOPPA
ROSEBUD
94
12
812
N
ONE STATUTE MILE
Modern Data: 1968-1984
Outbound: No Outbound
Return: Clark-July 28, 1806
EXPLORATIONS OF LEWIS AND CLARK 1804 - 1806
CARTOGRAPHIC RECONSTRUCTION
COAL BLUFFS
UTM ZONE 13
MAP NUMBER 471
Montana
CONTOUR INTERVAL 100 FEET

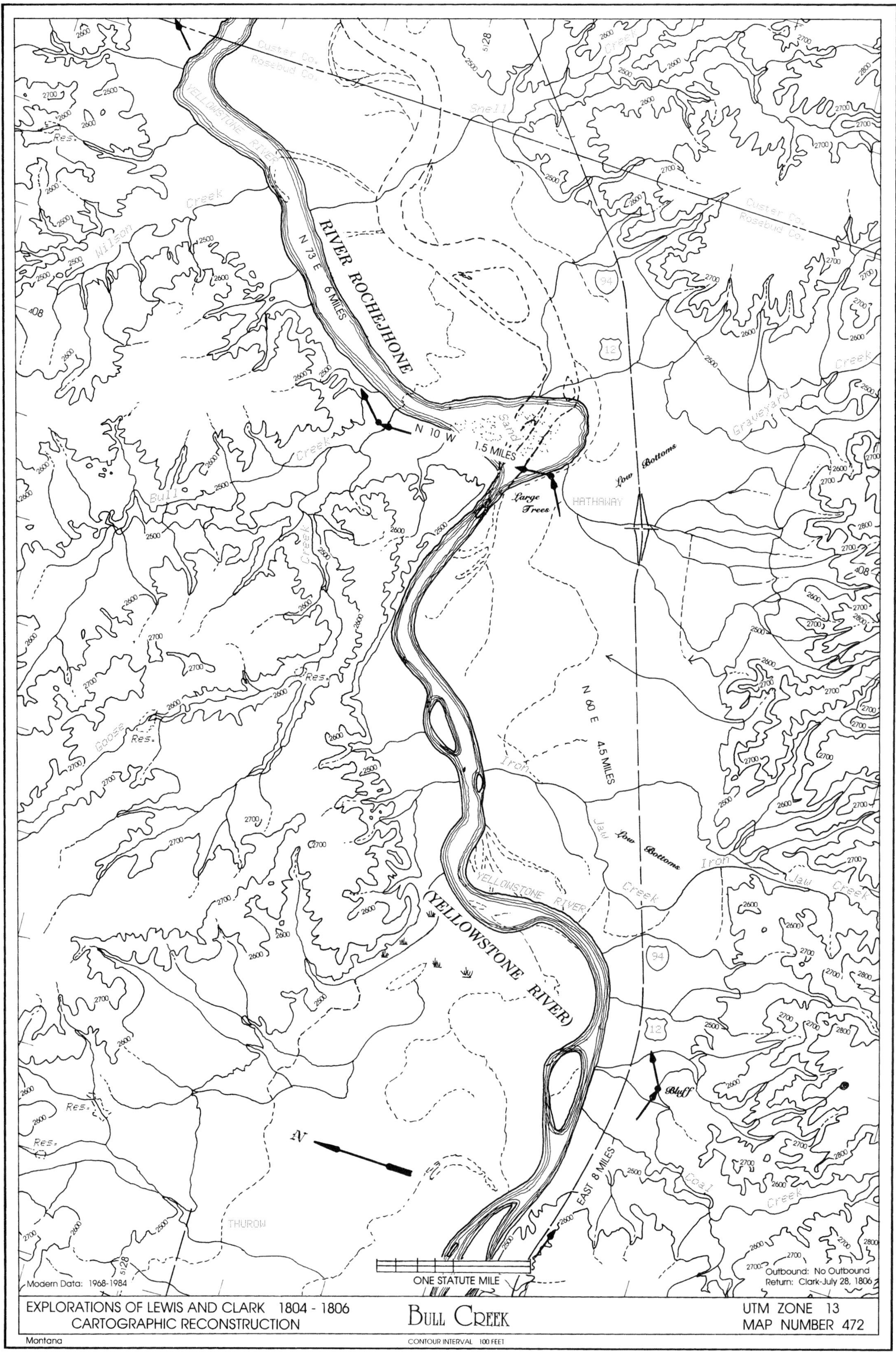
RIVER ROCHEJHONE
N 73 E 6 MILES
N 10 W
1.5 MILES
Large Trees
Low Bottoms
HATHAWAY
N 60 E 4.5 MILES
Low Bottoms
(YELLOWSTONE RIVER)
YELLOWSTONE RIVER
Bluff
EAST 8 MILES
N
ONE STATUTE MILE
Modern Data: 1968-1984
Outbound: No Outbound
Return: Clark-July 28, 1806
EXPLORATIONS OF LEWIS AND CLARK 1804 - 1806
CARTOGRAPHIC RECONSTRUCTION
BULL CREEK
UTM ZONE 13
MAP NUMBER 472
Montana
CONTOUR INTERVAL 100 FEET

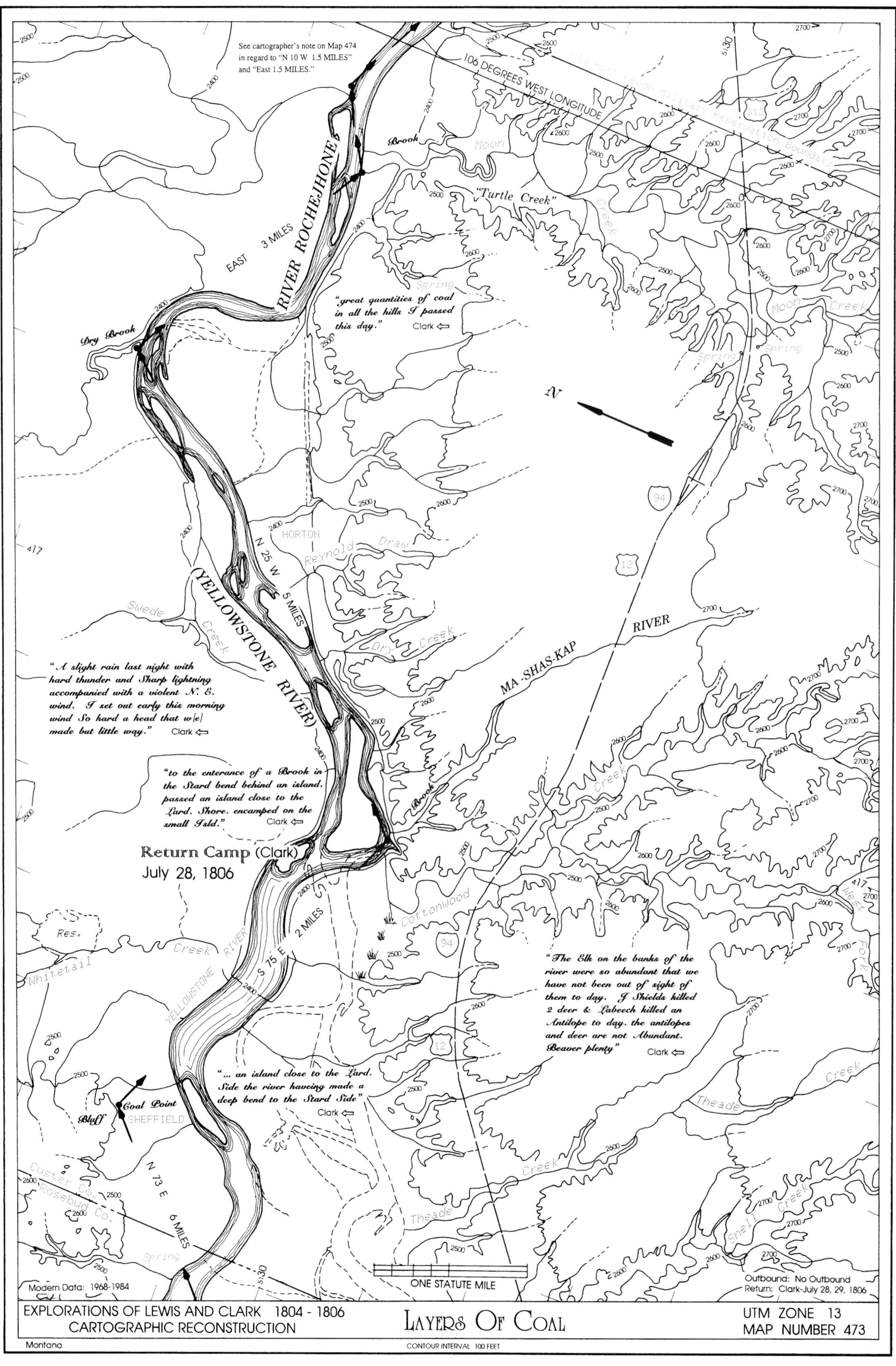

See cartographer's note on Map 474 in regard to "N 10 W 1.5 MILES" and "East 1.5 MILES."
106 DEGREES WEST LONGITUDE
RIVER ROCHEJHONE
EAST 3 MILES
Brook
Moon
"Turtle Creek"
"great quantities of coal in all the hills I passed this day." Clark ⇦
Dry Brook
N
HORTON
Reynold Draw
N 25 W 5 MILES
(YELLOWSTONE RIVER)
Swede Creek
Dry Creek
MA-SHAS-KAP RIVER
"A slight rain last night with hard thunder and Sharp lightning accompanied with a violent N. E. wind. I set out early this morning wind So hard a head that w[e] made but little way." Clark ⇦
"to the enterance of a Brook in the Stard bend behind an island. passed an island close to the Lard. Shore. encamped on the small Isld." Clark ⇦
Return Camp (Clark)
July 28, 1806
Cottonwood
Res.
Whitetail Creek
YELLOWSTONE RIVER
S 75 E 2 MILES
"The Elk on the banks of the river were so abundant that we have not been out of sight of them to day. J Shields killed 2 deer & Labeech killed an Antilope to day. the antilopes and deer are not Abundant. Beaver plenty" Clark ⇦
"... an island close to the Lard. Side the river haveing made a deep bend to the Stard Side" Clark ⇦
Coal Point
Bluff
SHEFFIELD
N 73 E 6 MILES
Custer Co.
Rosebud Co.
Theade Creek
Shell Creek
Modern Data: 1968-1984
ONE STATUTE MILE
Outbound: No Outbound
Return: Clark-July 28, 29, 1806
EXPLORATIONS OF LEWIS AND CLARK 1804 - 1806
CARTOGRAPHIC RECONSTRUCTION
LAYERS OF COAL
UTM ZONE 13
MAP NUMBER 473
Montana
CONTOUR INTERVAL 100 FEET

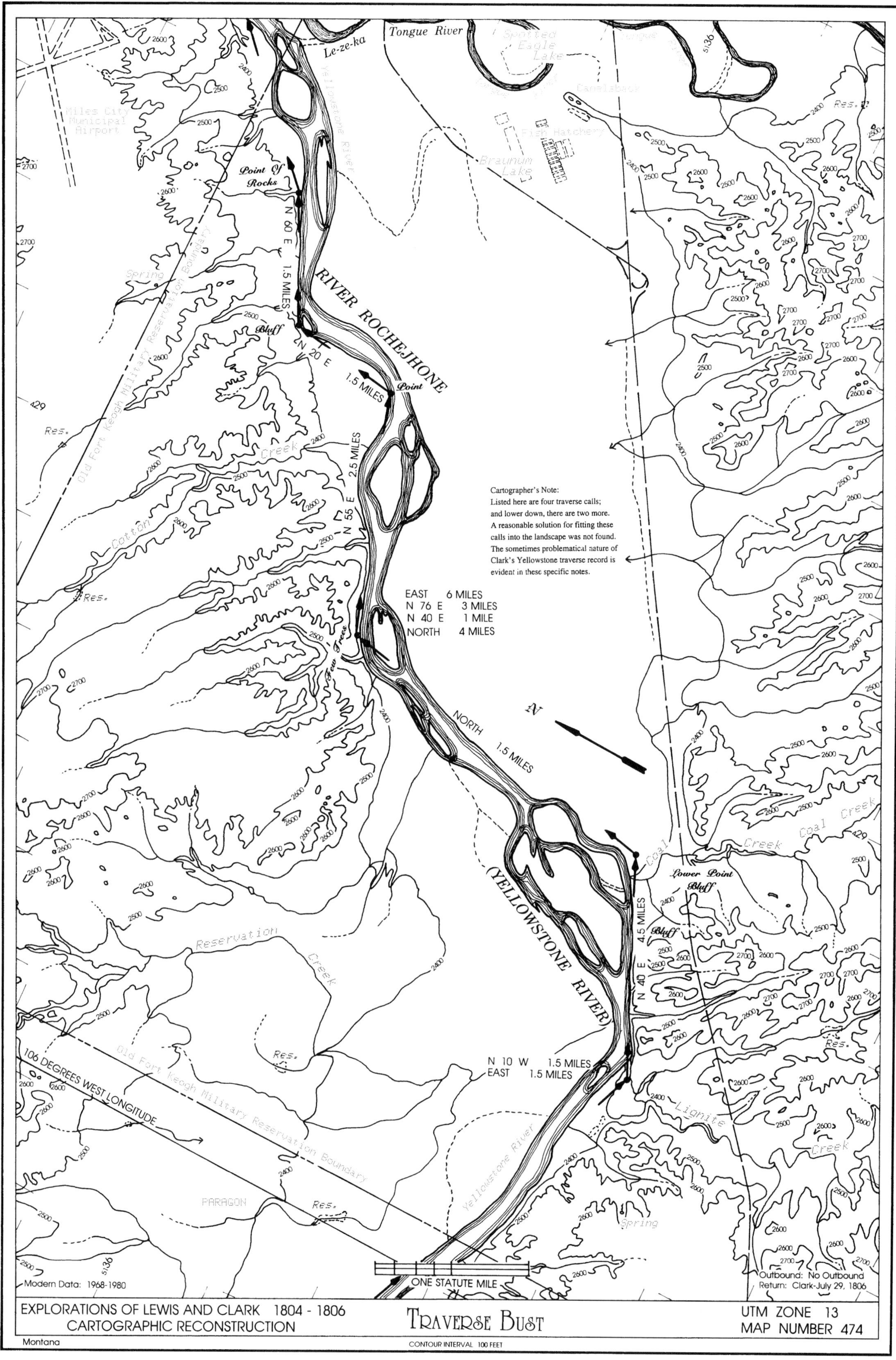
Tongue River
Le-ze-ka
Yellowstone River
Point Of Rocks
N 60 E 1.5 MILES
Bluff
N 20 E
RIVER ROCHEJHONE
1.5 MILES
Point
2.5 MILES
N 55 E
Cartographer's Note:
Listed here are four traverse calls; and lower down, there are two more. A reasonable solution for fitting these calls into the landscape was not found. The sometimes problematical nature of Clark's Yellowstone traverse record is evident in these specific notes.
EAST 6 MILES
N 76 E 3 MILES
N 40 E 1 MILE
NORTH 4 MILES
Yew Trees
NORTH 1.5 MILES
N
YELLOWSTONE RIVER
Lower Point
Bluff
Bluff
N 40 E 4.5 MILES
N 10 W 1.5 MILES
EAST 1.5 MILES
106 DEGREES WEST LONGITUDE
ONE STATUTE MILE
Modern Data: 1968-1980
Outbound: No Outbound
Return: Clark-July 29, 1806
EXPLORATIONS OF LEWIS AND CLARK 1804 - 1806
CARTOGRAPHIC RECONSTRUCTION
TRAVERSE BUST
UTM ZONE 13
MAP NUMBER 474
Montana
CONTOUR INTERVAL 100 FEET

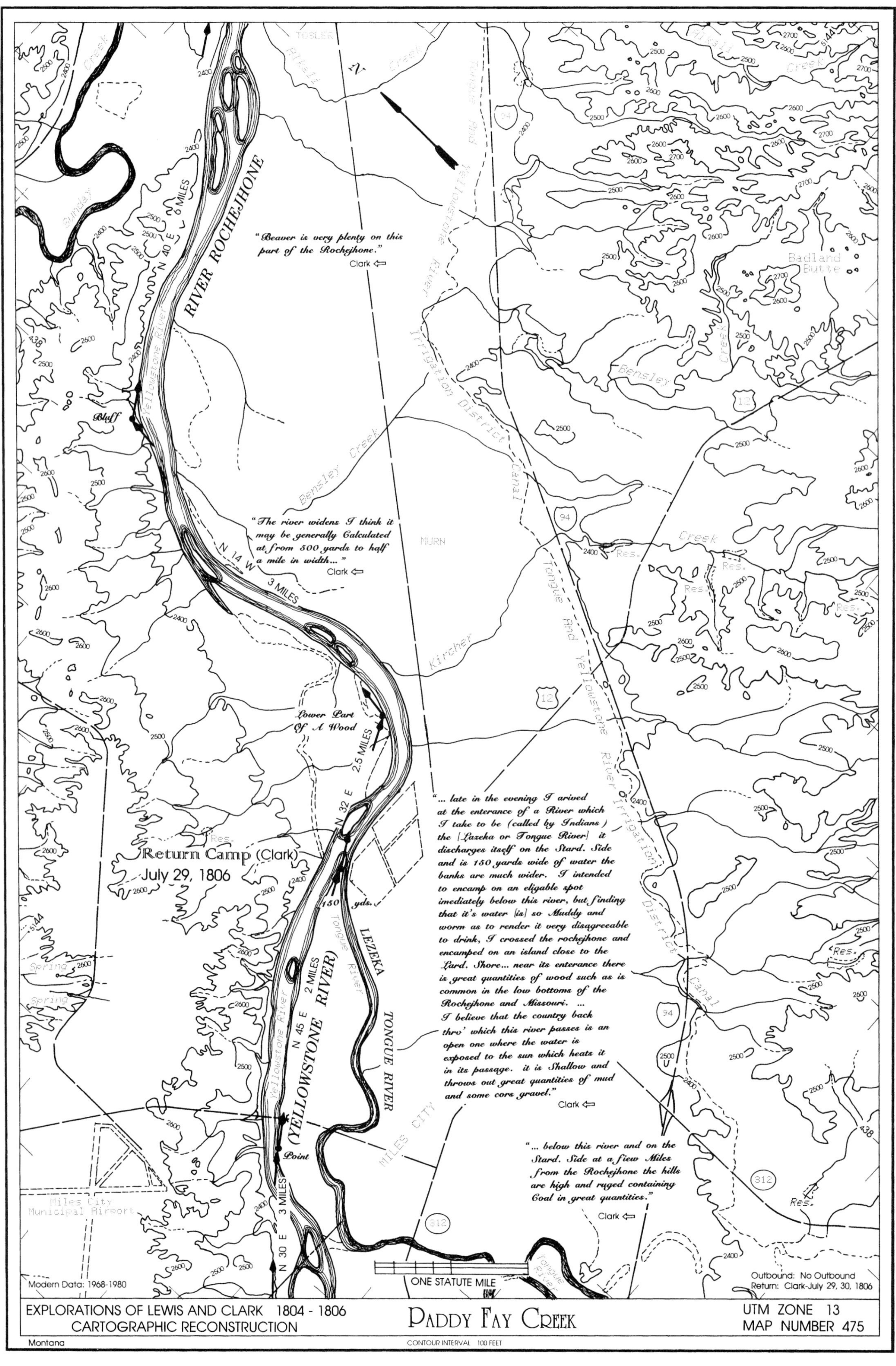

"Beaver is very plenty on this part of the Rochejhone."
Clark ⇦
RIVER ROCHEJHONE
"The river widens I think it may be generally Calculated at from 500 yards to half a mile in width..."
Clark ⇦
Bluff
Lower Part Of A Wood
Return Camp (Clark)
July 29, 1806
(YELLOWSTONE RIVER)
LEZEKA
TONGUE RIVER
Point
"... late in the evening I arived at the enterance of a River which I take to be (called by Indians) the [Lazeka or Tongue River] it discharges itself on the Stard. Side and is 150 yards wide of water the banks are much wider. I intended to encamp on an eligable spot imediately below this river, but finding that it's water [is] so Muddy and worm as to render it very disagreeable to drink, I crossed the rochejhone and encamped on an island close to the Lard. Shore... near its enterance there is great quantities of wood such as is common in the low bottoms of the Rochejhone and Missouri. ... I believe that the country back thro' which this river passes is an open one where the water is exposed to the sun which heats it in its passage. it is Shallow and throws out great quantities of mud and some cors gravel."
Clark ⇦
"... below this river and on the Stard. Side at a fiew Miles from the Rochejhone the hills are high and ruged containing Coal in great quantities."
Clark ⇦
Badland Butte
Miles City Municipal Airport
MILES CITY
ONE STATUTE MILE
Modern Data: 1968-1980
Outbound: No Outbound
Return: Clark-July 29, 30, 1806
EXPLORATIONS OF LEWIS AND CLARK 1804 - 1806
CARTOGRAPHIC RECONSTRUCTION
PADDY FAY CREEK
UTM ZONE 13
MAP NUMBER 475
Montana
CONTOUR INTERVAL 100 FEET

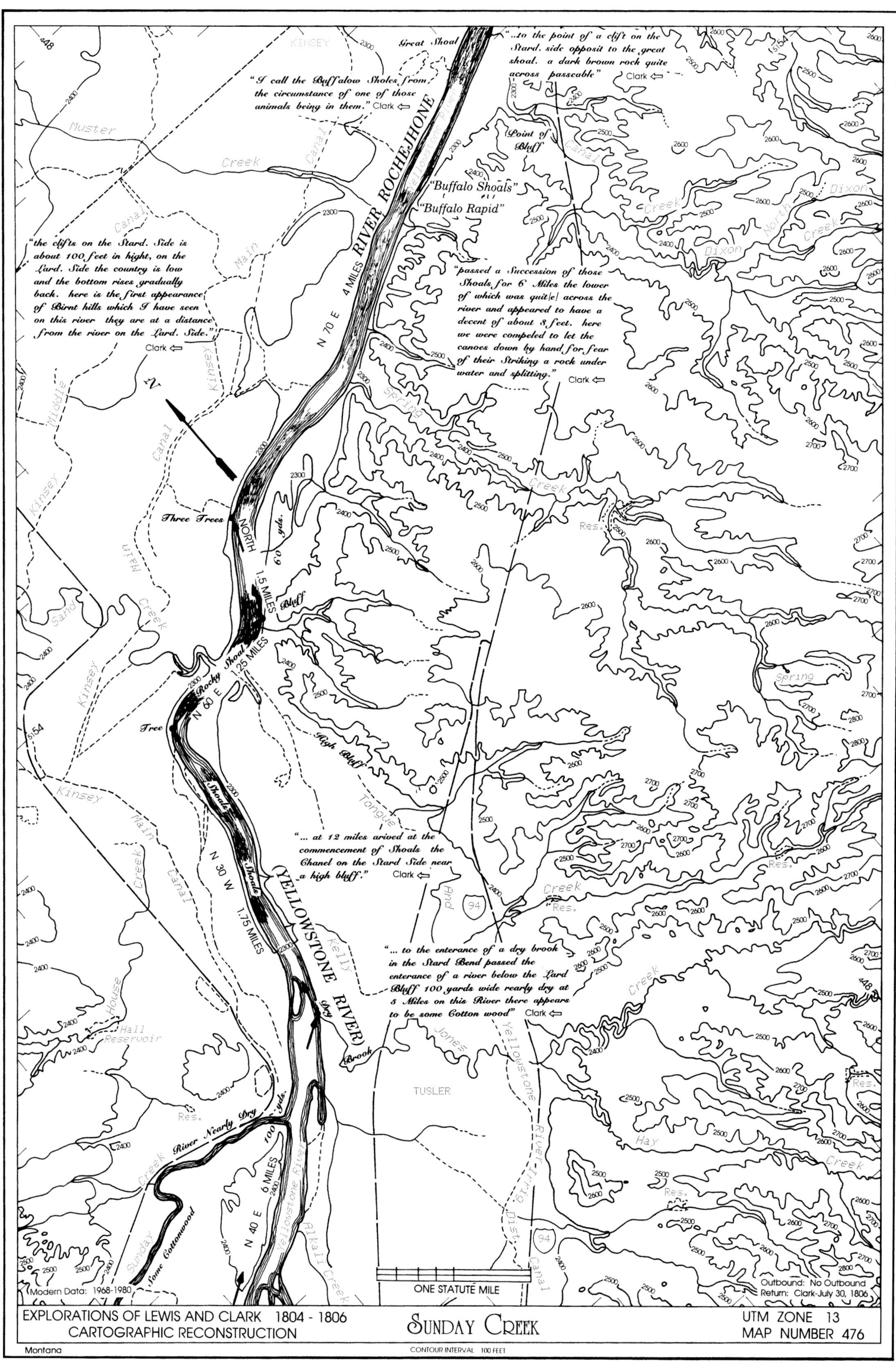
"..to the point of a clift on the Stard. side opposit to the great shoal. a dark brown rock quite across passcable" Clark
Great Shoal
"I call the Buffalow Sholes from the circumstance of one of those animals being in them." Clark
Point of Bluff
"Buffalo Shoals"
"Buffalo Rapid"
RIVER ROCHEJHONE
N 70 E 4 MILES
"the clifts on the Stard. Side is about 100 feet in hight, on the Lard. Side the country is low and the bottom rises gradually back. here is the first appearance of Birnt hills which I have seen on this river they are at a distance from the river on the Lard. Side." Clark
"passed a Succession of those Shoals for 6' Miles the lower of which was quit[e] across the river and appeared to have a decent of about 3 feet. here we were compeled to let the canoes down by hand for fear of their Striking a rock under water and splitting." Clark
Three Trees
NORTH
60 yds.
1.5 MILES
Bluff
Rocky Shoal
1.25 MILES
N 60 E
Tree
High Bluff
Shoals
N 30 W
1.75 MILES
(YELLOWSTONE RIVER)
"... at 12 miles arived at the commencement of Shoals the Chanel on the Stard Side near a high bluff." Clark
"... to the enterance of a dry brook in the Stard Bend passed the enterance of a river below the Lard Bluff 100 yards wide nearly dry at 5 Miles on this River there appears to be some Cotton wood" Clark
Dry Brook
TUSLER
River Nearly Dry
100 yds.
N 40 E
6 MILES
Some Cottonwood
ONE STATUTE MILE
Modern Data: 1968-1980
Outbound: No Outbound
Return: Clark-July 30, 1806
EXPLORATIONS OF LEWIS AND CLARK 1804 - 1806
CARTOGRAPHIC RECONSTRUCTION
SUNDAY CREEK
UTM ZONE 13
MAP NUMBER 476
Montana
CONTOUR INTERVAL 100 FEET

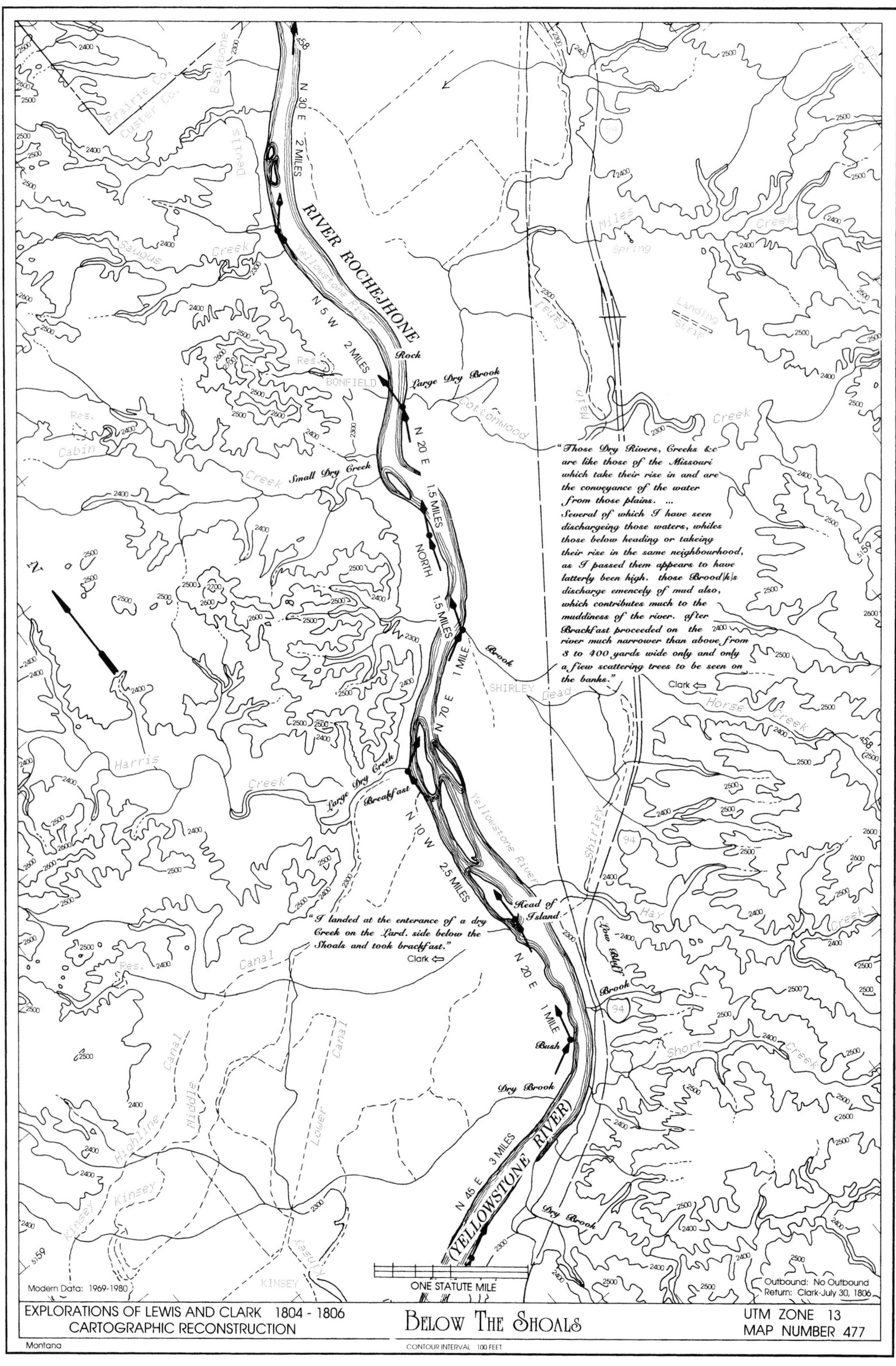
RIVER ROCHEJHONE
(YELLOWSTONE RIVER)
"Those Dry Rivers, Creeks &c are like those of the Missouri which take their rise in and are the conveyance of the water from those plains. ... Several of which I have seen dischargeing those waters, whiles those below heading or takeing their rise in the same neighbourhood, as I passed them appears to have latterly been high. those Brood(k)s discharge emencely of mud also, which contributes much to the muddiness of the river. after Brackfast proceeded on the river much narrower than above from 3 to 400 yards wide only and only a fiew scattering trees to be seen on the banks."
Clark ⇦
"I landed at the enterance of a dry Creek on the Lard. side below the Shoals and took brackfast."
Clark ⇦
Rock
Large Dry Brook
Small Dry Creek
Brook
Large Dry Creek
Breakfast
Head of Island
Low Bluff
Brook
Bush
Dry Brook
Dry Brook
N 30 E 2 MILES
N 5 W 2 MILES
N 20 E 1.5 MILES
NORTH 1.5 MILES
N 70 E 1 MILE
N 10 W 2.5 MILES
N 20 E 1 MILE
N 45 E 3 MILES
BONFIELD
SHIRLEY
KINSEY
Yellowstone River
Prairie Co.
Custer Co.
ONE STATUTE MILE
Modern Data: 1969-1980
Outbound: No Outbound
Return: Clark-July 30, 1806
EXPLORATIONS OF LEWIS AND CLARK 1804 - 1806
CARTOGRAPHIC RECONSTRUCTION
BELOW THE SHOALS
UTM ZONE 13
MAP NUMBER 477
Montana
CONTOUR INTERVAL 100 FEET

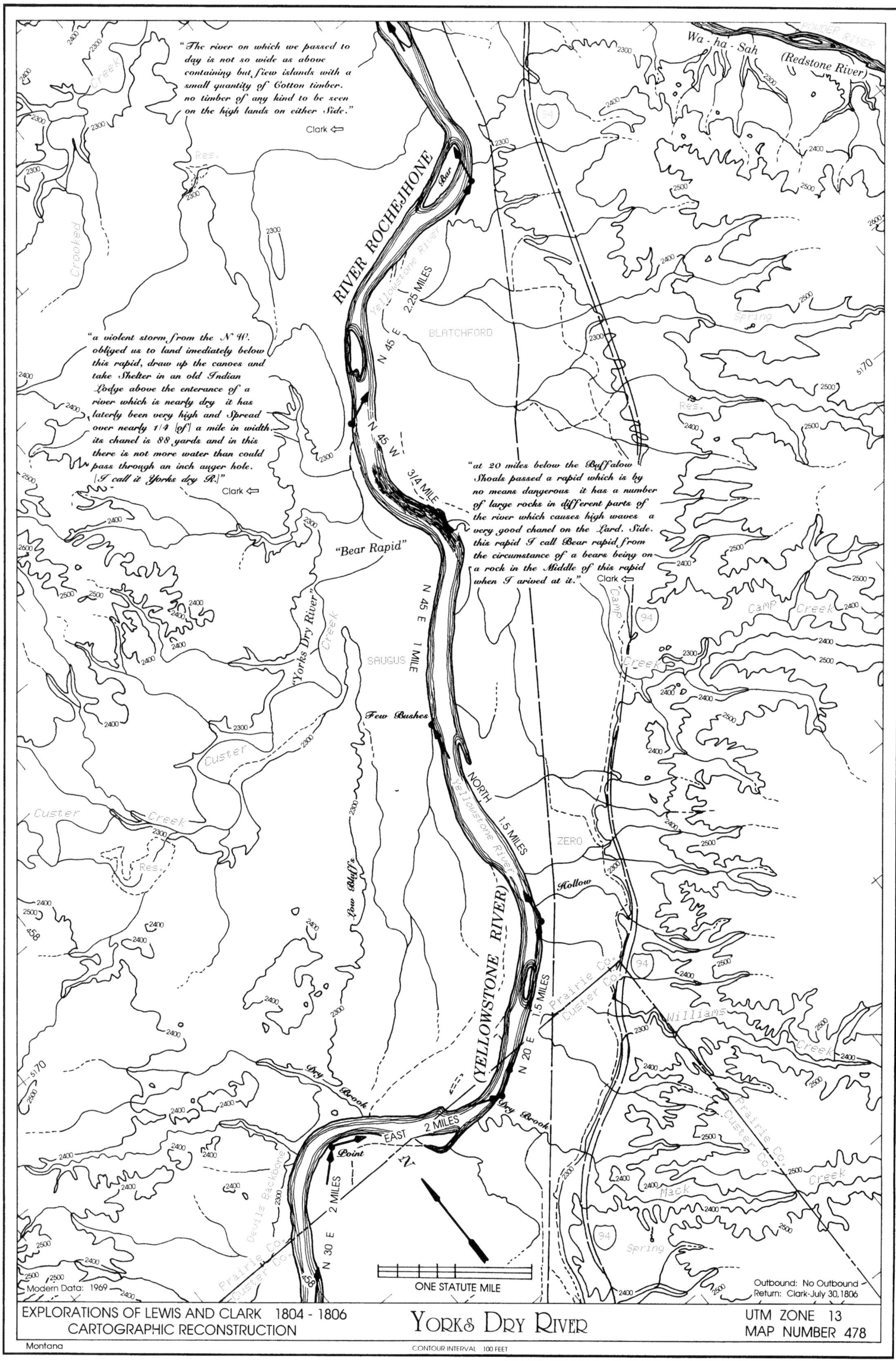
"The river on which we passed to day is not so wide as above containing but fiew islands with a small quantity of Cotton timber. no timber of any kind to be seen on the high lands on either Side."
Clark
"a violent storm from the N W. obliged us to land imediately below this rapid, draw up the canoes and take Shelter in an old Indian Lodge above the enterance of a river which is nearly dry it has laterly been very high and Spread over nearly 1/4 [of] a mile in width. its chanel is 88 yards and in this there is not more water than could pass through an inch auger hole. [I call it Yorks dry R.]"
Clark
"at 20 miles below the Buffalow Shoals passed a rapid which is by no means dangerous it has a number of large rocks in different parts of the river which causes high waves a very good chanel on the Lard. Side. this rapid I call Bear rapid from the circumstance of a bears being on a rock in the Middle of this rapid when I arived at it."
Clark
RIVER ROCHEJHONE
(YELLOWSTONE RIVER)
Wa - ha - Sah
(Redstone River)
POWDER RIVER
"Bear Rapid"
"Yorks Dry River"
Few Bushes
Low Bluffs
Hollow
Dry Brook
Point
Bar
N 45 E
2.25 MILES
N 45 W
3/4 MILE
1 MILE
NORTH
1.5 MILES
N 20 E
EAST
2 MILES
N 30 E
BLATCHFORD
SAUGUS
ZERO
Crooked
Custer Creek
Camp Creek
Williams Creek
Mack Creek
Devils Backbone
Prairie Co.
Custer Co.
Modern Data: 1969
ONE STATUTE MILE
Outbound: No Outbound
Return: Clark-July 30, 1806
EXPLORATIONS OF LEWIS AND CLARK 1804 - 1806
CARTOGRAPHIC RECONSTRUCTION
YORKS DRY RIVER
UTM ZONE 13
MAP NUMBER 478
Montana
CONTOUR INTERVAL 100 FEET

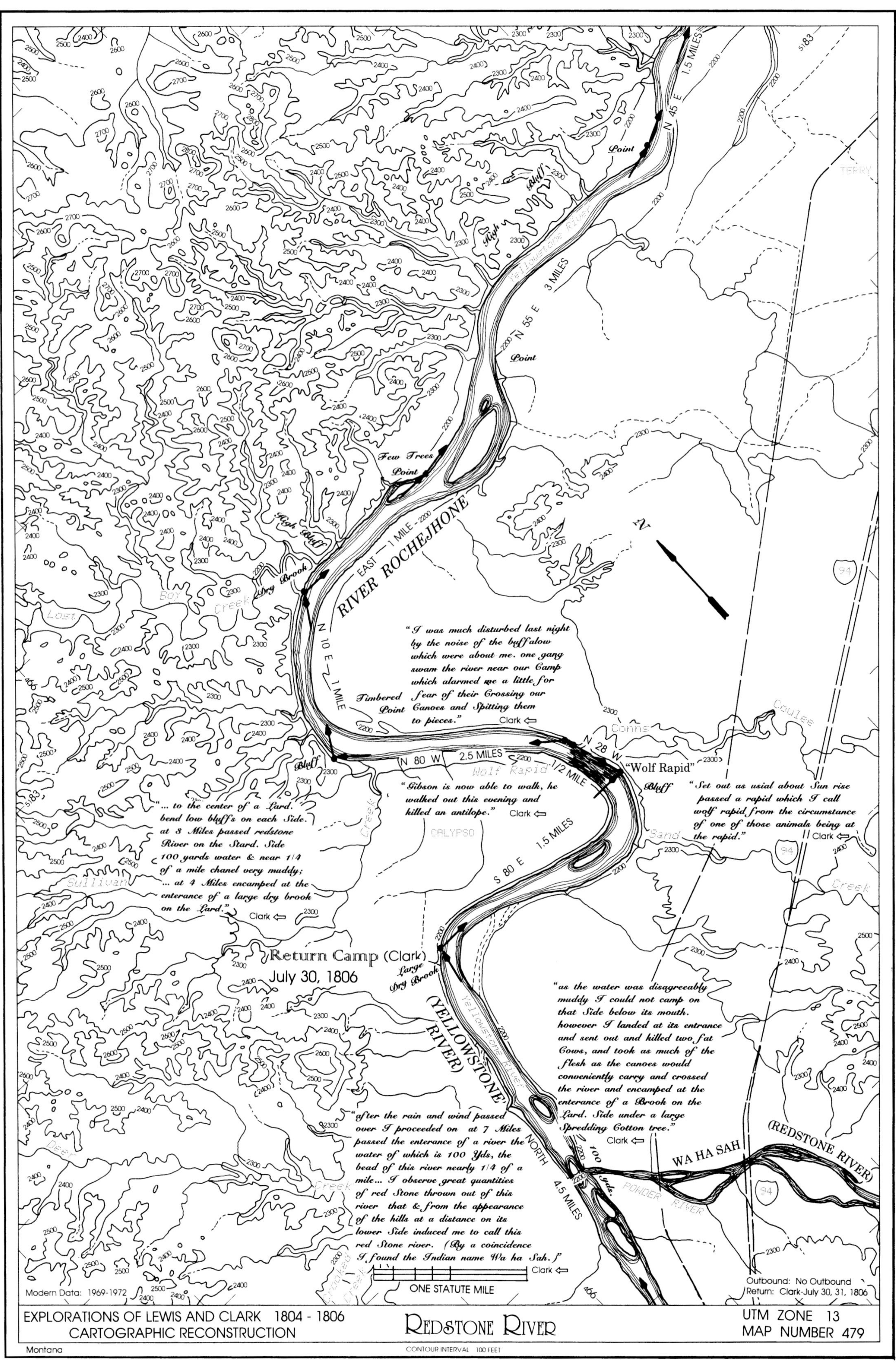

RIVER ROCHEJHONE
Yellowstone River
N 45 E 1.5 MILES
N 55 E 3 MILES
EAST 1 MILE
N 10 E 1 MILE
N 80 W 2.5 MILES
N 28 W 1/2 MILE
S 80 E 1.5 MILES
NORTH 4.5 MILES
Point
High Bluff
Few Trees Point
Dry Brook
Timbered Point
Bluff
"Wolf Rapid"
Wolf Rapid
Large Dry Brook
"I was much disturbed last night by the noise of the buffalow which were about me. one gang swam the river near our Camp which alarmed me a little for fear of their Crossing our Canoes and Spitting them to pieces." Clark
"Gibson is now able to walk, he walked out this evening and killed an antilope." Clark
"Set out as usial about Sun rise passed a rapid which I call wolf rapid from the circumstance of one of those animals being at the rapid." Clark
"... to the center of a Lard. bend low bluffs on each Side. at 3 Miles passed redstone River on the Stard. Side 100 yards water & near 1/4 of a mile chanel very muddy; ... at 4 Miles encamped at the enterance of a large dry brook on the Lard." Clark
Return Camp (Clark)
July 30, 1806
"as the water was disagreeably muddy I could not camp on that Side below its mouth. however I landed at its entrance and sent out and killed two fat Cows, and took as much of the flesh as the canoes would conveniently carry and crossed the river and encamped at the enterance of a Brook on the Lard. Side under a large Spredding Cotton tree." Clark
"after the rain and wind passed over I proceeded on at 7 Miles passed the enterance of a river the water of which is 100 Yds, the bead of this river nearly 1/4 of a mile... I observe great quantities of red Stone thrown out of this river that & from the appearance of the hills at a distance on its lower Side induced me to call this red Stone river. (By a coincidence I found the Indian name Wa ha Sah.)" Clark
(YELLOWSTONE RIVER)
WA HA SAH
(REDSTONE RIVER)
POWDER RIVER
100 yds.
Boy Creek
Lost
Conns Coulee
Sand Creek
CALYPSO
Sullivan
Deer Creek
Crooked Creek
TERRY
N
94
ONE STATUTE MILE
Modern Data: 1969-1972
Outbound: No Outbound
Return: Clark-July 30, 31, 1806
EXPLORATIONS OF LEWIS AND CLARK 1804 - 1806
CARTOGRAPHIC RECONSTRUCTION
REDSTONE RIVER
UTM ZONE 13
MAP NUMBER 479
Montana
CONTOUR INTERVAL 100 FEET

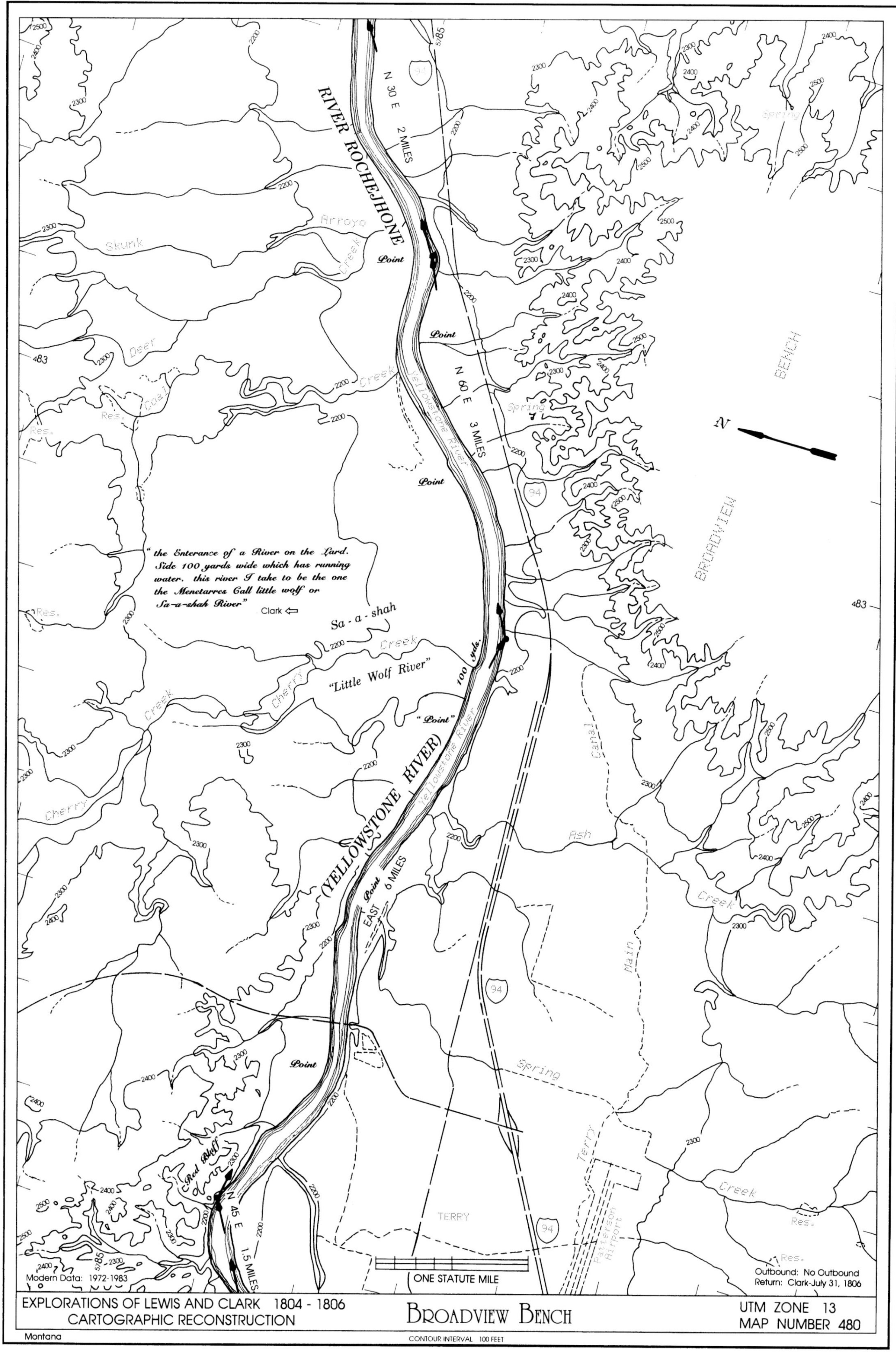
RIVER ROCHEJHONE
N 30 E 2 MILES
N 60 E 3 MILES
Point
Point
Point
Skunk
Arroyo
Creek
Deer
Coal
Creek
Yellowstone River
BROADVIEW BENCH
"the Enterance of a River on the Lard. Side 100 yards wide which has running water. this river I take to be the one the Menetarres Call little wolf or Sa-a-shah River" Clark
Sa - a - shah
Creek
Cherry
Creek
"Little Wolf River"
"Point"
100 yds.
(YELLOWSTONE RIVER)
Yellowstone River
Cherry
Ash
Creek
Canal
Main
Point
EAST 6 MILES
Point
Spring
Red Bluff
N 45 E
1.5 MILES
TERRY
Terry
Patterson Airport
Creek
Res.
Modern Data: 1972-1983
ONE STATUTE MILE
Outbound: No Outbound
Return: Clark-July 31, 1806
EXPLORATIONS OF LEWIS AND CLARK 1804 - 1806
CARTOGRAPHIC RECONSTRUCTION
BROADVIEW BENCH
UTM ZONE 13
MAP NUMBER 480
Montana
CONTOUR INTERVAL 100 FEET

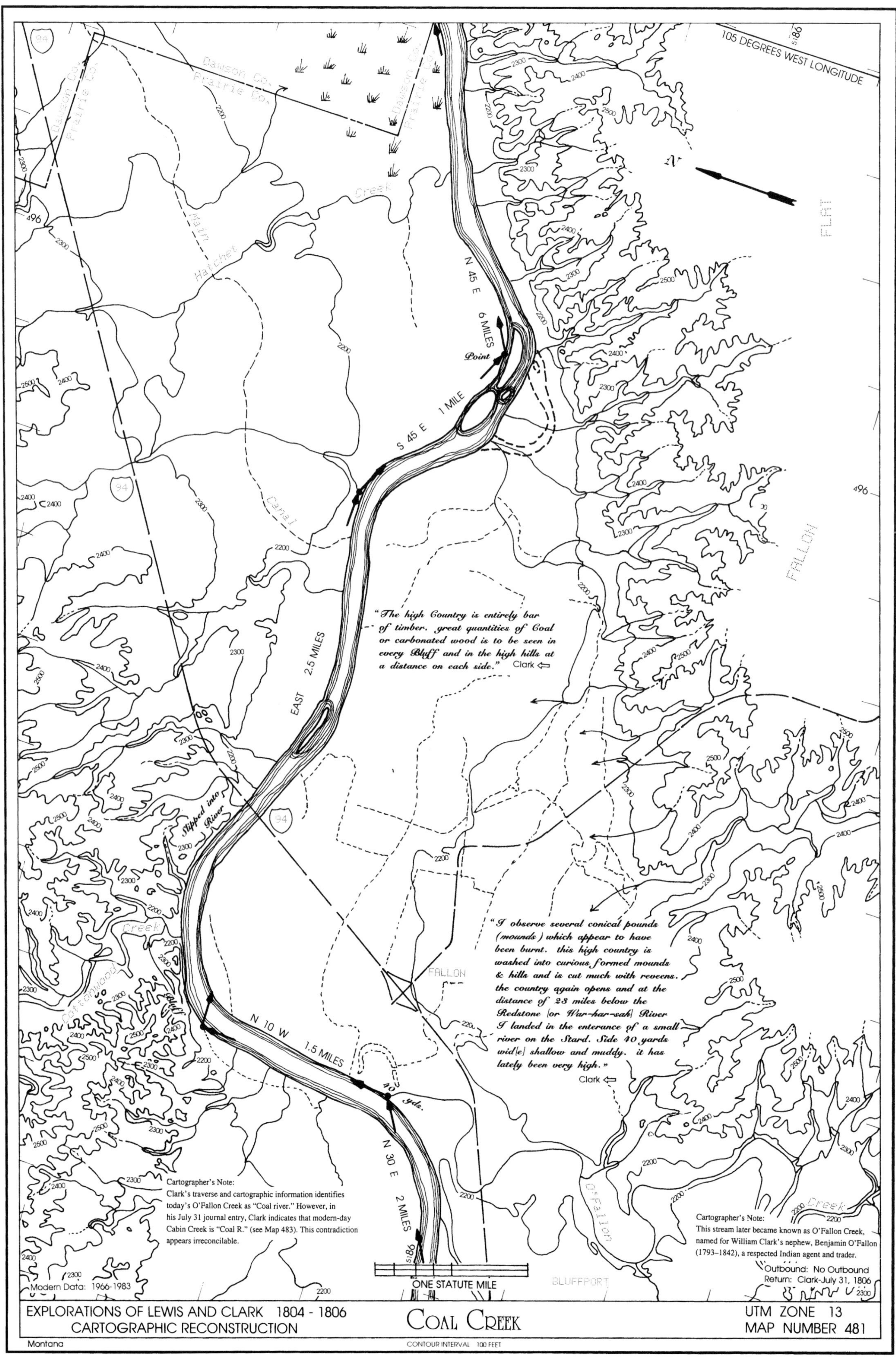
"The high Country is entirely bar of timber. great quantities of Coal or carbonated wood is to be seen in every Bluff and in the high hills at a distance on each side." Clark
"I observe several conical pounds (mounds) which appear to have been burnt. this high country is washed into curious formed mounds & hills and is cut much with reveens. the country again opens and at the distance of 23 miles below the Redstone [or War-har-sah] River I landed in the enterance of a small river on the Stard. Side 40 yards wid[e] shallow and muddy. it has lately been very high." Clark
Cartographer's Note:
Clark's traverse and cartographic information identifies today's O'Fallon Creek as "Coal river." However, in his July 31 journal entry, Clark indicates that modern-day Cabin Creek is "Coal R." (see Map 483). This contradiction appears irreconcilable.
Cartographer's Note:
This stream later became known as O'Fallon Creek, named for William Clark's nephew, Benjamin O'Fallon (1793–1842), a respected Indian agent and trader.
Outbound: No Outbound
Return: Clark-July 31, 1806
Modern Data: 1966-1983
ONE STATUTE MILE
105 DEGREES WEST LONGITUDE
N 45 E 6 MILES
Point
S 45 E 1 MILE
EAST 2.5 MILES
Slipped into River
N 10 W 1.5 MILES
40 yds.
N 30 E 2 MILES
Bluff
Dawson Co.
Prairie Co.
Main Hatchet Creek
Canal
Cottonwood Creek
O'Fallon Creek
FLAT
FALLON
BLUFFPORT
EXPLORATIONS OF LEWIS AND CLARK 1804 - 1806
CARTOGRAPHIC RECONSTRUCTION
COAL CREEK
UTM ZONE 13
MAP NUMBER 481
Montana
CONTOUR INTERVAL 100 FEET

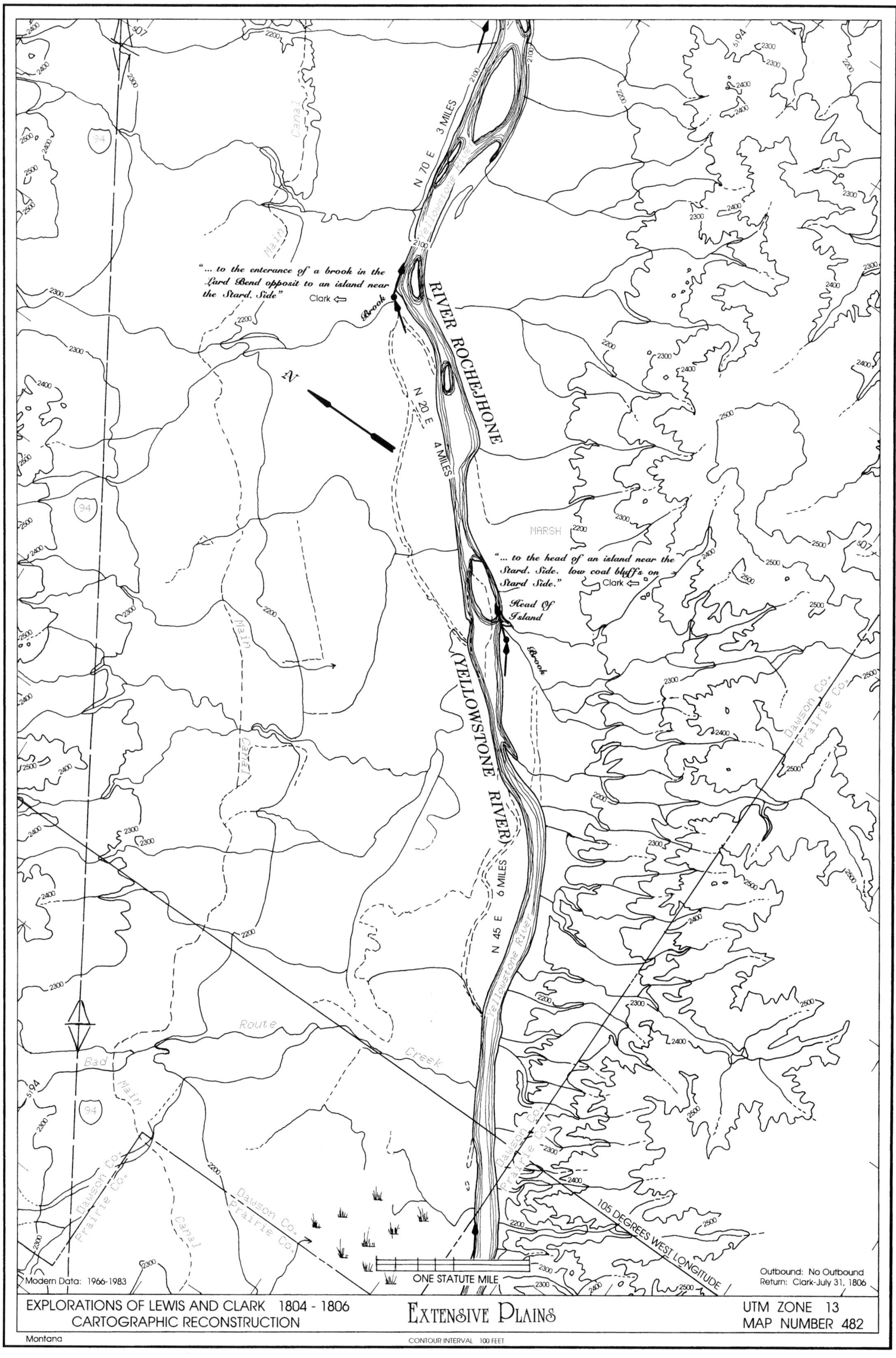
"... to the enterance of a brook in the Lard Bend opposit to an island near the Stard. Side"
Clark
Brook
RIVER ROCHEJHONE
(YELLOWSTONE RIVER)
N 70 E 3 MILES
N 20 E 4 MILES
N 45 E 6 MILES
N
MARSH
"... to the head of an island near the Stard. Side. low coal bluffs on Stard Side."
Clark
Head Of Island
Brook
Yellowstone River
Main Canal
Route
Bad
Creek
Dawson Co.
Prairie Co.
105 DEGREES WEST LONGITUDE
ONE STATUTE MILE
Modern Data: 1966-1983
Outbound: No Outbound
Return: Clark-July 31, 1806
EXPLORATIONS OF LEWIS AND CLARK 1804 - 1806
CARTOGRAPHIC RECONSTRUCTION
EXTENSIVE PLAINS
UTM ZONE 13
MAP NUMBER 482
Montana
CONTOUR INTERVAL 100 FEET

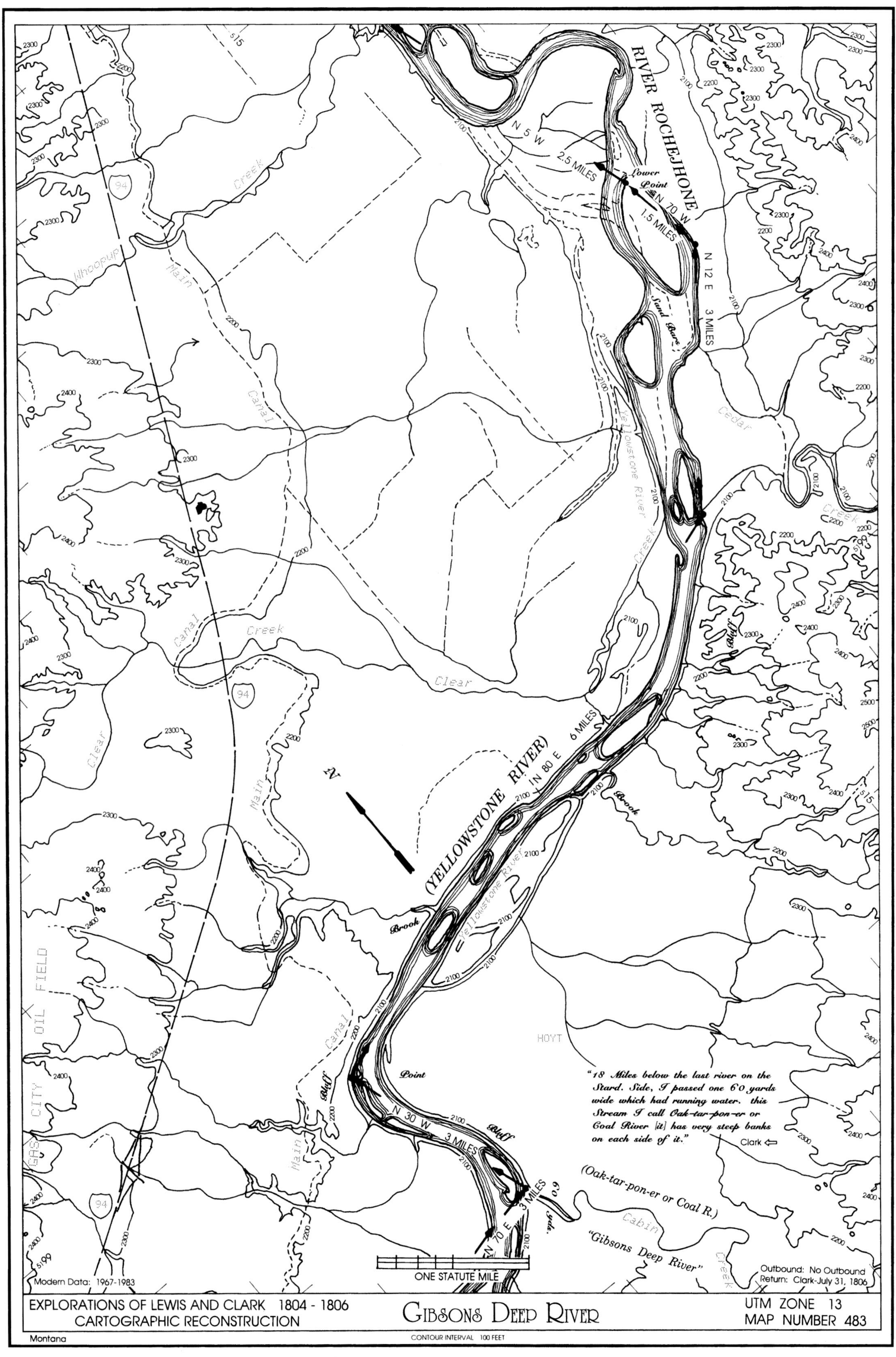
RIVER ROCHEJHONE
(YELLOWSTONE RIVER)
"18 Miles below the last river on the Stard. Side, I passed one 60 yards wide which had running water. this Stream I call Oak-tar-pon-er or Coal River [it] has very steep banks on each side of it."
Clark ⇦
(Oak-tar-pon-er or Coal R.)
"Gibsons Deep River"
Outbound: No Outbound
Return: Clark-July 31, 1806
Modern Data: 1967-1983
ONE STATUTE MILE
EXPLORATIONS OF LEWIS AND CLARK 1804 - 1806
CARTOGRAPHIC RECONSTRUCTION
GIBSONS DEEP RIVER
UTM ZONE 13
MAP NUMBER 483
Montana
CONTOUR INTERVAL 100 FEET

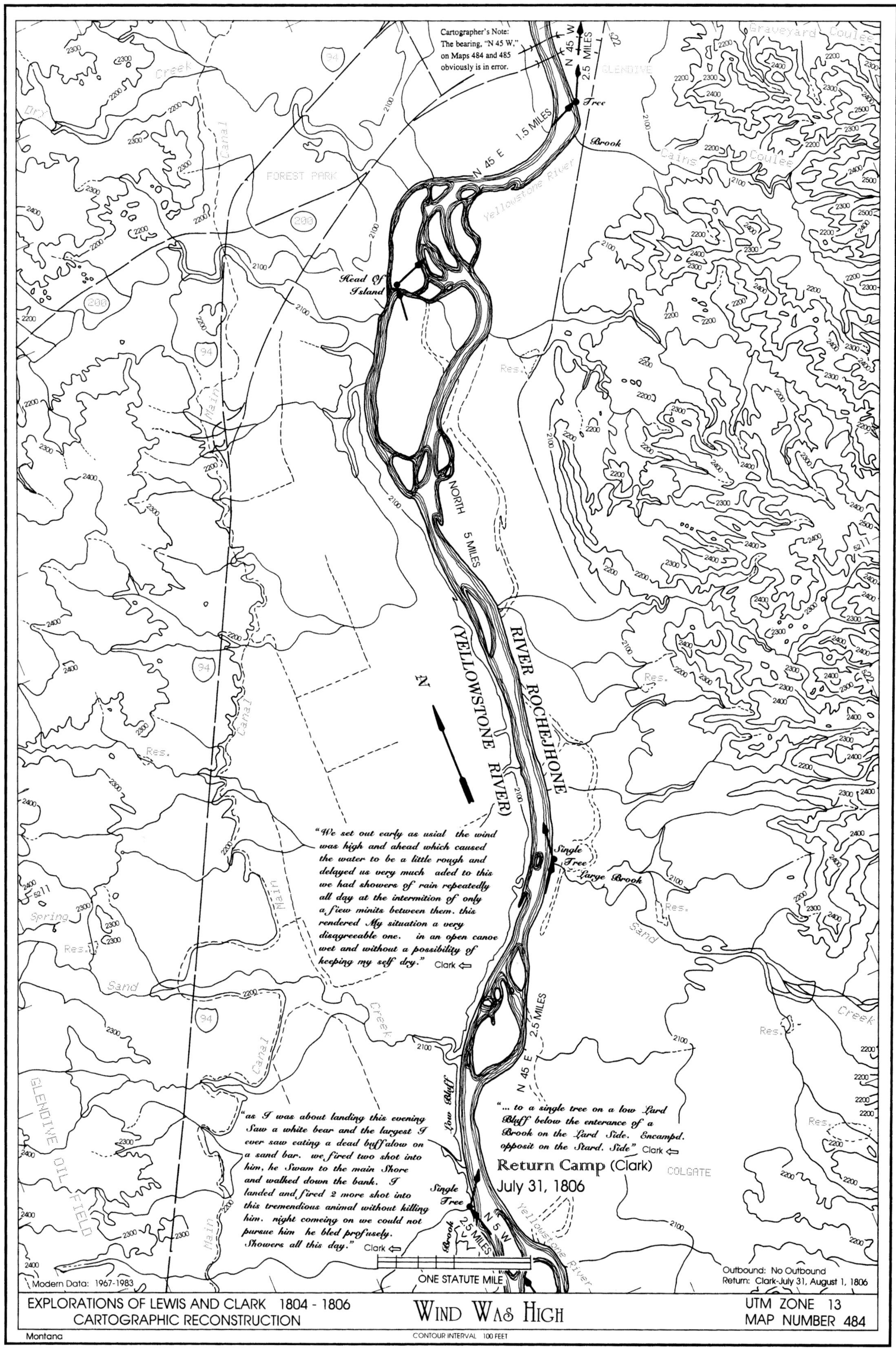
Cartographer's Note:
The bearing, "N 45 W,"
on Maps 484 and 485
obviously is in error.
N 45 W
2.5 MILES
Tree
Brook
N 45 E 1.5 MILES
Head Of Island
NORTH 5 MILES
RIVER ROCHEJHONE
(YELLOWSTONE RIVER)
N
"We set out early as usial the wind was high and ahead which caused the water to be a little rough and delayed us very much aded to this we had showers of rain repeatedly all day at the intermition of only a fiew minits between them. this rendered My situation a very disagreeable one. in an open canoe wet and without a possibility of keeping my self dry." Clark ⇦
Single Tree
Large Brook
N 45 E 2.5 MILES
Low Bluff
"... to a single tree on a low Lard Bluff below the enterance of a Brook on the Lard Side. Encampd. opposit on the Stard. Side" Clark ⇦
Return Camp (Clark)
July 31, 1806
"as I was about landing this evening Saw a white bear and the largest I ever saw eating a dead buffalow on a sand bar. we fired two shot into him, he Swam to the main Shore and walked down the bank. I landed and fired 2 more shot into this tremendious animal without killing him. night comeing on we could not pursue him he bled profusely. Showers all this day." Clark ⇦
Single Tree
Brook
N 5 W 2.5 MILES
ONE STATUTE MILE
Modern Data: 1967-1983
Outbound: No Outbound
Return: Clark-July 31, August 1, 1806
EXPLORATIONS OF LEWIS AND CLARK 1804 - 1806
CARTOGRAPHIC RECONSTRUCTION
WIND WAS HIGH
UTM ZONE 13
MAP NUMBER 484
Montana
CONTOUR INTERVAL 100 FEET

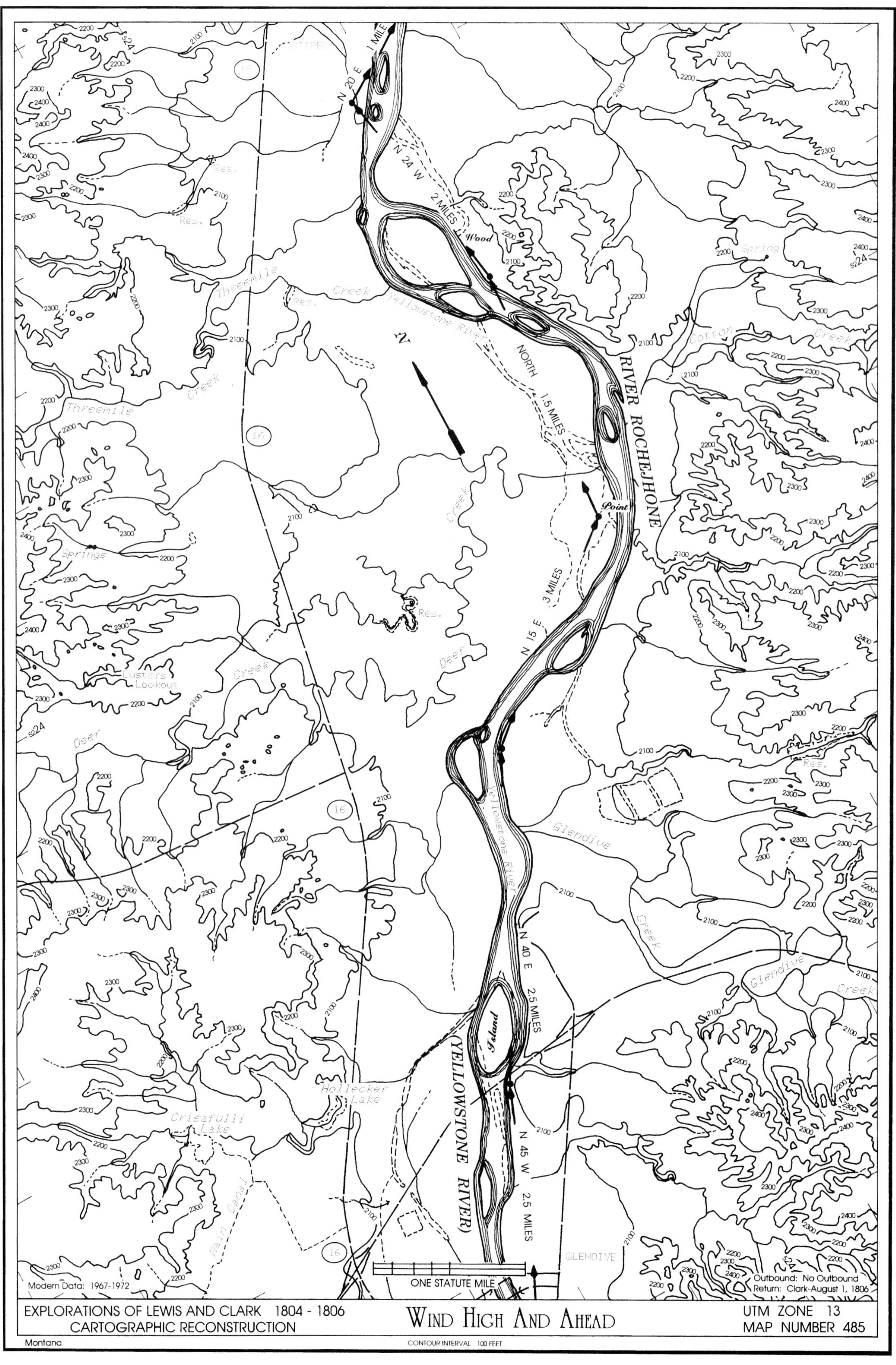
EXPLORATIONS OF LEWIS AND CLARK 1804 - 1806
CARTOGRAPHIC RECONSTRUCTION
WIND HIGH AND AHEAD
UTM ZONE 13
MAP NUMBER 485
Montana
CONTOUR INTERVAL 100 FEET
Modern Data: 1967-1972
ONE STATUTE MILE
Outbound: No Outbound
Return: Clark-August 1, 1806
RIVER ROCHEJHONE
(YELLOWSTONE RIVER)
Yellowstone River
N 20 E 1 MILE
N 24 W 2 MILES
Wood
NORTH 1.5 MILES
Point
N 15 E 3 MILES
N 40 E 2.5 MILES
Island
N 45 W 2.5 MILES
Threemile Creek
Cotton Creek
Spring
Springs
Deer Creek
Custers Lookout
Glendive Creek
Hollecker Lake
Crisafulli Lake
Main Canal
GLENDIVE
Res.

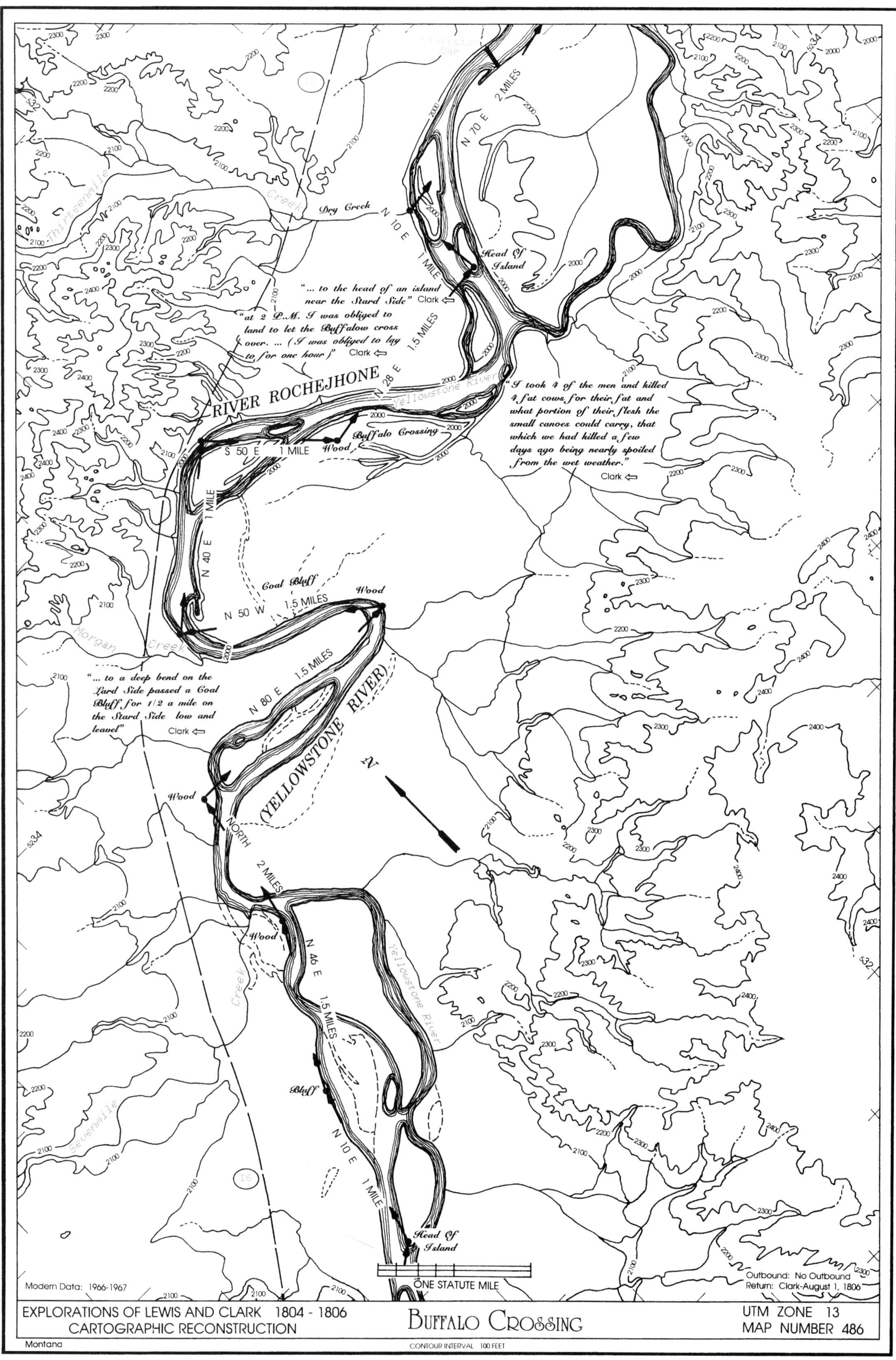
"... to the head of an island near the Stard Side" Clark
"at 2 P.M. I was obliged to land to let the Buffalow cross over. ... (I was obliged to lay to for one hour)" Clark
"I took 4 of the men and killed 4 fat cows for their fat and what portion of their flesh the small canoes could carry, that which we had killed a few days ago being nearly spoiled from the wet weather." Clark
"... to a deep bend on the Lard Side passed a Coal Bluff for 1/2 a mile on the Stard Side low and leavel" Clark
RIVER ROCHEJHONE
(YELLOWSTONE RIVER)
Yellowstone River
Buffalo Crossing
Coal Bluff
Dry Creek
Head Of Island
Wood
Bluff
Morgan Creek
Thirteenmile
Sevenmile
Creek
N 70 E 2 MILES
N 10 E 1 MILE
N 28 E 1.5 MILES
S 50 E 1 MILE
N 40 E 1 MILE
N 50 W 1.5 MILES
N 80 E 1.5 MILES
NORTH 2 MILES
N 46 E 1.5 MILES
N 10 E 1 MILE
N
ONE STATUTE MILE
Modern Data: 1966-1967
Outbound: No Outbound
Return: Clark-August 1, 1806
EXPLORATIONS OF LEWIS AND CLARK 1804 - 1806
CARTOGRAPHIC RECONSTRUCTION
BUFFALO CROSSING
UTM ZONE 13
MAP NUMBER 486
Montana
CONTOUR INTERVAL 100 FEET

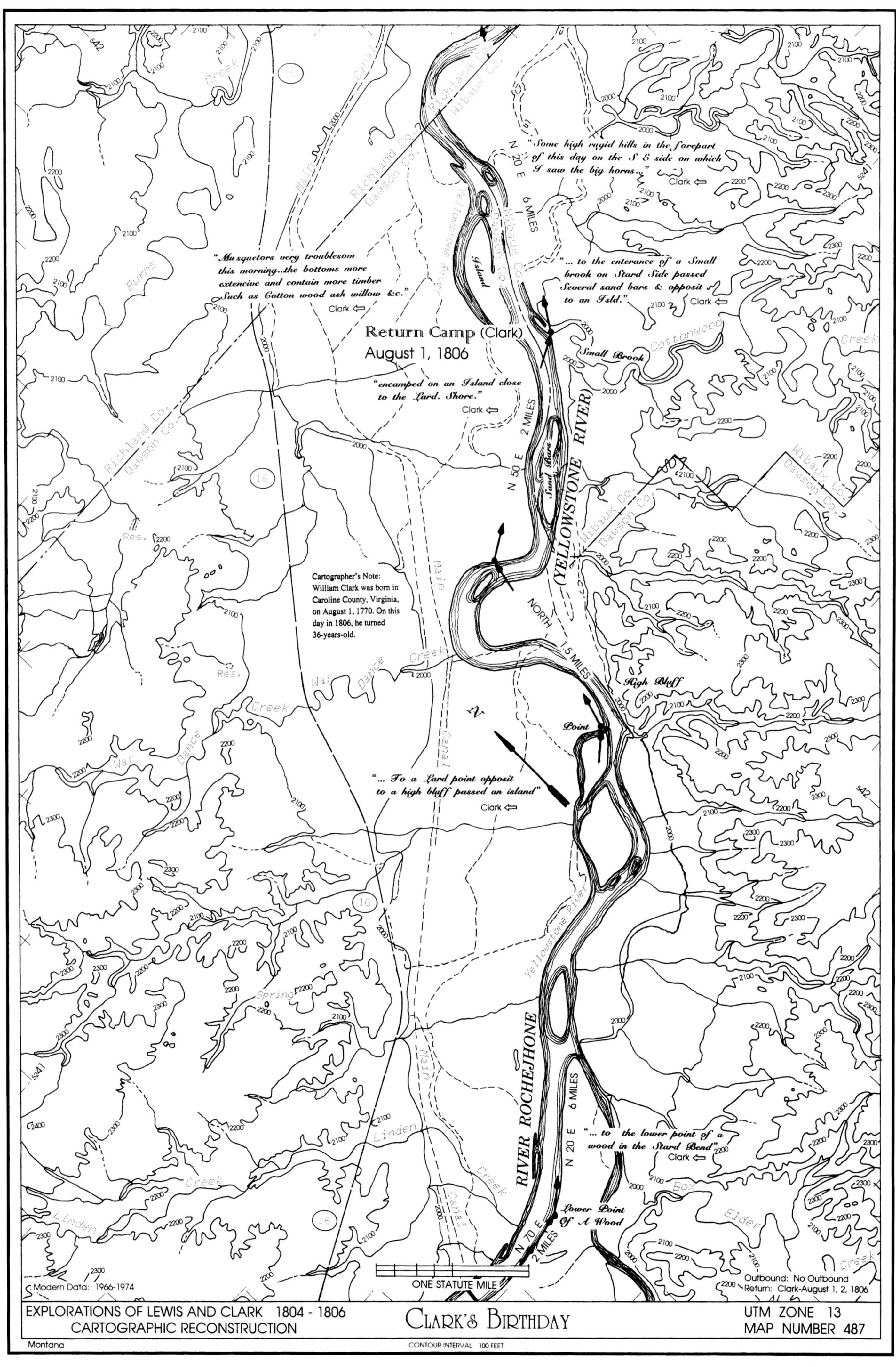
"Some high rugid hills in the forepart of this day on the S E side on which I saw the big horns..."
Clark ⇦
"Musquetors very troublesom this morning..the bottoms more extencive and contain more timber Such as Cotton wood ash willow &c."
Clark ⇦
"... to the enterance of a Small brook on Stard Side passed Several sand bars & opposit to an Isld."
Clark ⇦
Return Camp (Clark)
August 1, 1806
"encamped on an Island close to the Lard. Shore."
Clark ⇦
Small Brook
Island
Sand Bars
(YELLOWSTONE RIVER)
Cartographer's Note: William Clark was born in Caroline County, Virginia, on August 1, 1770. On this day in 1806, he turned 36-years-old.
High Bluff
Point
"... To a Lard point opposit to a high bluff passed an island"
Clark ⇦
RIVER ROCHEJHONE
"... to the lower point of a wood in the Stard Bend"
Clark ⇦
Lower Point Of A Wood
ONE STATUTE MILE
Modern Data: 1966-1974
Outbound: No Outbound
Return: Clark-August 1, 2, 1806
EXPLORATIONS OF LEWIS AND CLARK 1804 - 1806
CARTOGRAPHIC RECONSTRUCTION
CLARK'S BIRTHDAY
UTM ZONE 13
MAP NUMBER 487
Montana
CONTOUR INTERVAL 100 FEET

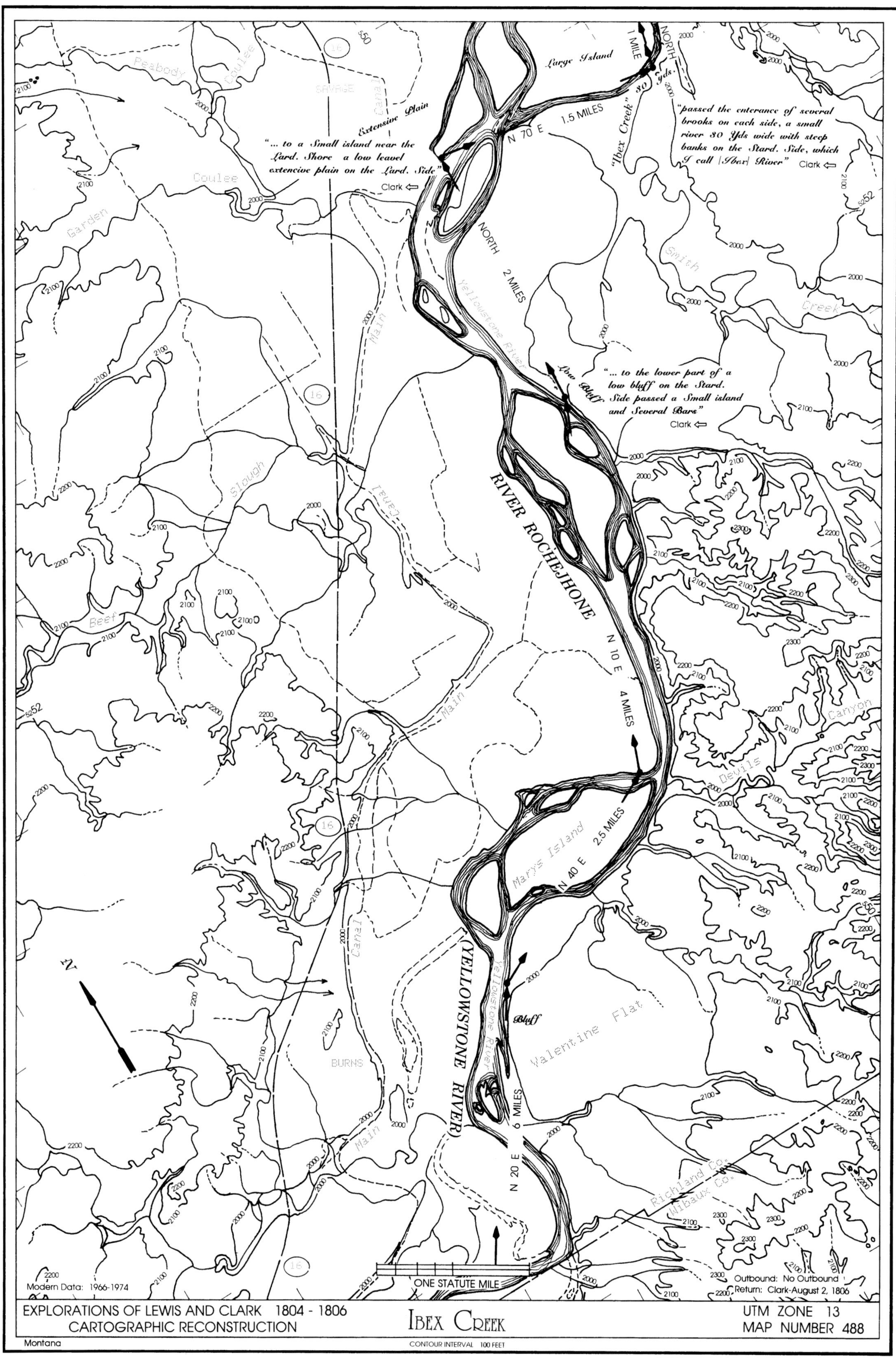
Large Island
"Ibex Creek" 80 yds.
"passed the enterance of several brooks on each side, a small river 80 Yds wide with steep banks on the Stard. Side, which I call [Ibex] River" Clark
"... to a Small island near the Lard. Shore a low leavel extencive plain on the Lard. Side" Clark
Extensive Plain
"... to the lower part of a low bluff on the Stard. Side passed a Small island and Several Bars" Clark
Low Bluff
RIVER ROCHEJHONE
YELLOWSTONE RIVER
Yellowstone River
Marys Island
Valentine Flat
Bluff
Peabody Coulee
Garden Coulee
Smith Creek
Slough
Beef
Devils Canyon
Main Canal
SAVAGE
BURNS
Richland Co.
Wibaux Co.
N 70 E 1.5 MILES
NORTH 1 MILE
NORTH 2 MILES
N 10 E 4 MILES
N 40 E 2.5 MILES
N 20 E 6 MILES
ONE STATUTE MILE
Modern Data: 1966-1974
Outbound: No Outbound
Return: Clark-August 2, 1806
EXPLORATIONS OF LEWIS AND CLARK 1804 - 1806
CARTOGRAPHIC RECONSTRUCTION
IBEX CREEK
UTM ZONE 13
MAP NUMBER 488
Montana
CONTOUR INTERVAL 100 FEET

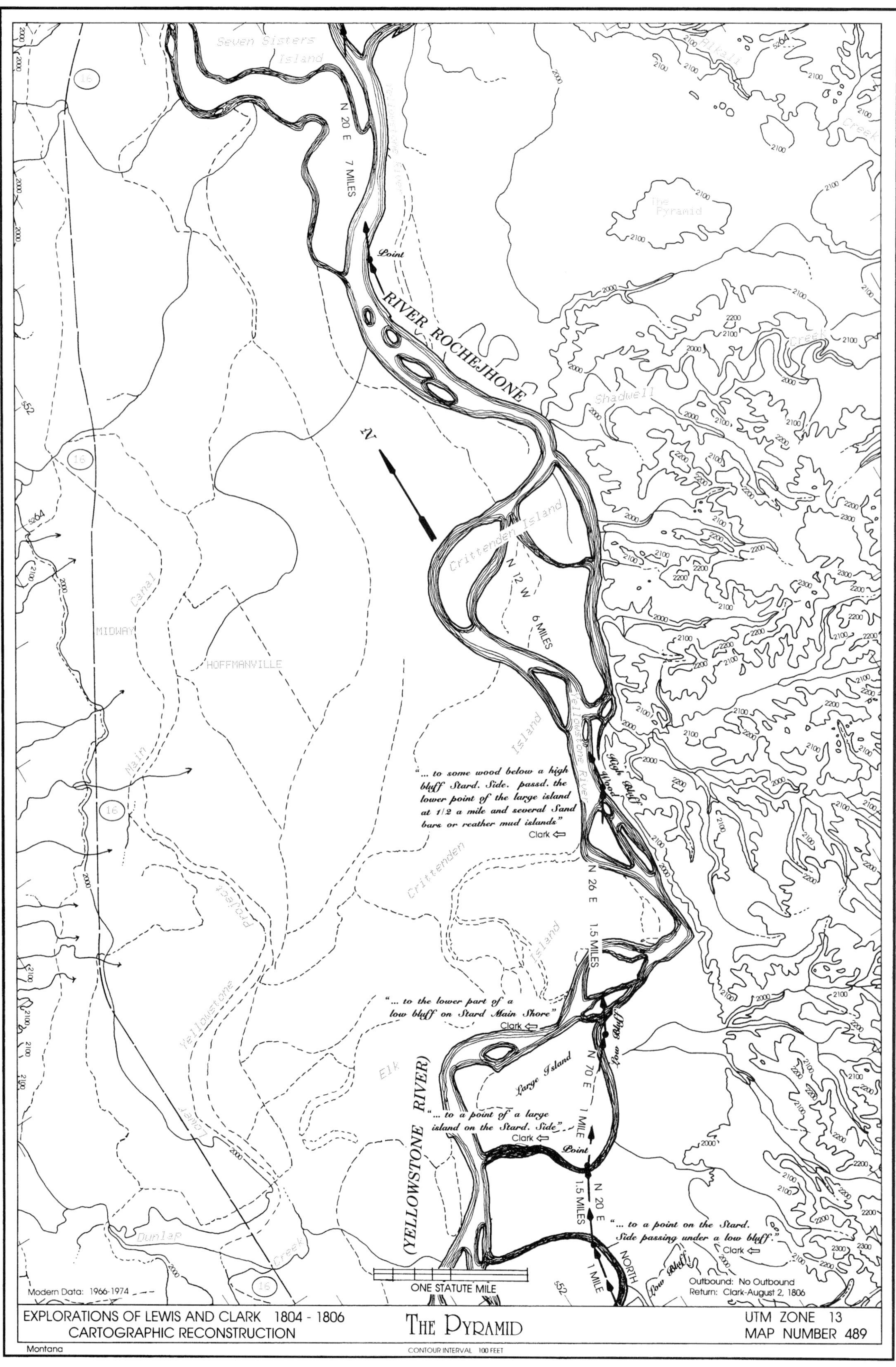
RIVER ROCHEJHONE
(YELLOWSTONE RIVER)
N 20 E 7 MILES
Point
N 12 W 6 MILES
N 26 E 1.5 MILES
N 70 E 1 MILE
N 20 E 1.5 MILES
NORTH 1 MILE
High Bluff
Wood
Low Bluff
Large Island
"... to some wood below a high bluff Stard. Side. passd. the lower point of the large island at 1/2 a mile and several Sand bars or reather mud islands"
Clark ⇦
"... to the lower part of a low bluff on Stard Main Shore"
Clark ⇦
"... to a point of a large island on the Stard. Side"
Clark ⇦
"... to a point on the Stard. Side passing under a low bluff."
Clark ⇦
Seven Sisters Island
The Pyramid
Alkali Creek
Shadwell Creek
Crittenden Island
Yellowstone River
Island
Crittenden Island
Elk
MIDWAY
HOFFMANVILLE
Main Canal
Project
Yellowstone
Lower
Dunlap Creek
ONE STATUTE MILE
Modern Data: 1966-1974
Outbound: No Outbound
Return: Clark-August 2, 1806
EXPLORATIONS OF LEWIS AND CLARK 1804 - 1806
CARTOGRAPHIC RECONSTRUCTION
THE PYRAMID
UTM ZONE 13
MAP NUMBER 489
Montana
CONTOUR INTERVAL 100 FEET

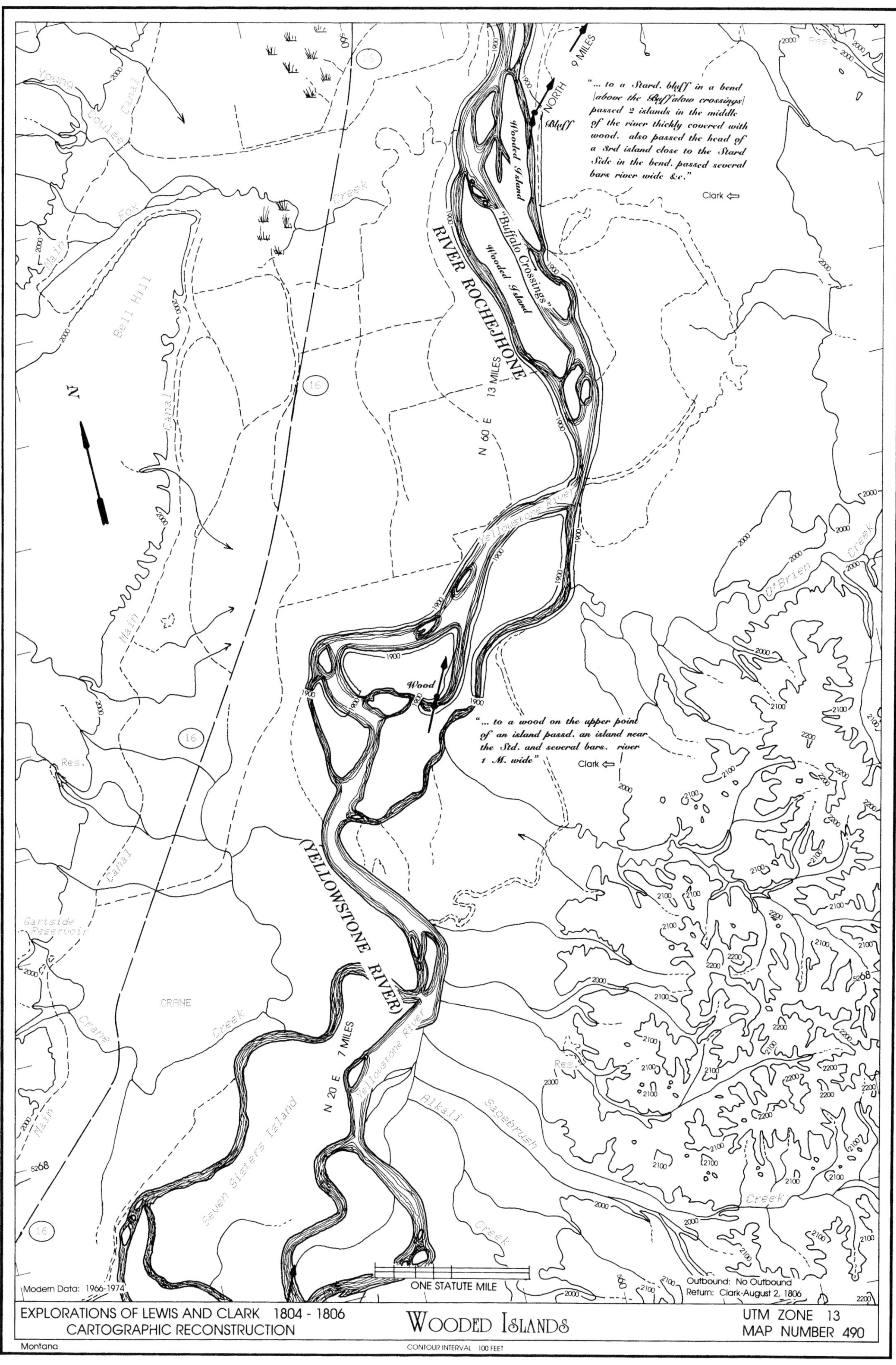
"... to a Stard. bluff in a bend [above the Buffalow crossings] passed 2 islands in the middle of the river thickly covered with wood. also passed the head of a 3rd island close to the Stard Side in the bend. passed several bars river wide &c."
Clark
Bluff
NORTH
9 MILES
Wooded Island
"Buffalo Crossings"
Wooded Island
RIVER ROCHEJHONE
N 60 E 13 MILES
Wood
"... to a wood on the upper point of an island passd. an island near the Std. and several bars. river 1 M. wide"
Clark
(YELLOWSTONE RIVER)
N 20 E 7 MILES
Yellowstone River
Young Coulee
Canal
Fox
Creek
Main
Bell Hill
Main
Canal
Res.
Gartside Reservoir
CRANE
Crane
Creek
Seven Sisters Island
Alkali
Sagebrush
Creek
O'Brien
Creek
Modern Data: 1966-1974
ONE STATUTE MILE
Outbound: No Outbound
Return: Clark-August 2, 1806
EXPLORATIONS OF LEWIS AND CLARK 1804 - 1806
CARTOGRAPHIC RECONSTRUCTION
WOODED ISLANDS
UTM ZONE 13
MAP NUMBER 490
Montana
CONTOUR INTERVAL 100 FEET

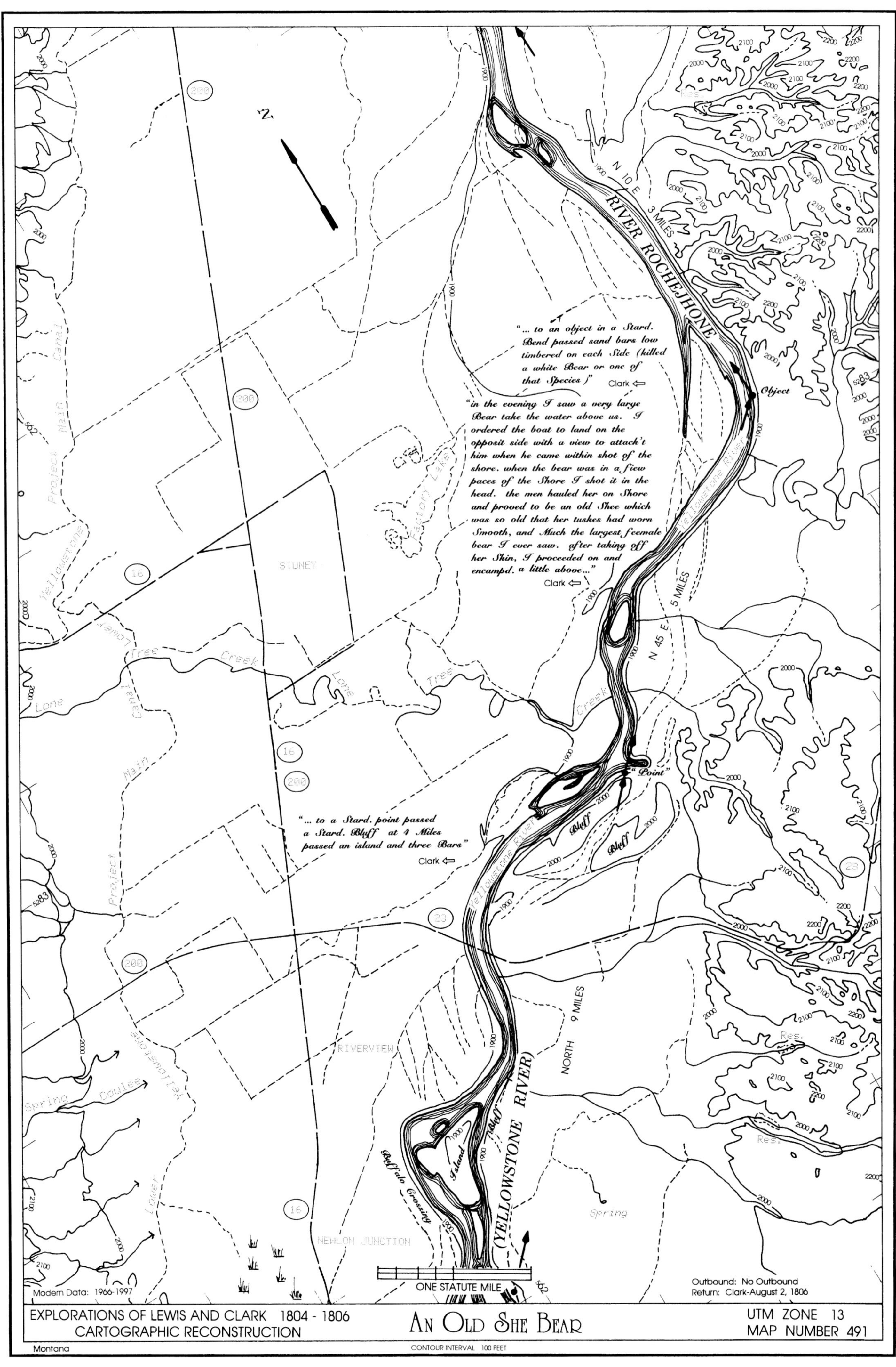
"... to an object in a Stard. Bend passed sand bars low timbered on each Side (killed a white Bear or one of that Species)"
Clark ⇦
"in the evening I saw a very large Bear take the water above us. I ordered the boat to land on the opposit side with a view to attack't him when he came within shot of the shore. when the bear was in a fiew paces of the Shore I shot it in the head. the men hauled her on Shore and proved to be an old Shee which was so old that her tuskes had worn Smooth, and Much the largest feemale bear I ever saw. after taking off her Skin, I proceeded on and encampd. a little above..."
Clark ⇦
"... to a Stard. point passed a Stard. Bluff at 4 Miles passed an island and three Bars"
Clark ⇦
RIVER ROCHEJHONE
N 10 E 3 MILES
Object
Yellowstone River
N 45 E 5 MILES
"Point"
Bluff
Bluff
NORTH 9 MILES
(YELLOWSTONE RIVER)
Island
Bluff
Buffalo Crossing
SIDNEY
RIVERVIEW
NEWLON JUNCTION
Factory Lake
Lone Tree Creek
Project Main Canal
Lower Yellowstone
Spring Coulee
Spring
Res.
ONE STATUTE MILE
Modern Data: 1966-1997
Outbound: No Outbound
Return: Clark-August 2, 1806
EXPLORATIONS OF LEWIS AND CLARK 1804 - 1806
CARTOGRAPHIC RECONSTRUCTION
AN OLD SHE BEAR
UTM ZONE 13
MAP NUMBER 491
Montana
CONTOUR INTERVAL 100 FEET

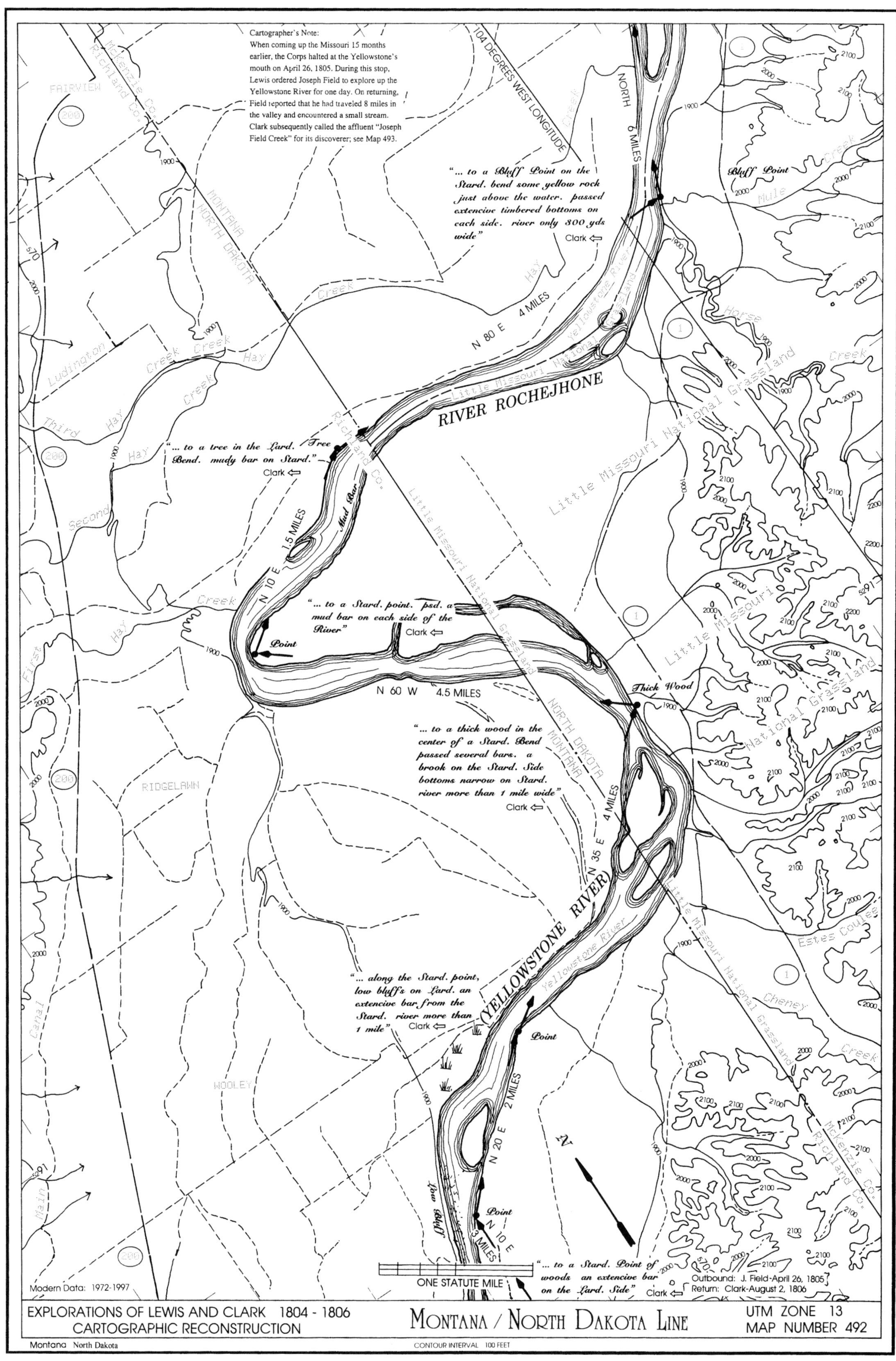
Cartographer's Note:
When coming up the Missouri 15 months earlier, the Corps halted at the Yellowstone's mouth on April 26, 1805. During this stop, Lewis ordered Joseph Field to explore up the Yellowstone River for one day. On returning, Field reported that he had traveled 8 miles in the valley and encountered a small stream. Clark subsequently called the affluent "Joseph Field Creek" for its discoverer; see Map 493.
"... to a Bluff Point on the Stard. bend some yellow rock just above the water. passed extencive timbered bottoms on each side. river only 300 yds wide" Clark
"... to a tree in the Lard. Tree Bend. mudy bar on Stard." Clark
"... to a Stard. point. psd. a mud bar on each side of the River" Clark
"... to a thick wood in the center of a Stard. Bend passed several bars. a brook on the Stard. Side bottoms narrow on Stard. river more than 1 mile wide" Clark
"... along the Stard. point, low bluffs on Lard. an extencive bar from the Stard. river more than 1 mile" Clark
"... to a Stard. Point of woods an extencive bar on the Lard. Side" Clark
RIVER ROCHEJHONE
YELLOWSTONE RIVER
104 DEGREES WEST LONGITUDE
MONTANA
NORTH DAKOTA
McKenzie Co.
Richland Co.
Little Missouri National Grassland
Yellowstone River
FAIRVIEW
RIDGELAWN
WOOLEY
Bluff Point
Tree
Mud Bar
Point
Thick Wood
Low Bluff
NORTH 6 MILES
N 80 E 4 MILES
N 10 E 1.5 MILES
N 60 W 4.5 MILES
N 35 E 4 MILES
N 20 E 2 MILES
N 10 E 3 MILES
ONE STATUTE MILE
Outbound: J. Field-April 26, 1805
Return: Clark-August 2, 1806
Modern Data: 1972-1997
EXPLORATIONS OF LEWIS AND CLARK 1804 - 1806
CARTOGRAPHIC RECONSTRUCTION
MONTANA / NORTH DAKOTA LINE
UTM ZONE 13
MAP NUMBER 492
Montana North Dakota
CONTOUR INTERVAL 100 FEET

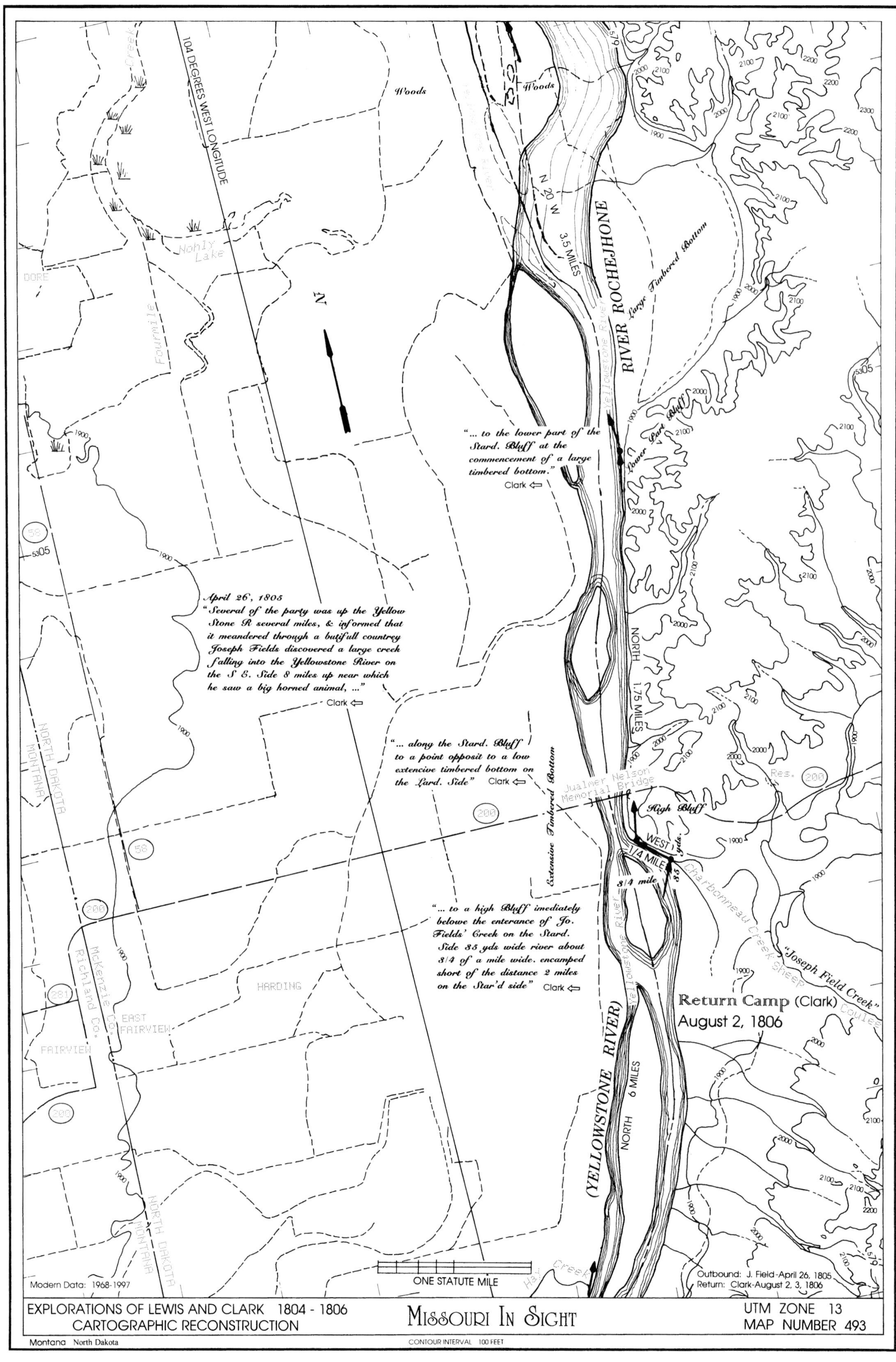

104 DEGREES WEST LONGITUDE
Woods
Woods
N 20 W
3.5 MILES
RIVER ROCHEJHONE
Large Timbered Bottom
Yellowstone River
Nohly Lake
Fourmile
N
"... to the lower part of the Stard. Bluff at the commencement of a large timbered bottom."
Clark
Lower Part Bluff
April 26, 1805
"Several of the party was up the Yellow Stone R several miles, & informed that it meandered through a butifull countrey Joseph Fields discovered a large creek falling into the Yellowstone River on the S E. Side 8 miles up near which he saw a big horned animal, ..."
Clark
NORTH
1.75 MILES
"... along the Stard. Bluff to a point opposit to a low extencive timbered bottom on the Lard. Side"
Clark
Extensive Timbered Bottom
Jualmer Nelson Memorial Bridge
High Bluff
WEST
1/4 MILE
3/4 mile
85 yds.
Charbonneau Creek
"... to a high Bluff imediately belowe the enterance of Jo. Fields' Creek on the Stard. Side 85 yds wide river about 3/4 of a mile wide. encamped short of the distance 2 miles on the Star'd side"
Clark
"Joseph Field Creek"
Sheep Coulee
Return Camp (Clark)
August 2, 1806
(YELLOWSTONE RIVER)
6 MILES
NORTH
NORTH DAKOTA
MONTANA
HARDING
McKenzie Co.
Richland Co.
EAST FAIRVIEW
FAIRVIEW
Res.
Modern Data: 1968-1997
ONE STATUTE MILE
Outbound: J. Field-April 26, 1805
Return: Clark-August 2, 3, 1806
EXPLORATIONS OF LEWIS AND CLARK 1804 - 1806
CARTOGRAPHIC RECONSTRUCTION
MISSOURI IN SIGHT
UTM ZONE 13
MAP NUMBER 493
Montana North Dakota
CONTOUR INTERVAL 100 FEET

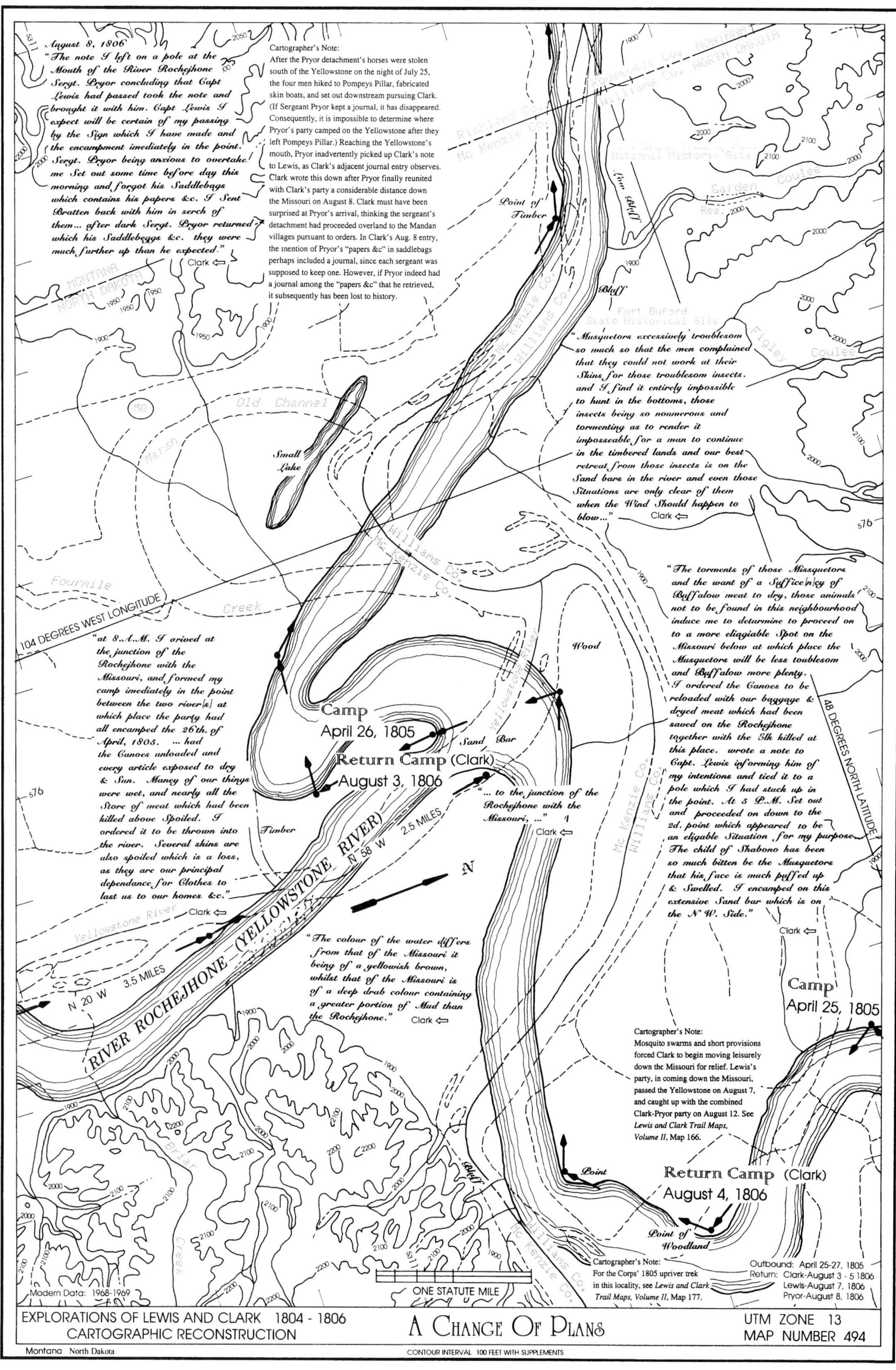

August 8, 1806
"The note I left on a pole at the Mouth of the River Rochejhone Sergt. Pryor concluding that Capt Lewis had passed took the note and brought it with him. Capt Lewis I expect will be certain of my passing by the Sign which I have made and the encampment imediately in the point. Sergt. Pryor being anxious to overtake me Set out some time before day this morning and forgot his Saddlebags which contains his papers &c. I Sent Bratten back with him in serch of them... after dark Sergt. Pryor returned with his Saddlebeggs &c. they were much further up than he expected."
Clark ⇦
Cartographer's Note:
After the Pryor detachment's horses were stolen south of the Yellowstone on the night of July 25, the four men hiked to Pompeys Pillar, fabricated skin boats, and set out downstream pursuing Clark. (If Sergeant Pryor kept a journal, it has disappeared. Consequently, it is impossible to determine where Pryor's party camped on the Yellowstone after they left Pompeys Pillar.) Reaching the Yellowstone's mouth, Pryor inadvertently picked up Clark's note to Lewis, as Clark's adjacent journal entry observes. Clark wrote this down after Pryor finally reunited with Clark's party a considerable distance down the Missouri on August 8. Clark must have been surprised at Pryor's arrival, thinking the sergeant's detachment had proceeded overland to the Mandan villages pursuant to orders. In Clark's Aug. 8 entry, the mention of Pryor's "papers &c" in saddlebags perhaps included a journal, since each sergeant was supposed to keep one. However, if Pryor indeed had a journal among the "papers &c" that he retrieved, it subsequently has been lost to history.
Roosevelt Co. MONTANA
Williams Co. NORTH DAKOTA
Richland Co.
Mc Kenzie Co.
MONTANA
NORTH DAKOTA
Fort Union Trading Post National Historic Site
Garden Coulee
Res.
Point of Timber
low Bluff
Bluff
Fort Buford State Historical Site
Figley Coulee
"Musquetors excessively troublesom so much so that the men complained that they could not work at their Skins for those troublesom insects. and I find it entirely impossible to hunt in the bottoms, those insects being so noumerous and tormenting as to render it imposseable for a man to continue in the timbered lands and our best retreat from those insects is on the Sand bars in the river and even those Situations are only clear of them when the Wind Should happen to blow..."
Clark ⇦
Old Channel
Marsh
Small Lake
58
Fourmile Creek
104 DEGREES WEST LONGITUDE
Williams Co.
Mc Kenzie Co.
"The torments of those Missquetors and the want of a Suffice[n]cy of Buffalow meat to dry, those animals not to be found in this neighbourhood induce me to deturmine to proceed on to a more eligiable Spot on the Missouri below at which place the Musquetors will be less toublesom and Buffalow more plenty. I ordered the Canoes to be reloaded with our baggage & dryed meat which had been saved on the Rochejhone together with the Elk killed at this place. wrote a note to Capt. Lewis informing him of my intentions and tied it to a pole which I had stuck up in the point. At 5 P.M. Set out and proceeded on down to the 2d. point which appeared to be an eligable Situation for my purpose. The child of Shabono has been so much bitten be the Musquetors that his face is much puffed up & Swelled. I encamped on this extensive Sand bar which is on the N.W. Side."
Clark ⇦
48 DEGREES NORTH LATITUDE
"at 8 A.M. I arived at the junction of the Rochejhone with the Missouri, and formed my camp imediately in the point between the two river[s] at which place the party had all encamped the 26th. of April, 1805. ...had the Canoes unloaded and every article exposed to dry & Sun. Maney of our things were wet, and nearly all the Store of meat which had been killed above Spoiled. I ordered it to be thrown into the river. Several skins are also spoiled which is a loss, as they are our principal dependance for Clothes to last us to our homes &c."
Clark ⇦
Wood
Yellowstone River
Camp
April 26, 1805
Sand Bar
Return Camp (Clark)
August 3, 1806
"... to the junction of the Rochejhone with the Missouri, ..."
Clark ⇦
Timber
N 58 W 2.5 MILES
N
RIVER ROCHEJHONE (YELLOWSTONE RIVER)
Yellowstone River
N 20 W 3.5 MILES
"The colour of the water differs from that of the Missouri it being of a yellowish brown, whilst that of the Missouri is of a deep drab colour containing a greater portion of Mud than the Rochejhone."
Clark ⇦
Camp
April 25, 1805
Cartographer's Note:
Mosquito swarms and short provisions forced Clark to begin moving leisurely down the Missouri for relief. Lewis's party, in coming down the Missouri, passed the Yellowstone on August 7, and caught up with the combined Clark-Pryor party on August 12. See Lewis and Clark Trail Maps, Volume II, Map 166.
Briar Creek
Bluff
Point
Return Camp (Clark)
August 4, 1806
Point of Woodland
Williams Co.
Mc Kenzie Co.
Cartographer's Note:
For the Corps' 1805 upriver trek in this locality, see Lewis and Clark Trail Maps, Volume II, Map 177.
Outbound: April 25-27, 1805
Return: Clark-August 3 - 5 1806
Lewis-August 7, 1806
Pryor-August 8, 1806
Modern Data: 1968-1969
ONE STATUTE MILE
EXPLORATIONS OF LEWIS AND CLARK 1804 - 1806
CARTOGRAPHIC RECONSTRUCTION
A CHANGE OF PLANS
UTM ZONE 13
MAP NUMBER 494
Montana North Dakota
CONTOUR INTERVAL 100 FEET WITH SUPPLEMENTS

Alternate Return Route No. 7

MAPS 495 - 527

LEWIS PARTY, UPPER MARIAS, MONTANA

Lewis contingent, upper Marias exploration, July 16–28, 1806—
- Capt. Meriwether Lewis (journalist)
- George Drouillard (interpreter)
- Pvt. Joseph Field
- Pvt. Reuben Field
- 6 horses

Gass contingent, Great Falls portage, July 11–28, 1806—
- Sgt. Patrick Gass (journalist)
- Pvt. Robert Frazer
- Pvt. Silas Goodrich
- Pvt. Hugh McNeal
- Pvt. John Thompson
- Pvt. William Werner
- 4 horses

Ordway contingent from Three Forks, joins Gass contingent at Great Falls portage, July 19–28, 1806—
- Sgt. John Ordway (journalist)
- Pvt. John Collins
- Pvt. John Colter
- Pvt. Pierre Cruzatte
- Pvt. Thomas Howard
- Pvt. Baptiste Lepage
- Pvt. John Potts
- Pvt. Joseph Whitehouse
- Pvt. Alexander Willard
- Pvt. Peter Wiser
- 6 canoes (1 broken up for firewood)

THE MOST NORTHERN POINT

During Lewis's short stay at the Great Falls, July 11–16, 1806, warriors ran off with 7 of his party's 17 horses. With only 10 mounts left, Lewis regretfully reduced the size of his upper Marias exploring party to 4 men and 6 horses, instead of 7 mounted men as originally planned.

The other 4 horses would be left to necessarily assist the 6 men portaging canoes and baggage around the Great Falls. The 10 men of Sergeant Ordway's canoe party, paddling downstream from the Three Forks, soon would join them.

President Thomas Jefferson's instructions to Meriwether Lewis had directed the Corps of Discovery to attempt to identify the most northerly point in the Missouri watershed and its relationship to Canadian rivers. This information could be critical in boundary negotiations regarding British and American possessions in western North America.

The captains had decided that the source of the Marias River likely was the most northerly Missouri tributary (in actuality, the Milk River headwaters extend considerably farther north). With his reduced party—including only Lewis, Drouillard, and the Field brothers—Lewis intended to ride north from the Great Falls to the Marias River and follow its tributaries to the most northern point to take latitude and longitude sightings. When this was accomplished, Lewis planned to return by August 5, 1806, to the Gass-Ordway portage group waiting for him below the Great Falls at the mouth of the Marias River. The combined party then would descend the Missouri as quickly as possible to rejoin Clark's party at the mouth of the Yellowstone River.

As it turned out, however, Lewis's detachment skirmished with an 8-man Blackfeet party on Two Medicine River, an upper Marias tributary. Fearing pursuit by larger bands of warriors, Lewis and his three men desperately rode southeast 120 miles (by Lewis's count) in slightly more than 24 hours, proceeding much of the way under moonlight and lightning flashes from thunderheads hanging on the horizon. On the morning of July 28, they were fortunate to quickly reunite with the portage party proceeding downstream toward the Marias. Turning the horses loose, the combined party of 20 men in 5 canoes and 1 pirogue proceeded down the Missouri.

A cartographic reconstruction of the Marias exploration is complicated by the fact that Lewis's traverse calls often are excessively long and overly simplified. For example, only a single call of "20 MILES" was recorded for the entire first day's travel from the Great Falls to their encampment on the Teton River. Another long call of "20 MILES" was recorded on July 19, "28 MILES" on July 20, "15 MILES" on July 21, and so forth. Lewis's extremely brief 120-mile compilation for the headlong July 27–28 ride obviously has its complications. Such long calls are not reassuring when mapping. No doubt, the course of Lewis's party along his traverse calls occasionally diverted at times to bypass gullies or other difficult terrain features. Consequently, their actual line of travel sometimes may have varied somewhat off the edges of MAPS 495–527. However, such key places as the first Marias River campsite, "Camp Disappointment," the fight site, the July 27 return camp near the Missouri, and some other locations can be identified with reasonable assurance or approximation.

RETURN TO ST. LOUIS

Maps 528, 529, 530 at the end of this section depict the return of the Corps of Discovery to the Mississippi River on September 23, 1806, and their arrival at St. Louis. With these final map pages, the *Lewis and Clark Trail Maps, A Cartographic Reconstruction, Volumes I, II, & III* are concluded.

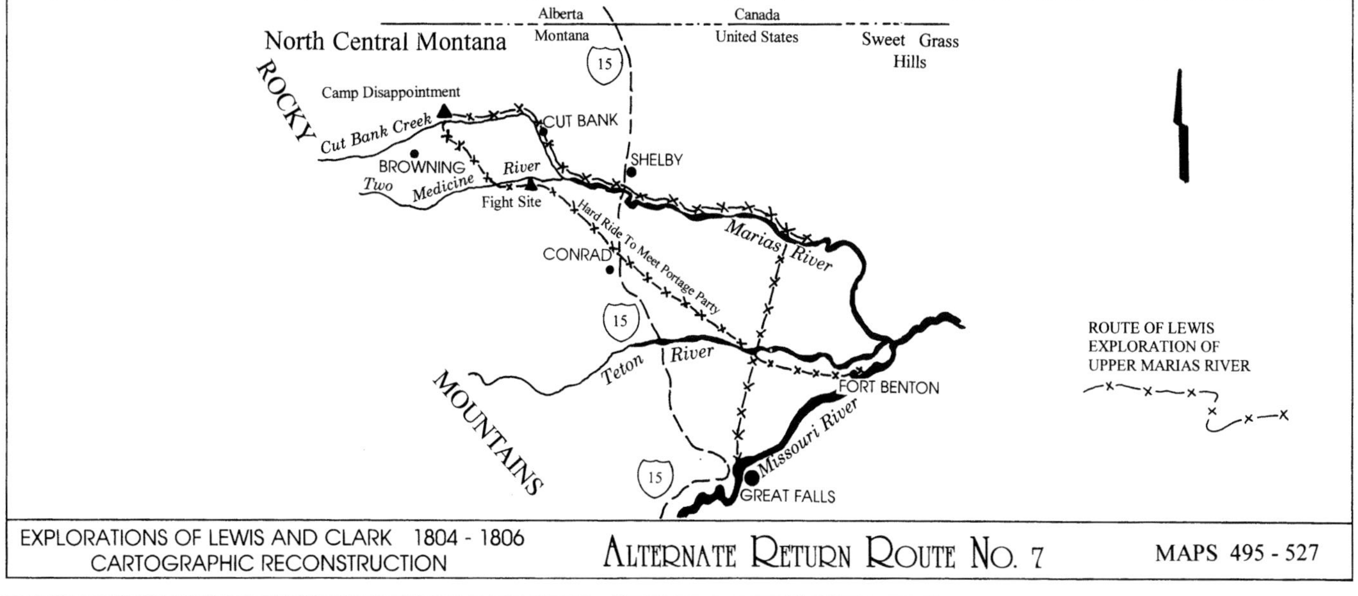

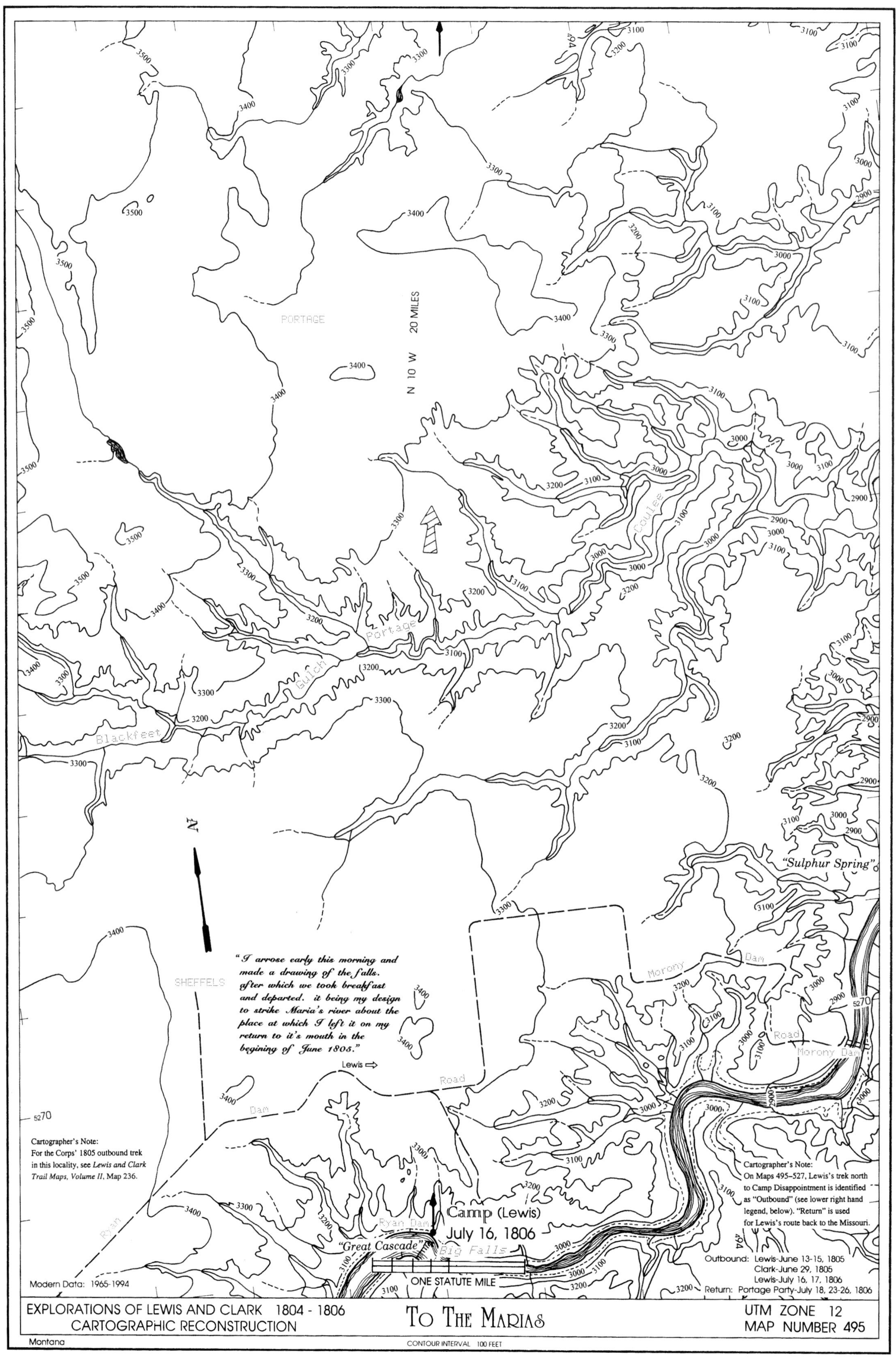

PORTAGE
N 10 W 20 MILES
Coulee
Portage
Gulch
Blackfeet
SHEFFELS
"I arrose early this morning and made a drawing of the falls. after which we took breakfast and departed. it being my design to strike Maria's river about the place at which I left it on my return to it's mouth in the begining of June 1805."
Lewis ⇨
Road
Dam
Morony
Dam
Road
Morony Dam
"Sulphur Spring"
Ryan
Ryan Dam
Camp (Lewis)
July 16, 1806
"Great Cascade"
Big Falls
ONE STATUTE MILE
Cartographer's Note:
For the Corps' 1805 outbound trek in this locality, see Lewis and Clark Trail Maps, Volume II, Map 236.
Cartographer's Note:
On Maps 495–527, Lewis's trek north to Camp Disappointment is identified as "Outbound" (see lower right hand legend, below). "Return" is used for Lewis's route back to the Missouri.
Outbound: Lewis-June 13-15, 1805
Clark-June 29, 1805
Lewis-July 16, 17, 1806
Return: Portage Party-July 18, 23-26, 1806
Modern Data: 1965-1994
EXPLORATIONS OF LEWIS AND CLARK 1804 - 1806
CARTOGRAPHIC RECONSTRUCTION
TO THE MARIAS
UTM ZONE 12
MAP NUMBER 495
Montana
CONTOUR INTERVAL 100 FEET

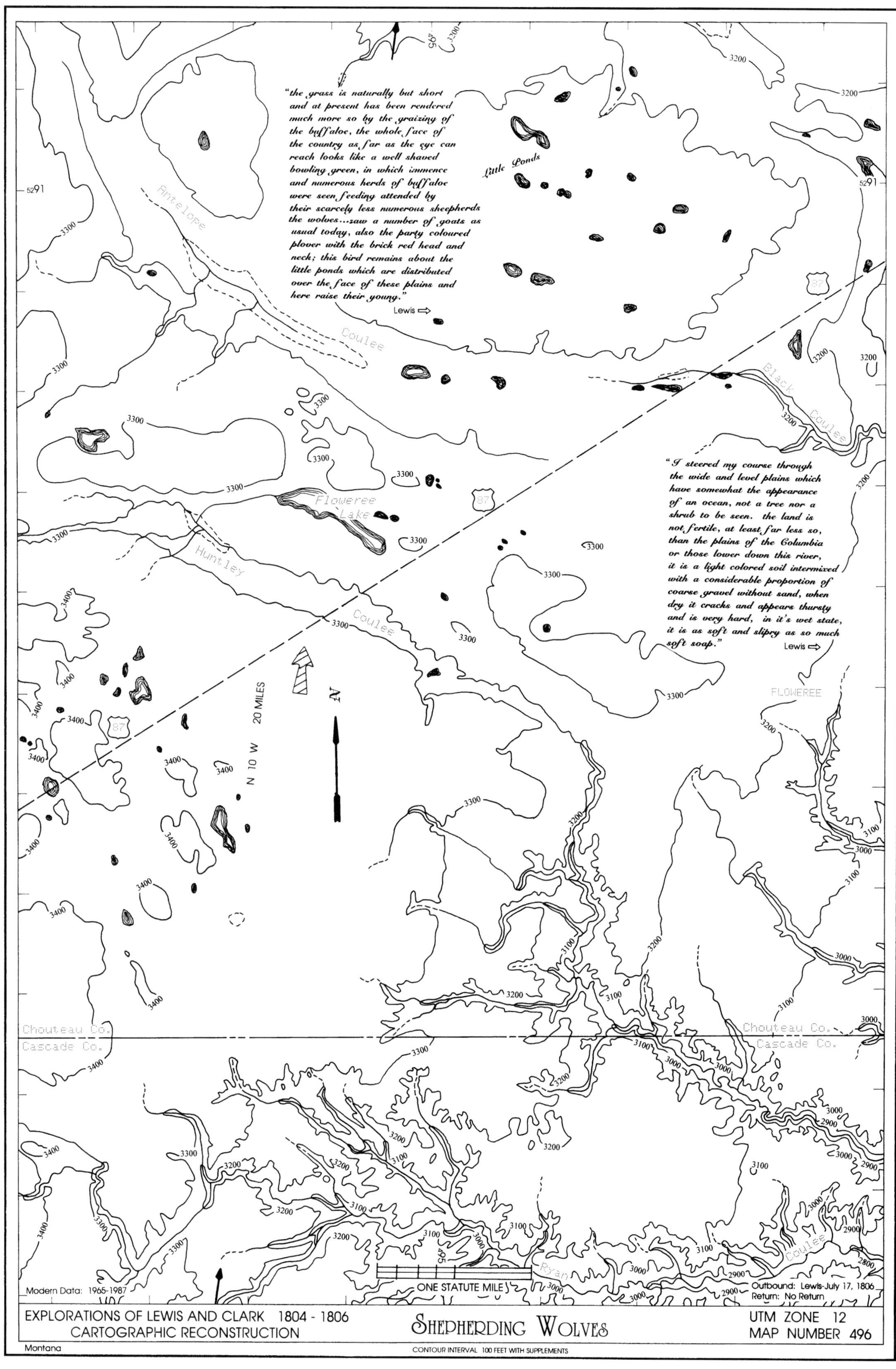
"the grass is naturally but short and at present has been rendered much more so by the graizing of the buffaloe, the whole face of the country as far as the eye can reach looks like a well shaved bowling green, in which immence and numerous herds of buffaloe were seen feeding attended by their scarcely less numerous sheepherds the wolves...saw a number of goats as usual today, also the party coloured plover with the brick red head and neck; this bird remains about the little ponds which are distributed over the face of these plains and here raise their young."
Lewis ⇨
Little Ponds
Antelope Coulee
Black Coulee
Floweree Lake
Huntley Coulee
"I steered my course through the wide and level plains which have somewhat the appearance of an ocean, not a tree nor a shrub to be seen. the land is not fertile, at least far less so, than the plains of the Columbia or those lower down this river, it is a light colored soil intermixed with a considerable proportion of coarse gravel without sand, when dry it cracks and appears thursty and is very hard, in it's wet state, it is as soft and slipry as so much soft soap."
Lewis ⇨
FLOWEREE
N 10 W 20 MILES
N
Chouteau Co.
Cascade Co.
Ryan Coulee
Modern Data: 1965-1987
ONE STATUTE MILE
Outbound: Lewis-July 17, 1806
Return: No Return
EXPLORATIONS OF LEWIS AND CLARK 1804 - 1806
CARTOGRAPHIC RECONSTRUCTION
SHEPHERDING WOLVES
UTM ZONE 12
MAP NUMBER 496
Montana
CONTOUR INTERVAL 100 FEET WITH SUPPLEMENTS

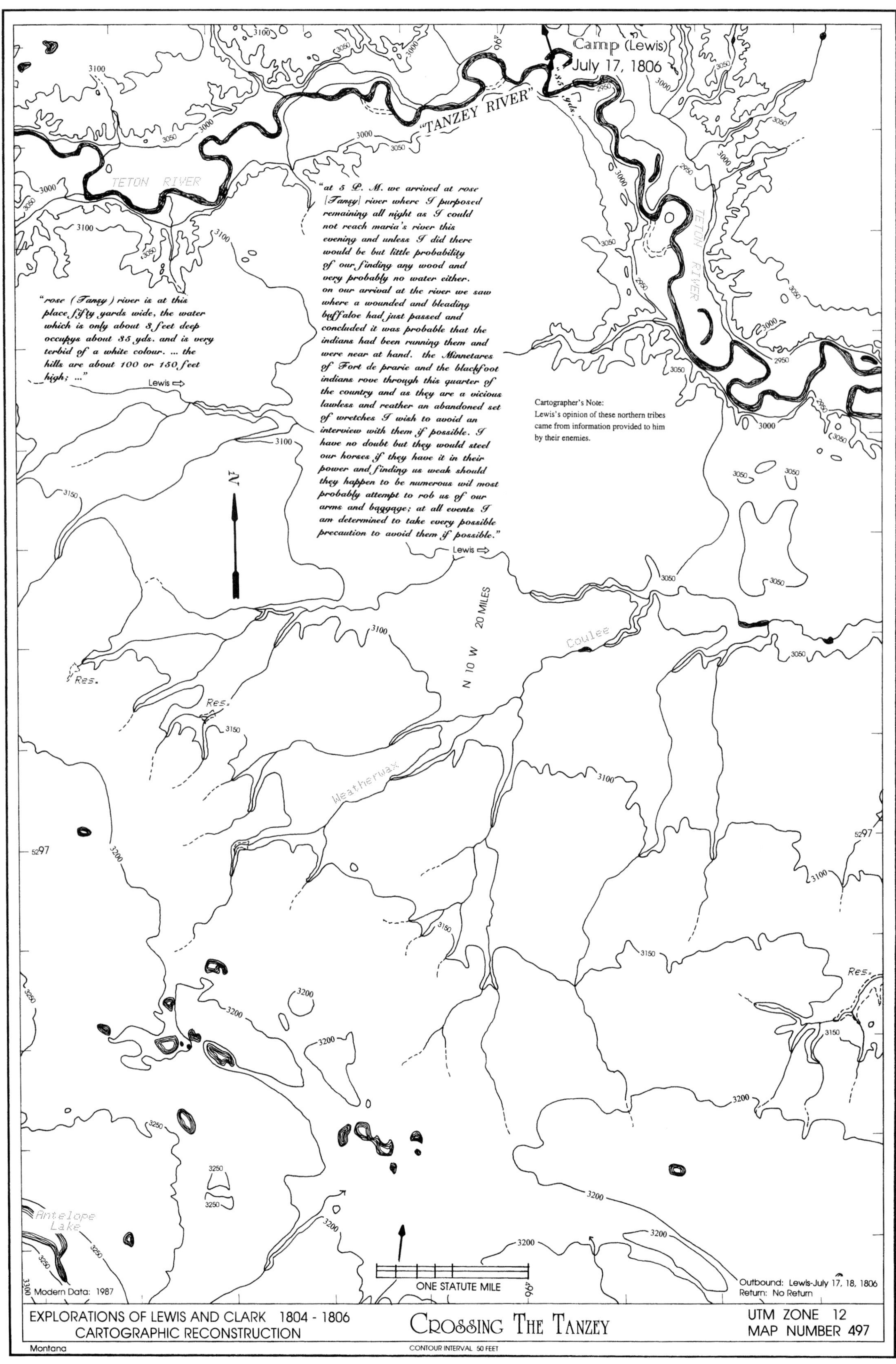

Camp (Lewis)
July 17, 1806
"TANZEY RIVER"
TETON RIVER
"at 5 P. M. we arrived at rose [Tansy] river where I purposed remaining all night as I could not reach maria's river this evening and unless I did there would be but little probability of our finding any wood and very probably no water either. on our arrival at the river we saw where a wounded and bleading buffaloe had just passed and concluded it was probable that the indians had been running them and were near at hand. the Minnetares of Fort de prarie and the blackfoot indians rove through this quarter of the country and as they are a vicious lawless and reather an abandoned set of wretches I wish to avoid an interview with them if possible. I have no doubt but they would steel our horses if they have it in their power and finding us weak should they happen to be numerous wil most probably attempt to rob us of our arms and baggage; at all events I am determined to take every possible precaution to avoid them if possible."
"rose (Tansy) river is at this place fifty yards wide, the water which is only about 3 feet deep occupys about 35 yds. and is very terbid of a white colour. ... the hills are about 100 or 150 feet high; ..."
Lewis ⇨
Lewis ⇨
Cartographer's Note:
Lewis's opinion of these northern tribes came from information provided to him by their enemies.
N
N 10 W 20 MILES
Coulee
Res.
Res.
Res.
Weatherwax
Antelope Lake
ONE STATUTE MILE
Modern Data: 1987
Outbound: Lewis-July 17, 18, 1806
Return: No Return
EXPLORATIONS OF LEWIS AND CLARK 1804 - 1806
CARTOGRAPHIC RECONSTRUCTION
CROSSING THE TANZEY
UTM ZONE 12
MAP NUMBER 497
Montana
CONTOUR INTERVAL 50 FEET

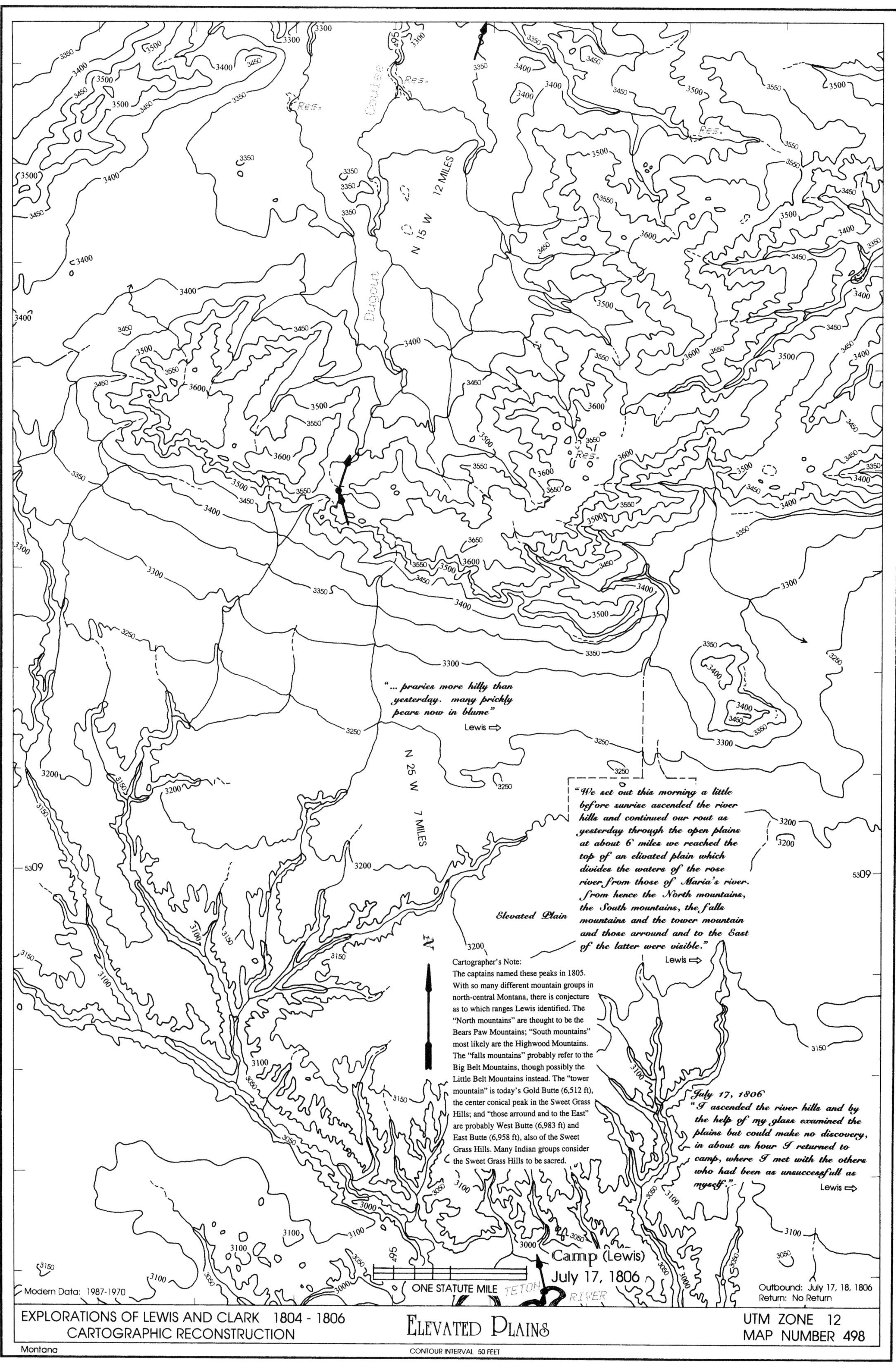
Coulee
Dugout
Res.
N 15 W 12 MILES
N 25 W 7 MILES
"... praries more hilly than yesterday. many prickly pears now in blume"
Lewis ⇨
"We set out this morning a little before sunrise ascended the river hills and continued our rout as yesterday through the open plains at about 6 miles we reached the top of an elivated plain which divides the waters of the rose river from those of Maria's river. from hence the North mountains, the South mountains, the falls mountains and the tower mountain and those arround and to the East of the latter were visible."
Lewis ⇨
Elevated Plain
Cartographer's Note:
The captains named these peaks in 1805. With so many different mountain groups in north-central Montana, there is conjecture as to which ranges Lewis identified. The "North mountains" are thought to be the Bears Paw Mountains; "South mountains" most likely are the Highwood Mountains. The "falls mountains" probably refer to the Big Belt Mountains, though possibly the Little Belt Mountains instead. The "tower mountain" is today's Gold Butte (6,512 ft), the center conical peak in the Sweet Grass Hills; and "those arround and to the East" are probably West Butte (6,983 ft) and East Butte (6,958 ft), also of the Sweet Grass Hills. Many Indian groups consider the Sweet Grass Hills to be sacred.
July 17, 1806
"I ascended the river hills and by the help of my glass examined the plains but could make no discovery, in about an hour I returned to camp, where I met with the others who had been as unsuccessfull as myself."
Lewis ⇨
Camp (Lewis)
July 17, 1806
TETON RIVER
ONE STATUTE MILE
Modern Data: 1987-1970
Outbound: July 17, 18, 1806
Return: No Return
EXPLORATIONS OF LEWIS AND CLARK 1804 - 1806
CARTOGRAPHIC RECONSTRUCTION
ELEVATED PLAINS
UTM ZONE 12
MAP NUMBER 498
Montana
CONTOUR INTERVAL 50 FEET

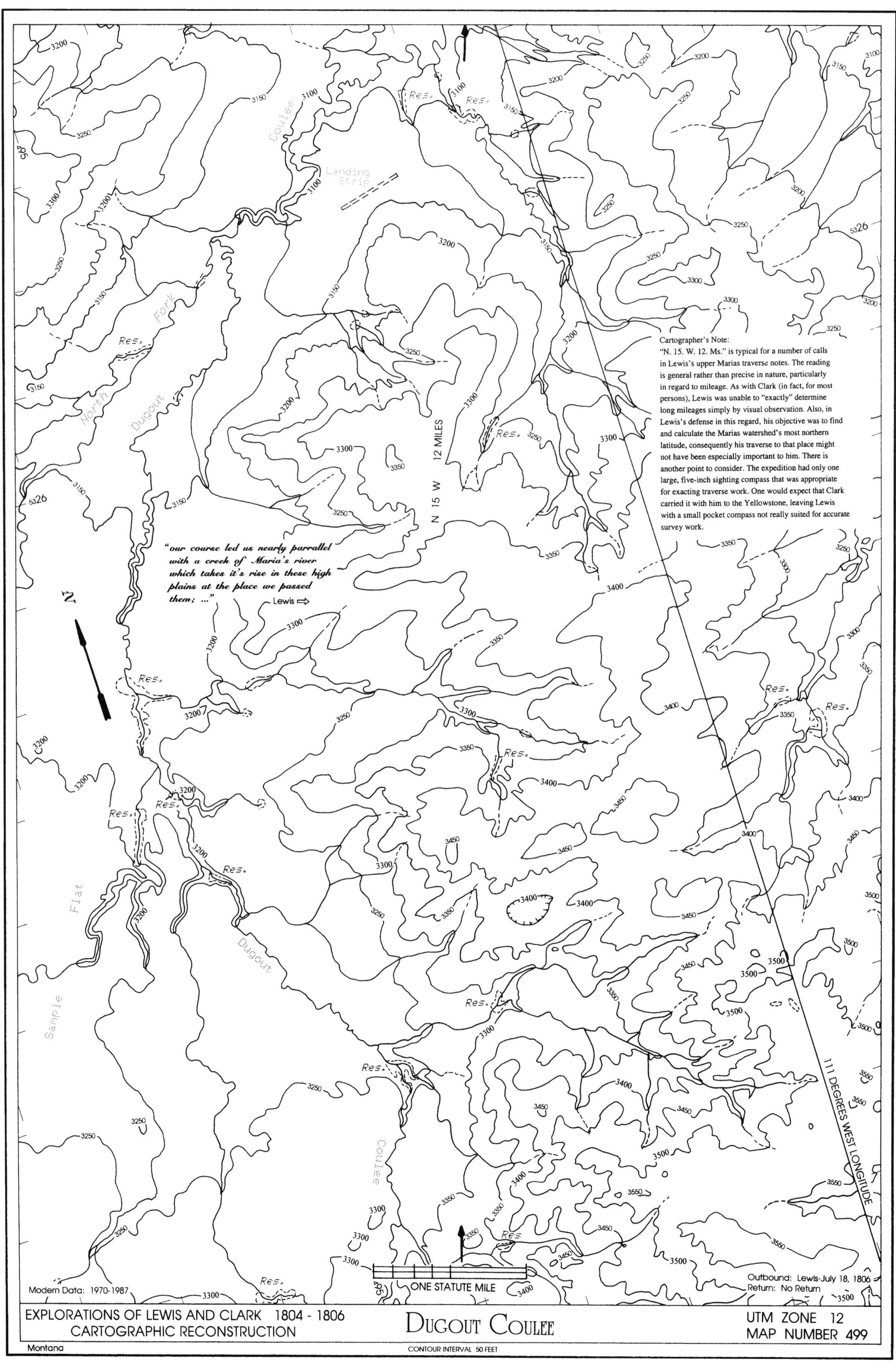
Cartographer's Note:
"N. 15. W. 12. Ms." is typical for a number of calls in Lewis's upper Marias traverse notes. The reading is general rather than precise in nature, particularly in regard to mileage. As with Clark (in fact, for most persons), Lewis was unable to "exactly" determine long mileages simply by visual observation. Also, in Lewis's defense in this regard, his objective was to find and calculate the Marias watershed's most northern latitude, consequently his traverse to that place might not have been especially important to him. There is another point to consider. The expedition had only one large, five-inch sighting compass that was appropriate for exacting traverse work. One would expect that Clark carried it with him to the Yellowstone, leaving Lewis with a small pocket compass not really suited for accurate survey work.
"our course led us nearly parrallel with a creek of Maria's river which takes it's rise in these high plains at the place we passed them; ..."
Lewis ⇨
N 15 W 12 MILES
111 DEGREES WEST LONGITUDE
Landing Strip
Coulee
North Fork Dugout
Dugout
Flat
Sample
Dugout Coulee
Res.
N
ONE STATUTE MILE
Modern Data: 1970-1987
Outbound: Lewis-July 18, 1806
Return: No Return
EXPLORATIONS OF LEWIS AND CLARK 1804 - 1806
CARTOGRAPHIC RECONSTRUCTION
DUGOUT COULEE
UTM ZONE 12
MAP NUMBER 499
Montana
CONTOUR INTERVAL 50 FEET

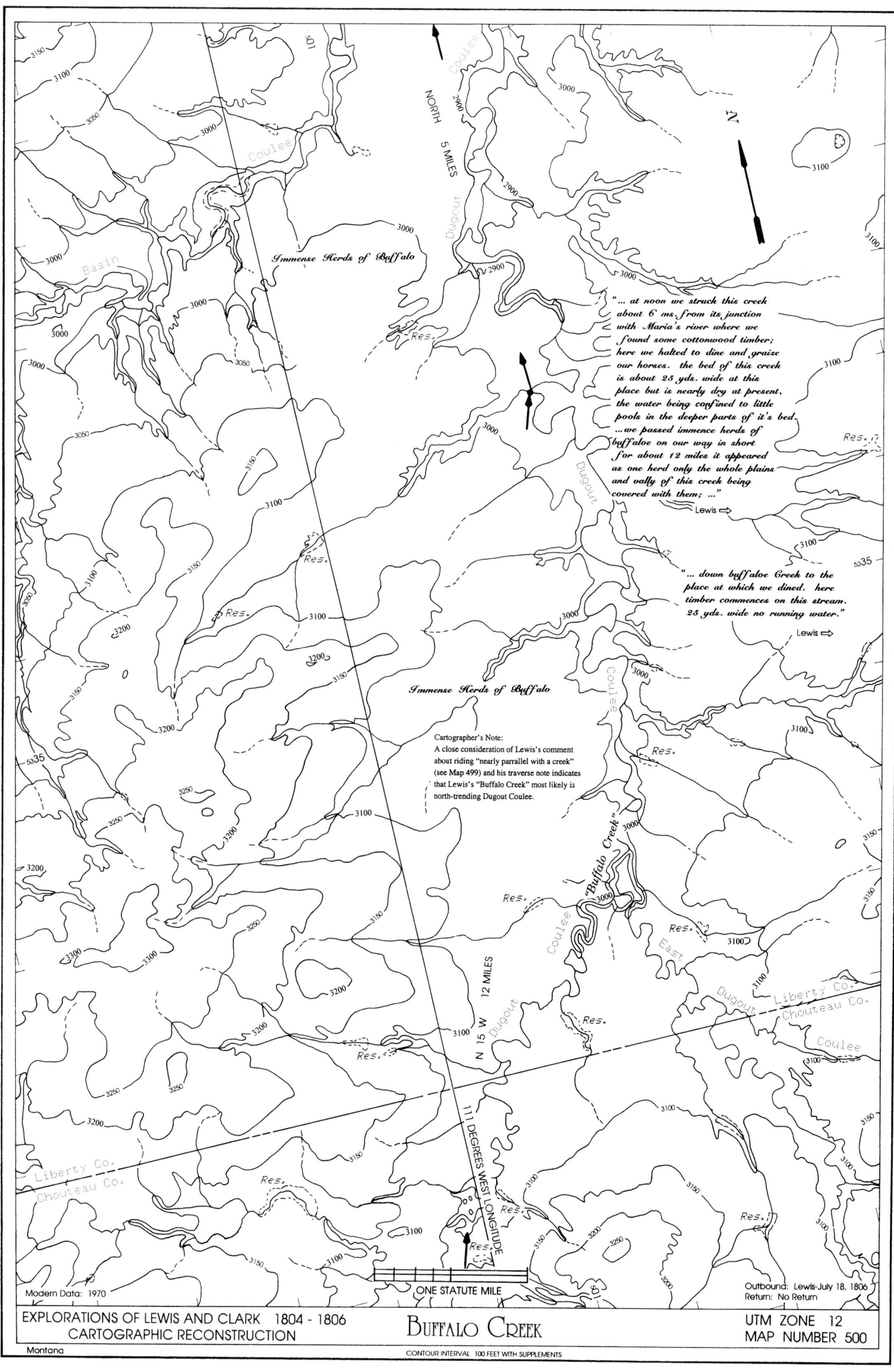

Immense Herds of Buffalo
"... at noon we struck this creek about 6 ms. from its junction with Maria's river where we found some cottonwood timber; here we halted to dine and graize our horses. the bed of this creek is about 25 yds. wide at this place but is nearly dry at present, the water being confined to little pools in the deeper parts of it's bed. ...we passed immence herds of buffaloe on our way in short for about 12 miles it appeared as one herd only the whole plains and vally of this creek being covered with them; ..."
Lewis
"... down buffaloe Creek to the place at which we dined. here timber commences on this stream. 25 yds. wide no running water."
Lewis
Immense Herds of Buffalo
Cartographer's Note:
A close consideration of Lewis's comment about riding "nearly parrallel with a creek" (see Map 499) and his traverse note indicates that Lewis's "Buffalo Creek" most likely is north-trending Dugout Coulee.
"Buffalo Creek"
NORTH 5 MILES
N 15 W 12 MILES
111 DEGREES WEST LONGITUDE
Dugout Coulee
East Dugout Coulee
Basin
Coulee
Res.
Liberty Co.
Chouteau Co.
ONE STATUTE MILE
Modern Data: 1970
Outbound: Lewis-July 18, 1806
Return: No Return
EXPLORATIONS OF LEWIS AND CLARK 1804 - 1806
CARTOGRAPHIC RECONSTRUCTION
BUFFALO CREEK
UTM ZONE 12
MAP NUMBER 500
Montana
CONTOUR INTERVAL 100 FEET WITH SUPPLEMENTS

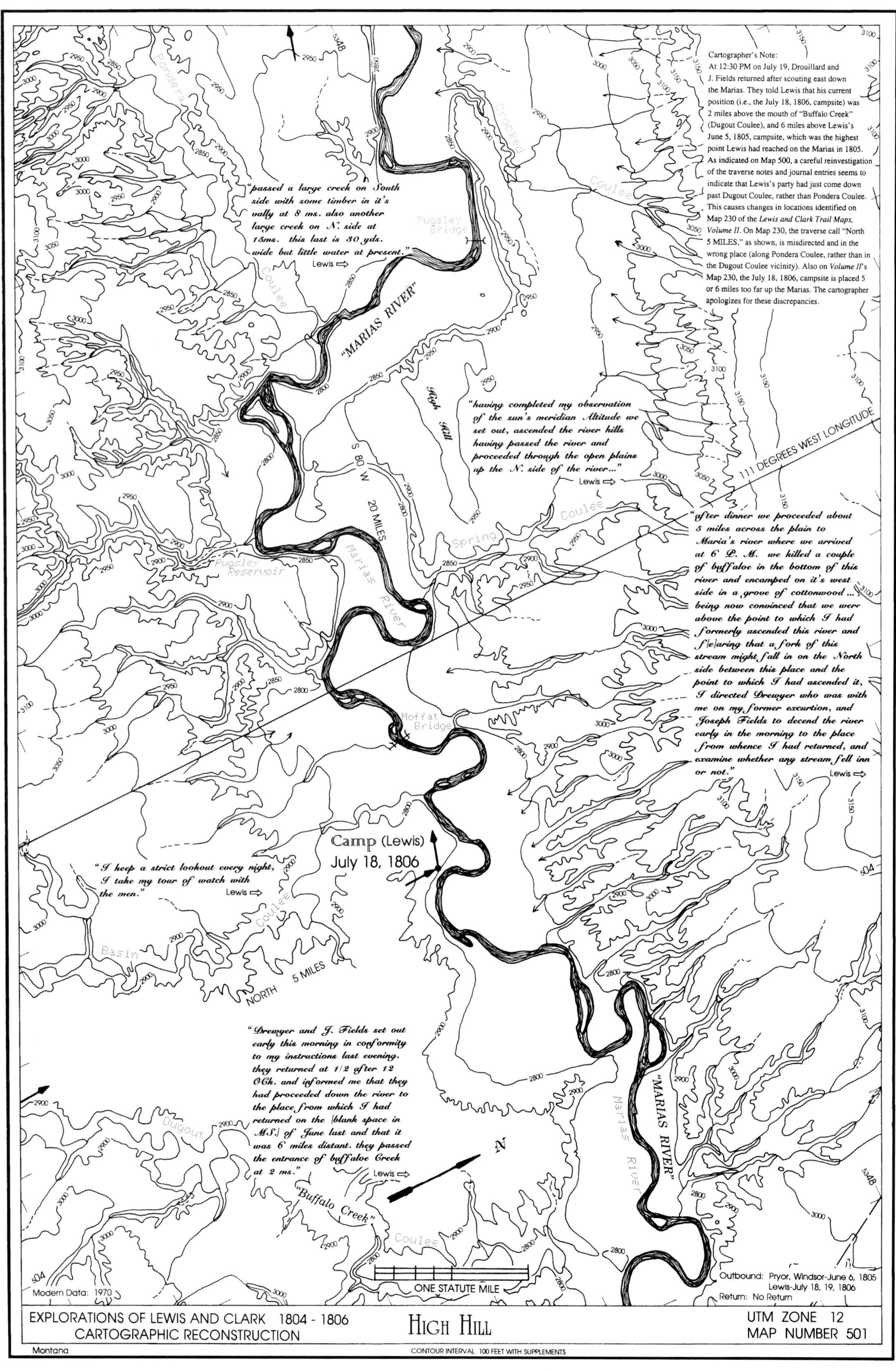

Cartographer's Note:
At 12:30 PM on July 19, Drouillard and J. Fields returned after scouting east down the Marias. They told Lewis that his current position (i.e., the July 18, 1806, campsite) was 2 miles above the mouth of "Buffalo Creek" (Dugout Coulee), and 6 miles above Lewis's June 5, 1805, campsite, which was the highest point Lewis had reached on the Marias in 1805. As indicated on Map 500, a careful reinvestigation of the traverse notes and journal entries seems to indicate that Lewis's party had just come down past Dugout Coulee, rather than Pondera Coulee. This causes changes in locations identified on Map 230 of the Lewis and Clark Trail Maps, Volume II. On Map 230, the traverse call "North 5 MILES," as shown, is misdirected and in the wrong place (along Pondera Coulee, rather than in the Dugout Coulee vicinity). Also on Volume II's Map 230, the July 18, 1806, campsite is placed 5 or 6 miles too far up the Marias. The cartographer apologizes for these discrepancies.
"passed a large creek on South side with some timber in it's vally at 8 ms. also another large creek on N. side at 15ms. this last is 30 yds. wide but little water at present."
Lewis ⇨
"having completed my observation of the sun's meridian Altitude we set out, ascended the river hills having passed the river and proceeded through the open plains up the N. side of the river..."
Lewis ⇨
"after dinner we proceeded about 5 miles across the plain to Maria's river where we arrived at 6 P. M. we killed a couple of buffaloe in the bottom of this river and encamped on it's west side in a grove of cottonwood ... being now convinced that we were above the point to which I had formerly ascended this river and f[e]aring that a fork of this stream might fall in on the North side between this place and the point to which I had ascended it, I directed Drewyer who was with me on my former excurtion, and Joseph Fields to decend the river early in the morning to the place from whence I had returned, and examine whether any stream fell inn or not."
Lewis ⇨
"I keep a strict lookout every night, I take my tour of watch with the men."
Lewis ⇨
"Drewyer and J. Fields set out early this morning in conformity to my instructions last evening. they returned at 1/2 after 12 OCk. and informed me that they had proceeded down the river to the place from which I had returned on the [blank space in MS.] of June last and that it was 6 miles distant. they passed the entrance of buffaloe Creek at 2 ms."
Lewis ⇨
Camp (Lewis)
July 18, 1806
"MARIAS RIVER"
Marias River
Pondera Coulee
Crooked Coulee
Pugsley Bridge
Pugsley Reservoir
Spring Coulee
Moffat Bridge
High Hill
S 80 W
20 MILES
NORTH 5 MILES
111 DEGREES WEST LONGITUDE
Basin Coulee
Dugout Coulee
"Buffalo Creek"
N
ONE STATUTE MILE
Modern Data: 1970
Outbound: Pryor, Windsor-June 6, 1805
Lewis-July 18, 19, 1806
Return: No Return
EXPLORATIONS OF LEWIS AND CLARK 1804 - 1806
CARTOGRAPHIC RECONSTRUCTION
HIGH HILL
UTM ZONE 12
MAP NUMBER 501
Montana
CONTOUR INTERVAL 100 FEET WITH SUPPLEMENTS

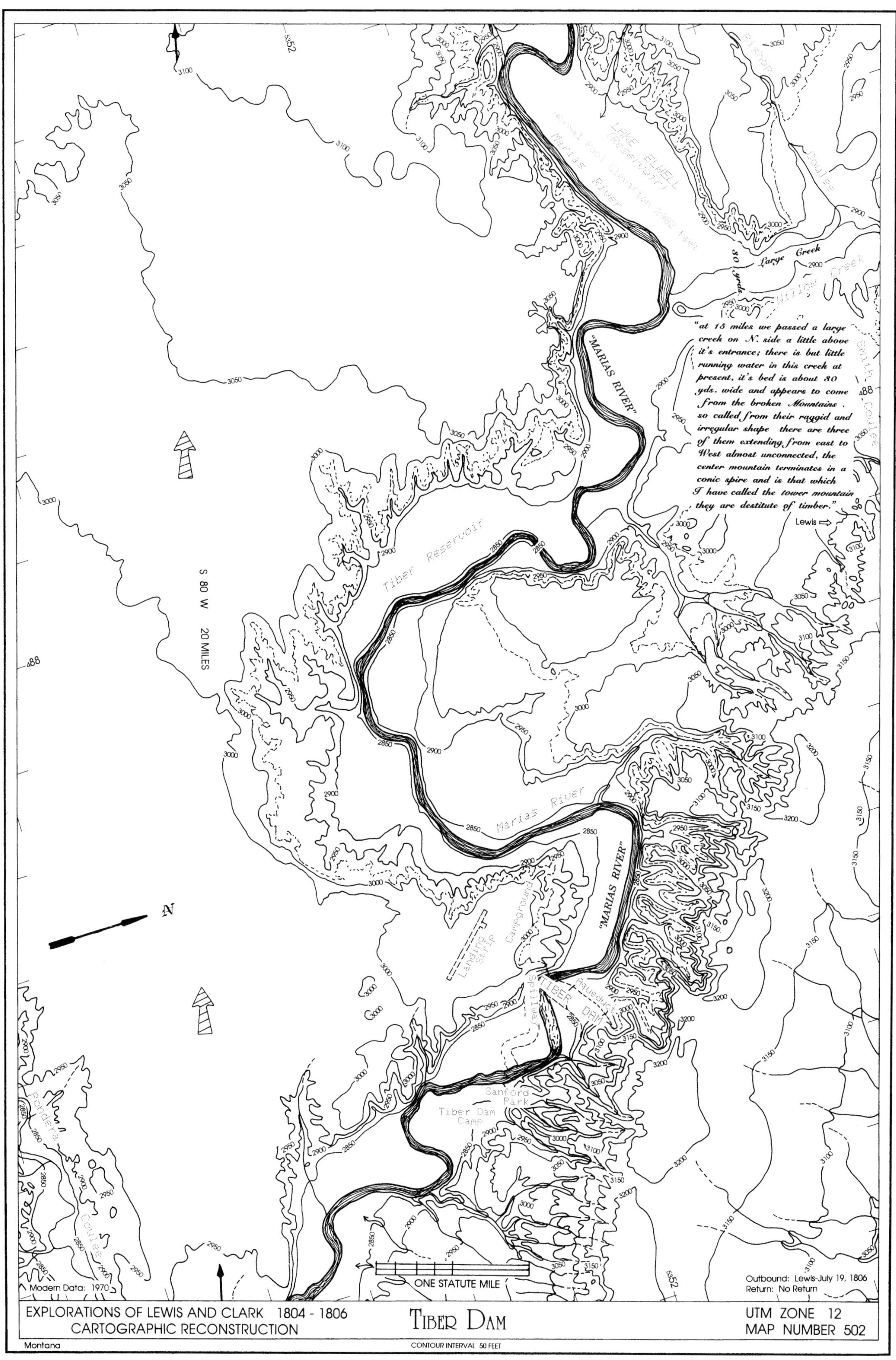

LAKE ELWELL (Reservoir)
Normal Pool Elevation 2982 feet
Marias River
Large Creek
80 yds.
Willow Creek
Bishop Coulee
Smith Coulee
"MARIAS RIVER"
"at 15 miles we passed a large creek on N. side a little above it's entrance; there is but little running water in this creek at present, it's bed is about 80 yds. wide and appears to come from the broken Mountains. so called from their raggid and irregular shape there are three of them extending from east to West almost unconnected, the center mountain terminates in a conic spire and is that which I have called the tower mountain they are destitute of timber."
Lewis
Tiber Reservoir
S 80 W 20 MILES
Marias River
"MARIAS RIVER"
N
Landing Strip
Campground
Spillway
TIBER DAM
Aqueduct
Sanford Park
Tiber Dam Camp
Pondera Coulee
ONE STATUTE MILE
Modern Data: 1970
Outbound: Lewis-July 19, 1806
Return: No Return
EXPLORATIONS OF LEWIS AND CLARK 1804 - 1806
CARTOGRAPHIC RECONSTRUCTION
TIBER DAM
UTM ZONE 12
MAP NUMBER 502
Montana
CONTOUR INTERVAL 50 FEET

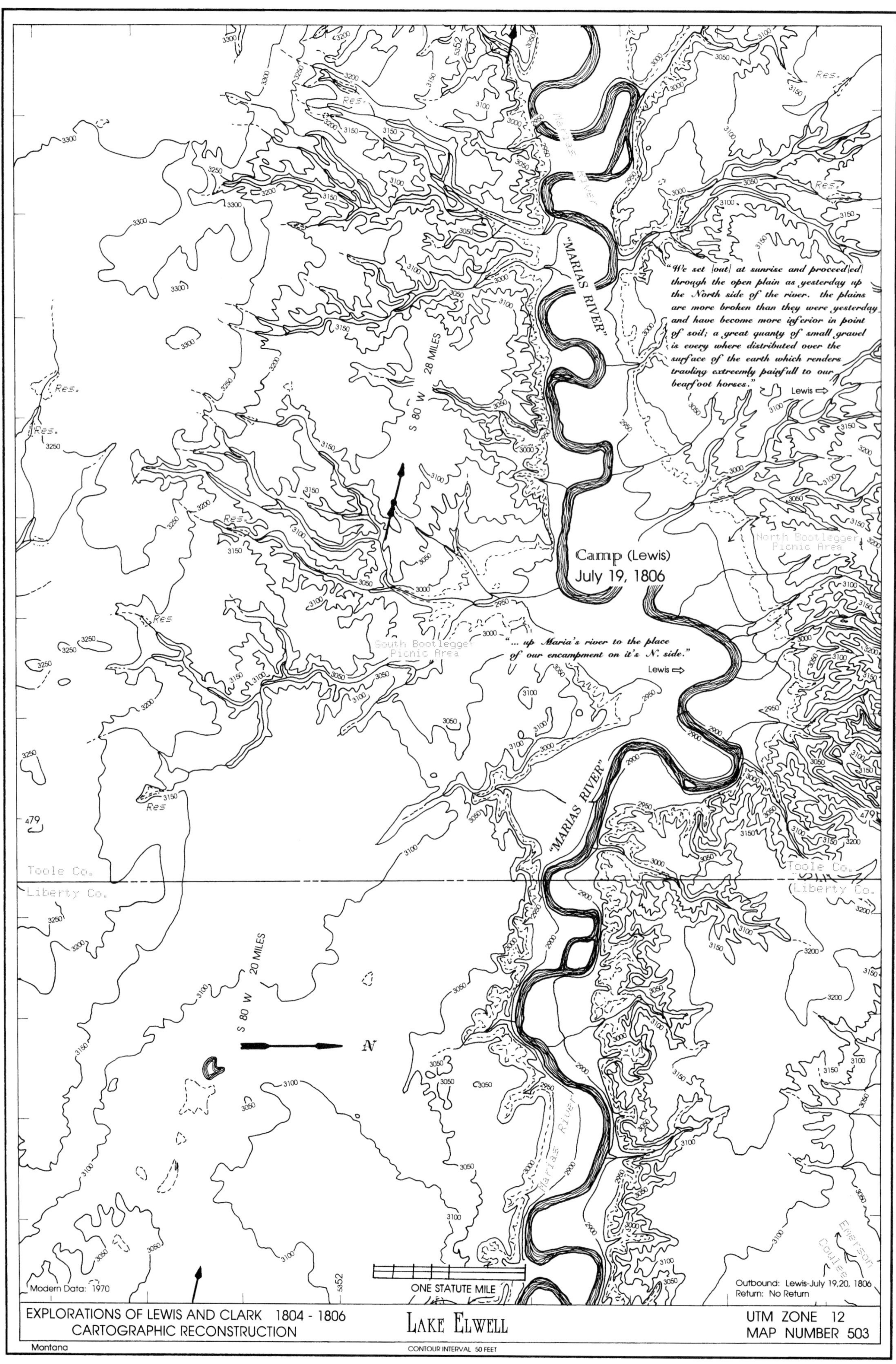
"We set [out] at sunrise and proceed[ed] through the open plain as yesterday up the North side of the river. the plains are more broken than they were yesterday and have become more inferior in point of soil; a great quanty of small gravel is every where distributed over the the surface of the earth which renders traveling extreemly painfull to our bearfoot horses."
"... up Maria's river to the place of our encampment on it's N. side."
Camp (Lewis)
July 19, 1806
"MARIAS RIVER"
Marias River
S 80 W 28 MILES
S 80 W 20 MILES
North Bootlegger Picnic Area
South Bootlegger Picnic Area
Toole Co.
Liberty Co.
Emerson Coulee
Lewis
ONE STATUTE MILE
Modern Data: 1970
Outbound: Lewis-July 19,20, 1806
Return: No Return
EXPLORATIONS OF LEWIS AND CLARK 1804 - 1806
CARTOGRAPHIC RECONSTRUCTION
LAKE ELWELL
UTM ZONE 12
MAP NUMBER 503
Montana
CONTOUR INTERVAL 50 FEET

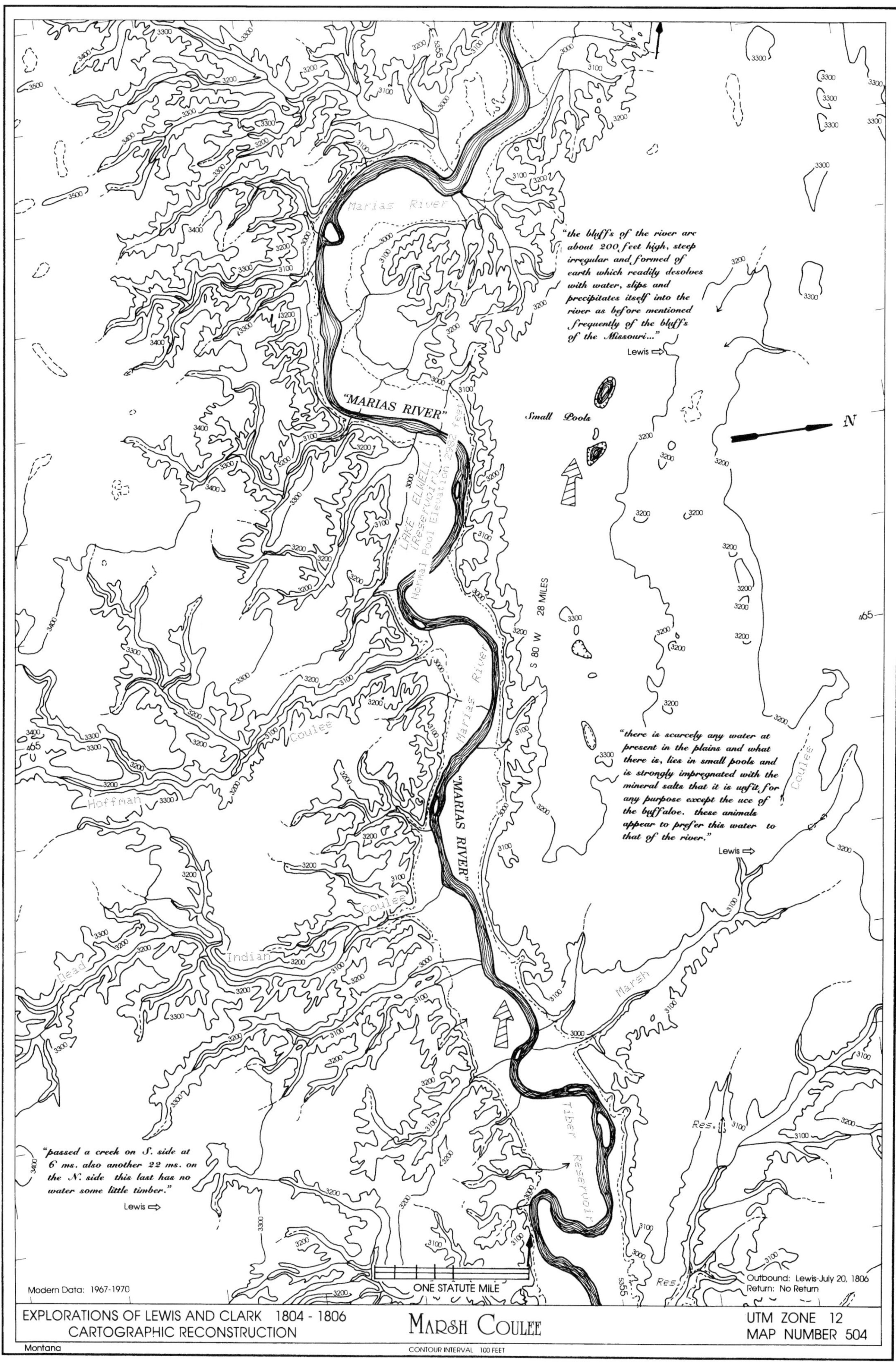
Marias River
"the bluffs of the river are about 200 feet high, steep irregular and formed of earth which readily desolves with water, slips and precipitates itself into the river as before mentioned frequently of the bluffs of the Missouri..."
Lewis ⇨
"MARIAS RIVER"
Small Pools
N
LAKE ELWELL (Reservoir) Normal Pool Elevation 2992 feet
S 80 W 28 MILES
Marias River
Coulee
Hoffman
"MARIAS RIVER"
"there is scarcely any water at present in the plains and what there is, lies in small pools and is strongly impregnated with the mineral salts that it is unfit for any purpose except the uce of the buffaloe. these animals appear to prefer this water to that of the river."
Lewis ⇨
Coulee
Coulee
Dead
Indian
Marsh
Tiber Reservoir
Res.
"passed a creek on S. side at 6 ms. also another 22 ms. on the N. side this last has no water some little timber."
Lewis ⇨
Modern Data: 1967-1970
ONE STATUTE MILE
Res.
Outbound: Lewis-July 20, 1806
Return: No Return
EXPLORATIONS OF LEWIS AND CLARK 1804 - 1806
CARTOGRAPHIC RECONSTRUCTION
MARSH COULEE
UTM ZONE 12
MAP NUMBER 504
Montana
CONTOUR INTERVAL 100 FEET

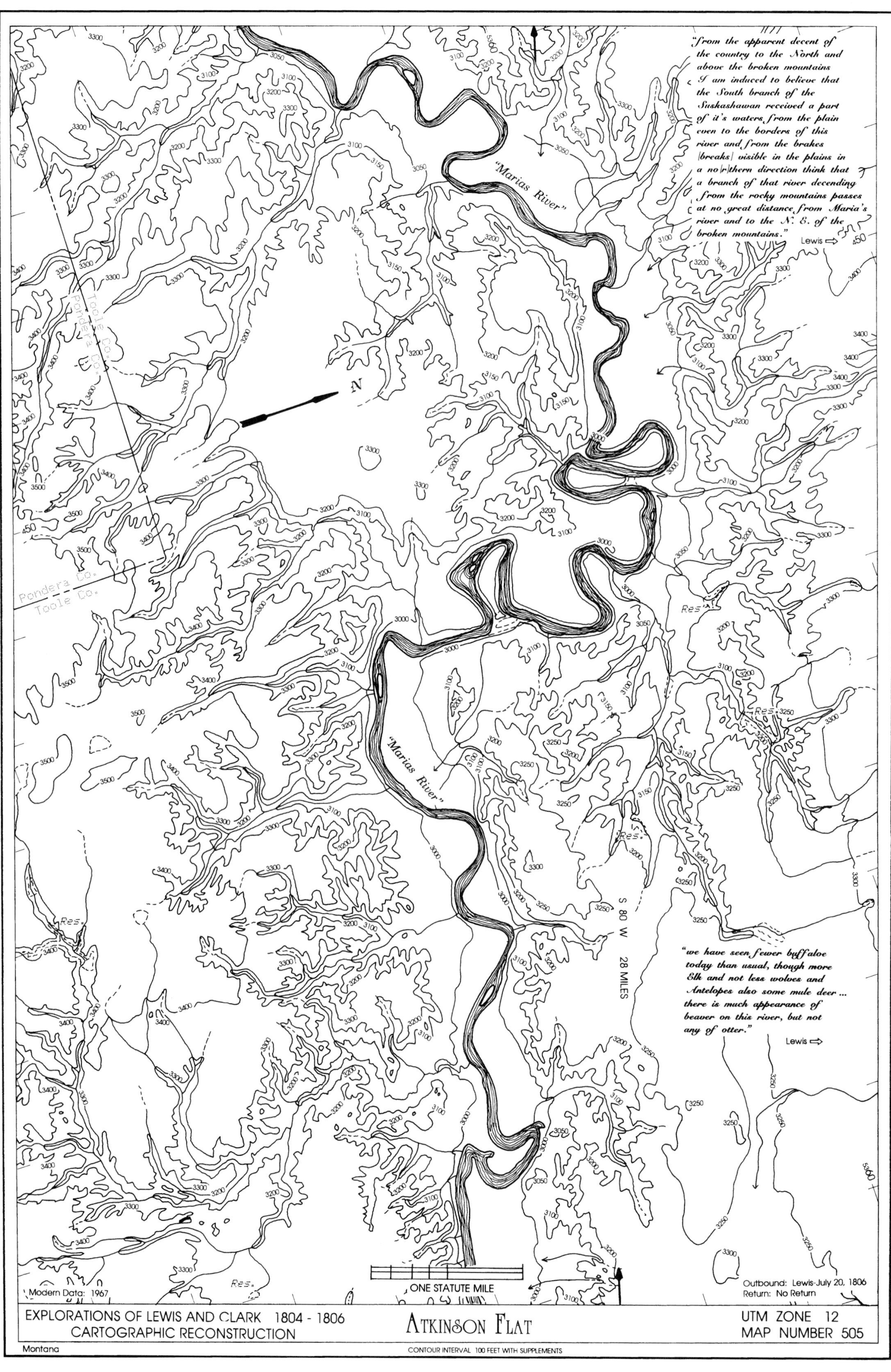
"from the apparent decent of the country to the North and above the broken mountains I am induced to believe that the South branch of the Suskashawan received a part of it's waters from the plain even to the borders of this river and from the brakes [breaks] visible in the plains in a no[r]thern direction think that a branch of that river decending from the rocky mountains passes at no great distance from Maria's river and to the N. E. of the broken mountains."
Lewis ⇨
"Marias River"
"Marias River"
N
Toole Co.
Pondera Co.
S 80 W 28 MILES
"we have seen fewer buffaloe today than usual, though more Elk and not less wolves and Antelopes also some mule deer ... there is much appearance of beaver on this river, but not any of otter."
Lewis ⇨
ONE STATUTE MILE
Modern Data: 1967
Outbound: Lewis-July 20, 1806
Return: No Return
EXPLORATIONS OF LEWIS AND CLARK 1804 - 1806
CARTOGRAPHIC RECONSTRUCTION
ATKINSON FLAT
UTM ZONE 12
MAP NUMBER 505
Montana
CONTOUR INTERVAL 100 FEET WITH SUPPLEMENTS

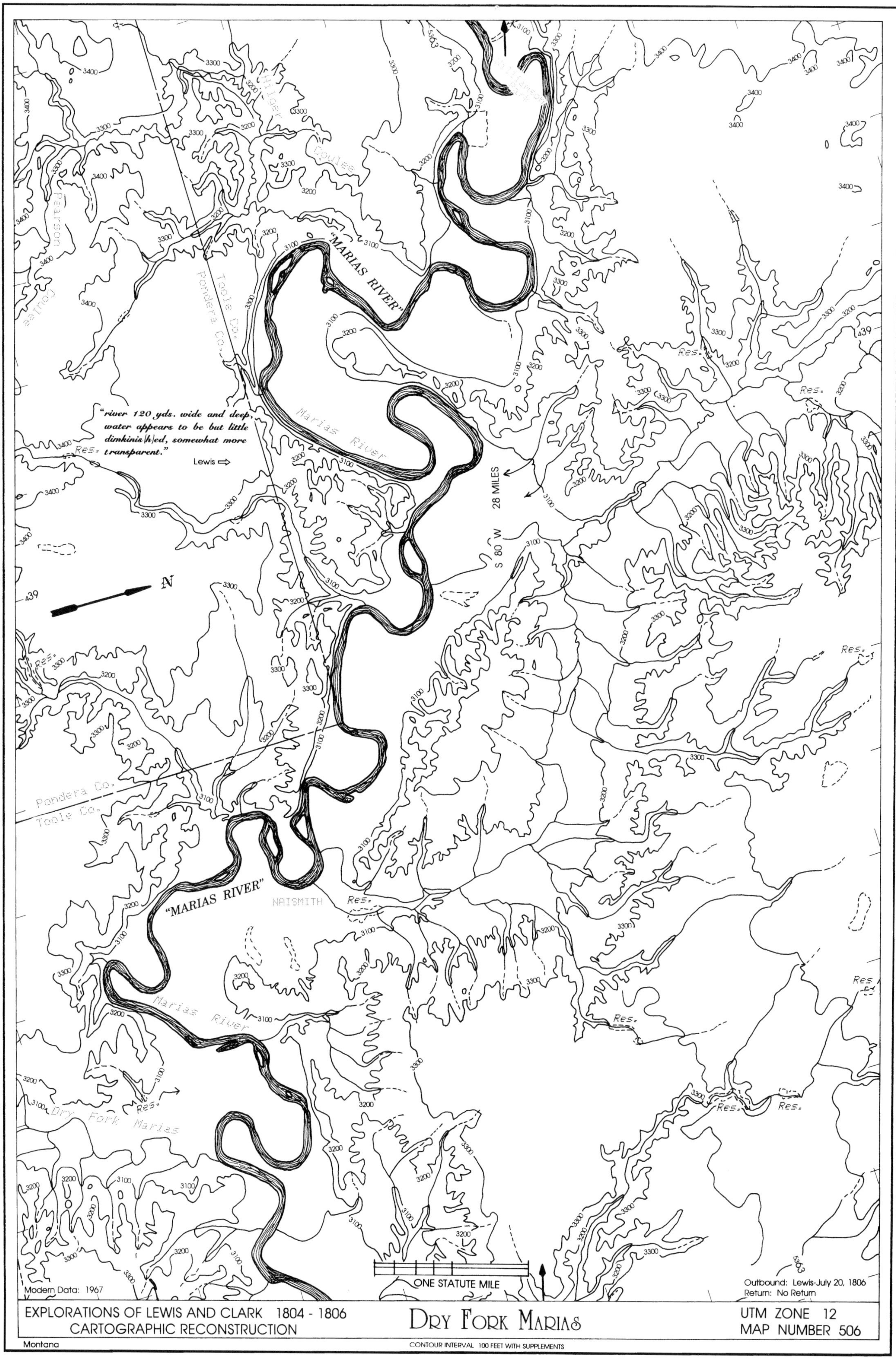
"MARIAS RIVER"
"river 120 yds. wide and deep, water appears to be but little dimkinis[h]ed, somewhat more transparent."
Lewis ⇨
28 MILES
S 80 W
N
"MARIAS RIVER"
NAISMITH
Modern Data: 1967
ONE STATUTE MILE
Outbound: Lewis-July 20, 1806
Return: No Return
EXPLORATIONS OF LEWIS AND CLARK 1804 - 1806
CARTOGRAPHIC RECONSTRUCTION
DRY FORK MARIAS
UTM ZONE 12
MAP NUMBER 506
Montana
CONTOUR INTERVAL 100 FEET WITH SUPPLEMENTS

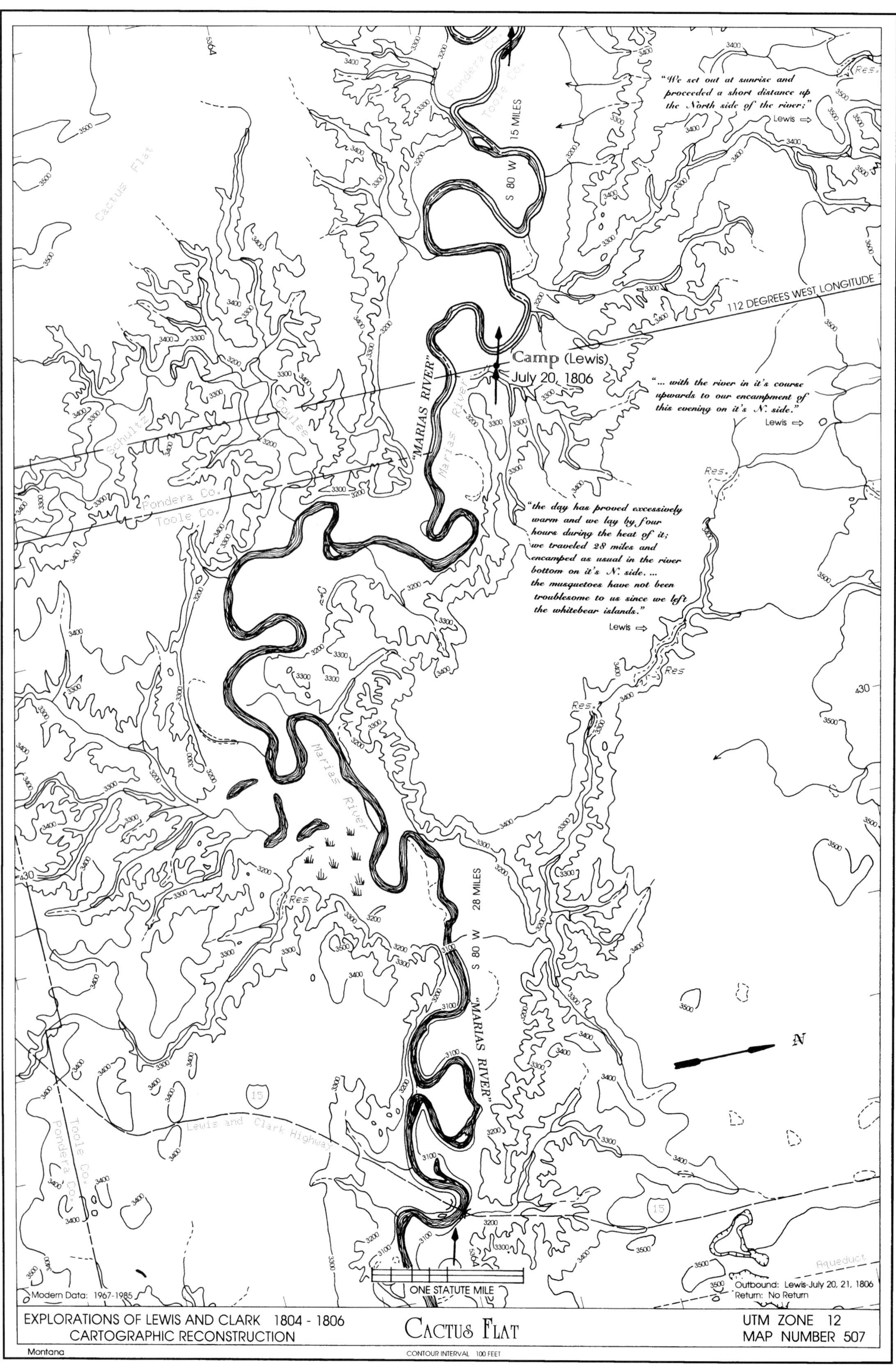
"We set out at sunrise and proceeded a short distance up the North side of the river;"
Lewis ⇨
Camp (Lewis)
July 20, 1806
"... with the river in it's course upwards to our encampment of this evening on it's N. side."
Lewis ⇨
"the day has proved excessively warm and we lay by four hours during the heat of it; we traveled 28 miles and encamped as usual in the river bottom on it's N. side. ... the musquetoes have not been troublesome to us since we left the whitebear islands."
Lewis ⇨
112 DEGREES WEST LONGITUDE
S 80 W 15 MILES
S 80 W 28 MILES
"MARIAS RIVER"
Marias River
Cactus Flat
Schultz Coulee
Pondera Co.
Toole Co.
Lewis and Clark Highway
Aqueduct
Res.
N
ONE STATUTE MILE
Modern Data: 1967-1985
Outbound: Lewis-July 20, 21, 1806
Return: No Return
EXPLORATIONS OF LEWIS AND CLARK 1804 - 1806
CARTOGRAPHIC RECONSTRUCTION
CACTUS FLAT
UTM ZONE 12
MAP NUMBER 507
Montana
CONTOUR INTERVAL 100 FEET

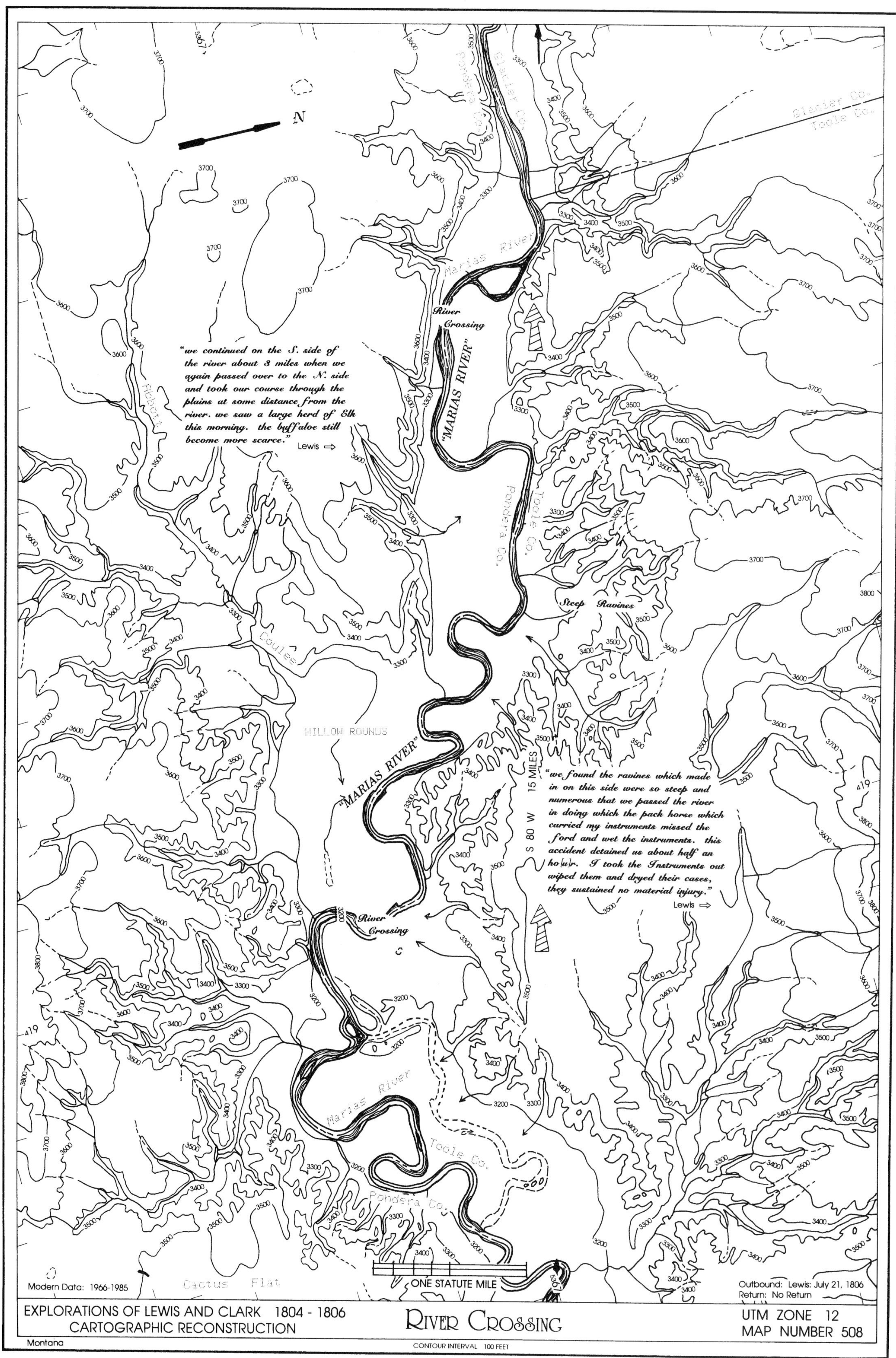
N
Pondera Co.
Glacier Co.
Glacier Co.
Toole Co.
Marias River
River Crossing
"MARIAS RIVER"
"we continued on the S. side of the river about 3 miles when we again passed over to the N. side and took our course through the plains at some distance from the river. we saw a large herd of Elk this morning. the buffaloe still become more scarce."
Lewis ⇨
Abbott
Pondera Co.
Toole Co.
Steep Ravines
Coulee
WILLOW ROUNDS
"MARIAS RIVER"
S 80 W 15 MILES
"we found the ravines which made in on this side were so steep and numerous that we passed the river in doing which the pack horse which carried my instruments missed the ford and wet the instruments. this accident detained us about half an ho[u]r. I took the Instruments out wiped them and dryed their cases, they sustained no material injury."
Lewis ⇨
River Crossing
Marias River
Toole Co.
Pondera Co.
Cactus Flat
Modern Data: 1966-1985
ONE STATUTE MILE
Outbound: Lewis: July 21, 1806
Return: No Return
EXPLORATIONS OF LEWIS AND CLARK 1804 - 1806
CARTOGRAPHIC RECONSTRUCTION
RIVER CROSSING
UTM ZONE 12
MAP NUMBER 508
Montana
CONTOUR INTERVAL 100 FEET

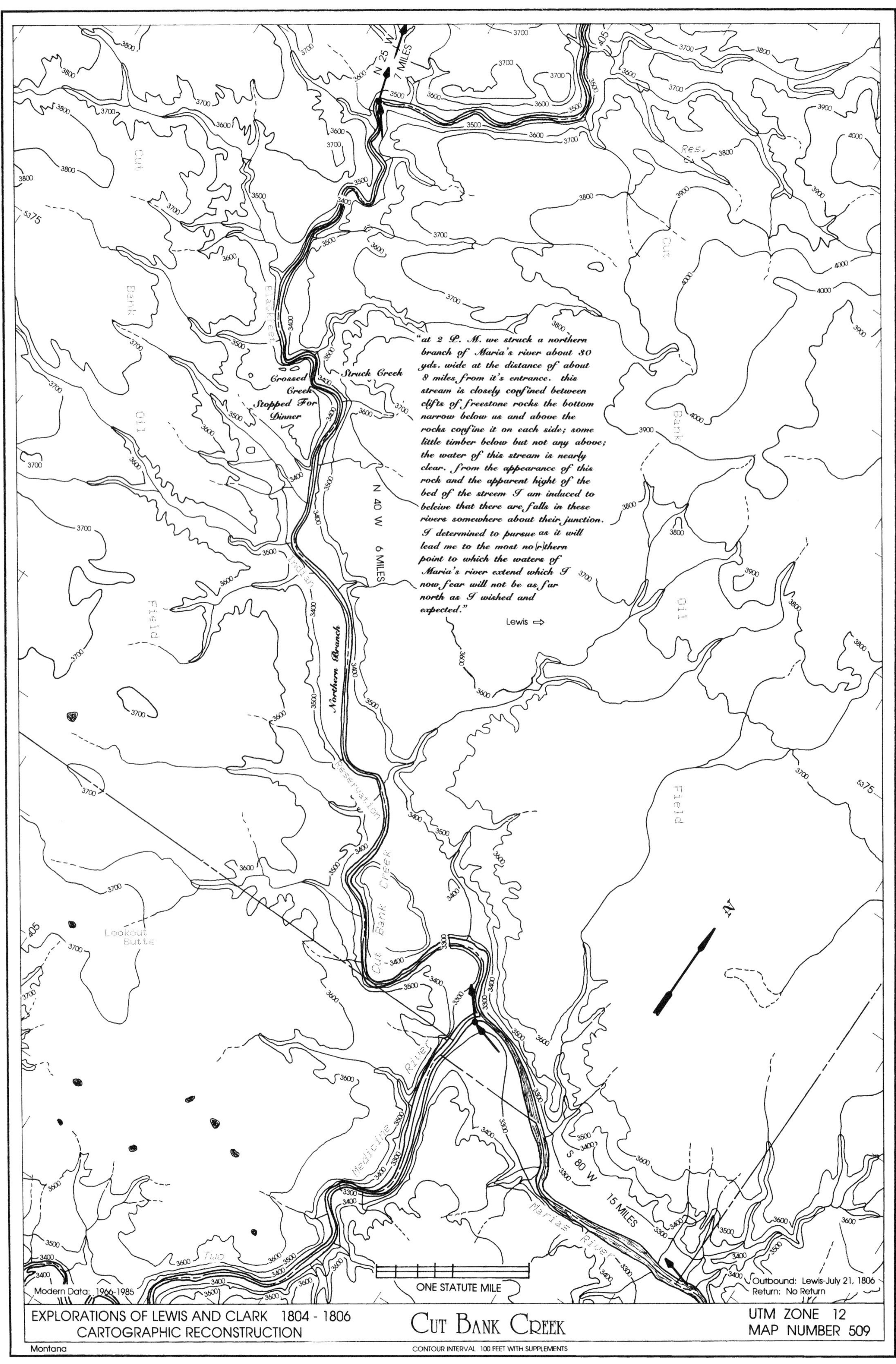

N 25 W
7 MILES
"at 2 P. M. we struck a northern
branch of Maria's river about 30
yds. wide at the distance of about
8 miles from it's entrance. this
stream is closely confined between
clifts of freestone rocks the bottom
narrow below us and above the
rocks confine it on each side; some
little timber below but not any above;
the water of this stream is nearly
clear. from the appearance of this
rock and the apparent hight of the
bed of the streem I am induced to
beleive that there are falls in these
rivers somewhere about their junction.
I determined to pursue as it will
lead me to the most no[r]thern
point to which the waters of
Maria's river extend which I
now fear will not be as far
north as I wished and
expected."
Lewis ⇨
Struck Creek
Crossed Creek
Stopped For Dinner
N 40 W
6 MILES
Northern Branch
Blackfeet
Indian
Reservation
Cut Bank Oil Field
Lookout Butte
Cut Bank Creek
Medicine River
Two
Maria's River
S 80 W
15 MILES
N
ONE STATUTE MILE
Modern Data: 1966-1985
Outbound: Lewis-July 21, 1806
Return: No Return
EXPLORATIONS OF LEWIS AND CLARK 1804 - 1806
CARTOGRAPHIC RECONSTRUCTION
CUT BANK CREEK
UTM ZONE 12
MAP NUMBER 509
Montana
CONTOUR INTERVAL 100 FEET WITH SUPPLEMENTS

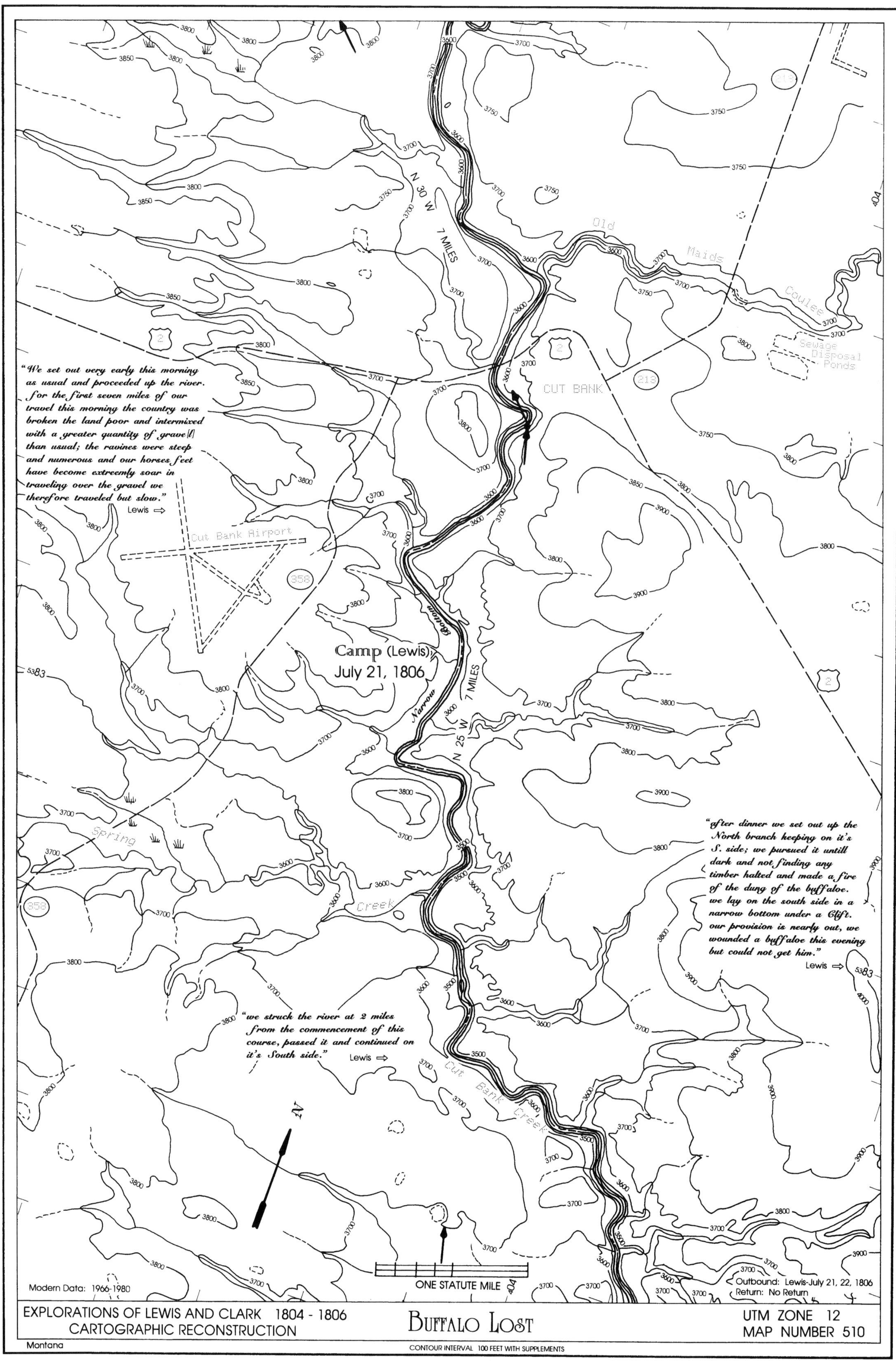
"We set out very early this morning as usual and proceeded up the river. for the first seven miles of our travel this morning the country was broken the land poor and intermixed with a greater quantity of grave[l] than usual; the ravines were steep and numerous and our horses feet have become extreemly soar in traveling over the gravel we therefore traveled but slow."
Lewis ⇨
Camp (Lewis); July 21, 1806
Narrow Bottom
N 30 W 7 MILES
N 25 W 7 MILES
CUT BANK
Old Maids Coulee
Sewage Disposal Ponds
Cut Bank Airport
Spring Creek
Cut Bank Creek
"after dinner we set out up the North branch keeping on it's S. side; we pursued it untill dark and not finding any timber halted and made a fire of the dung of the buffaloe. we lay on the south side in a narrow bottom under a Clift. our provision is nearly out, we wounded a buffaloe this evening but could not get him."
Lewis ⇨
"we struck the river at 2 miles from the commencement of this course, passed it and continued on it's South side."
Lewis ⇨
N
ONE STATUTE MILE
Modern Data: 1966-1980
Outbound: Lewis-July 21, 22, 1806
Return: No Return
EXPLORATIONS OF LEWIS AND CLARK 1804 - 1806
CARTOGRAPHIC RECONSTRUCTION
BUFFALO LOST
UTM ZONE 12
MAP NUMBER 510
Montana
CONTOUR INTERVAL 100 FEET WITH SUPPLEMENTS

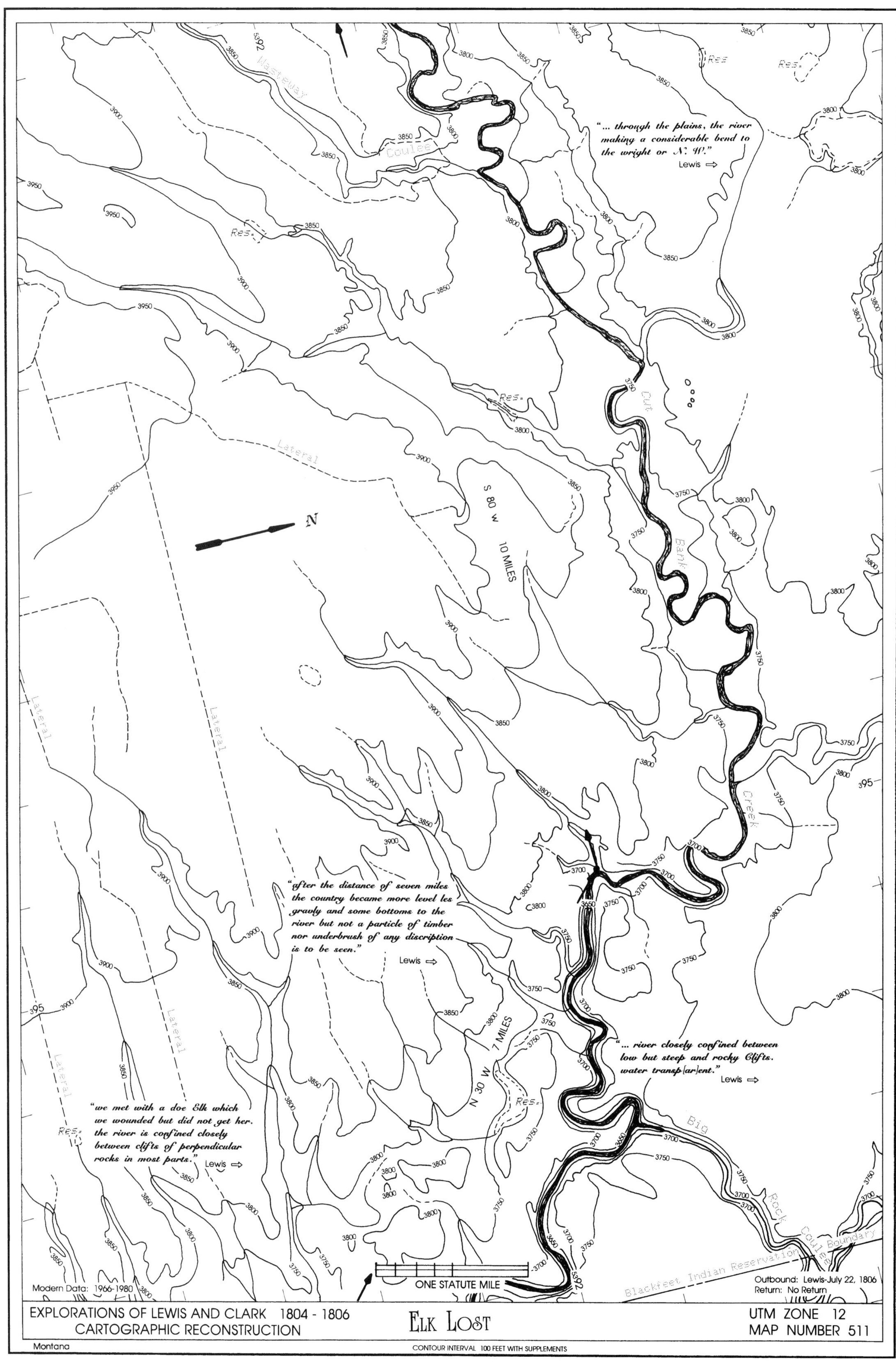
"... through the plains, the river making a considerable bend to the wright or N. W."
Lewis ⇨
"after the distance of seven miles the country became more level les gravly and some bottoms to the river but not a particle of timber nor underbrush of any discription is to be seen."
Lewis ⇨
"... river closely confined between low but steep and rocky Clifts. water transp[ar]ent."
Lewis ⇨
"we met with a doe Elk which we wounded but did not get her. the river is confined closely between clifts of perpendicular rocks in most parts."
Lewis ⇨
S 80 W 10 MILES
N 30 W 7 MILES
N
Wasteway
Coulee
Res.
Lateral
Cut Bank Creek
Big Rock Coulee
Blackfeet Indian Reservation Boundary
ONE STATUTE MILE
Modern Data: 1966-1980
Outbound: Lewis-July 22, 1806
Return: No Return
EXPLORATIONS OF LEWIS AND CLARK 1804 - 1806
CARTOGRAPHIC RECONSTRUCTION
Elk Lost
UTM ZONE 12
MAP NUMBER 511
Montana
CONTOUR INTERVAL 100 FEET WITH SUPPLEMENTS

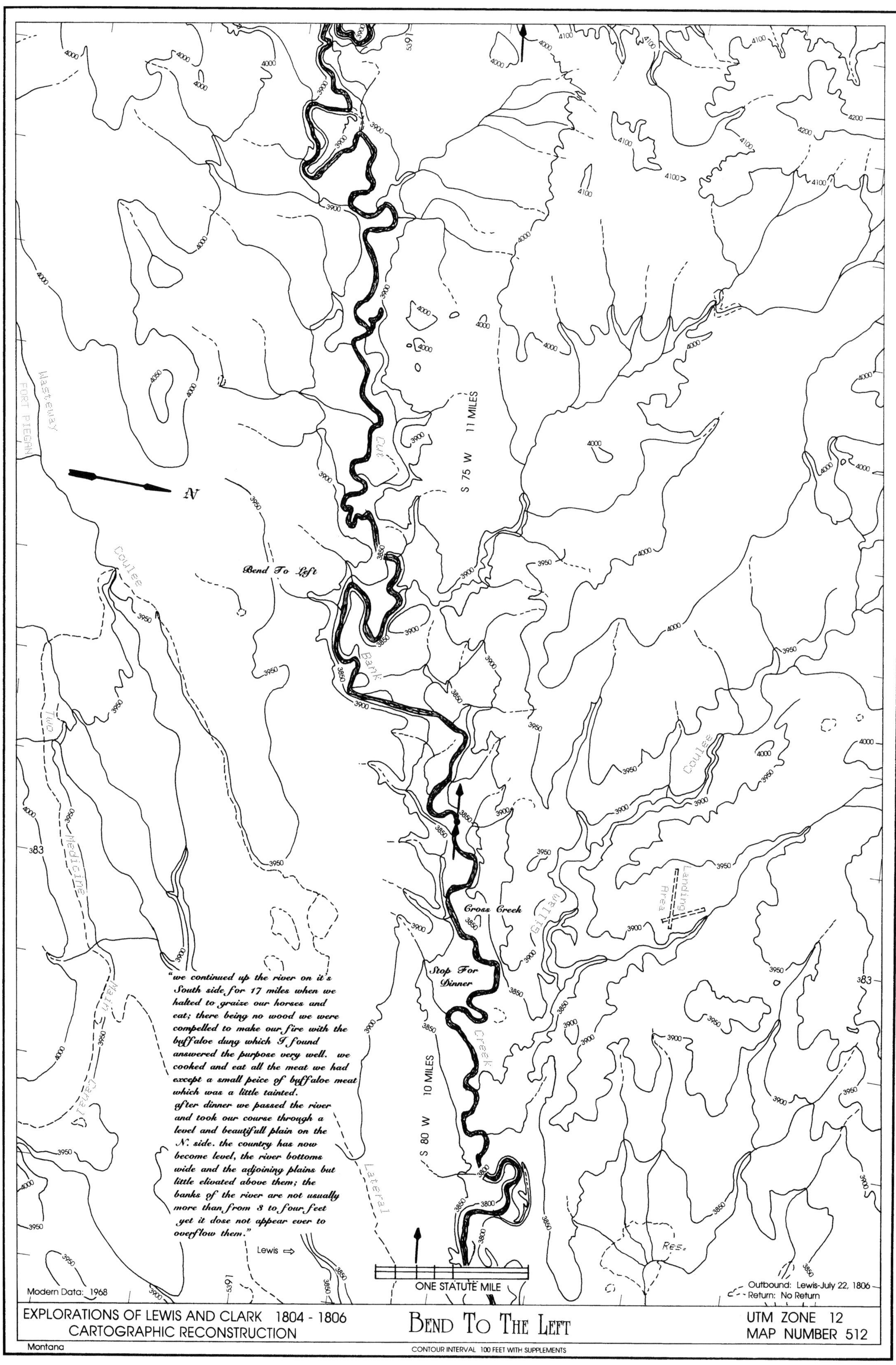

FORT PIEGAN
Wasteway
N
Coulee
Cut
Bank
Creek
S 75 W 11 MILES
Bend To Left
Two
Medicine
Main
Canal
Lateral
Cross Creek
Gilliam
Coulee
Landing
Area
Stop For
Dinner
S 80 W 10 MILES
Res.
"we continued up the river on it's
South side for 17 miles when we
halted to graize our horses and
eat; there being no wood we were
compelled to make our fire with the
buffaloe dung which I found
answered the purpose very well. we
cooked and eat all the meat we had
except a small peice of buffaloe meat
which was a little tainted.
after dinner we passed the river
and took our course through a
level and beautifull plain on the
N. side. the country has now
become level, the river bottoms
wide and the adjoining plains but
little elivated above them; the
banks of the river are not usually
more than from 3 to four feet
yet it dose not appear ever to
overflow them."
Lewis ⇨
ONE STATUTE MILE
Modern Data: 1968
Outbound: Lewis-July 22, 1806
Return: No Return
EXPLORATIONS OF LEWIS AND CLARK 1804 - 1806
CARTOGRAPHIC RECONSTRUCTION
BEND TO THE LEFT
UTM ZONE 12
MAP NUMBER 512
Montana
CONTOUR INTERVAL 100 FEET WITH SUPPLEMENTS

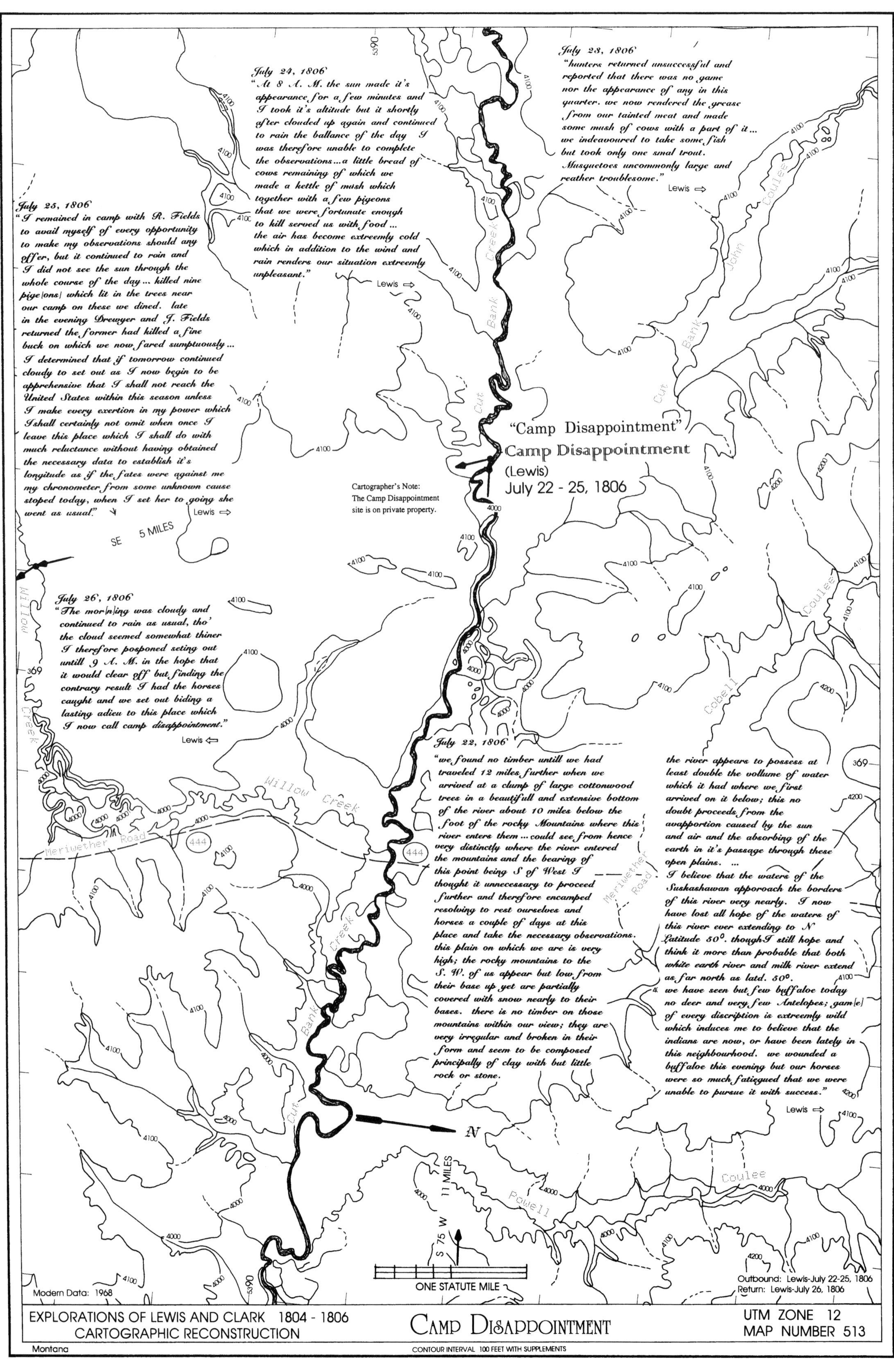
July 24, 1806
"At 8 A. M. the sun made it's appearance for a few minutes and I took it's altitude but it shortly after clouded up again and continued to rain the ballance of the day I was therefore unable to complete the observations...a little bread of cows remaining of which we made a kettle of mush which together with a few pigeons that we were fortunate enough to kill served us with food... the air has become extreemly cold which in addition to the wind and rain renders our situation extreemly unpleasant."
Lewis ⇨
July 23, 1806
"hunters returned unsuccessful and reported that there was no game nor the appearance of any in this quarter. we now rendered the grease from our tainted meat and made some mush of cows with a part of it... we indeavoured to take some fish but took only one smal trout. Musquetoes uncommonly large and reather troublesome."
Lewis ⇨
July 25, 1806
"I remained in camp with R. Fields to avail myself of every opportunity to make my observations should any offer, but it continued to roin and I did not see the sun through the whole course of the day... killed nine pige[ons] which lit in the trees near our camp on these we dined. late in the evening Drewyer and J. Fields returned the former had killed a fine buck on which we now fared sumptuously... I determined that if tomorrow continued cloudy to set out as I now begin to be apprehensive that I shall not reach the United States within this season unless I make every exertion in my power which I shall certainly not omit when once I leave this place which I shall do with much reluctance without having obtained the necessary data to establish it's longitude as if the fates were against me my chronometer from some unknown cause stoped today, when I set her to going she went as usual."
Lewis ⇨
"Camp Disappointment"
Camp Disappointment
(Lewis)
July 22 - 25, 1806
Cartographer's Note:
The Camp Disappointment site is on private property.
SE 5 MILES
July 26, 1806
"The mor[n]ing was cloudy and continued to rain as usual, tho' the cloud seemed somewhat thiner I therefore posponed seting out untill 9 A. M. in the hope that it would clear off but finding the contrary result I had the horses caught and we set out biding a lasting adieu to this place which I now call camp disappointment."
Lewis ⇦
July 22, 1806
"we found no timber untill we had traveled 12 miles further when we arrived at a clump of large cottonwood trees in a beautifull and extensive bottom of the river about 10 miles below the foot of the rocky Mountains where this river enters them...could see from hence very distinctly where the river entered the mountains and the bearing of this point being S of West I thought it unnecessary to proceed further and therefore encamped resolving to rest ourselves and horses a couple of days at this place and take the necessary observations. this plain on which we are is very high; the rocky mountains to the S. W. of us appear but low from their base up yet are partially covered with snow nearly to their bases. there is no timber on those mountains within our view; they are very irregular and broken in their form and seem to be composed principally of clay with but little rock or stone.
the river appears to possess at least double the vollume of water which it had where we first arrived on it below; this no doubt proceeds from the avapportion caused by the sun and air and the absorbing of the earth in it's passage through these open plains. ...
I believe that the waters of the Suskashawan apporoach the borders of this river very nearly. I now have lost all hope of the waters of this river ever extending to N Latitude 50°. though I still hope and think it more than probable that both white earth river and milk river extend as far north as latd. 50°.
we have seen but few buffaloe today no deer and very few Antelopes; gam[e] of every discription is extreemly wild which induces me to believe that the indians are now, or have been lately in this neighbourhood. we wounded a buffaloe this evening but our horses were so much fatigued that we were unable to pursue it with success."
Lewis ⇨
Cut Bank Creek
John Coulee
Cut Bank Coulee
Cobell Coulee
Willow Creek
Meriwether Road
Powell Coulee
444
369
5390
N
S 75 W
11 MILES
ONE STATUTE MILE
Modern Data: 1968
Outbound: Lewis-July 22-25, 1806
Return: Lewis-July 26, 1806
EXPLORATIONS OF LEWIS AND CLARK 1804 - 1806
CARTOGRAPHIC RECONSTRUCTION
Montana
CAMP DISAPPOINTMENT
CONTOUR INTERVAL 100 FEET WITH SUPPLEMENTS
UTM ZONE 12
MAP NUMBER 513

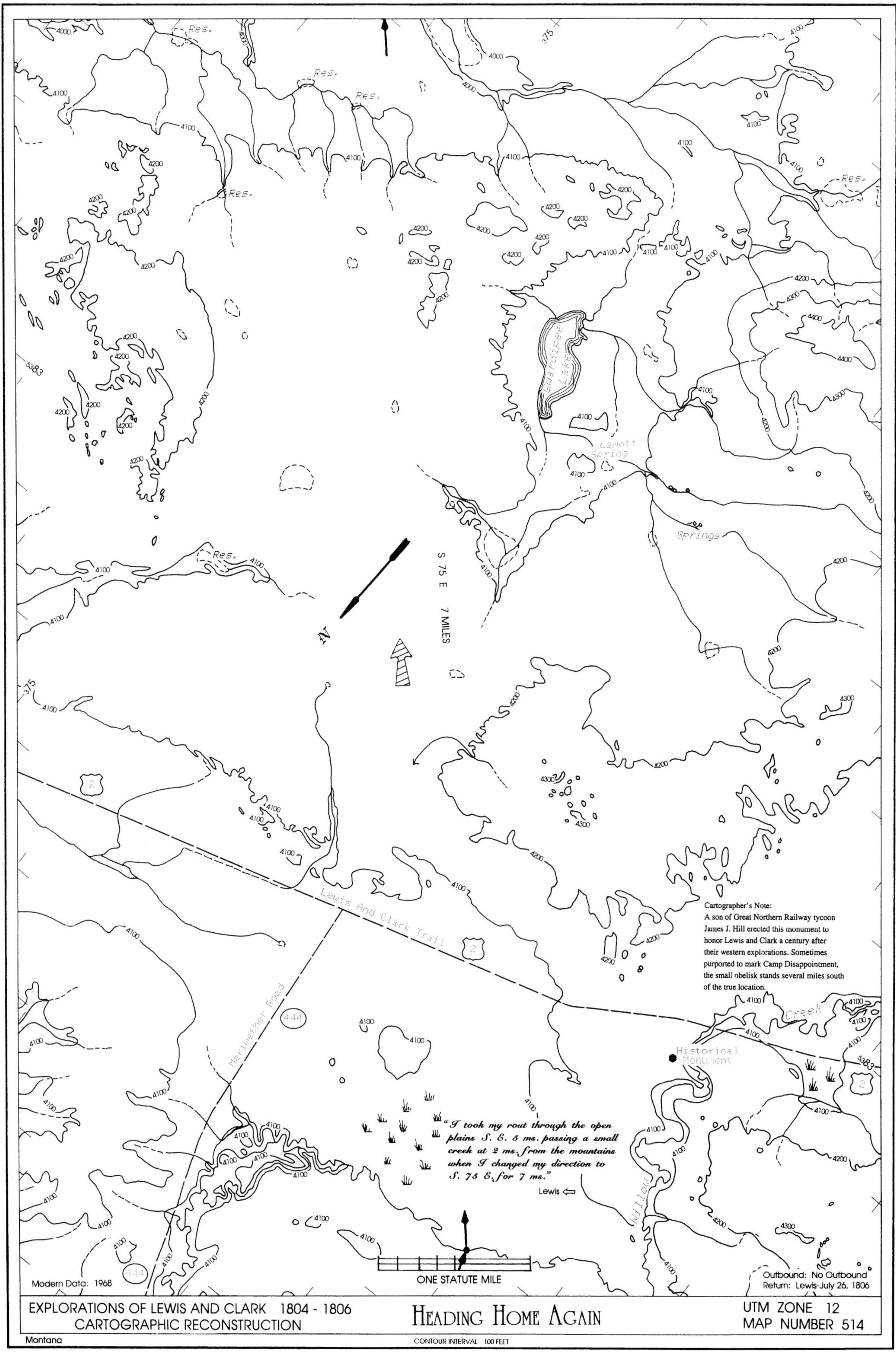
Cartographer's Note:
A son of Great Northern Railway tycoon James J. Hill erected this monument to honor Lewis and Clark a century after their western explorations. Sometimes purported to mark Camp Disappointment, the small obelisk stands several miles south of the true location.
"I took my rout through the open plains S. E. 5 ms. passing a small creek at 2 ms. from the mountains when I changed my direction to S. 75 E. for 7 ms."
Lewis
S 75 E 7 MILES
Guardipee Lake
Lamott Spring
Springs
Res.
Lewis And Clark Trail
Meriwether Road
Historical Monument
Creek
Willow
ONE STATUTE MILE
Modern Data: 1968
Outbound: No Outbound
Return: Lewis-July 26, 1806
EXPLORATIONS OF LEWIS AND CLARK 1804 - 1806
CARTOGRAPHIC RECONSTRUCTION
HEADING HOME AGAIN
UTM ZONE 12
MAP NUMBER 514
Montana
CONTOUR INTERVAL 100 FEET

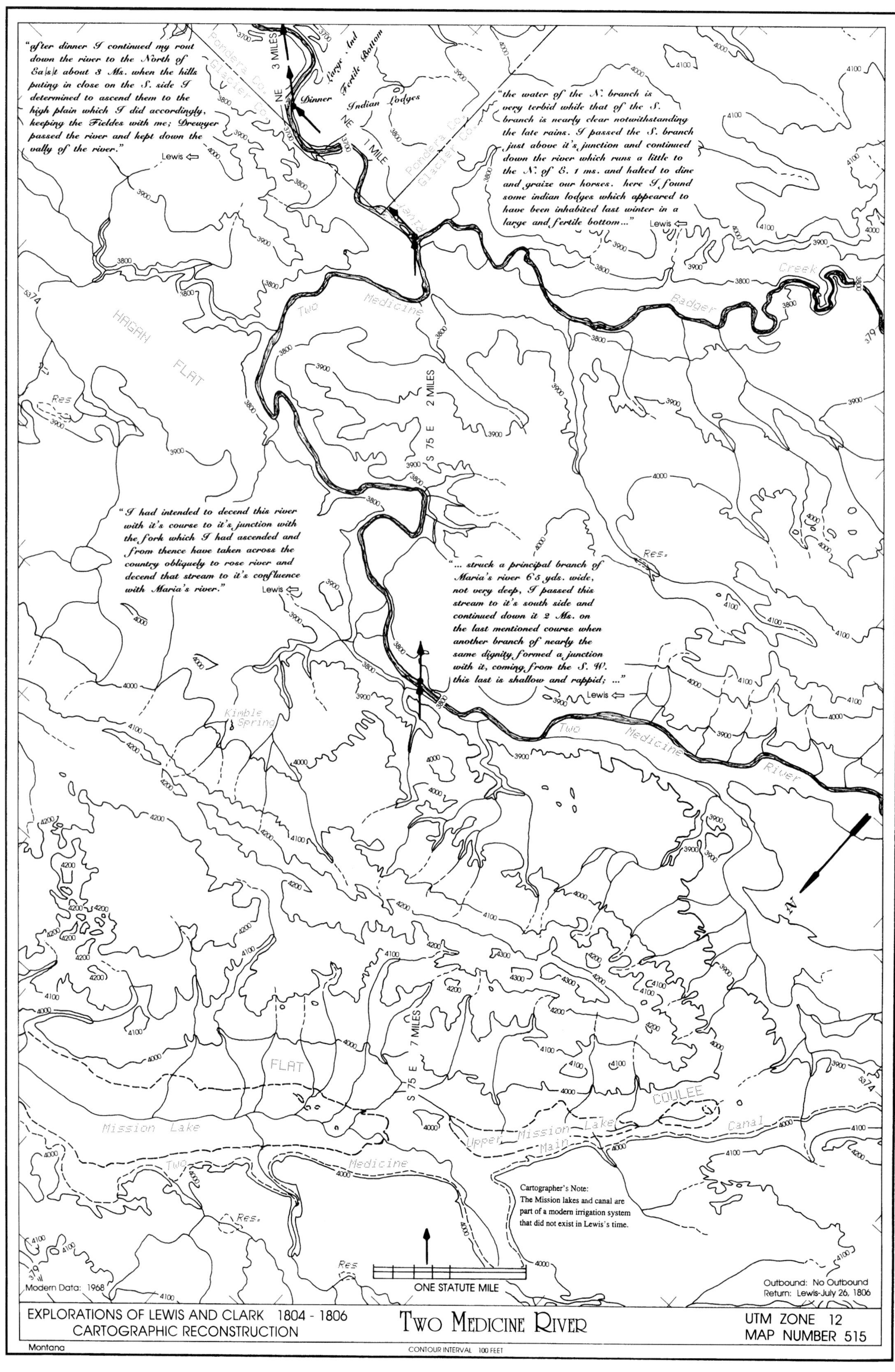
"after dinner I continued my rout down the river to the North of Ea[s]t about 3 Ms. when the hills puting in close on the S. side I determined to ascend them to the high plain which I did accordingly, keeping the Fieldes with me; Drewyer passed the river and kept down the vally of the river."
Lewis
"the water of the N. branch is very terbid while that of the S. branch is nearly clear notwithstanding the late rains. I passed the S. branch just above it's junction and continued down the river which runs a little to the N. of E. 1 ms. and halted to dine and graize our horses. here I found some indian lodges which appeared to have been inhabited last winter in a large and fertile bottom..."
Lewis
"I had intended to decend this river with it's course to it's junction with the fork which I had ascended and from thence have taken across the country obliquely to rose river and decend that stream to it's confluence with Maria's river."
Lewis
"... struck a principal branch of Maria's river 65 yds. wide, not very deep, I passed this stream to it's south side and continued down it 2 Ms. on the last mentioned course when another branch of nearly the same dignity formed a junction with it, coming from the S. W. this last is shallow and rappid; ..."
Lewis
NE 3 MILES
Dinner
Large And Fertile Bottom
Indian Lodges
NE 1 MILE
Pondera Co.
Glacier Co.
HAGAN FLAT
Two Medicine
Badger Creek
S 75 E 2 MILES
Kimble Spring
Two Medicine River
Res.
S 75 E 7 MILES
FLAT
COULEE
Mission Lake
Upper Mission Lake Main Canal
Two Medicine
Cartographer's Note: The Mission lakes and canal are part of a modern irrigation system that did not exist in Lewis's time.
Modern Data: 1968
ONE STATUTE MILE
Outbound: No Outbound
Return: Lewis-July 26, 1806
EXPLORATIONS OF LEWIS AND CLARK 1804 - 1806
CARTOGRAPHIC RECONSTRUCTION
TWO MEDICINE RIVER
UTM ZONE 12
MAP NUMBER 515
Montana
CONTOUR INTERVAL 100 FEET

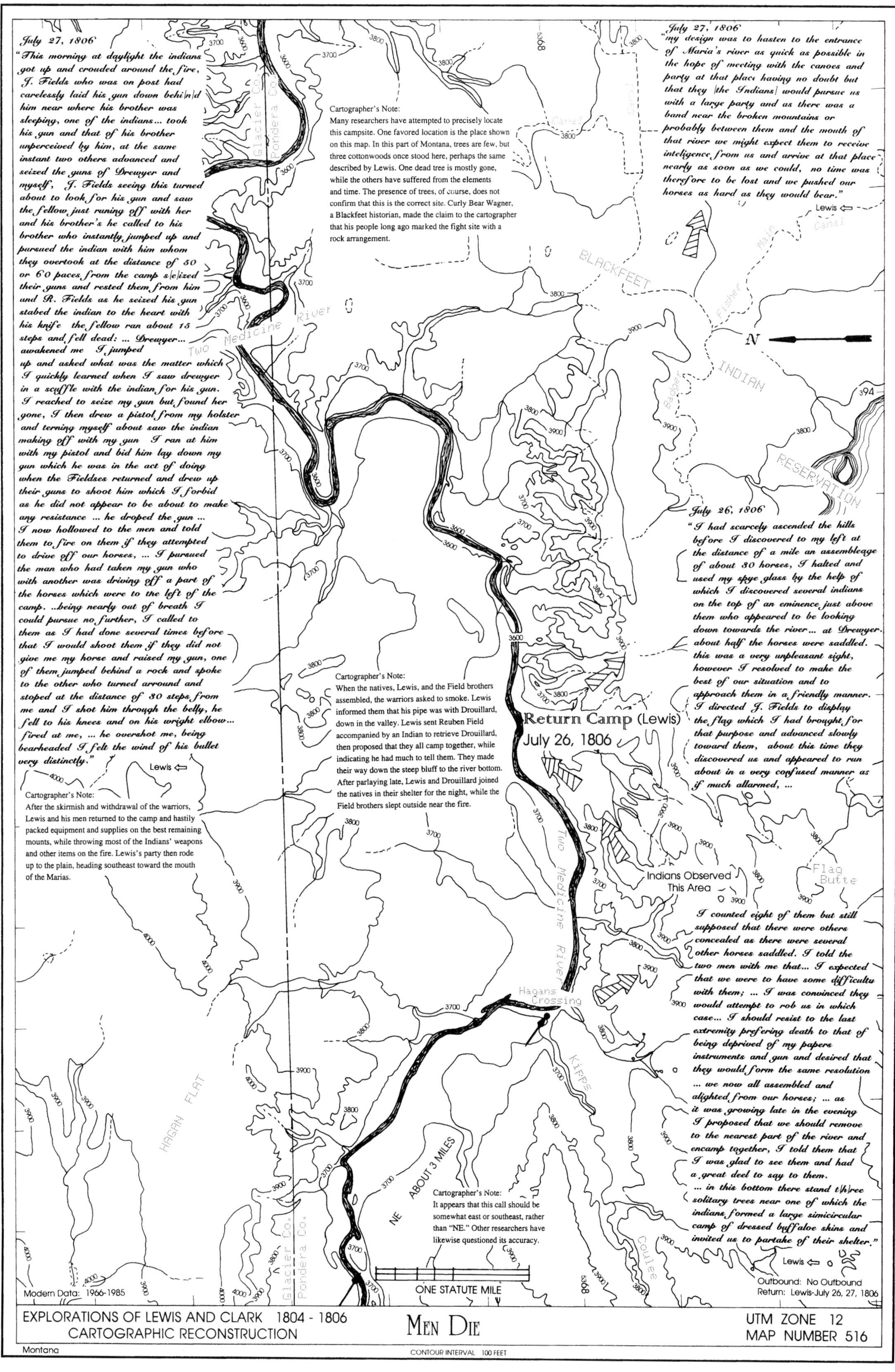

July 27, 1806
"This morning at daylight the indians got up and crouded around the fire, J. Fields who was on post had carelessly laid his gun down behi[n]d him near where his brother was sleeping, one of the indians... took his gun and that of his brother unperceived by him, at the same instant two others advanced and seized the guns of Drewyer and myself, J. Fields seeing this turned about to look for his gun and saw the fellow just runing off with her and his brother's he called to his brother who instantly jumped up and pursued the indian with him whom they overtook at the distance of 50 or 60 paces from the camp s[e]ized their guns and rested them from him and R. Fields as he seized his gun stabed the indian to the heart with his knife the fellow ran about 15 steps and fell dead: ... Drewyer... awakened me I jumped up and asked what was the matter which I quickly learned when I saw drewyer in a scuffle with the indian for his gun. I reached to seize my gun but found her gone, I then drew a pistol from my holster and terning myself about saw the indian making off with my gun I ran at him with my pistol and bid him lay down my gun which he was in the act of doing when the Fieldses returned and drew up their guns to shoot him which I forbid as he did not appear to be about to make any resistance ... he droped the gun ... I now hollowed to the men and told them to fire on them if they attempted to drive off our horses, ... I pursued the man who had taken my gun who with another was driving off a part of the horses which were to the left of the camp. ..being nearly out of breath I could pursue no further, I called to them as I had done several times before that I would shoot them if they did not give me my horse and raised my gun, one of them jumped behind a rock and spoke to the other who turned arround and stoped at the distance of 30 steps from me and I shot him through the belly, he fell to his knees and on his wright elbow... fired at me, ... he overshot me, being bearheaded I felt the wind of his bullet very distinctly."
Lewis ⇦
Cartographer's Note:
After the skirmish and withdrawal of the warriors, Lewis and his men returned to the camp and hastily packed equipment and supplies on the best remaining mounts, while throwing most of the Indians' weapons and other items on the fire. Lewis's party then rode up to the plain, heading southeast toward the mouth of the Marias.
Cartographer's Note:
Many researchers have attempted to precisely locate this campsite. One favored location is the place shown on this map. In this part of Montana, trees are few, but three cottonwoods once stood here, perhaps the same described by Lewis. One dead tree is mostly gone, while the others have suffered from the elements and time. The presence of trees, of course, does not confirm that this is the correct site. Curly Bear Wagner, a Blackfeet historian, made the claim to the cartographer that his people long ago marked the fight site with a rock arrangement.
July 27, 1806
"my design was to hasten to the entrance of Maria's river as quick as possible in the hope of meeting with the canoes and party at that place having no doubt but that they [the Indians] would pursue us with a large party and as there was a band near the broken mountains or probably between them and the mouth of that river we might expect them to receive inteligence from us and arrive at that place nearly as soon as we could, no time was therefore to be lost and we pushed our horses as hard as they would bear."
Lewis ⇦
Canal
Main
Fisher
BLACKFEET
INDIAN
RESERVATION
Badger
N
Glacier Co.
Pondera Co.
Two Medicine River
Cartographer's Note:
When the natives, Lewis, and the Field brothers assembled, the warriors asked to smoke. Lewis informed them that his pipe was with Drouillard, down in the valley. Lewis sent Reuben Field accompanied by an Indian to retrieve Drouillard, then proposed that they all camp together, while indicating he had much to tell them. They made their way down the steep bluff to the river bottom. After parlaying late, Lewis and Drouillard joined the natives in their shelter for the night, while the Field brothers slept outside near the fire.
Return Camp (Lewis)
July 26, 1806
July 26, 1806
"I had scarcely ascended the hills before I discovered to my left at the distance of a mile an assembleage of about 30 horses, I halted and used my spye glass by the help of which I discovered several indians on the top of an eminence just above them who appeared to be looking down towards the river... at Drewyer. about half the horses were saddled. this was a very unpleasant sight, however I resolved to make the best of our situation and to approach them in a friendly manner. I directed J. Fields to display the flag which I had brought for that purpose and advanced slowly toward them, about this time they discovered us and appeared to run about in a very confused manner as if much allarmed, ...
Indians Observed This Area
Flag Butte
I counted eight of them but still supposed that there were others concealed as there were several other horses saddled. I told the two men with me that... I expected that we were to have some difficulty with them; ... I was convinced they would attempt to rob us in which case... I should resist to the last extremity prefering death to that of being deprived of my papers instruments and gun and desired that they would form the same resolution ... we now all assembled and alighted from our horses; ... as it was growing late in the evening I proposed that we should remove to the nearest part of the river and encamp together, I told them that I was glad to see them and had a great deel to say to them. ... in this bottom there stand t[h]ree solitary trees near one of which the indians formed a large simicircular camp of dressed buffaloe skins and invited us to partake of their shelter."
Lewis ⇦
Hagans Crossing
Kipps
Coulee
HAGAN FLAT
NE ABOUT 3 MILES
Cartographer's Note:
It appears that this call should be somewhat east or southeast, rather than "NE." Other researchers have likewise questioned its accuracy.
ONE STATUTE MILE
Modern Data: 1966-1985
Outbound: No Outbound
Return: Lewis-July 26, 27, 1806
EXPLORATIONS OF LEWIS AND CLARK 1804 - 1806
CARTOGRAPHIC RECONSTRUCTION
MEN DIE
UTM ZONE 12
MAP NUMBER 516
Montana
CONTOUR INTERVAL 100 FEET

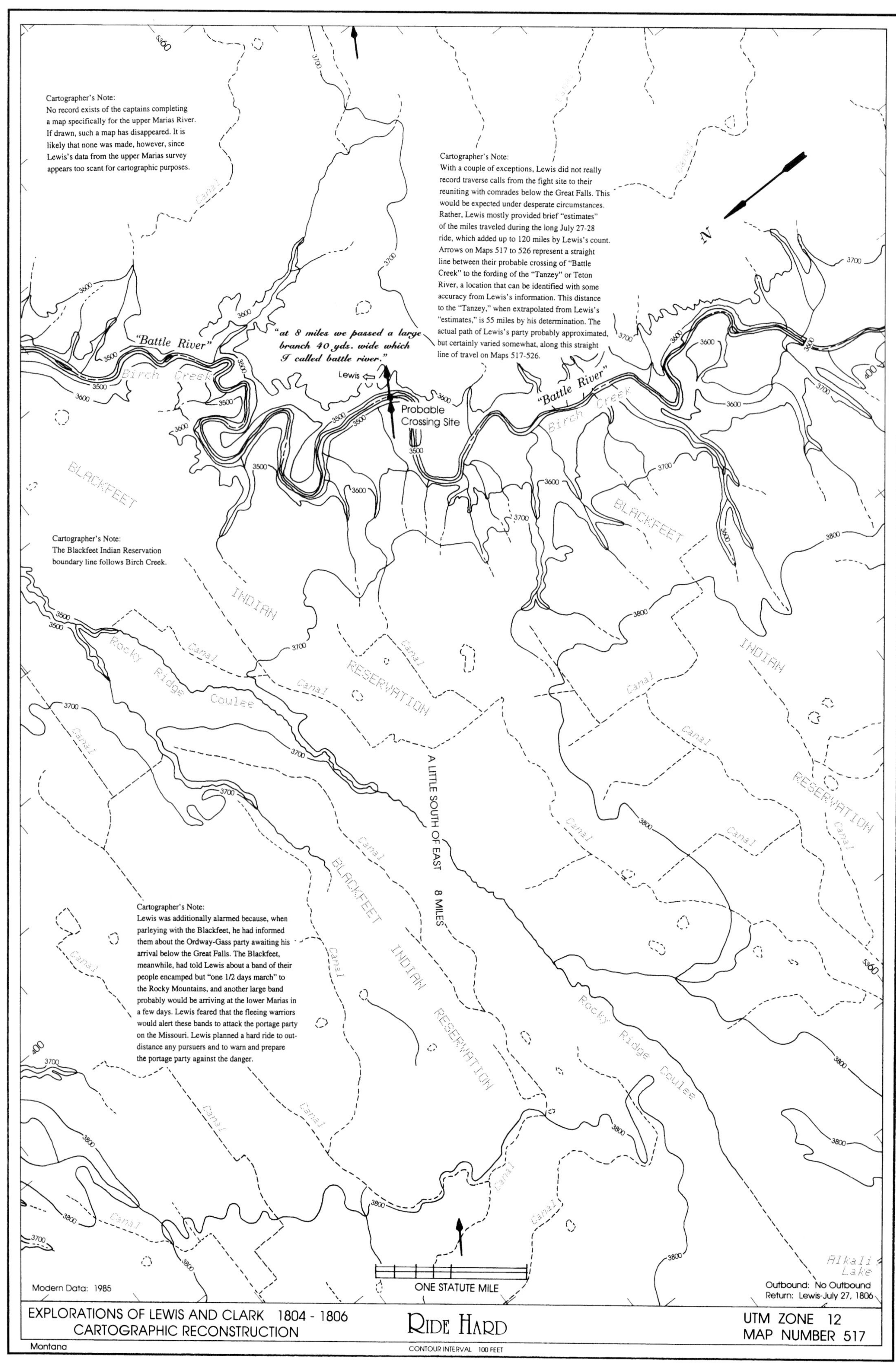
Cartographer's Note:
No record exists of the captains completing a map specifically for the upper Marias River. If drawn, such a map has disappeared. It is likely that none was made, however, since Lewis's data from the upper Marias survey appears too scant for cartographic purposes.
Cartographer's Note:
With a couple of exceptions, Lewis did not really record traverse calls from the fight site to their reuniting with comrades below the Great Falls. This would be expected under desperate circumstances. Rather, Lewis mostly provided brief "estimates" of the miles traveled during the long July 27-28 ride, which added up to 120 miles by Lewis's count. Arrows on Maps 517 to 526 represent a straight line between their probable crossing of "Battle Creek" to the fording of the "Tanzey" or Teton River, a location that can be identified with some accuracy from Lewis's information. This distance to the "Tanzey," when extrapolated from Lewis's "estimates," is 55 miles by his determination. The actual path of Lewis's party probably approximated, but certainly varied somewhat, along this straight line of travel on Maps 517-526.
N
"Battle River"
Birch Creek
"at 8 miles we passed a large branch 40 yds. wide which I called battle river."
Lewis
Probable Crossing Site
"Battle River"
Birch Creek
BLACKFEET
BLACKFEET
Cartographer's Note:
The Blackfeet Indian Reservation boundary line follows Birch Creek.
INDIAN
RESERVATION
INDIAN
RESERVATION
Rocky Ridge Coulee
Canal
A LITTLE SOUTH OF EAST 8 MILES
BLACKFEET
INDIAN
RESERVATION
Cartographer's Note:
Lewis was additionally alarmed because, when parleying with the Blackfeet, he had informed them about the Ordway-Gass party awaiting his arrival below the Great Falls. The Blackfeet, meanwhile, had told Lewis about a band of their people encamped but "one 1/2 days march" to the Rocky Mountains, and another large band probably would be arriving at the lower Marias in a few days. Lewis feared that the fleeing warriors would alert these bands to attack the portage party on the Missouri. Lewis planned a hard ride to outdistance any pursuers and to warn and prepare the portage party against the danger.
Rocky Ridge Coulee
Alkali Lake
Modern Data: 1985
ONE STATUTE MILE
Outbound: No Outbound
Return: Lewis-July 27, 1806
EXPLORATIONS OF LEWIS AND CLARK 1804 - 1806
CARTOGRAPHIC RECONSTRUCTION
RIDE HARD
UTM ZONE 12
MAP NUMBER 517
Montana
CONTOUR INTERVAL 100 FEET

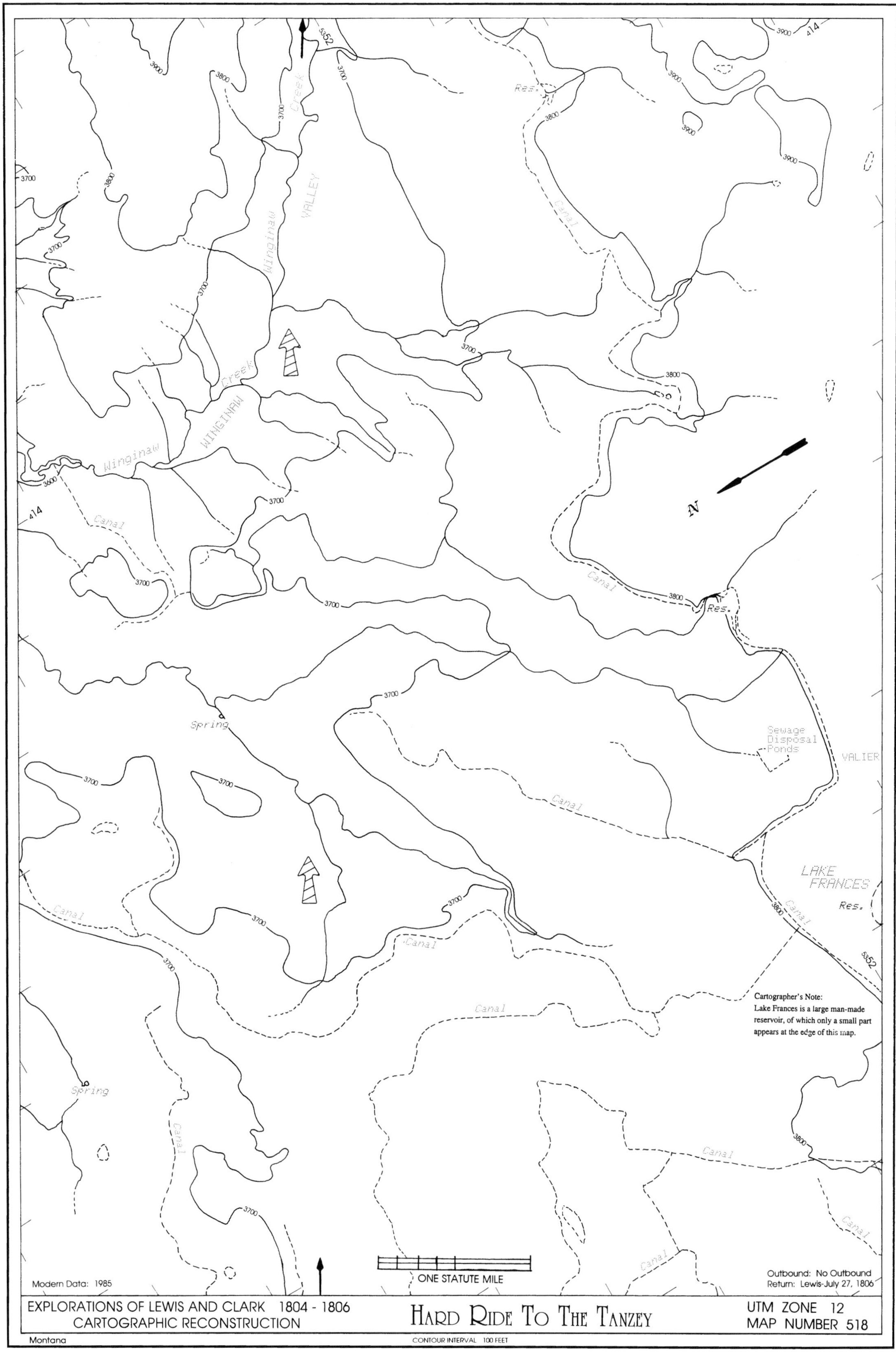

Cartographer's Note:
Lake Frances is a large man-made reservoir, of which only a small part appears at the edge of this map.
Modern Data: 1985
ONE STATUTE MILE
Outbound: No Outbound
Return: Lewis-July 27, 1806
EXPLORATIONS OF LEWIS AND CLARK 1804 - 1806
CARTOGRAPHIC RECONSTRUCTION
HARD RIDE TO THE TANZEY
UTM ZONE 12
MAP NUMBER 518
Montana
CONTOUR INTERVAL 100 FEET
LAKE FRANCES
VALIER
Sewage Disposal Ponds
Wingingaw
VALLEY
Canal
Spring
Res.

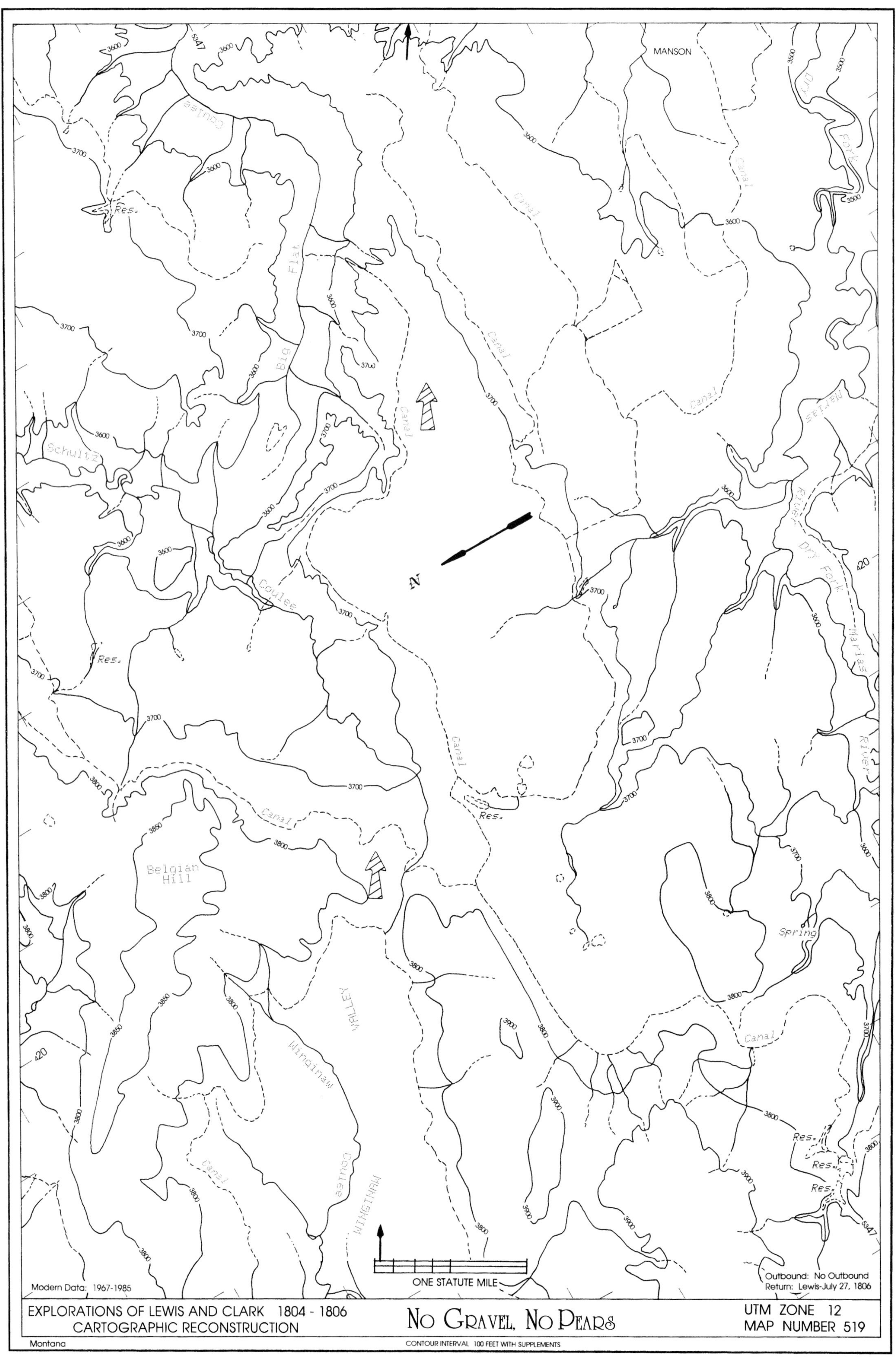
MANSON
Big Flat Coulee
Schultz
Res.
Canal
Dry Fork
Marias
Marias River
Dry Fork Marias River
Coulee
Belgian Hill
Spring
VALLEY
Wingirnaw
Coulee
WINGINAW
N
Modern Data: 1967-1985
ONE STATUTE MILE
Outbound: No Outbound
Return: Lewis-July 27, 1806
EXPLORATIONS OF LEWIS AND CLARK 1804 - 1806
CARTOGRAPHIC RECONSTRUCTION
NO GRAVEL, NO PEARS
UTM ZONE 12
MAP NUMBER 519
Montana
CONTOUR INTERVAL 100 FEET WITH SUPPLEMENTS

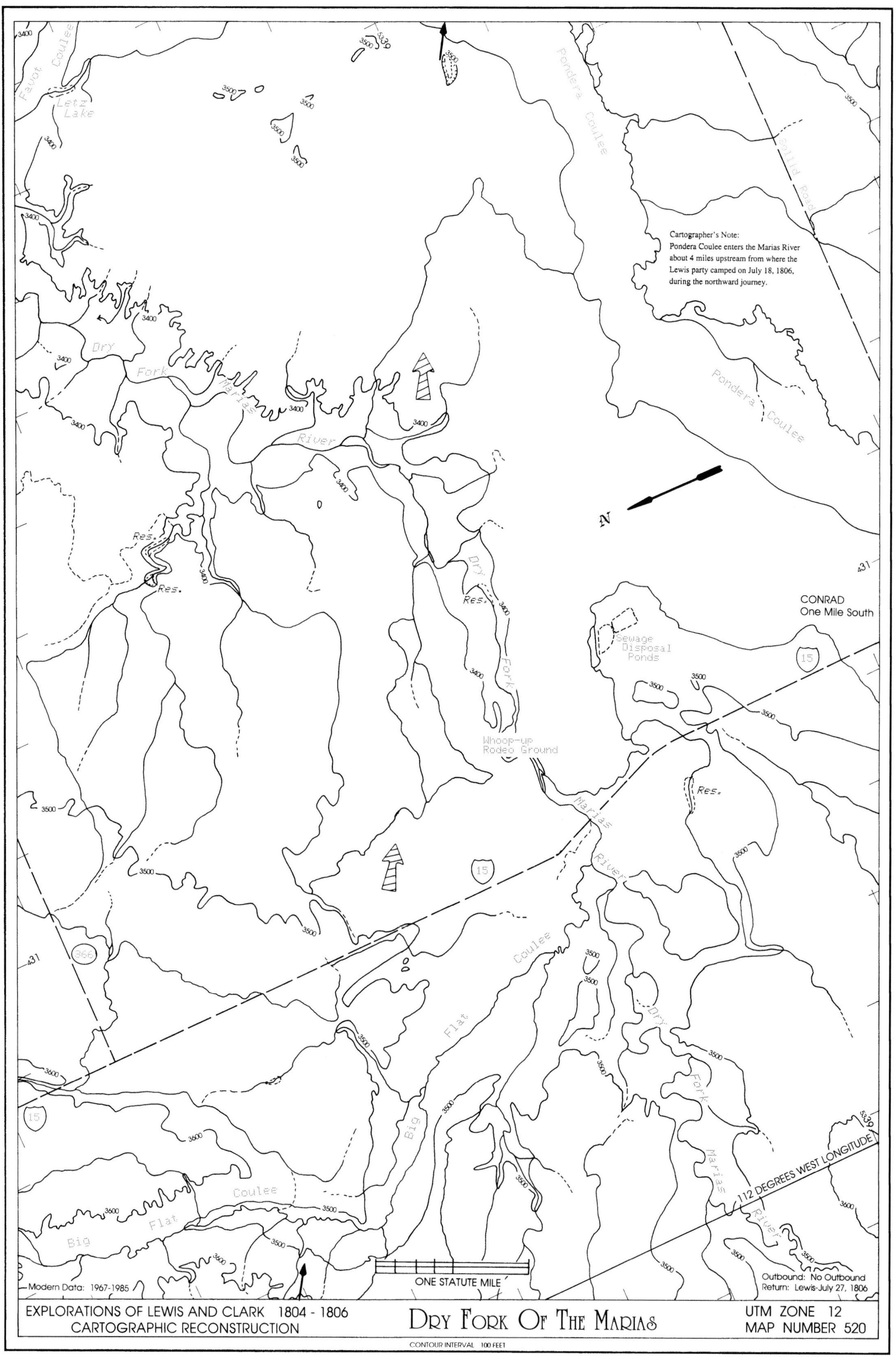
Cartographer's Note:
Pondera Coulee enters the Marias River about 4 miles upstream from where the Lewis party camped on July 18, 1806, during the northward journey.
CONRAD
One Mile South
Modern Data: 1967-1985
ONE STATUTE MILE
Outbound: No Outbound
Return: Lewis-July 27, 1806
EXPLORATIONS OF LEWIS AND CLARK 1804 - 1806
CARTOGRAPHIC RECONSTRUCTION
DRY FORK OF THE MARIAS
UTM ZONE 12
MAP NUMBER 520
CONTOUR INTERVAL 100 FEET
112 DEGREES WEST LONGITUDE

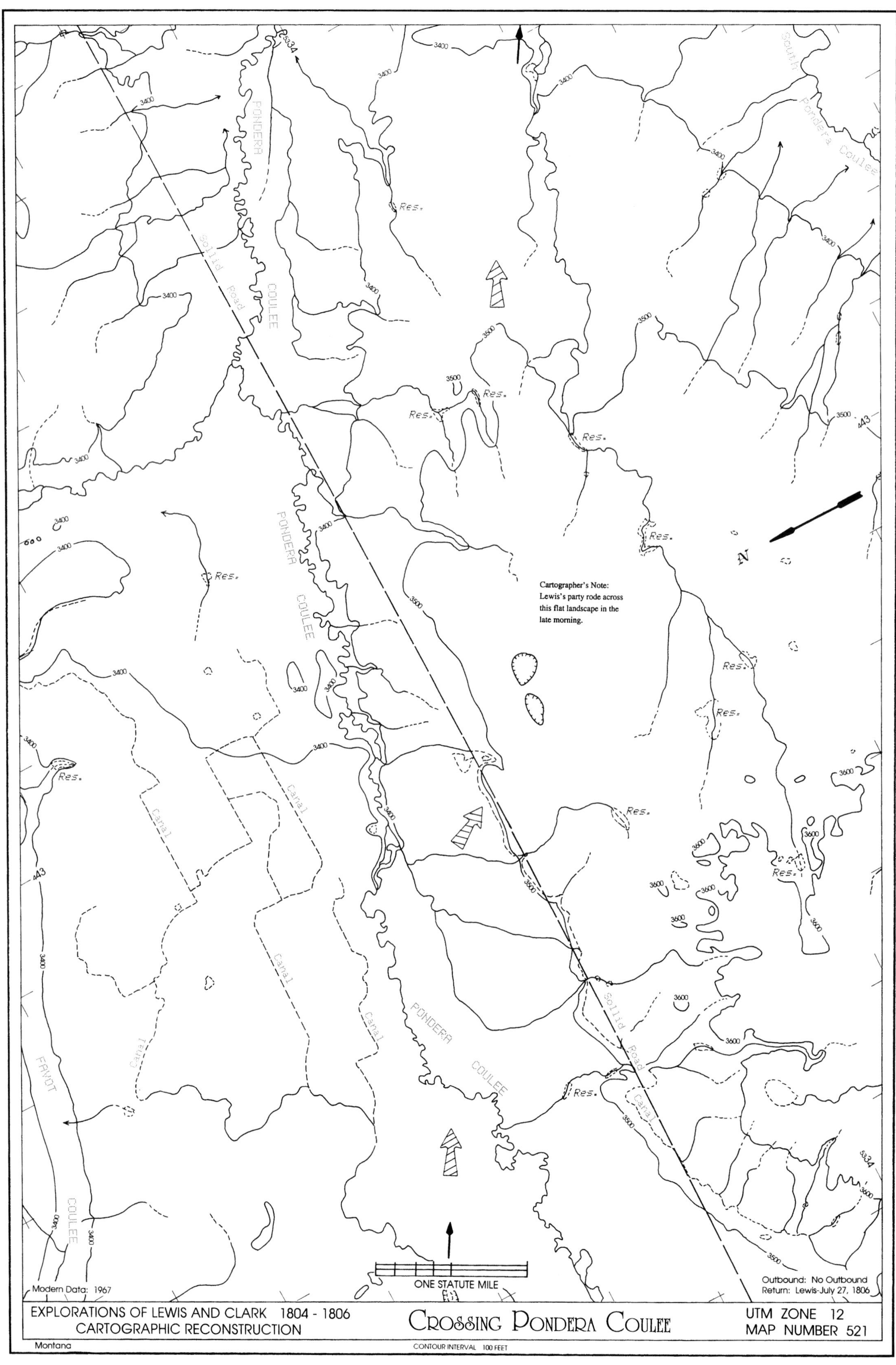
Cartographer's Note:
Lewis's party rode across this flat landscape in the late morning.
PONDERA COULEE
South Pondera Coulee
Sollid Road
Canal
Res.
FAVOT COULEE
N
ONE STATUTE MILE
Modern Data: 1967
Outbound: No Outbound
Return: Lewis-July 27, 1806
EXPLORATIONS OF LEWIS AND CLARK 1804 - 1806
CARTOGRAPHIC RECONSTRUCTION
CROSSING PONDERA COULEE
UTM ZONE 12
MAP NUMBER 521
Montana
CONTOUR INTERVAL 100 FEET

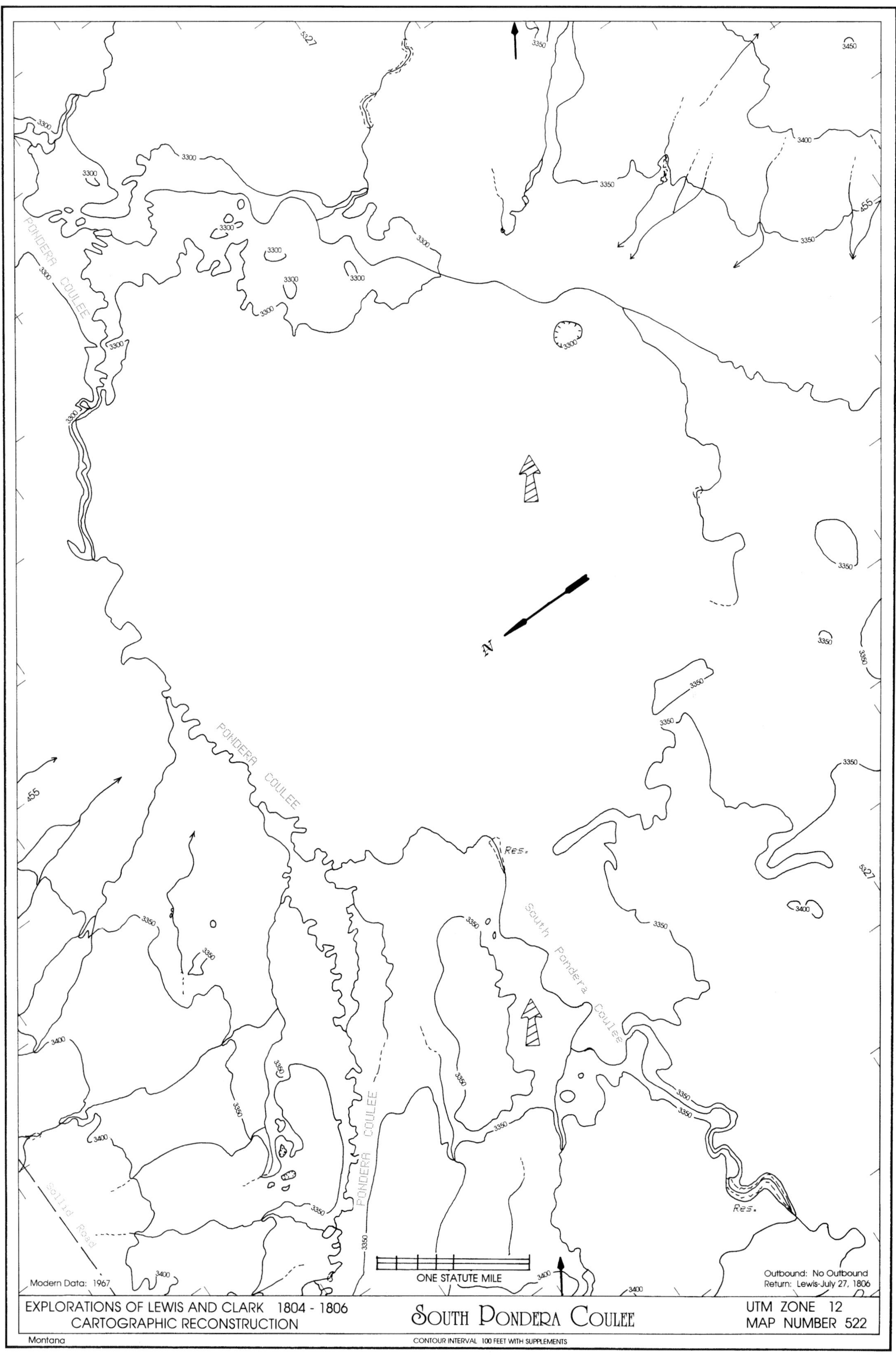
PONDERA COULEE
PONDERA COULEE
South Pondera Coulee
PONDERA COULEE
Res.
Res.
Solid Road
ONE STATUTE MILE
Modern Data: 1967
Outbound: No Outbound
Return: Lewis-July 27, 1806
EXPLORATIONS OF LEWIS AND CLARK 1804 - 1806
CARTOGRAPHIC RECONSTRUCTION
SOUTH PONDERA COULEE
UTM ZONE 12
MAP NUMBER 522
Montana
CONTOUR INTERVAL 100 FEET WITH SUPPLEMENTS

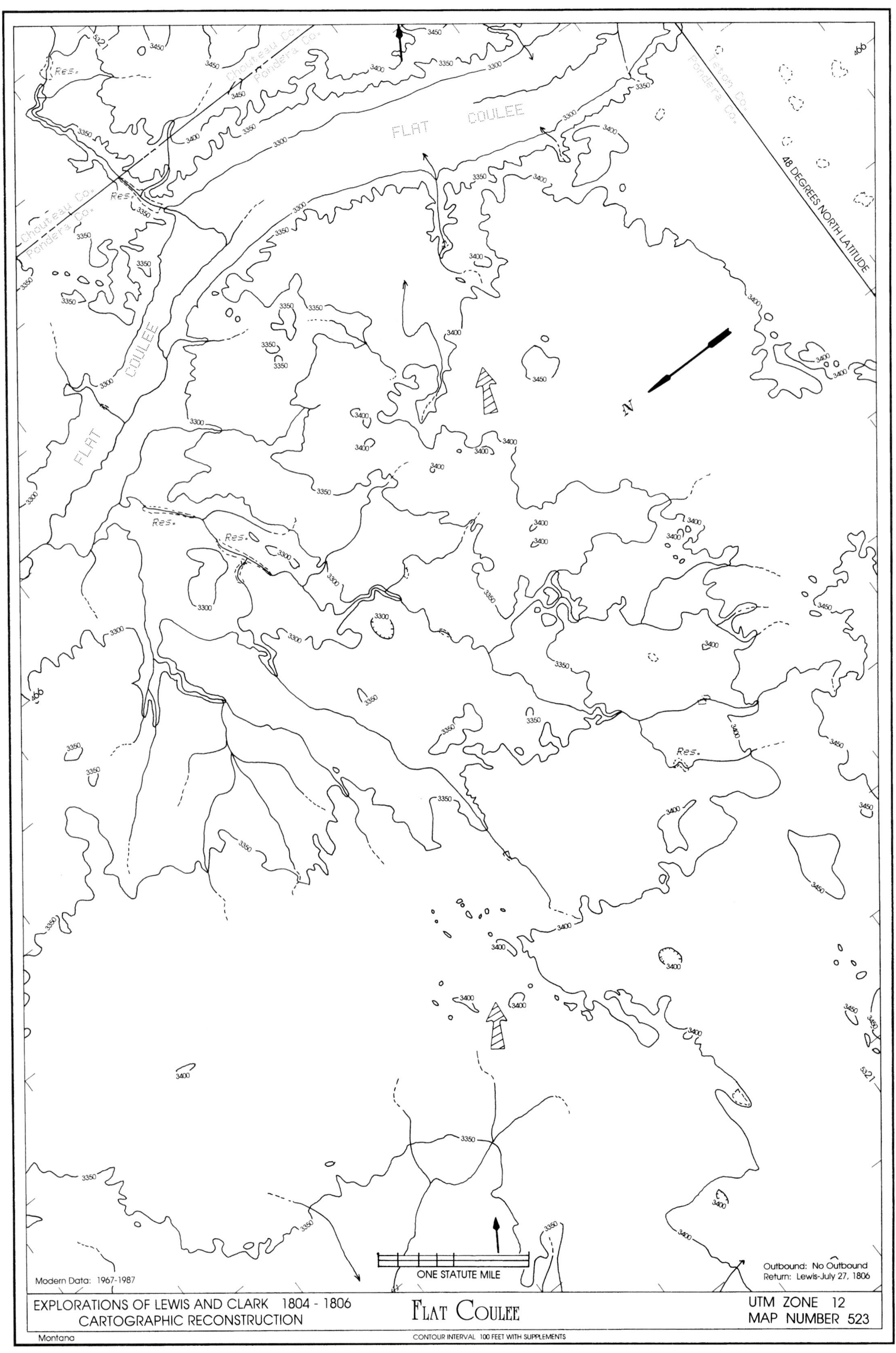

Chouteau Co.
Pondera Co.
Teton Co.
Pondera Co.
FLAT COULEE
48 DEGREES NORTH LATITUDE
Res.
N
Modern Data: 1967-1987
ONE STATUTE MILE
Outbound: No Outbound
Return: Lewis-July 27, 1806
EXPLORATIONS OF LEWIS AND CLARK 1804 - 1806
CARTOGRAPHIC RECONSTRUCTION
FLAT COULEE
UTM ZONE 12
MAP NUMBER 523
Montana
CONTOUR INTERVAL 100 FEET WITH SUPPLEMENTS

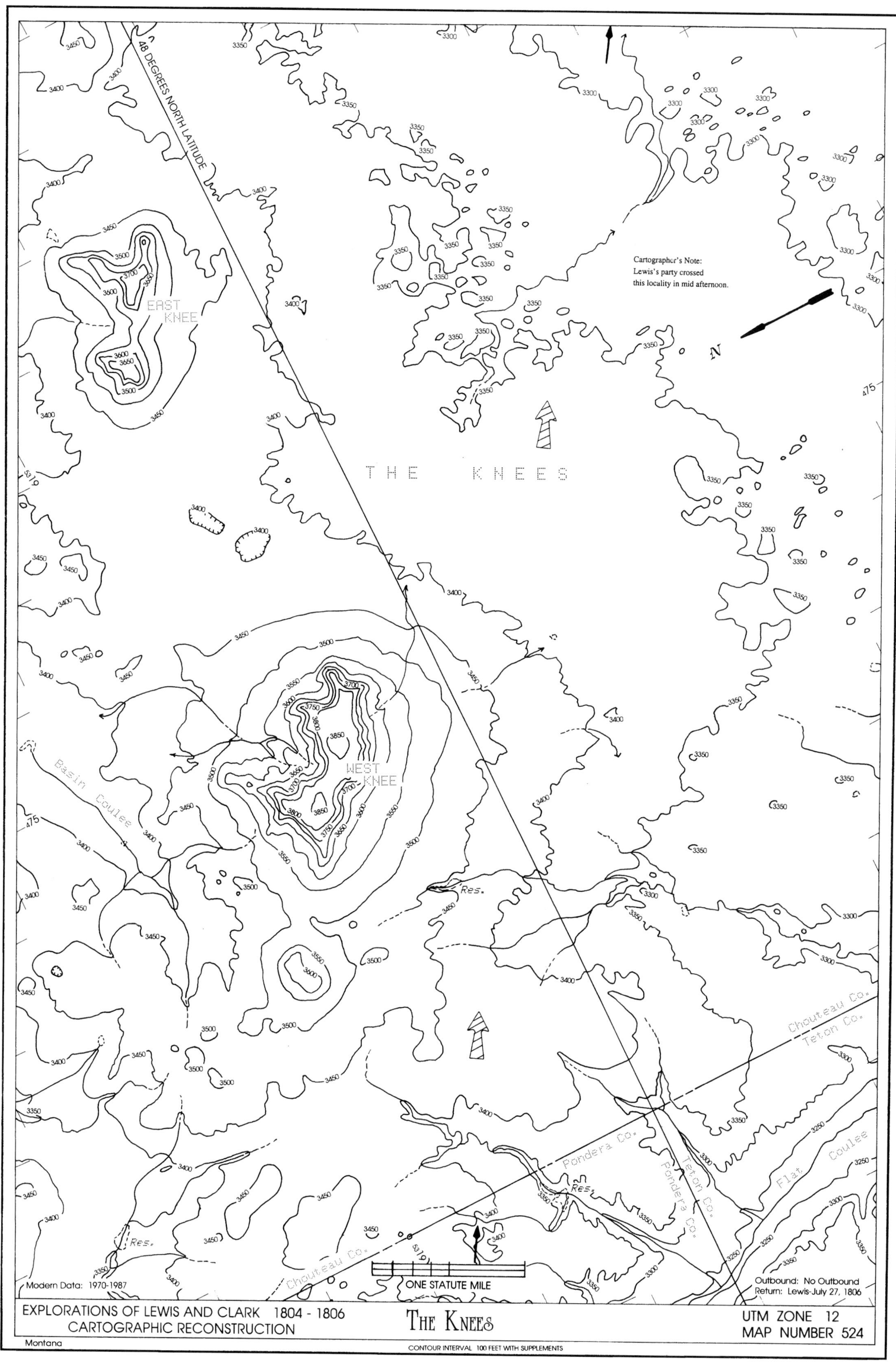

48 DEGREES NORTH LATITUDE
EAST KNEE
WEST KNEE
THE KNEES
Cartographer's Note:
Lewis's party crossed
this locality in mid afternoon.
N
Basin Coulee
Res.
Chouteau Co.
Teton Co.
Pondera Co.
Flat Coulee
ONE STATUTE MILE
Modern Data: 1970-1987
Outbound: No Outbound
Return: Lewis-July 27, 1806
EXPLORATIONS OF LEWIS AND CLARK 1804 - 1806
CARTOGRAPHIC RECONSTRUCTION
THE KNEES
UTM ZONE 12
MAP NUMBER 524
Montana
CONTOUR INTERVAL 100 FEET WITH SUPPLEMENTS

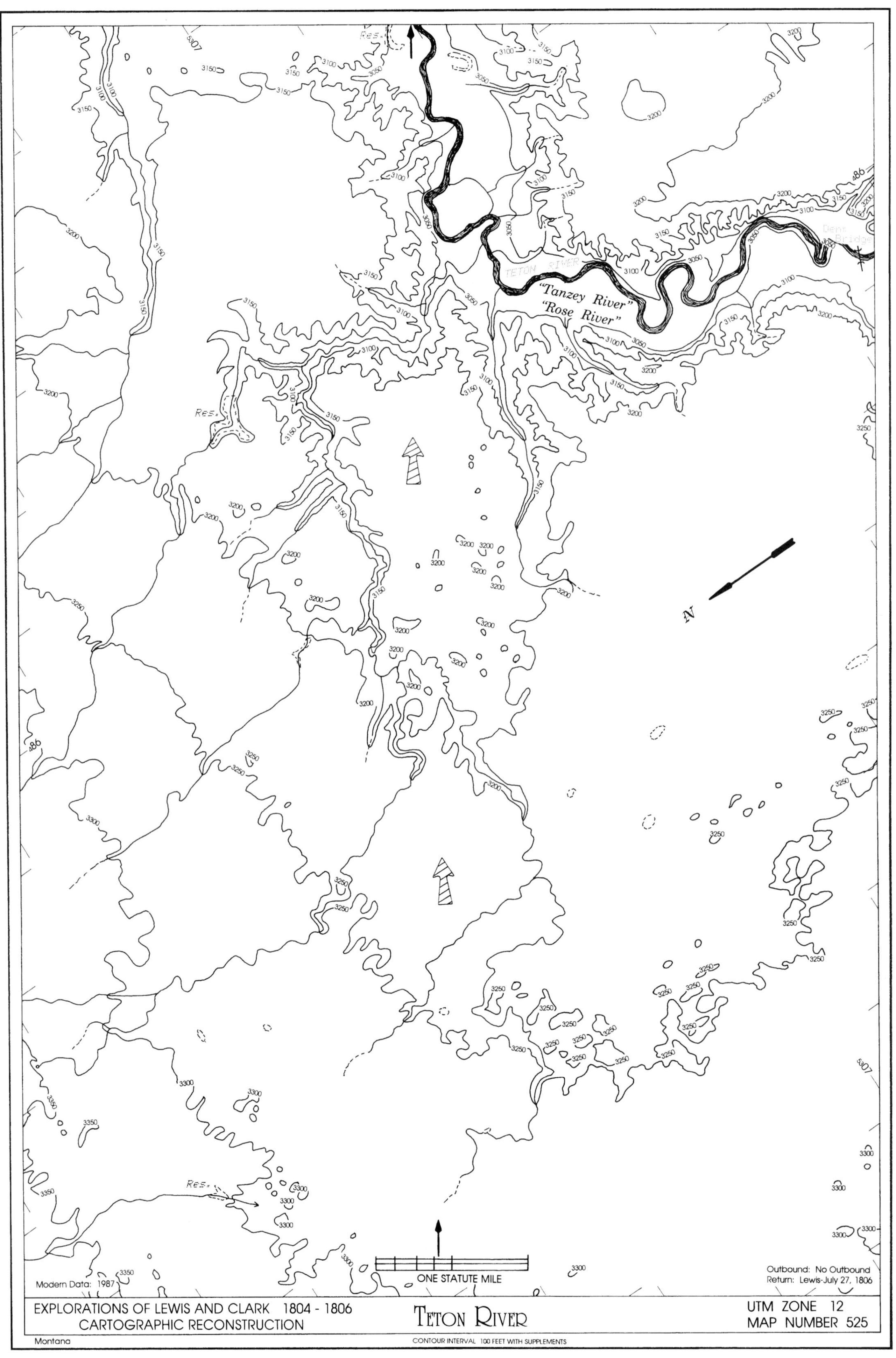
"Tanzey River"
"Rose River"
TETON RIVER
ONE STATUTE MILE
Modern Data: 1987
Outbound: No Outbound
Return: Lewis-July 27, 1806
EXPLORATIONS OF LEWIS AND CLARK 1804 - 1806
CARTOGRAPHIC RECONSTRUCTION
TETON RIVER
UTM ZONE 12
MAP NUMBER 525
Montana
CONTOUR INTERVAL 100 FEET WITH SUPPLEMENTS

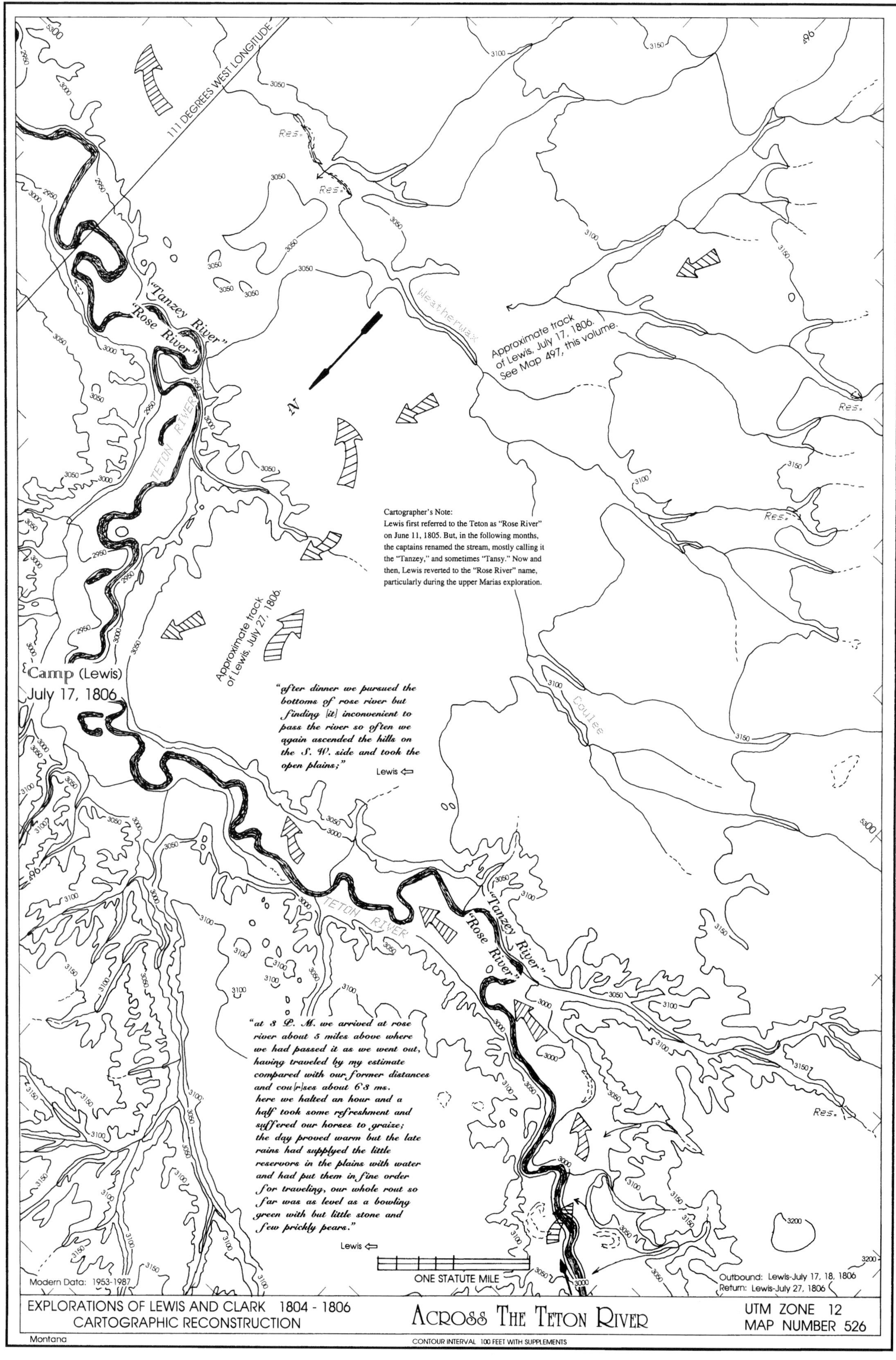

111 DEGREES WEST LONGITUDE
"Tanzey River"
"Rose River"
TETON RIVER
Weatherwax
Coulee
Approximate track of Lewis, July 17, 1806. See Map 497, this volume.
Approximate track of Lewis, July 27, 1806.
Camp (Lewis) July 17, 1806
Cartographer's Note:
Lewis first referred to the Teton as "Rose River" on June 11, 1805. But, in the following months, the captains renamed the stream, mostly calling it the "Tanzey," and sometimes "Tansy." Now and then, Lewis reverted to the "Rose River" name, particularly during the upper Marias exploration.
"after dinner we pursued the bottoms of rose river but finding [it] inconvenient to pass the river so often we again ascended the hills on the S. W. side and took the open plains;"
Lewis ⇦
"at 3 P. M. we arrived at rose river about 5 miles above where we had passed it as we went out, having traveled by my estimate compared with our former distances and cou[r]ses about 63 ms. here we halted an hour and a half took some refreshment and suffered our horses to graize; the day proved warm but the late rains had supplyed the little reservoirs in the plains with water and had put them in fine order for traveling, our whole rout so far was as level as a bowling green with but little stone and few prickly pears."
Lewis ⇦
ONE STATUTE MILE
Modern Data: 1953-1987
Outbound: Lewis-July 17, 18, 1806
Return: Lewis-July 27, 1806
EXPLORATIONS OF LEWIS AND CLARK 1804 - 1806
CARTOGRAPHIC RECONSTRUCTION
ACROSS THE TETON RIVER
UTM ZONE 12
MAP NUMBER 526
Montana
CONTOUR INTERVAL 100 FEET WITH SUPPLEMENTS

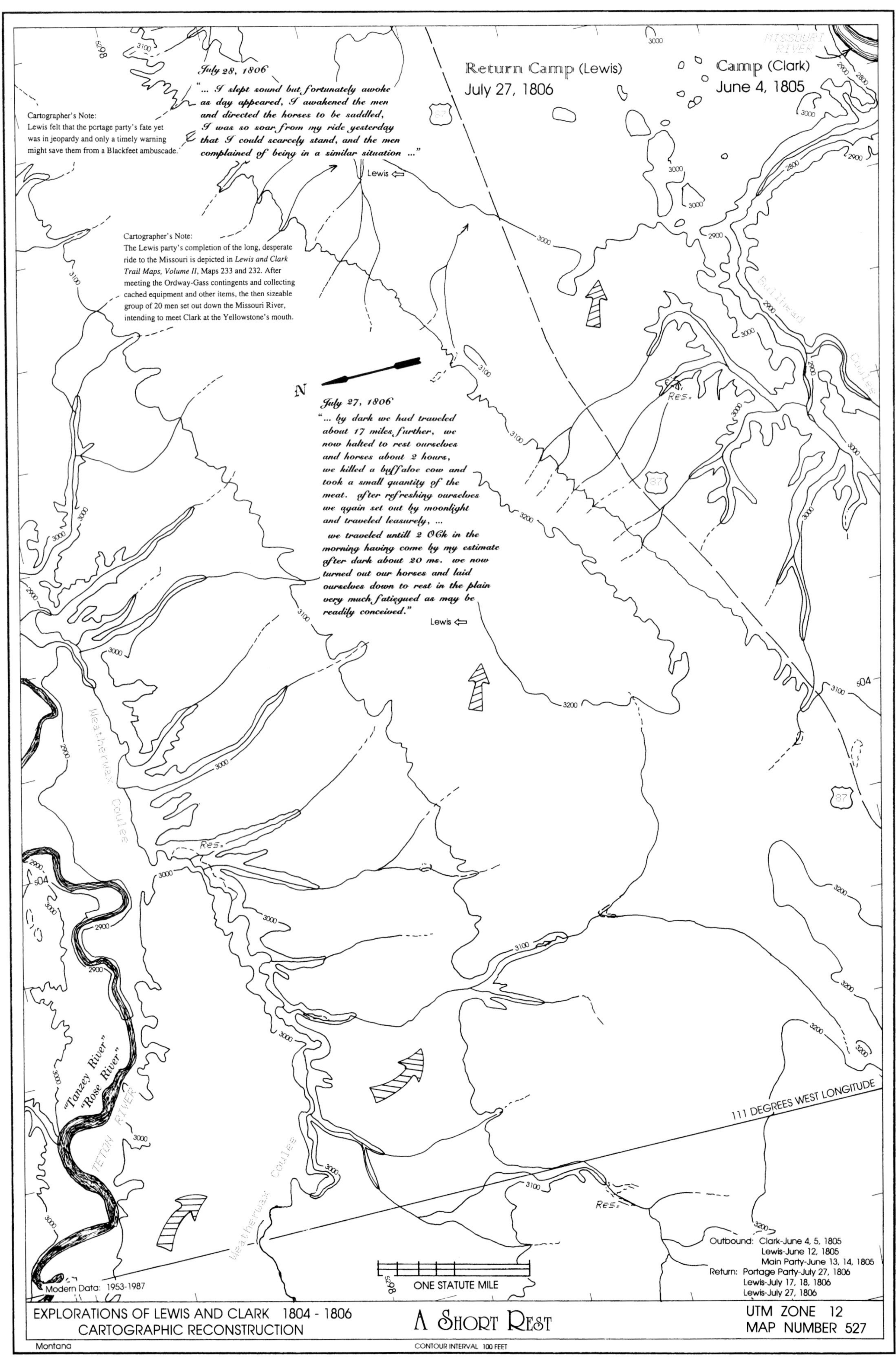

July 28, 1806
"... I slept sound but fortunately awoke as day appeared, I awakened the men and directed the horses to be saddled, I was so soar from my ride yesterday that I could scarcely stand, and the men complained of being in a similar situation ..."
Cartographer's Note:
Lewis felt that the portage party's fate yet was in jeopardy and only a timely warning might save them from a Blackfeet ambuscade.
Return Camp (Lewis)
July 27, 1806
Camp (Clark)
June 4, 1805
MISSOURI RIVER
Cartographer's Note:
The Lewis party's completion of the long, desperate ride to the Missouri is depicted in Lewis and Clark Trail Maps, Volume II, Maps 233 and 232. After meeting the Ordway-Gass contingents and collecting cached equipment and other items, the then sizeable group of 20 men set out down the Missouri River, intending to meet Clark at the Yellowstone's mouth.
Bullhead Coulee
N
July 27, 1806
"... by dark we had traveled about 17 miles further, we now halted to rest ourselves and horses about 2 hours, we killed a buffaloe cow and took a small quantity of the meat. after refreshing ourselves we again set out by moonlight and traveled leasurely, ... we traveled untill 2 OCk in the morning having come by my estimate after dark about 20 ms. we now turned out our horses and laid ourselves down to rest in the plain very much fatigued as may be readily conceived."
Lewis
Weatherwax Coulee
Res.
"Tanzey River"
"Rose River"
TETON RIVER
111 DEGREES WEST LONGITUDE
ONE STATUTE MILE
Outbound: Clark-June 4, 5, 1805
Lewis-June 12, 1805
Main Party-June 13, 14, 1805
Return: Portage Party-July 27, 1806
Lewis-July 17, 18, 1806
Lewis-July 27, 1806
Modern Data: 1953-1987
EXPLORATIONS OF LEWIS AND CLARK 1804 - 1806
CARTOGRAPHIC RECONSTRUCTION
A SHORT REST
UTM ZONE 12
MAP NUMBER 527
Montana
CONTOUR INTERVAL 100 FEET

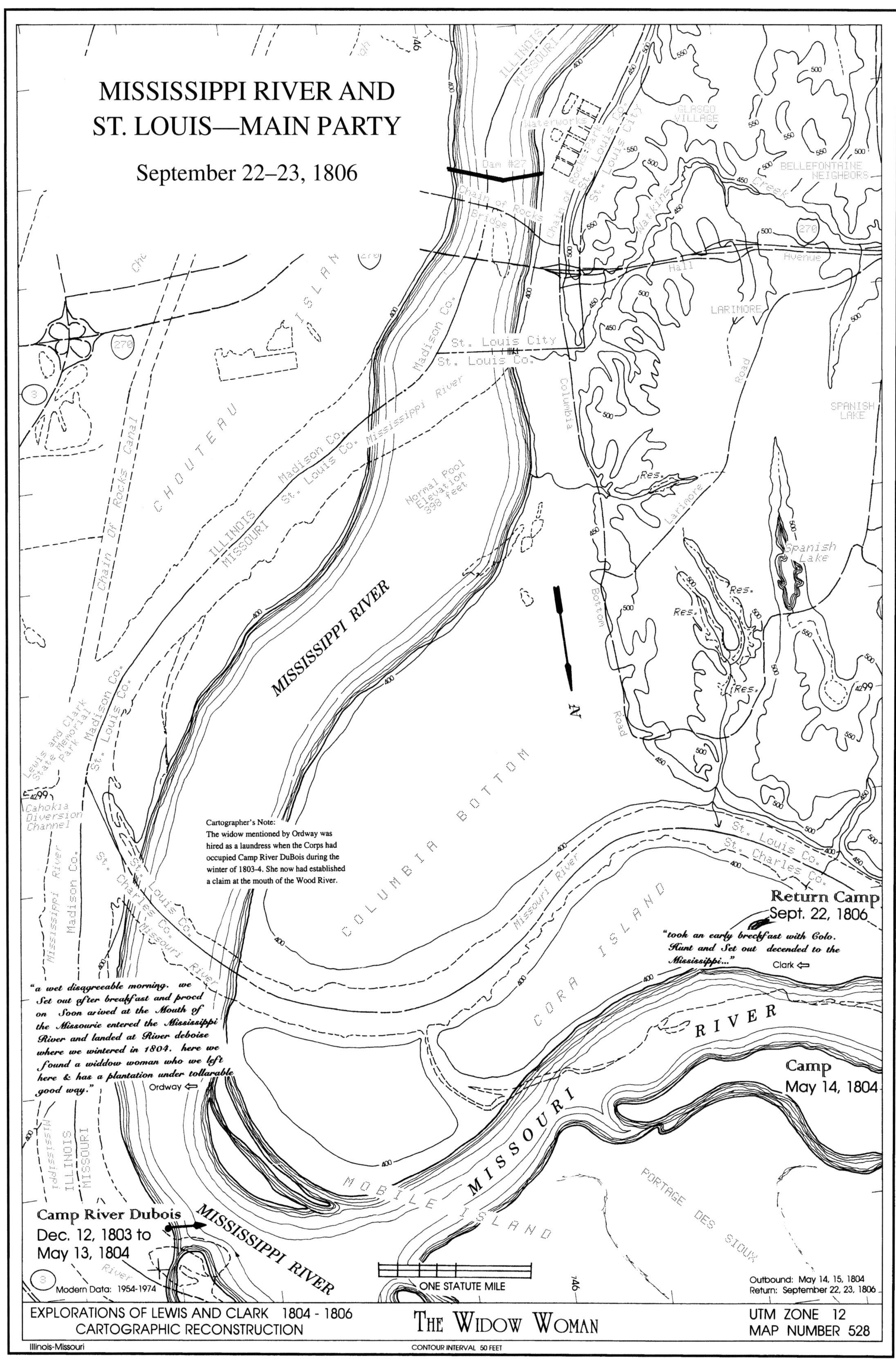
MISSISSIPPI RIVER AND
ST. LOUIS—MAIN PARTY
September 22–23, 1806
Cartographer's Note:
The widow mentioned by Ordway was hired as a laundress when the Corps had occupied Camp River DuBois during the winter of 1803-4. She now had established a claim at the mouth of the Wood River.
"a wet disagreeable morning. we Set out after breakfast and procd on Soon arived at the Mouth of the Missourie entered the Mississippi River and landed at River deboise where we wintered in 1804. here we found a widdow woman who we left here & has a plantation under tollarable good way."
Ordway ⇦
"took an early breckfast with Colo. Hunt and Set out decended to the Mississippi..."
Clark ⇦
Return Camp
Sept. 22, 1806
Camp
May 14, 1804
Camp River Dubois
Dec. 12, 1803 to
May 13, 1804
MISSISSIPPI RIVER
MISSOURI RIVER
COLUMBIA BOTTOM
CORA ISLAND
MOBILE ISLAND
CHOUTEAU ISLAND
PORTAGE DES SIOUX
Dam #27
Chain of Rocks Bridge
Chain of Rocks Canal
Waterworks
GLASGO VILLAGE
BELLEFONTAINE NEIGHBORS
SPANISH LAKE
Spanish Lake
LARIMORE
Normal Pool Elevation 398 feet
Lewis and Clark State Memorial Park
Cahokia Diversion Channel
Madison Co.
St. Louis Co.
St. Louis City
St. Charles Co.
ILLINOIS
MISSOURI
N
ONE STATUTE MILE
Modern Data: 1954-1974
Outbound: May 14, 15, 1804
Return: September 22, 23, 1806
EXPLORATIONS OF LEWIS AND CLARK 1804 - 1806
CARTOGRAPHIC RECONSTRUCTION
THE WIDOW WOMAN
UTM ZONE 12
MAP NUMBER 528
Illinois-Missouri
CONTOUR INTERVAL 50 FEET

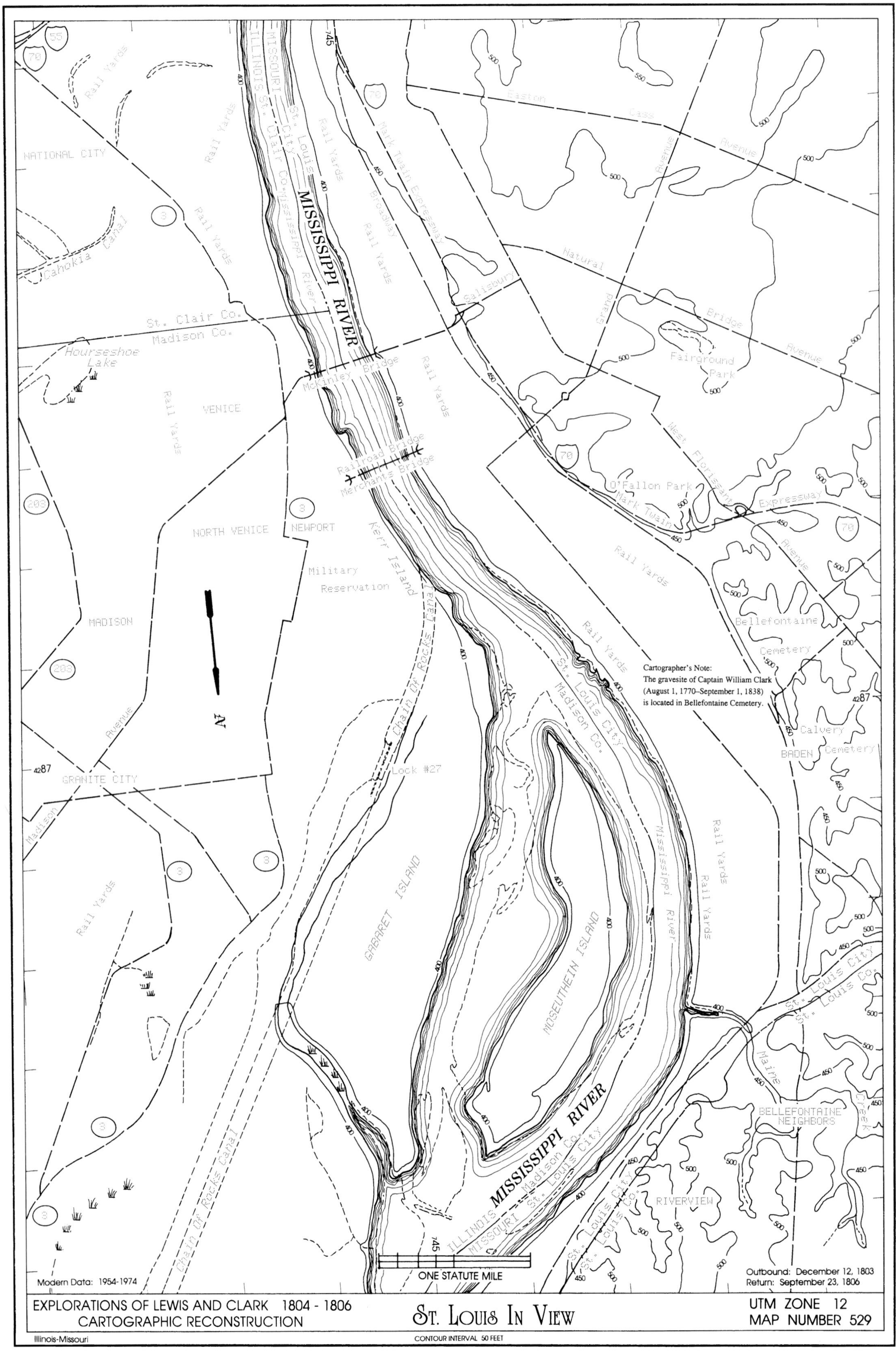
Cartographer's Note:
The gravesite of Captain William Clark (August 1, 1770–September 1, 1838) is located in Bellefontaine Cemetery.
MISSISSIPPI RIVER
NATIONAL CITY
VENICE
NORTH VENICE
NEWPORT
MADISON
GRANITE CITY
GABARET ISLAND
MOSEUTHEIN ISLAND
BELLEFONTAINE NEIGHBORS
RIVERVIEW
BADEN
Bellefontaine Cemetery
Calvery Cemetery
Fairground Park
O'Fallon Park
Military Reservation
Kerr Island
Chain Of Rocks Canal
Lock #27
Cahokia Canal
Hourseshoe Lake
McKinley Bridge
Railroad Bridge
Merchants Bridge
Mark Twain Expressway
Natural Bridge Avenue
Easton Cass Avenue
West Florissant
Grand Avenue
Rail Yards
ONE STATUTE MILE
Modern Data: 1954-1974
Outbound: December 12, 1803
Return: September 23, 1806
EXPLORATIONS OF LEWIS AND CLARK 1804 - 1806
CARTOGRAPHIC RECONSTRUCTION
St. Louis In View
UTM ZONE 12
MAP NUMBER 529
Illinois-Missouri
CONTOUR INTERVAL 50 FEET

RETURN TO ST. LOUIS

CORPS OF DISCOVERY

Departure from Camp River Dubois—May 14, 1804
Pacific Ocean at Station Camp—November 15–24, 1805
Arrival at St. Louis—September 23, 1806

Sergeant John Ordway's final journal words, September 23, 1806—

"... about 12 oClock we arived in Site of St Louis fired three Rounds as we approached the... Town and landed oppocit the center of the Town, the people gathred on the Shore and Hizzared three cheers. we unloaded the canoes and carried the baggage all up to a store house in Town drew out the canoes then the party all considerable much rejoiced that we have the Expedition Completed and now we look for boarding in Town and wait for our Settlement and then we entend to return to our native homes to See our parents once more as we have been so long from them.—finis."

Sergeant Patrick Gass's last journal note, September 23, 1806—

"we arrived on the 23rd and were received with great kindness and marks of friendship by the inhabitants, after an absence of two years, four months and ten days."

William Clark's concluding journal entries, September 23–26, 1806—

September 23, 1806
"...decended to the Mississippi and down that river to St. Louis at which place we arived about 12 oClock. we Suffered the party to fire off their pieces as a Salute to the Town. we were met by all the village and received a harty welcom from it's inhabitants &c. here I found my old acquaintance Majr. W. Christy who had settled in this town in a public line as a Tavern Keeper. he furnished us with store rooms for our baggage and we accepted of the invitation of Mr. Peter Choteau and took a room in his house. we payed a friendly visit to Mr. August Chotau and some of our old friends this evening. as the post had departed from St. Louis Capt. Lewis wrote a note to Mr. Hay in Kahoka [Cahokia] to detain the post at that place untill 12 tomorrow which was reather later than his usial time of leaving it."*

September 24, 1806
"I sleped but little last night however we rose early and commenc[e]d wrighting our letters Capt Lewis wrote one to the presidend and I wrote Govr. Harrison & my friends in Kentucky and Sent of[f] George Drewyer with those letters to Kohoka & delivered them to Mr. Hays &c. we dined with Mr. Chotoux to day, and after dinner went to a store and purchased some clothes, which we gave to a Tayler and derected to be made. Capt. Lewis in opening his trunk found all his papers wet, and some seeds spoiled."

September 25, 1806
"had all our skins &c. suned and stored away in a store-room of Mr. Caddy Choteau. payed some visits of form, to the gentlemen of St. Louis. in the evening a dinner & Ball" **

September 26, 1806
"a fine morning we commenced wrighting &c."

* John Hay was in charge of the federal mails in Cahokia.

** "[A]t William Christy's tavern. Eighteen toasts were drunk, starting with one to President Jefferson, 'The friend of science, the polar star of *discovery*, the philosopher and the patriot,' and ending with 'Captains Lewis and Clark—Their perilous services endear them to every American heart.' Ronda (SLW)." From Gary E. Moulton, ed., *The Journals of the Lewis & Clark Expedition*, Vol. 8 (Lincoln: University of Nebraska Press, 1993), p. 372 (fn. 2).

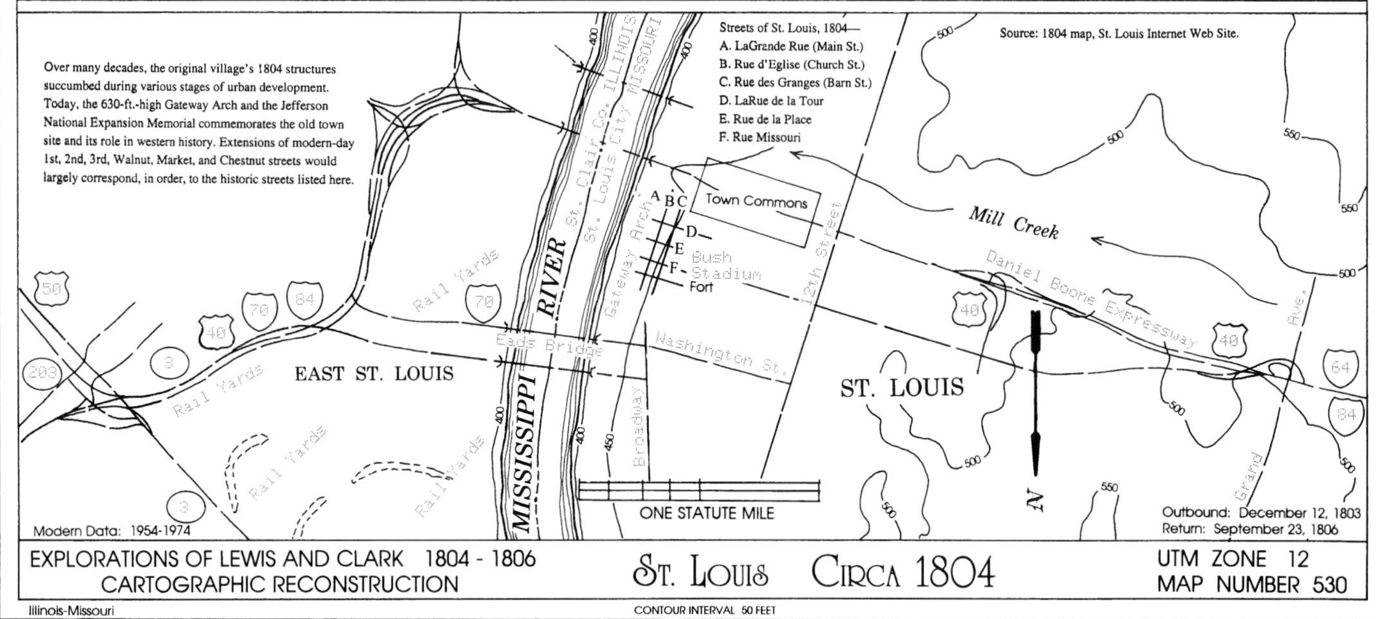

Index to the Lewis and Clark Trail Maps, Volume III

Outbound, Winter Activity, and Return Camps

Outbound, Winter Activity, and Return Camps

Outbound Camps—Columbia River (October 16 to November 24, 1805)

Winter Activity Camps—Mouth of Columbia River (November 25, 1805 to March 22, 1806)

Return Camps—Columbia River (March 23 to April 28, 1806)

Alternate Return Route No. 1—Nez Perce Trail (April 29 to May 4, 1806)

Alternate Return Route No. 2—Clearwater River (May 8 to June 14, 1806)

Alternate Return Route No. 3—Lolo Pass (June 28–29, 1806)

Clark Alternate Return Route No. 4—Big Hole (July 5–8, 1806)

Lewis Alternate Return Route No. 5—Best Route (July 3–12, 1806)

Clark Alternate Return Route No. 6—Yellowstone River (July 13 to August 3, 1806)

Lewis Alternate Return Route No. 7—Upper Marias River (July 16–28, 1806)

Return to St. Louis (September 22–23, 1806)

(The longer length, more extensive map coverage, and other considerations regarding *Volume III* precluded detailed indexes as presented in *Volumes I & II*.)

Errata

Volume I—

P. xiii, 2nd column, line 6: "Bismark" should read "Bismarck."

P. 11, 1st column, 2nd paragraph: see revised layout of Fort Clatsop in *Volume III,* p. 73.

P. 18, Index Map 7: turn north arrow 90 degrees counter-clockwise; "95 East Longitude" should read "95 West Longitude."

P. 21, Index Map 20: "Bismark" should read "Bismarck" (twice)

P. 169, Map 145: "Bismark" should read "Bismarck" (thrice).

Volume II—

Dedication page, line 6: Captain and Mrs. J.P. Brooks reside in "Camas, Washington," not "Washougal, Washington."

P. xiii, 2nd column, 4th paragraph: according to recent investigators, The Eye of the Needle may have succumbed to a natural cause, not vandalism.

P. xiii, 2nd column, line 34: ""Washougal" should read "Camas."

P. xv, "Harpers Ferry" note: doubts have arisen about Lewis's involvement in modifying Army issue rifles.

P. 3, map legend: "27. Little Muddy Creek" and "28. Little Dry Creek" should read "27. Big Muddy Creek" and "28. Big Dry Creek."

P. 44, Map 177: add "Return Camp (Clark) August 4, 1806" at inside of Missouri bend at bottom of map.

P. 88, Map 221: according to recent investigators, The Eye of the Needle may have succumbed to a natural cause, not vandalism.

PP. 100–1, Maps 233, 234: "Camp (Clark) June 4, 1805" indicated at two different locations due to vague journal data.

P. 109, Map 242: the name "Tower Rock" should be added for an isolated hill at lower right.

P. 122, Map 255: relocate "Fort Rock" between the Gallatin and Madison rivers. Correctly located in *Volume III*, Map 435.

P. 165, Map 295: UTM values misplaced; can be corrected by switching values from north and south lines to east and west lines, and vice versa.

P. 166, Map 296: UTM values misplaced; can be corrected by switching values from north and south lines to east and west lines, and vice versa.

P. 167, Map 298: UTM values misplaced; can be corrected by switching values from north and south lines to east and west lines, and vice versa.

P. 180, Map 310: turn north arrow 90 degrees counter-clockwise.

P. 202, Map 332: add "S 28 W 6.5 MILES" for last lower Snake River traverse call; actual mile indicators are missing; Clark's mile indicators missing on upper Columbia leg; missing island at the Snake's mouth. Much of this corrected in *Volume III*, Map 334.

P. 203, Map 333: Clark's and the actual mile indicators are missing.